Nanostructured Materials

NATO ASI Series

Advanced Science Institutes Series

A Series presenting the results of activities sponsored by the NATO Science Committee, which aims at the dissemination of advanced scientific and technological knowledge, with a view to strengthening links between scientific communities.

The Series is published by an international board of publishers in conjunction with the NATO Scientific Affairs Division

A Life Sciences	Plenum Publishing Corporation
B Physics	London and New York
C Mathematical and Physical Sciences	Kluwer Academic Publishers
D Behavioural and Social Sciences	Dordrecht, Boston and London
E Applied Sciences	
F Computer and Systems Sciences	Springer-Verlag
G Ecological Sciences	Berlin, Heidelberg, New York, London,
H Cell Biology	Paris and Tokyo
I Global Environmental Change	

PARTNERSHIP SUB-SERIES

1. Disarmament Technologies	Kluwer Academic Publishers
2. Environment	Springer-Verlag / Kluwer Academic Publishers
3. High Technology	Kluwer Academic Publishers
4. Science and Technology Policy	Kluwer Academic Publishers
5. Computer Networking	Kluwer Academic Publishers

The Partnership Sub-Series incorporates activities undertaken in collaboration with NATO's Cooperation Partners, the countries of the CIS and Central and Eastern Europe, in Priority Areas of concern to those countries.

NATO-PCO-DATA BASE

The electronic index to the NATO ASI Series provides full bibliographical references (with keywords and/or abstracts) to more than 50000 contributions from international scientists published in all sections of the NATO ASI Series.
Access to the NATO-PCO-DATA BASE is possible in two ways:

– via online FILE 128 (NATO-PCO-DATA BASE) hosted by ESRIN,
Via Galileo Galilei, I-00044 Frascati, Italy.

– via CD-ROM "NATO-PCO-DATA BASE" with user-friendly retrieval software in English, French and German (© WTV GmbH and DATAWARE Technologies Inc. 1989).

The CD-ROM can be ordered through any member of the Board of Publishers or through NATO-PCO, Overijse, Belgium.

3. High Technology – Vol. 50

Nanostructured Materials
Science & Technology

edited by

Gan-Moog Chow

Materials Science and Technology Division,
Naval Research Laboratory,
Washington, U.S.A.

and

Nina Ivanovna Noskova

Institute of Metal Physics,
Ural Division of Russian Academy of Sciences,
Ekaterinburg, Russia

Springer-Science+Business Media, B.V.

Proceedings of the NATO Advanced Study Institute on
Nanostructured Materials: Science & Technology
St. Petersburg, Russia
August 10–20, 1997

A C.I.P. Catalogue record for this book is available from the Library of Congress.

ISBN 978-94-010-6100-1   ISBN 978-94-011-5002-6 (eBook)
DOI 10.1007/978-94-011-5002-6

TABLE OF CONTENTS

Preface vii

Chapter 1. Nanoparticle Synthesis: A Key Process in the Future of Nano-
technology 1
    J.R. Brock

Chapter 2. Nanoparticles and Nanostructures: Aerosol Synthesis and
Characterization 15
    Richard C. Flagan

Chapter 3. Chemical Synthesis and Processing of Nanostructured Particles
and Coatings 31
    G.M. Chow

Chapter 4. Nanostructured Materials produced by High-Energy Mechanical
Milling and Electrodeposition 47
    Michel L. Trudeau

Chapter 5. Perspective on Nanoparticle Manufacturing Research 71
    M.C. Roco

Chapter 6. The Nanocrystalline Alloys: The Structure and Properties 93
    N.I. Noskova

Chapter 7. Processing and Properties of Nanostructured Materials prepared
by Severe Plastic Deformation 121
    R.Z. Valiev, I.V. Alexandrov and R.K. Islamgaliev

Chapter 8. Changes in Structure and Properties associated with the Transition
from the Amorphous to the Nanocrystalline State 143
    A.L. Greer

Chapter 9. Melt Quenched Nanocrystals 163
    A. Glezer

Chapter 10. Quasiperiodic and Disordered Interfaces in Nanostructured
Materials 183
    I.A. Ovid'ko

Chapter 11. Micromechanisms of Defects in Nanostructured Materials 207
    Alexei E. Romanov

Chapter 12. The Properties of Fe-Ni FCC Alloys having a Nanostructure
produced by Deformation, Irradiation and Cyclic Phase 243
  V.V. Sagaradze

Chapter 13. The State-of-the-Art of Nanostructured High Melting Point
Compound-Based Materials 263
  R.A. Andrievski

Chapter 14. Thermal Spray Processing of Nanocrystalline Materials 283
  E.J. Lavernia, M.L. Lau and H.G. Jiang

Chapter 15. The Surface Characterization of Nanosized Powders: Relevance
of the FTIR Surface Spectrometry 303
  Marie-Isabelle Baraton

Chapter 16. Enhanced Transformation and Sintering of Transitional Alumina
through Mechanical Seeding 319
  Martin L. Panchula and Jackie Y. Ying

Chapter 17. Nanostructured Materials for Gas-Reactive Applications 335
  V. Provenzano

Chapter 18. Nanocrystalline Ceramics for Structural Applications:
Processing and Properties 361
  M.J. Mayo

Chapter 19. Features of Nanocrystalline Structure Formation on Sintering
of Ultra-Fine Powders 387
  V.V. Skorokhod and A.V. Ragulya

Chapter 20. Consolidation of Nanocrystalline Materials at High Pressures 405
  V.S. Urbanovich

Chapter 21. Magnetic State, Transport Properties and Structure of Granular
Nanophased Systems: 1. Mechanically Alloyed Cu-20%Co System. 2. Hydro-
genated $Pr(Cu,Co)_5$ Intermetallics 425
  A.Ye. Yermakov, M.A. Uimin, N.V. Mushnikov, N.K. Zajkov,
  V.V. Serikov, A.Yu. Korobejnikov, N.M. Kleinerman and A.K. Shtolz

Chapter 22. Metallic Superlatives with Governed Non-Collinear Magnetic
Ordering: Atomic Structure, Interlayer Exchange and Magnetotransport
Properties 441
  V.V. Ustinov and E.A. Kravtsov

Index 457

Preface

The NATO Advanced Study Institute on Nanostructured Materials (St. Petersburg, August 10-21, 1997), the first of this kind held in Russia, successfully provided a critical, up-to-date tutorial review and discussion on science and technology of nanostructured metallic and ceramic materials. Focuses were placed on synthesis and processing of nanoparticles, assembly and stability of nanostructures, characterization and properties, and applications. It was well recognized that many progresses have been made in the synthesis of nanoparticles, using vairous approaches via vapour phase deposition, solution chemistry and solid state attrition. The fundamentals of various approaches were discussed. Although different methods can be used to make a material, the properties and the impurities of nanoparticles significantly depend on the chosen synthesis method. The agglomeration of nanoparticles, which may lead to deterioration of properties and difficulty in subsequent processing, has been addressed in both vapour phase and solution phase synthetic approaches by controlling the kinetics and interfacial interactions.

It was shown that growing interest is placed in the processing of nanoparticles into consolidated bulk materials and coatings. The metastability of nanoparticles may lead to undesirable grain growth during thermally assisted consolidation or other processing routes, and the retention of nanostructures in a processed part or component with desirable density continues to draw much attention. Currently, significant activities are found in the deposition of nanostructured coatings using established thermal spray technologies and wet chemistry methods. Naturally existing or artificially synthesized templates with unique structures and morphologies have been used to fabricate nanostructured materials with the same structural and morphological characteristics of the templates.

Many recent advances in nanoscale characterization techniques have provided useful information on the structure, the surface and bulk chemistry of nanoparticles, and the structure and chemistry of exposed surfaces and buried interfaces of nanostructured coatings. It was recognized that future advances in the field of nanostructured masterials research require increasing interactions from various disciplines in physics, chemistry, biology, and materials science and engineering. Research areas such as the modeling of processing, structure and interfaces, and the fundamental understanding of properties of nanostructured materials were identified to warrant more attention. Progresses in structural, electronic, magnetic, and catalytic applications were reported and discussed. The many fruitful discussions between the lecturers and participants during the ASI led to the development of many collavorative projects and more future global interactions.

This NATO ASI publication includes most of the invited lecture papers presented at the meeting. All the published papers were peer-reviewed.

We would like to acknowledge the significant contribution of the organizing committee: R. Andrievski (Russia), A.L. Greer (UK), L. Kabacoff (USA), I.A. Ovid'ko (Russia) and M.C. Roco (USA). We gratefully acknowledge the major financial support of this ASI by the NATO Science Committee and the NATO High Technology Area

Program, and the co-sponsorships of the Office of Naval Research (USA), the National Science Foundation (USA), the Russian Academy of Science, and the Russian Foundation for Basic Research. We also thank the American Society for Engineering Education and the Advanced Materials and Technologies Foundation (Russia) for their organizational and technical support, and many others who provided their valuable assistance in the preparation and conduction of the meeting. Finally, G.M. Chow would like to specially thank his family for their support and encouragement in organizing this meeting.

Gan-Moog Chow   and   Nina Ivanovna Noskova
ASI director   ASI co-director
Naval Research Laboratory   Russian Academy of Science
Washington, D.C., USA   Ekaterinburg, Russia

February 3, 1998

NANOPARTICLE SYNTHESIS: A KEY PROCESS IN THE FUTURE OF NANOTECHNOLOGY

J. R. BROCK
Chemical Engineering Department
University of Texas
Austin, Texas   78712-1062

1. Introduction

The unusual properties of nanoscale particles have sparked commercial interest in diverse fields including magnetics, pharmaceutics, aerospace and microelectronics.

It is established that for a given volume of a material, as we shrink that volume, so the dimensions are of the order of the de Broglie wavelength of the carriers, the carrier movement becomes quantized; that volume is now said to be of "quantum scale". Such quantum scale objects on semiconductors are currently the subject of worldwide intensive research (particularly in Europe) and are suggested[1] to have dimensions of a few 100 nm.. The limit in reduction in scale is reached at what has been termed a quantum dot where all the dimensions have been reduced to the quantum regime, approximately 10 nm for GaAs.

As suggested some time ago by the great Japanese physicist, R. Kubo quantum dots can be regarded as artificial atoms. A single quantum dot can emit an extremely sharp line (energy level~200μeV). If the number of atoms in a quantum dot is increased by as little as one atom, then inhomogeneous broadening of emission lines can occur[2]. Therefore if such quantum dots can be synthesized commercially with precise control of particle size and composition, the door is open to new materials and phenomena no longer constrained by the periodic table of elements. In this sense, nanoparticle synthesis yielding particles with the specified properties may be said to be a key process in the future of nanotechnology. Quantum dot lasers promise very low threshold currents and other attractive qualities.

Now, it is a matter of manufacturing engineering and economics-- one must be able to produce large quantities of quantum dots with identical properties cheaply and reliably.

Both quantum dots and nanoparticles are alike in that they represent quantum confinement with zero dimension. In one application, the term quantum dot implies a fabricated volume on a semiconductor that might function as a semiconductor laser; of course, one can suppose that nanoparticles could play the role of quantum dots in such applications, although this has not yet been demonstrated. The importance of quantum dots is indicated in a recent book[3] whose thesis is that quantum dots will play a major role in the future of microelectronics.

It has been widely suggested that future computer architectures might use suitable arrays of quantum dots.

1

G.M. Chow and N.I. Noskova (eds.), Nanostructured Materials, 1–14.
© 1998 *Kluwer Academic Publishers.*

The density of states in a quantum dot is given as a summation of delta functions:

$$\rho_{0D} = \frac{2}{s_x s_y s_z} \sum_{l,m,n} \delta(E - E_{l,m,n}) \tag{1}$$

where s_x, $s_y s_z$ are the sides of a cube representing the quantum dot.

For a fine metallic particle in the single electron approximation, the energy levels are given by Kubo:

$$\Delta E \sim \frac{\zeta}{N} \tag{2}$$

where ζ is the Fermi potential and N the number of atoms in the particle.

In order to realize the promise of nanoparticles, the nature of the technical challenge in nanoparticle synthesis is daunting and clear: synthesis at low cost of large quantities of monodisperse nanoparticles with specified diameter(sufficiently small to achieve quantum confinement), exceptional purity, and exact composition.

In this chapter, synthesis methods are briefly surveyed in the light of the criteria noted here. First, the dynamics of gas phase synthesis methods are discussed from the standpoint of achieving the desired criteria. Finally, the synthesis of np's by laser ablation is examined in detail, because the method shows commercial possibilities.

2. Gas Phase Dynamics of Particle Formation

The dynamics of particle formation in the gas phase will be discussed here, although, of course, liquid phase methods are also important, but a discussion of liquid phase methods are beyond the scope of this review. However the similarities with the gas phase processes give an idea of the liquid phase dynamics. Here several questions will be addressed for gas phase synthesis: How can very small particles be synthesized ? How are monodisperse (or near-monodisperse) size distributions to be produced? How can complex stoichiometry of some source material be reproduced faithfully in the nanoparticles synthesized ? How can all this be accomplished in large volumes of particles ?

2.1 INERT-GAS EVAPORATIONx

Some years ago, Granqvist and Buhrman[4].suggested that inert-gas evaporation was the best method for making ultrafine metal particles. However the great Russian aerosol scientist, N. A. Fuchs has pointed out the drawbacks of such method[5]--namely, difficulties arising from presence of more volatile impurities (oxides) in the system being heated that will issue as the first particles formed   Beyond such difficulties, if one wishes to prepare nanoparticles of some highly refractory material, inert-gas evaporation becomes more problematic, as, at the high temperatures required, impurities arising from evaporation of the containment apparatus might be difficult to suppress. .

Beyond these difficulties, there is the question of synthesizing nanoparticles of a given chemical composition. The evaporation method has not been demonstrated to preserve stoichiometry; it is reasonable to suppose that the less volatile species will evaporate first so that one will have ultimately separation of the components by a sort of batch distillation process. In many studies using the inert-gas evaporation method, it has been necessary to minimize, or eliminate, agglomeration of the condensed particles; this is often done by introducing some surfactant. Another approach is to dilute rapidly the condensing vapor by a jet of some gas. This requires that the particles be dispersed in times less than the agglomeration time, t_a, This can be estimated by assuming a constant agglomeration coefficient, b. Then the required dispersion time, t_d, must be much less than t_a, which is: t_d ~ 1/(bN), where N is the total number density of particles formed. For a density of 10^{16} particles/cc.(a desirable density to be achieved), this time would be approx. <10 nsec., possibly difficult to achieve by simple fluid dispersion. Of course this estimate doesn't account for the increasing value of b as agglomeration proceeds..

2.1.1 *Dynamics Favoring Ultrafine Particle Formation.*
In order to produce such ultrafine particles as nanoparticles, an estimate is possible of conditions that favor such small particles by inert-gas evaporation method. The basis is very simple: given a mass,M, of vapor that condenses by homogeneous nucleation to yield N particles(or nuclei), then the mean particle mass, <m>, of particles formed would be <m>~M/N . According to the classical theory of homogeneous nucleation, the rate of formation of new particles, dN/dt, increases faster than the first power of vapor supersaturation, while the rate of condensation of vapor on existing particles is nearly proportional to supersaturation. Therefore with sufficiently rapid cooling of the vapor (or rapid increase of vapor supersaturation), new particle formation is favored over condensation growth of existing particles. Put simply, high rates of vapor cooling lead to high nucleation rates, yielding larger N, and hence smaller particles without a proportionate increase of particle size by condensation.

2.1.2 *Synthesis of Monodisperse Distributions*
The conditions for forming a monodisperse particle size distribution have been known for some time.   The so-called Sinclair -LaMer monodisperse aerosol generator has been well described[6].
This method relies on the experimental observation that a distribution subjected to condensation growth becomes more monodisperse. This can be demonstrated quantitatively using the evolution equation for condensation growth of some density function of an aerosol, n(r) where r is particle radius :

$$\partial n / \partial t = -(\partial / \partial r)(f(r)n) \qquad (3)$$

where f(r) is the growth rate for a particle of radius r. As an approximation, in the free molecule regime, f(r) ~a, a constant . In this case, it is simple to show that the ratio of the standard deviation of the distribution, to the mean particle radius, γ, has the value:

$$\frac{\sigma}{\gamma} = \frac{\sigma_0}{(\gamma_0 + at)} \qquad (4)$$

where subscript o denotes initial values. Fuchs[7] proposed a criterion for monodispersity--namely, the ratio, $\alpha = \sigma/\gamma$. Fuchs suggested that the criterion for practical monodispersity was, of course, arbitrary, but as a definition of monodispersity, then the ratio, $\alpha < ~0.2$ could serve as a working definition; here we choose to regard a distribution with $\alpha = 0.2$ as being near-monodisperse. According to Eq. (4) $\alpha \rightarrow 0$ as condensation continues with time and thus the distribution becomes ever more monodisperse. A similar result can be demonstrated for condensation in the continuum regime. Of course, condensation will increase particle radius in the distribution, so this method is perhaps not practical for synthesis of monodisperse nanoparticle distributions. One must consider, as stated in Section 1, that increasing the size of a quantum dot by as much as *one atom* will broaden the emission line. Here indeed is a requirement of monodispersity that goes far beyond present synthesis technology, at least for large volume synthesis.

2.1.3 *Preservation of Stoichiometry*

If the source material is composed of some complex mixture, it may be most desirable that nanoparticles formed from this material preserve the stoichiometry of the source material. For generation of thin films of complex stoichiometry, pulsed laser deposition appears to be the method of choice[7] in which reproducible preparation of smooth $YBa_2Cu_3O_{7-\delta}$ thin films with good superconducting properties has been demonstrated. Inert-gas evaporation method fails with complex stoichiometry, for reasons already discussed.

2.1.4 *Reduction Methods* .

When it was discovered that drugs delivered in nanoparticle state had remarkable properties in concentrating in certain organs of the body, great interest was sparked in synthesis of drugs in the nanoparticle state. The first approach was to use mills to grind the material to the nanoparticle state. This turned out to have decided drawbacks in that on grinding, metallic residues from the milling apparatus became concentrated in the drug nanoparticles. The same drawbacks are to be found in all synthesis methods that start with large samples and then seek to reduce their size--spray pyrolysis, for example. If one begins with a 1.0μm droplet and then evaporates it to size 1.0 nm, then if the impurities are involatile, such impurities will be concentrated in the final nanoparticle by a factor of 10^9 !

2.1.5 *Comparison of Nanoparticle Synthesis Methods* . Various desirable characteristics of nanoparticle synthesis methods have been discussed in Section 2. For example, it is desirable that nanoparticles be synthesized with no agglomeration and with high purity; the distribution should be monodisperse, or near-monodisperse, the method should have high volume capability and preserve stoichiometry. These criteria represent oversimplifications and should be weighed with the previous discussion. A comparison, by these criteria, for various synthesis methods is given in Table 2.1. It was not possible to cover all synthesis methods; there have been many and some quite ingenious. In some of the listed methods, agglomeration can be minimized, but such minimization may be problematic. Likewise there is some ambiguity in assigning particular methods' potential for large volume synthesis. Nevertheless, at risk of controversy.

Table 2.1 CHARACTERISTICS OF IMPORTANT NANOPARTICLE SYNTHESIS METHODS

METHOD	agglomeration	high purity	large volume syn.	near-monodispersity	stoichiometry preserve.
vacuum syn.	yes	poss.	maybe	no	no
i.gas cond.	yes	no	yes	poss.	no
sputt.	no	yes	no	yes	poss.
laser ablatn.	likely	yes	poss.	no	yes
flame hydrol	yes	no	yes	no	no
gas prec. pyrolysis	yes	poss.	yes	no	poss.
cond. syn.	likely	no	yes	no	no
soln.sol-gel	yes	no	yes	not likely	not likely
part. laser abltn.	no	yes	yes	yes	yes

3.1 Laser Ablation of Microparticles

For formation of high quality thin films, Pulsed Laser Deposition (PLD) is commonly used, principally because it preserves stoichiometry. This is vital in synthesis of high quality high temperature superconducting films (such as the $YBa_2Cu_3O_{7-\delta}$ family). However, PLD has a decided drawback[8], because the method causes deposition of very large particles, or "droplets"[9] that result in poor film morphology. Fig. 3-1 below indicates this schematically

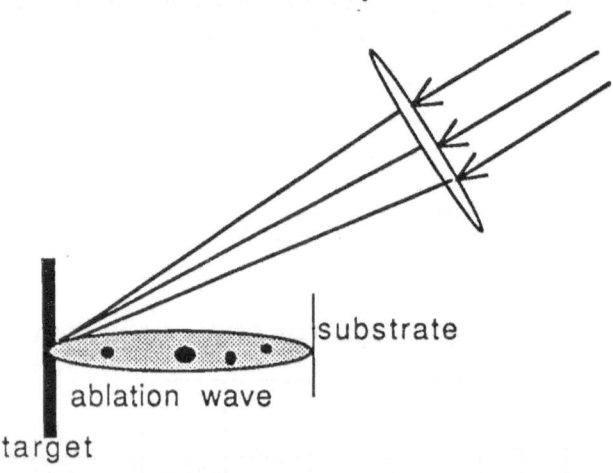

Fig. 3-1. Schematic of PLD for thin film synthesis. Note indicated large particles, "droplets", found in ablation wave.

We have discovered a new method that avoids these problems, but appears to preserve the advantages of PLD[10]. We[11] have termed this method Laser Ablation of Microparticles (LAM). LAM produces nanoparticles that are near- monodisperse and preserves stoichiometry. Table 3-1 compares statistics for glass particles formed by PLD and by LAM.

SOURCE	Fluence	Threshold fluenc J/cm^2	N, # of particles	mean diam, nm	min., nm	max., nm	s,std dev.., nm	mono disper sity index s/dia.
Slide	10.0	7.2	224	332	32	930	198	0.6
Spheres	10.5	0.5	560	74	32	327	23	0.31

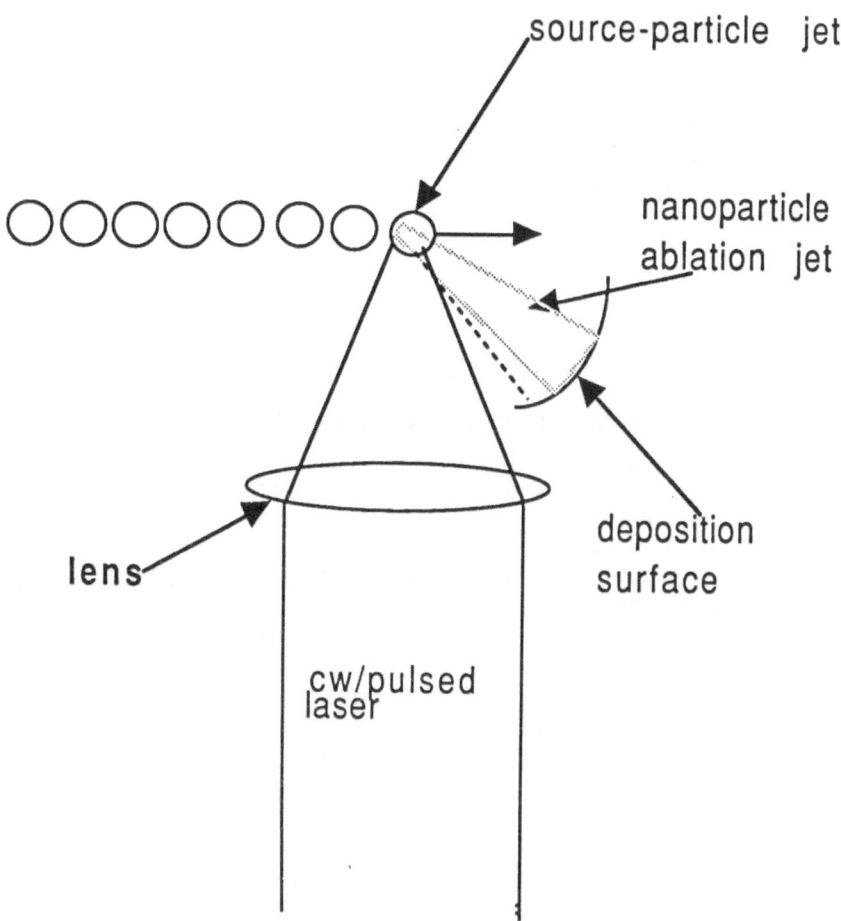

Fig. 3-2. SCHEMATIC diagram of continuous LAM process.

Fig. 3-2 indicates the LAM synynthesis mode used in obtaining the data discussed in this section, although there was some variation in the system used. In some cases (for silica) source particles were coated onto a quartz plate and the sessile source particles were then ablated. The continuous production method suggested makes possible high volume production rates; a second laserbeam detects particles and triggers the ablating beam. Based on experimental results of conversion efficiency of microparticle source particles (ranging from 60-90%), it is estimated that production rates in excess of 1 kg/hr. of microparticles are possible. Of course, such high production rates obviously will involve changes in the continuous process that have already been put into effect.

Note in Table 3-1 that the threshold fluence for laser induced breakdown of the glass particles is more than an order of magnitude less than that required for the glass slide--the analog of PLD in this case. Note also that the measure of monodispersity for particles, 0.31, is less than that for PLD, 0.6. According to N. A. Fuchs' criterion, near-monodispersity is present when the monodispersity index is around 0.2-0.25. Also, the largest particles formed in PLD are 930 nm in size, compared to the maximum size of 327 nm by LAM, and the mean particle size by PLD is five times that by LAM. This suggests that LAM could prove to be a replacement method for PLD in deposition of thin films.

Silica nanoparticles were synthesized by LAM. Results for this synthesis are reported here:

3.2 SYNTHESIS OF METAL NANOPARTICLES

If LAM is to be useful, it must be versatile for synthesis of metal nanoparticles, as well as for dielectrics, discussed above for silica. As LAM depends on the optical properties of the source microparticles, the synthesis of conductive metal nanoparticles is a large departure from dielectrics. The interaction of metal microparticles with pulsed laser light shows marked difference from that of dielectrics. Metal particles are conductors and hence, unlike dielectrics. have no internal electric fields under illumination by light. From our observations, it appears that maxima in the source function occur in the skin, or in the gas phase just outside surface of the metal sphere; the exact position of breakdown has not yet been determined but it is believed to correspond to the maxima in the source function--that is, in the skin. For dielectric spheres, on the other hand, for silica, breakdown appears to occur inside the microsphere near the shadow face, where there is a large maximum in the source function. We have synthesized nanoparticles from gold microsphere source particles. More recently, nanoparticles of silver have been synthesized from spherical silver microparticles with diameter of 8μm,. The size histogram from this synthesis is shown immediately below: The total number of particles in the size histogram is around 300. Ablation of the 8μm silver microparticles was done in air at 248 nm excimer laser beam wavelength at 2.2 J /cm^2. No agglomeration was observed between the silver nanoparticles collected on TEM grids.; all the nanoparticles represented in the figure below were separated spherical particles.

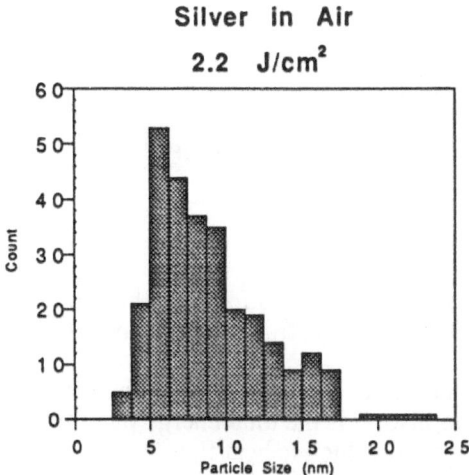

Silver in Air

2.2 J/cm²

Silver nanoparticles produced by ablation in air of 8 μm spherical silver source particles. at 248 nm wavelength and 2.2 J /cm².

3-3. Dynamics of Breakdown Process

With modern lasers, irradiances high enough to cause dielectric breakdown have been possible. This phenomenon has received much attention; laser induced breakdown in clean gases is well understood; Razier[12] provides an excellent book on this subject. It is well known that the presence of small particles in a gas lowers substantially the breakdown threshold of a gas-- ~10^6 W/cm² in air in presence of particles for 10.6 μm radiation and 10^9 W/cm² for clean air at same conditions. It is believed that breakdown probably occurs in the region where the source function is enhanced; here multiphoton ionization is suggested, leading to formation of a region of dense plasma.

By means of a 1-D numerical model, Carls and Brock [13,14,15] have studied the hydrodynamics of the ablation process for microparticles. The study does not attempt to include the initiation of breakdown and plasma formation;it begins with an assumed breakdown region. It proceeds by numerical solution of the 1-D Euler equations that have the form for this problem:

$$\frac{\partial \rho_i}{\partial t} + \nabla \cdot \rho_i V = -\nabla \cdot j_i \qquad (5)$$

$$\frac{\partial \rho V}{\partial t} + \nabla \cdot \rho VV = -\nabla P - \nabla \cdot \tau \qquad (6)$$

$$\frac{\partial \rho E}{\partial t} + \nabla \cdot \rho EV = -\nabla \cdot PV + S_{rad}$$
$$(7)$$
$$-\nabla \cdot q_{con} - \nabla \cdot (\tau \cdot V)$$

where Eq. (5) is the mass conservation equation, (6) the momentum equation (7) the energy equation. In (7), S_{rad} is the term that couples

the laser energy into the fluid dynamics, N is the number of chemical species in the plasma; also, E is the total energy per unit mass, kinetic plus potential; q_{con} is the conductive heat flux.

For permalloy, $Ni_{81}Fe_{19}$, we have shown [17] that the source microparticles need not be spherical; the source micropartucles were highly irregular in shape; also, based on magnetic measurements, stoichiometry was evidently preserved.

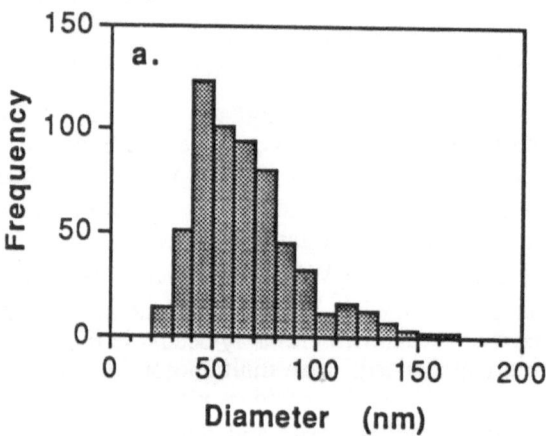

Fig. 3-2. Distribution of permalloy particles from LAM. produced at 3.1 J/cm^2.

Our calculations, using (5), (6) and (7) give the shock speeds as a function of laser fluence. For dielectric particles, the model is initiated by locating the initial location of breakdown at the internal maximum of the source function. This breakdown creates a dense plasma with extremely high pressures and temperatures. This plasma propagates through the particle, toward the laser principally by mechanical compression of the material; depending on the laser pulse, this shock can become an optical detonation wave, and finally on reaching the illuminated surface emerges as a blast wave propagating in the host gas toward the laser; by numerical studies, we have shown[16] that nanoparticles are formed by homogeneous nucleation immediately behind the outgoing shock in the host gas where there is a large drop in temperature.

We obtained the equation of state (E.O.S.) for our calculations from the SESAME Library maintained and originated by the Mechanics of Materials and E.O.S. groups at the Los Alamos National Laboratory . By comparison with Ref.[16], moment simulation of particle nucleation and growth gave mean diameters of nanoparticles formed that agreed well with experiment. [17]

An experimental observable in many investigations is the wave velocity. In our model simulations, three waves appeared--namely, a shadow side(the side shielded by the microparticle) blast wave, an illuminated side(facing laser beam), an internal liquid shock and the illuminated side blast wave. The calculation in our model begins with a plasma layer imbedded in the microparticle, close to the shadow side of the droplet. After this positioning of the plasma layer, immediately, two identical compression waves erupt from the plasma region. The two waves form when the high pressure initial plasma region explodes in both directions, compressing and ionizing the adjacent material. At early times these two waves are nearly identical, but the wave nearest the shadow face quickly leaves the droplet, causing rapid decompression and heating the surrounding gas. The shock accelerates into a blast wave as it enters the gas phase, forming the shadow side blast wave.

The second shock, moving through the droplet and toward the laser, also compresses, heats, and ionizes the material that it traverses. The shock strength determines the initial degree of ionization. As the material ionizes, it absorbs light, leading to further ionization. With sufficient ionization, the material becomes a plasma and absorbs strongly. Thus two ionization mechanisms are in operation-- so-called pressure ionization and laser beam heating. Strong laser beam heating can only occur after pressure ionization has ionized the material sufficiently. At very high laser beam intensities, the high intensity causes the shocked ionized material to absorb more energy than it uses compressing. The resultant higher ionization leads to higher absorption; the flow becomes self-accelerating. The shock grows strong enough to ionize the material completely. , and so the ionized material absorbs strongly, immediately after the shock passes, without any laser-assisted ionization. This is known as an optical detonation wave (ODW).

The table below compares velocities obtained by modeling water droplet breakdown in water vapor with those measured[18] during flurocarbon droplet breakdown in argon. The two systems are comparable because the breakdown mechanism is mechanical, through shock heating, and the two liquids have similar mechanical properties. Also, the plasma optics depend mostly on the free-electron density and not on the chemical nature of the positive ions. This table shows good agreement for the detonation and breakout waves, indicating that these waves are essentially planar in this case. This is probable because the

the laser beam forces a directional (planar) response of the fluids for these waves. Also, the narrow shock widths [~1nm] [19] minimize the distortions of the planar shocks by droplet geometry. Agreement is not so good for the plasma ejection velocity. Its motion is much less dominated by the laser beam and thus has a greater radial (more two-dimensional) component. The results from the model are encouraging and warrant more detailed experimental and computational studies.

Table 3-3. Comparison of Experimental Shock Wave Velocities with those Predicted by the Ablation Model

	Model H_2O(liq.)/ H_2O(vap.) (km/sec)	Experiment[18] C_8F_{18} (liq.)/ Ar (gas) (km/sec)
Detonation wave		
Medium beam intensity (10^{10} W/cm^2)	2.2	2.0
High beam intensity (10^{11} W/cm^2)	7.5	5-9
Plasma ejection		
Medium intensity	3	8
High intensity	8	6
Breakout wave	24	15-20

3-4. Advantages of the LAM Process

The LAM process permits easy control of nanoparticle mean size by choosing among a number of control variables: fluence, beam wavelength, and others. In addition for LAM, synthesis of nanoparticles with high purity is possible and it is sufficiently versatile to permit synthesis of nanoparticles of many diverse materials, including highly refractory materials; as support for this, it is noted that initial breakdown temperature is ~10^5 K. Although we have not demonstrated this in numerous experiments, we can state that LAM is an ablation process and therefore should share the property of ablation processes in preserving stoichiometry of its source material. Based on laboratory results to date, LAM has a high mass efficiency, in the range 70-90 % conversion of the source microparticles. This implies that high nanoparticle production rates are possible, something around ~1 kg/hr. We have made estimates of production costs of LAM; these are in the range, $4.50-6.00/kg of nanoparticles   There are many assumptions in this estimate. True costs can only be determined in the marketplace. Based on experiments performed, synthesized nanoparticles typically can have monodispersity index, <~0.25. The process is environmentally benign, depending, of course , on toxicity of source material.

In the process under development, all source material is retained and there should be no adverse emissions to the environment.

ACKNOWLEDGMENT

Work reported here was supported by National Science Foundation under grant ECS 9119043.

References

[1] Bolton , F.andRossler, U.(1993)Classical model of a Wigner Crystal in a Quantum Dot Superlattices and Microstructures 13,139-144

[2] Fafard, S.(1997) , Quantum Dots, *Photonics Spectra*, May ,pp.160-163.

[3] Turton,R .(1995),*The quantum dot: a Journey into the future of microelectronics*, R. , John Publisher, NY.

[4] Granqvist ,C.G. and Buhrman,R.A.(1976),Ultrafine metal particles, J. Applied Physics 47, p.2200.

[5] Fuchs, N. A. and Sutugin,A.G. (1971)Formation and Methods of Generation of Highly Dispersed Aerosols, G. M. Hidy and J. R. Brock,Eds., *International Reviews of Aerosol Physics and Chemistry*, Pergamon Press, Oxford, vol.2, pp10-12.

[6] Green, H. L. and Lane, W. R.(1964) *Particulate Clouds: Dusts, Smokes and Mists, E.. &F.N. Spon Ltd., London,pp.18-21.*

[7] Fuchs, N. A. and Sutugin, A. G.,(1966),Generation and Use of Monodisperse Aerosols,in C. N. Dasvies, ed., *Aerosol Science, Academic Press, London,* pp. 1-27.

[8] Holzapfel,,B., Roas, B., Schultz,L. Bauer,P. and Saemann-Ischenko, G. (1992),Off-axis deposition of $YBa_2Cu_3O_7-\delta$ thin filmsAppl. Phys. Lett. 61 3178-3180.

[9] B. Holzapfel,B. Roas,L. Schultz,L. Bauer, and G. Saemann-Ischenko, G. (1992). Off-axis deposition of $YBa_2Cu_3O_7-\delta$ thin filmsAppl. Phys. Lett. 61 3178-3180.

10 M. Becker, ,J. Keto, and J. Brock, J.,C. Juang,, andH. Cai,(1994). Synthesis of nanometer glass particles by pulsed-laser ablation of microparticles, Appl. Phys. Lett.65,40.

11 Material given in Sect.3 is result of collaborative efforts, at Univ. of Texas/Austin, between J. Brock (CH.E.), J. Keto (Physics), M. Becker (E.C.E.), and graduate students under grant ECS9119043, National Science Foundation.

12 Y. P.Raizer(1977). Laser Induced Discharge Phenomena, PlenumPub., NY.

13 J. Carls, and J. Brock,(1987), Explosion of a water droplet by pulsed laser heating, Aerosol Sci. Tech 7, 79-90

14 J. Carls and J. Brock,(1988)Propagation of laser breakdown and detonation waves in transparent droplets, Optics Letters 13 273.

15. J. Carls, J. Brock,,andY. Seo,(1991). Laser-induced breakout and detonation waves in droplets: II. Model, J. Optical Soc. America B, 8,329-336.

16. Lee, J., Becker, M., Brock, J., Keto, J., and Walser, R.(1998). Permalloy nanoparticles generated by laser ablation, IEEE Trans. Magn,32, 4484-4486.

17. N. Chaudhary, (1996). Investigation of pulsed laser ablation of microparticles in nanoparticle synthesis, Ph. D. Dissertation, Univ. of Texas/Austin, pp.149-190.
18. W. F. Hsieh, J. B. Zheng, C. F. Wood,B. T. Chu and R. K. Chang((1987). Propagation velocity of laser-induced plasma inside and outside a transparent drop, Opt. Letters12,576.
19. G. A.. Lyzenga, , T. J. Ahrens, W. J. Nellis, and A,. C. Mitchell, (1982),The temperature of shock-compressed water,. J. Chem. Phys. 76, 6282

NANOPARTICLES AND NANOSTRUCTURES: AEROSOL SYNTHESIS AND CHARACTERIZATION

RICHARD C. FLAGAN
California Institute of Technology
Pasadena, California 91125
USA

Abstract

The remarkable properties of materials synthesized from nanometer-sized particles were discovered using particles that were formed as aerosols. Although other routes for nanoparticle synthesis have evolved, aerosol routes remain a major method. The original processes employed by Gleiter and coworkers entailed evaporation of a metal into a low density gas where the vapors formed nanoparticles by homogeneous nucleation. Those nanoparticles were then collected by thermophoresis for subsequent consolidation. In a study that predated the use of vapor condensation as a step in the production of consolidated nanostructures, Granqvist and Buhrman provided empirical observations of the role of major control variables. A number of points of confusion remain in this method of nanoparticle synthesis, notably the distinctions between particle size, and the sizes of the microstructures that attract the attention of materials scientists.

This paper will examine the aerosol physics of nanoparticle synthesis with emphasis on unraveling this distinction. The physical processes that govern particle formation, growth, structure, and deposition will be examined. The on-line characterization of aerosol nanoparticles will also be probed, with a view toward monitoring of the synthesis process.

Introduction

Gleiter and coworkers[2] catalyzed the development of a new field of research with his reports of remarkable properties of materials that he synthesized by consolidation of nanometer-sized particles. Those particles had been synthesized by evaporation of a metal into a low density gas. Particles formed by homogeneous nucleation as the vapors diffused in the cold surrounding gas were collected on a cold finger and then consolidated in situ to prevent contamination by common atmospheric gases. This landmark work was by no means the first synthesis of such particulate material, for nanoparticle synthesis from the vapor phase is a well developed technology that has been used for over a millennium, and is now the basis for such commodity chemicals as carbon black which is produced by millions •of tons per year worldwide for structural reinforcement of polymers (a nanocomposite material), in pigments, in carbon resistors, and in many other applications[33]. That nanostructures of so simple a material as carbon can be commercially viable in such a wide range of common products clearly demonstrate the potential for aerosol synthesized nanomaterials to be produced economically at technologically significant rates. An understanding of the processes that govern the nature of aerosol synthesized nanoparticles will be extremely helpful in the development of viable technologies for nanoparticles production.

Optimization of nanoparticle synthesis is generally empirical, using off-line post-process characterization of collected materials to guide process development. Even in the

G.M. Chow and N.I. Noskova (eds.), Nanostructured Materials, 15–30.
© 1998 *Kluwer Academic Publishers.*

commercial synthesis of carbon black, off-line characterization is the rule. Only in those few research laboratories where process is the focus are on-line characterizations actively pursued. Yet methods are available that can provide near real-time characterization of the materials emanating from a nanoparticle synthesis reactor. Having been developed primarily for environmental research, these methods have not been fully optimized for the nanoparticle synthesis environment, but laboratory applications clearly demonstrate their potential.

This paper will examine the use of the methods of aerosol science in the development of aerosol processes for nanomaterials synthesis, with an emphasis on those processes that govern particle size and structure. We begin with a discussion of the nature of nanoparticles.

Nanoparticle Characteristization

The properties of materials produced by consolidation of nanoparticles depend on the distribution of sizes of the crystalites or primary particles that comprise the nanoparticles, on the chemical state of their surface which may be seriously degraded by contaminants, and on any factors that may limit the density of the compact. The latter issue is not well understood so much of the attention on nanoparticle synthesis has justifiably been focused on the former two issues. Hence, the most common descriptions of nanoparticles are the sizes of the "particles" as determined by electron microscopy. Figure 1(Left) shows such a micrograph. Although some information can be gleaned from this type of picture, it is woefully incomplete.

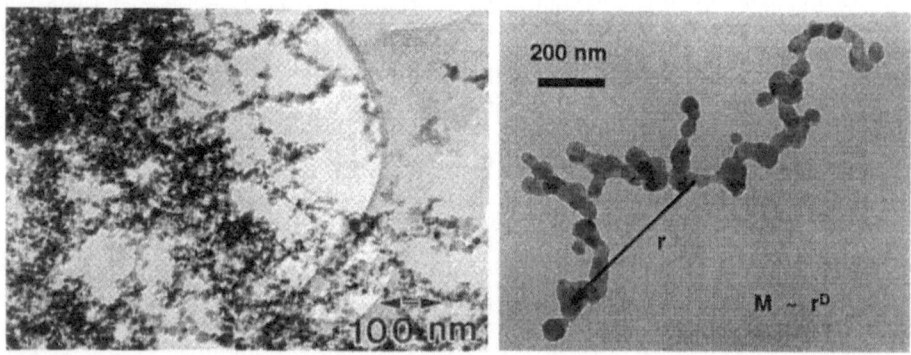

Figure 1. Transmission electron micrographs of nanoparticles synthesized from the vapor phase: (Left) Silicon nanoparticles produced by evaporation and condensation and collected on a holey carbon TEM grid. (Right) Transmission electron micrograph of a single agglomerate particle produced by silane pyrolysis in a nitrogen carrier gas. The particle was deposited thermophoretically directly on the TEM grid.

Although TEM pictures of collected samples reveal information that is relevant to the properties of the consolidated material, they do not give an accurate picture of the aerosol particles. To see the nature of the particles as they exist in the aerosol phase, it is necessary to deposit particles at very low surface coverage so that isolated particles can be seen. Fig. 1(Right) shows such a particle. The particle is a low density aggregate of small, roughly spherical primary particles. Such particles are formed by Brownian coagulation at temperatures that are two low for the particle that is produced when two particles collide to coalesce completely before the next agglomeration event occurs.

The different interpretations of particle size have caused considerable confusion. In an early study of aerosol synthesis of nanoparticles, Ulrich and coworkers[32, 35, 36, 34] probed the growth of SiO_2 particles produced by flame oxidation of $SiCl_4$ by BET gas adsorption. The uniform small size thus determined initially led them to infer that the probability that two particles would stick upon collision decreased abruptly. Later, this anomalous behavior was shown to be caused by slow coalescence. The inference of a low sticking probability was an artifact of using the primary particle size the characterize the growth process.

The structures of the agglomerates that is formed are remarkable similar for a wide range of materials and synthesis conditions. Forrest and Witten[7] observed that the structures were similar over a wide range of scales. This self-similarity was later described in terms of a power-law behavior, wherein the mass of material contained within a radius r from the center of a particle scales as

$$m \sim r^{D_f}. \tag{1}$$

The power-law exponent in Eq. (1) has come to be known as the mass-fractal dimension[25]. The power-law, or "fractal" scaling of particulate properties arises from the stochastic nature of coagulation events. Aerosol aggregation usually follows a process that has been labeled "cluster-cluster aggregation" which yields a mass-fractal dimension of about 1.8.

Although off-line measurements by TEM or BET gas adsorption have dominated the characterization of nanopowders synthesized by aerosol (or other) routes, the long analysis time makes them of limited value for process monitoring and control. Tools developed for on-line characterization of aerosol particles are capable of sizing particles in the nanometer size range in near-real-time. The differential mobility analyzer (DMA) [15, 40, 30] is the primary tool for nanoparticle characterization, providing size distributions on the aerosol particles in a few minutes or less depending on the mode of operation. The DMA is an aerodynamic analog of a mass spectrometer in which charged aerosol particles are caused to migrate across a clean, particle-free sheath flow as illustrated in Fig. 2. Only particles that migrate within a narrow interval of velocities are transmitted, as illustrated by transmission electron micrograph size analysis of the transmitted particles as illustrated in Fig. 3[3].

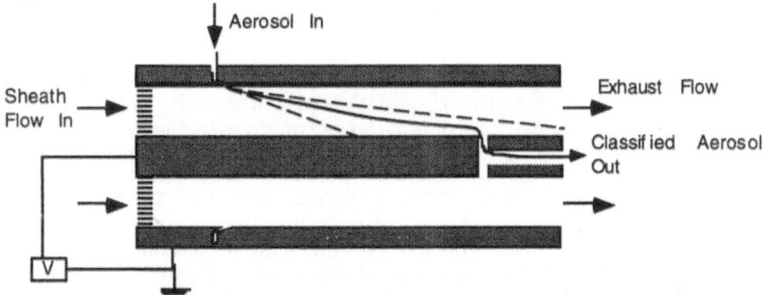

Figure 2. Schematic of coaxial-cylinder differential mobility analyzer.

The difference between the microscopic characterization of the crystallite size and the nature of the particles as they exist in the synthesis process environment can readily be seen by comparing different characterizations of the same particles, as shown in Fig. 4. Silicon nanoparticles were synthesized by pyrolytic decomposition of silane gas in a nitrogen carrier gas at atmospheric pressure. Collected particles were sized by transmission electron microscopy. In a related experiment, the BET gas adsorption method, has been shown to yield estimates of the mean particle size that agree well with TEM observations[27]. On the other hand, as shown in Fig. 4, the size determined by differential mobility analysis shows a very different trend of particle size with precursor concentration. The difference between these measures is that the former methods characterize the fine structures of the particles after collection, while the latter method measures the aerodynamic drag on the particle while it is still in the aerosol phase. For particles that are much smaller than the mean-free-path of the gas molecules, the drag force is proportional to the projected area of the particle. The ramified structure of the aerosol agglomerates exposes much of the total primary particle area to the flow of surrounding gas, so the drag force scales approximately with the total area of the agglomerate. Because measurements are made on singly charged particles (due to the low probability of a particle acquiring any charge when exposed to ambipolar gas ions produced by radioisotope decay in the aerosl charger), the mobility of

18

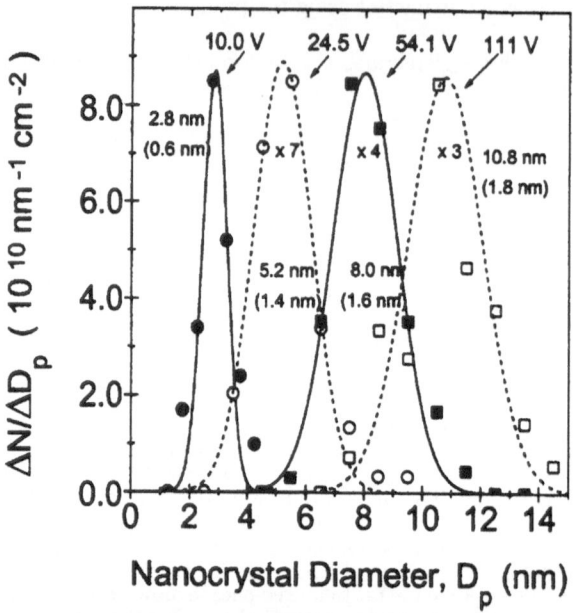

Figure 3. Size distributions measured by transmission electron microscopy on silicon nanoparticle fractions that were size classified with a radial differential mobility analyzer. (Adapted from Camata, 1997)

the agglomerate scales roughly as the inverse of the number of primary particles which comprise it (assuming a uniform primary particle size). In mobility analysis, the size of a particle is generally reported as the size of a dense sphere with the same mobility as the particle in question. This mobility-equivalent size, D_{mob}, shown in Fig. 4, decreases with increasing temperature as coalescence of smaller particles reduces the total projected area while increasing the primary particle size as determined by TEM or BET analysis.

From these observations, we may form a picture, illustrated in Fig. 5, of how particles evolve in synthesis from the vapor phase. The hot source, or reactions of a chemical precursor generates vapors. In the case of the system used by Gleiter, those vapors diffuse into the surrounding atmosphere. At some point as the vapors cool, the vapors nucleate to form small particles. The particles then grow by coagulation, initially coalescing to form dense particles that become the primary particles in later agglomeration. That the sizes of the primary particles tend to be fairly uniform may suggest that the transition from dense particle growth to agglomerate formation is relatively abrupt. To understand what determines the transition from one growth mechanism to another, we must examine the physical processes involved in some detail.

Particle Inception

Although particles are frequently produced by mechanical break-up of bulk material, the work required to generate the huge surface areas of nanoparticles is so large that it is impractical to employ such routes for nanoparticle synthesis. The vast majority of nanoparticle synthesis is, therefore, based on transformations from a fluid phase to a solid phase. The fluid phase may be liquid in the case of colloidal processing or vapor for aerosol processing. Phase transitions within large aerosol droplets can be used to produce nanostructures[28] by routes that are similar to colloidal processing. This paper focuses on synthesis from the vapor phase, the approach to aerosol synthesis that initiated the field of nanostructured materials.

Most synthesis of aerosol nanoparticles has been accomplished by physical condensation

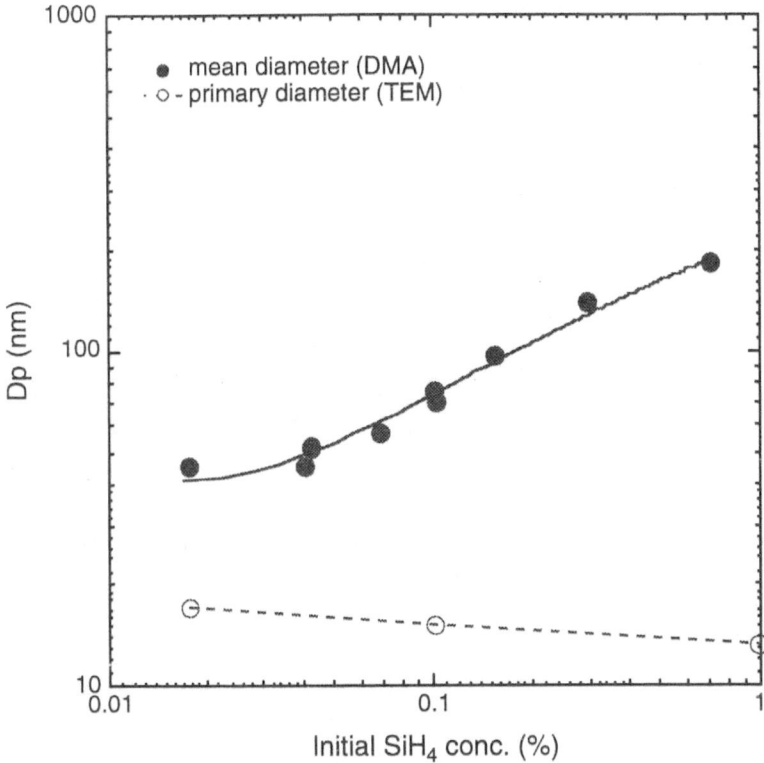

Figure 4. Variation of apparent particle size with precursor concentration in silicon nanoparticle synthesis by silane pyrolysis as inferred by TEM measurements and differential mobility analysis.

of vapors, a process that is initiated by homogeneous nucleation of the vapors. The theory of homogeneous nucleation of condensing vapors is well developed, although predictions of absolute nucleation rates frequently differ from experimentally measured values by many orders of magnitude[12]. It should be noted, however, that few measurements of the nucleation rate of the substances of interest to researchers investigating nanostructured materials. Substances that have been investigated tend to be materials that are volatile at room temperature, e.g., water, organics, etc. One metal that has been studied is mercury, for which deviations from experimental measurements are as high as 40 orders of magnitude. Nonetheless, recent theoretical investigations show promise toward resolving this uncertainty, at least for simple species.

The classical theory of homogeneous nucleation describes the growth of clusters by a

Particle Growth Processes

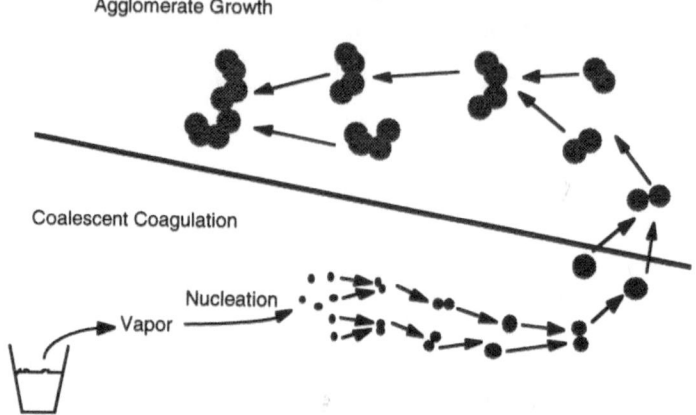

Agglomerate Growth

Coalescent Coagulation

Nucleation

Vapor

Figure 5. Schematic of the processes that contribute to particle growth in nanoparticle synthesis from the vapor phase.

sequence of monomer addition reactions, i.e.,

$$A + A \;\rightleftharpoons\; A_2$$
$$A_2 + A \;\rightleftharpoons\; A_3$$
$$\ldots$$
$$A_k + A \;\rightleftharpoons\; A_{k+1},$$

when the vapor is present at concentrations in excess of the equilibrium value. The rate of addition of monomers to small clusters is the product of the monomer impingement flux, β, which can be calculated from the kinetic theory of gases, the surface area of the cluster, and the sticking probability α_k. The net rate of cluster growth from size k to size $k+1$ is the difference between the rate of monomers additions and that for evaporation.

$$J_k = \beta s_k k_k N_k - E_{k+1} N_{k+1} \qquad (2)$$

The sticking probability is generally assumed to equal unity although the need for collisions of background gas molecules to rapidly extract excess collision energy will lead to a lower sticking probability on very small clusters. The principle of detailed balancing is usually applied to derive the evaporation rate coefficient E_k noting that, at full thermodynamic equilibrium, the forward and reverse fluxes must exactly balance[12].

To determine the evaporative flux using detailed balancing, one must know the distribution of clusters of different sizes at full thermodynamic equilibrium. Thermodynamic data are not generally available for molecular clusters. The free energy of a molecular cluster can be represented as the sum of the free energy of the same number of molecules of bulk liquid and a correction factor that accounts for the finite cluster size. The difficulty with the classical theory lies in assuming that the correction term is equal to a surface free energy contribution and that the applicable surface tension is that of the bulk liquid. Efforts to correct the classical theory have shown that inclusion of size dependence of the surface free energy lead to much closer agreement between theory and experiment. The most straightforward extension is to include the Tolman model for the size dependent surface tension[29]. More elaborate models have been developed as well. These have been shown to be different levels of treatment of the series approximation to the surface free energy of the clusters[23]. If, instead, the free energies of the small clusters are determined directly using the methods of computational chemistry, more realistic models of the nucleation process result[4, 31, 24]. For example, detailed calculations of the nucleation of mercury lead to rates that agree within a few orders of magnitude with experimental observations[24]. In this case, it has been found that a transition from metallic to nonmetallic behavior at a cluster size of 13 atoms accounts for the extreme discrepencies when the classical theory is applied directly.

When reactive gases are used to generate nanoparticles rather than physical condensation, the rate of material addition to the surface of the growing cluster becomes a chemical process rather than simple physical vapor deposition. The theory of nucleation via chemical reactions is less well developed than is the classical nucleation theory, although some progress has been made[13]. Again, detailed knowledge of the thermodynamic properties of the small clusters is needed to develop quantitative predictions.

In either case, the conditions in particle synthesis systems usually lead to rapid formation of large numbers of small particles. The surface area of those particles is large enough that they then scavenge additional vapor or reactive species, reducing the supersaturation and quenching further nucleation. As a result of this competition between nucleation and vapor deposition, new particles are typically formed only during a brief burst of nucleation.

PARTICLE GROWTH

The nucleation burst generates large numbers of particles that then grow by continued vapor deposition, either physical condensation or chemical vapor deposition, or by Brownian coagulation. When high number concentrations are generated, during the nucleation burst, coagulation will usually dominate. Special circumstances exist, however, when vapor deposition continues through the growth process.

Growth by Vapor Deposition

If the production of condensible vapors is slow, the number concentration generated during the nucleation burst may be small enough that coagulation proceeds relatively slowly. Vapor deposition may then play a key role in determining the nature of the particle size distribution. Alam and Flagan[?] demonstrated that gradual acceleration of the rate of reaction of a gaseous precursor could be used to grow particles to far larger sizes than can be achieved within reasonable reactor residence times in coagulation dominated particle growth. Okuyama et al.[?] subsequently demonstrated that the same method could be applied to controlled growth of aerosol nanoparticles. A tube furnace with multiple heating zones was used to produce different temperature profiles shown in Fig. 6. Reaction at constant, high temperature (circles) produced a broadly distributed, coagulation-dominated size distribution. Reaction with a gradually increasing temperature (squares) produced a narrower distribution, with a peak in the mobility distribution below 10 nm diameter. These results clearly indicate that a suitably designed reactor can produce nanoparticles that are not agglomerated to larger sizes prior to collection.

Coagulation Dominated Growth

Due to the very high number concentrations that are produced during the nucleation busrt in many synthesis environments, remaining vapors are rapidly depleted by deposition onto the surfaces of the nuclei. The particles then undergo rapid growth by Brownian coagulation. While the particles are small and the gases relatively hot, particles coalesce quickly after agglomeration. Later, however, due to decreasing mobility of the atoms in the small particles due to cooling and decreasing excess surface free energy driving forces, the coalescence slows, and the particles grow as low density agglomerates. Although the fine structure of the resulting particles may be on the desired nanometer scale, the particles are actually much larger and may be quite rigid, hindering consolidation of the powder. To understand the competing processes that govern the structure of nanoparticles, we must examine each of the processes involved in the particle growth.

The growth of aerosol particles by coagulation is described by an integro-differential

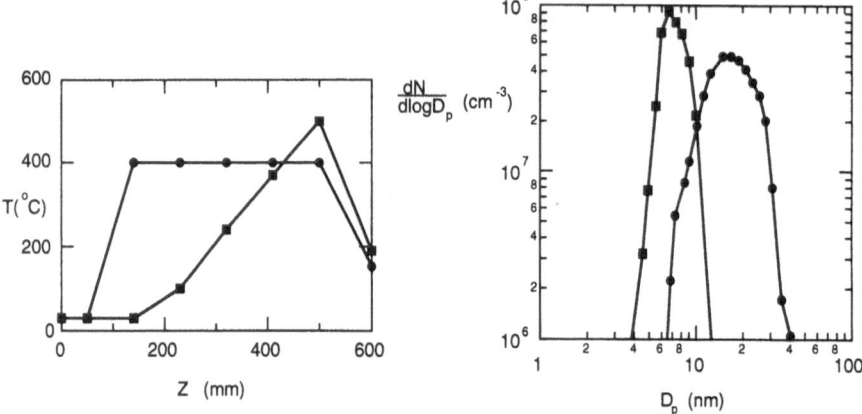

Figure 6. Experimental results of Okuyama et al. (1989) showing the influence of the temperature profile on the size distribution of particles produced by titanium tetraisopropoxide pyrolysis.

equation,

$$\left(\frac{\partial n(v,t)}{\partial t}\right)_{\text{coag}} = \frac{1}{2}\int_0^v \rho K(v-u,u)n(v-u,t)n(u,t)du - n(v,t)\int_0^\infty \rho K(v,u)n(u,t)du \quad (3)$$

where $K(v,u)$ is the rate coefficient for Brownian collisions between particles of volumes u and v and $d\mathcal{N} = n(v,t)dv$ is the number of particles with volums between v and dv per unit mass of gas, and ρ is the gas density. The first term on the right-hand-side describes the formation of particles of volume v by coagulation of smaller particles. The second term describes the loss of such particles by coagulation with particles of any size. Although the particle sizes of interest are small, the numbers of atoms involved are sufficiently large that the continuous representation of the particle size distribution ($n(v,t)$) is an excellent approximation that leads to important simplifications results for predicting the performance of nanoparticle synthesis systems.

Even at atmospheric pressure, nanoparticles remain small compared to the mean–free–path λ of the gas molecules throughout their growth. The appropriate collision frequency function for modeling nanoparticle synthesis is, therefore, that for coagulation of free molecular particles, i.e.,

$$K(v,u) = A_{\text{collision}}\frac{\bar{c}_{uv}}{4} \quad (4)$$

The mean relative thermal speed of the two particles is

$$\bar{c}_{uv} = \sqrt{\frac{8kT}{\pi\rho_p}}\left(\frac{1}{v}+\frac{1}{u}\right)^{\frac{1}{2}} \quad (5)$$

and $A_{\text{collision}}$ is the collision cross sectional area. Early in the growth process, the particles coalesce rapidly and may, therefore, be assumed to be spherical, so that the effective cross section for collision of a particle of volume v with one of volume u is

$$A_{\text{collision,s}} = 4\pi\left(\frac{3}{4\pi}\right)^{\frac{2}{3}}\left(v^{\frac{1}{3}}+u^{\frac{1}{3}}\right)^2. \quad (6)$$

Once coalescence slows, however, the apparent density of the particle decreases and the collision area increases. As a result, the coagulation process accelerates once agglomerates begin to form[5]. This observation suggests that, if dense particles are sought, agglomerate formation must be prevented altogether. This rather strong statement can be tempered somewhat once the nature of the transition from coalescent coagulation to agglomeration is understood. Before turning to that question, however, it will be useful to examine the coagulation process a little further.

Solution of the general dynamic equation (Eq. 3) is possible only in a few cases. One special case is when the collision frequency function is *homogeneous*. The collision frequency function is said to be homogeneous of degree γ if $K(\beta\eta, \beta\tilde{\eta}) = \beta^\gamma K(\eta, \tilde{\eta})$. The free molecular collision frequency functions for dense spheres and mass fractal agglomerates are homogeneous of degree $\gamma = 1/6$, and $\gamma = \nu/D_f - 1/2$, where $\nu = Max(D_f, 2)$, respectively.

When the collision frequency function is homogeneous, the particle size distribution of a coagulating aerosol asymptotically approaches a self–preserving form that can be expressed in terms of the dimensionless particle volume[21, 26, 22]

$$\eta = \frac{v}{\bar{v}(t)} \quad , \quad \psi(\eta) = \frac{n(v,t)}{\mathcal{N}(t)}\bar{v}(t) \tag{7}$$

where $\bar{v}(t) = V/\mathcal{N}(t)$ is the instantaneous mean paarticle volume, and V is the total volume of particulate material per unit mass. Substituting the self–preserving particle size distribution into Eq. (3), and integrating both sides over all particle volumes yields a rate equation for the total particle volume per unit mass of gas in the aerosol system,

$$\frac{d\mathcal{N}}{dt} = -2^{1-\nu}\alpha\kappa\rho\bar{v}_0^{\frac{2}{3}-\frac{\nu}{D_f}} V^{\frac{\nu}{D_f}-\frac{1}{2}} \mathcal{N}^{\frac{5}{2}-\frac{\nu}{D_f}} \tag{8}$$

where α is the dimensionless coagulation integral for the self–preserving particle size distribution function that must be determined by numerical solution of the coagulation equation. For free–molecular coagulation of spherical particles, $\alpha = 6.67$.[21] The self–preserving particle size distribution for free molecular coagulation of spherical particles is approximately log–normal, with a geometric standard deviation of $\sigma_G \approx 1.35$. The value of α for agglomerate particles should, at least, be of the same order of magnitude.

The robust nature of the self–preserving solution to the GDE can be seen by examining the size distribution of aerosols at very different sizes produced by Brownian coagulation. Figure 7 shows size distributions measured by differential mobility analysis for silicon aerosols that had been heat treated to to a temperature high enough to evaporate some of the material. The vapors then nucleated upon cooling, leading to a size distribution with the same shape, but a very different size from the coagulation aerosol from which the vapors were derived.

Assuming that the total particle volume per unit mass of gas V is constant throughout the coagulation growth process, the mean particle volume thus obeys

$$\frac{d\bar{v}}{dt} = 2^{1-\nu}\alpha\kappa\rho\bar{v}_0^{\frac{2}{3}-\frac{\nu}{D_f}} V\bar{v}^{\frac{\nu}{D_f}-\frac{1}{2}}. \tag{9}$$

Integrating Eq. (9) over time subject to an initial condition of $\bar{v}(0) = \bar{v}_0$, e.g., the size of particle produced by homogeneous nucleation, yields

$$\bar{v}(t) = \bar{v}_0 \left[1 + 2^{1-\nu}(\frac{3}{2} - \frac{\nu}{D_f})\frac{\alpha\kappa_0\rho_0 V}{\bar{v}_0^{\frac{5}{6}}} \int_0^t \sqrt{\frac{T}{T_0}}dt \right]^{\frac{1}{\frac{3}{2}-\frac{\nu}{D_f}}} \tag{10}$$

The integration over the temperature-time history of the aerosols arises from the dependence of the density and mean thermal speed on temperature.

Equation (10) shows that the rate of particle growth depends on the total volume of particulate material per unit volume of gas, ρV, the fractal dimension of the particles, D_f

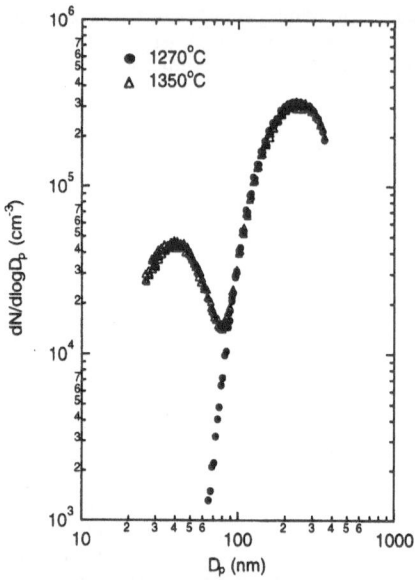

Figure 7. Size distributions of a silicon aerosol that was heat treated after growth by coagulation. Heat treatment at 1270°C maintained the original self-preserving particle size distribution. Heat treatment at 1350°C vaporized a small amount of material that nucleated upon cooling, producing a second self-preserving particle mode in the size distribution.

(for dense particles, $D_f = 3$), the sizes of the primary particles from which the agglomerates are assembled, and on the temperature-time history of the aerosol system. Because the particles of interest are much smaller than the mean-free-path of the gas molecules, there is no explicit dependence on the total pressure in the reactor. Instead, the dependence on $\rho V = P_v / R_{gas} T$ indicates a influence of the partial pressure of the vapor precursor, P_v.

Experimental observations in inert gas condensation reactors, on the other hand, show a strong dependence of the primary particle size on the system pressure[9]. This apparent contradiction can be resolved by examining the complex vapor source commonly used in vapor condensation systems. For a system with uniform composition and starting from thermal equilibrium with a hot vapor source, particle growth would be independent from the overall pressure. However, diffusion of the vapors from the hot source into the surrounding low density gas varies inversely with overall pressure. Hence, lowering the pressure increases vapor dispersion and dilution, thereby lowering the volume concentration of particulate material produced by condensation and slowing their growth by coagulation. The complex interplay between heat and mass transfer in the vicinity of the vapor source is beyond the scope of this paper. Rather than attempt to model this complex source in detail, the calculations presented in the remainder of this paper will be based upon a uniform gas composition. The resulting model captures the essential features of particle growth in these systems without the complexities of modeling the buoyancy-driven flows of present vapor condensation systems and is more applicable to forced flow systems that will probably dominate in large-scale nanoparticle production.

A number of important features of nanoparticle synthesis systems are revealed, when this simple model is applied to a homogeneous synthesis system. Consider the question of how large particles grow within the region of the reactor where temperatures are high enough to ensure rapid coalescence. As noted above, particle evolution depends on the partial pressure of the vapor precursor rather than the total system pressure. Figure 8 presents an operating map for spherical particle growth by coagulation, showing isopleths of constant final particle size and mass production rate as a function of the precursor partial pressure the time available for particle growth. Isopleths of constant mass production rate per unit of active reactor volume are estimated as the ratio of the initial vapor density to the mean residence time in the reactor region where particles grow by coalescent coagulation, i.e., ρ_v/τ. As one increases the precursor vapor pressure at constant growth time, both the size of the product particles and the mass throughput increase. Thus, increasing the system vapor pressure has two effects, one desirable and one undesirable for nanoparticle synthesis.

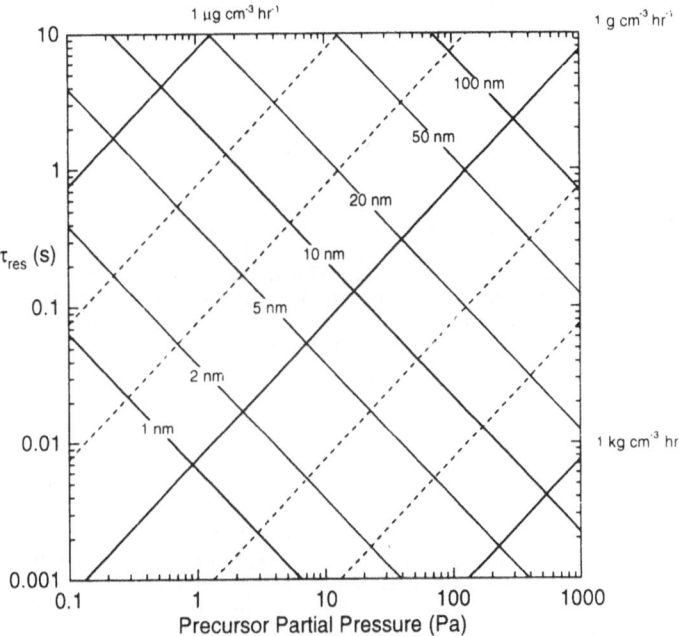

Figure 8. Operating map of isothermal nanoparticle synthesis reactor performance. Isopleths of constant mean particle size and reactor throughput are plotted as functions of pressure and residence time for spherical particle growth by Brownian coagulation at 1600K. The mass throughput per unit of active reactor volume is also indicated. (Flagan and Lunden, 1995)

The operating map shows that, to produce particles of a given size, the residence time must be increased as the pressure is decreased, leading to a decrease in the mass production rate. To increase the mass throughput of particles of a given size, the pressure could be increased, but the residence time would have to be decreased to limit growth. In an inert gas condensation reactor operated at 1000 Pa vapor pressure, i.e., under conditions similar to the Gleiter reactor, 5 nm radius particles are produced in about 0.1s at a rate of order 1 mg cm^{-3}hr^{-1}. A higher pressure source, such as a thermal plasma reactor operated at atmospheric pressure, could produce such particles in times as short as 1 ms. This thousand-fold reduction in the growth time is accompanied by a comparable increase in the mass production rate. Of course, neither of these systems is operated in an isothermal, homogeneous composition mode as we have modeled here, but the comparison is qualitatively valid. If one wants to increase the production rate of a reactor in which the primary growth mechanism is coalescent coagulation, operation at elevated pressures and short growth times is very beneficial. The time for growth of the primary particles is determined by the time during which the temperature is high enough for particles to coalesce completely. Once the temperature has dropped below that point, either hard or soft agglomerates may be generated, but primary particle growth will be severely limited.

To understand this transition, we must examine the competition between coagulation and coalescence. That comparison is conveniently made on the basis of characteristic times for the two processes. Since, as we have noted above, once agglomerates begin to form, coagulation accelerates. This generally happens at the same time that coalescence is slowing due to decreasing temperature and surface free energy driving force. Thus, the time scale relevant to this transition is that for coagulation of dense, spherical particles.

The time scale on which coagulation significantly decreases the particle concentration can readily be estimated from Eq. (8) as

$$\tau_{cs} = \frac{12 \ \bar{v}P^{5/6}}{5 \ \alpha\kappa\rho V} \tag{11}$$

for spherical particles. The characteristic time for agglomerate coagulation shorter by a factor that depends on the fractal dimension of the aggregates.

Particle Coalescence

A number of condensed phase transport mechanisms contribute to the coalescence of agglomerates produced by coagulation. Our interest is in the time required for complete coalescence of the bisphere that is produced when two spherical particles coagulate. The excess surface of the bisphere provides the driving force for interdiffusion of the two particles. Viscous flow and diffusion through the particle volume, grain boundaries, or surfaces may contribute to the coalescence depending on the material properties and growth conditions. The early phase of this process is described in classical sintering models[8, 20, 14, 11] in terms of the growth of the radius of the neck between the spheres relative to the sphere radius by a single mechanism, i.e.,

$$\left(\frac{x}{a_i}\right)^n = \frac{B(T)}{a_i^m} t \tag{12}$$

where a_i is the initial radius of the spherical particle and m and n depend upon the individual mechanism. These expressions are based upon simplified geometries for the growing neck that limit their applicability to a neck radius below about $x/a_i \approx 0.3$. Although they do not describe the complete coalescence that we seek to understand, the early stage sintering models do provide an estimate for the time required for sintering. That characteristic time for neck growth is

$$\tau_s = \frac{a_i^m}{B(T)} \tag{13}$$

A number of investigators have applied such models of neck growth to the description of the structural evolution of aerosol particles. Ulrich[32] observed that the sizes of silica particles produced in a flame synthesis process grew much more slowly than coagulation calculations would suggest. He later attributed the slow growth to slow viscous coalescence of agglomerate particles, and attempted to model the process by tracking the number of primary particles that make up the agglomerates Ulrich[36]. He characterized the rate of assimilation of one primary particle into another one in terms of the characteristic time for viscous coalescence[8],

$$\tau_{sv} = \frac{\mu a}{\gamma} \tag{14}$$

where γ is the surface free energy of the material and μ is its viscosity. More recently, Koch[17] expressed the relaxation of the particle structure in terms of its surface area. They suggested that a rate equation be written for the decay of the particle surface area and that the decay could be modeled as a first order process, i.e.,

$$\left(\frac{ds}{dt}\right)_{coalescence} = -\frac{s - s_s}{\tau_s} \tag{15}$$

where s_s is the surface area of the fully coalesced particle, i.e., that of a sphere with the same volume as that of the agglomerate, and τ_s is the characteristic time for coalescence that was estimated using classical sintering models. Although there is no rigorous basis for this model, calculations of Hiram [10] suggest that the approach surface area to that of a dense sphere size in viscous coalescence is approximately exponential for $t \gg \tau_s$.

Based on the long time limit, they estimated τ_s based on the final size of the fully densified spherical particle. Although the model was developed to describe nearly dense particles undergoing viscous coalescence, Koch[17] applied their model to the entire evolution of the aerosol, extending from early times when particles coalesce quickly to later times when large agglomerates are the dominant structures. From this beginning, both approximate and detailed predictions of the size and surface area distribution function have evolved.[17, 18, 37, 16, 19, 38, 39] The use of a simple parameter such as the agglomerate surface area or the number of primary parameters is cogent. However, although considerable effort has been invested in developing and solving representations of the coupled coagulation/sintering problem, remarkably little attention has been given to form of the sintering term itself.

For the present, we limit our analysis to the transition from coalescent coagulation to agglomerate formation by comparing the characteristic times for coagulation with that for coalescence. We note that, once τ_s/τ_c exceeds a value near unity, complete coalescence is unlikely. From that time on, particles may be expected to grow as agglomerates. Some sintering may continue, and this will generally lead to undesired neck growth resulting in hard agglomerates rather than complete coalescence. Only by increasing the temperature as the particles grow, thereby accelerating sintering and reducing τ_s, can particles continue to grow beyond this transition point. More commonly, the temperature will decrease as growth progresses, accelerating the transition. Thus, the key to the growth of dense, nonagglomerated particles by coagulation is to quench coagulation before the onset of agglomeration, i.e., while τ_s/τ_c is still less than unity. In the discussion that follows, particle growth is followed only until $\tau_s/\tau_c = 1$. Throughout the growth phase, the particles coalesce on a time scale that is short by comparison with the time between collision events. Hence, for the present purposes it is sufficient to consider only the coagulation of spherical particles.

For isothermal growth, the ratio of the sintering time to the coagulation time will increase only slowly with particle size, or even decrease depending on the growth mechanism. Thus, the transition would not generally occur within the residence time in the synthesis reactor. This is suggested by the characteristic times for isothermal particle growth in Fig.9 using a silicon aerosol and surface diffusion, $m = 4$, as the sintering mechanism. If the transition is passed, the slow increase in the coalescence time means that substantial neck growth can be expected following any coagulation events that subsequently occur. However, isothermal growth is an idealized case that would only be expected in carefully controlled laboratory experiments.

The inert gas condensation system and most other fine particle synthesis technologies employ a high temperature source, but allow substantial cooling of the aerosol after particles are formed. Indeed, in the inert gas condensation method and thermal plasma reactors, the temperature drop is responsible for particle formation in the first place. To examine the role of cooling on the transition from coalescent growth to agglomeration, we consider the particle dynamics when the aerosol is cooled at a constant rate.

Figure 9 shows the influence of the cooling rate on the variation of the characteristic time for coalescence of a silicon aerosol formed at an initial temperature of 1400 K and subjected to cooling rates of 10^2 K/s and 10^4 K/s. In both cases, cooling causes the characteristic time for coalescence to increase dramatically. The low cooling rate allows a longer time for growth before the coalescence time surpasses that for coagulation. The two characteristic times rapidly diverge, although the divergence is more rapid at the higher cooling rate. The divergence will be greater than indicated here because the more rapid growth of the agglomerate particles is not considered in Fig. 9.

The rapid reduction in the sintering rate essentially freezes the structure of the primary particles. Ideally, neck growth would be terminated abruptly by rapid cooling, so that agglomerates formed after the desired product particle size is reached would be held together only by Van der Waals forces, Unfortunately, necks may continue to grow between agglomerated particles after τ_s exceeds τ_c. This will depend upon the rate of divergence of the characteristic times. This divergence is not always as large as the ones shown in Fig. 9(left). The initial temperature also influences the transition from coalescent growth to agglomeration [5]. If the initial growth temperature is low enough that the coalescence time is only slightly smaller than that for coagulation, the rate of divergence is reduced dramatically as shown in Fig. 7 (right). The lower divergence rate will allow greater neck formation of agglomerates formed shortly after the onset of agglomeration. Reactor

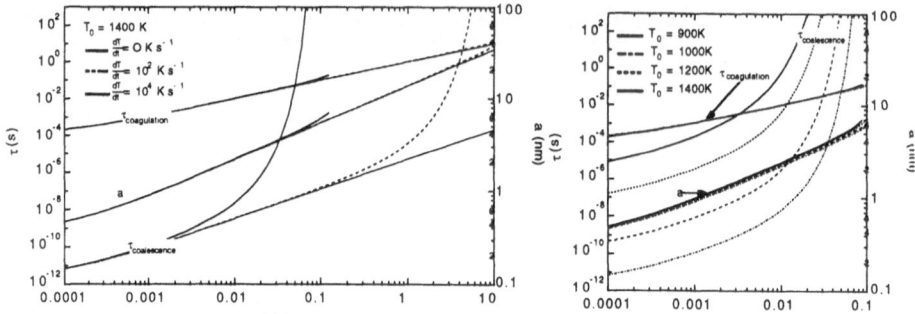

Figure 9. Influence of the cooling rate on the variation of the characteristic time for coagulation and coalescence of silicon nanoparticles. The product particle radius is also shown. (Flagan and Lunden, 1995)

operation with a higher temperature in the primary growth region will, therefore, inhibit the formation of strongly necked agglomerates.

Nanoparticle Separations

Once nanoparticles are formed, they must be separated from the gas in which they are carried. Small-scale laboratory synthesis systems have collected particles on a cold surface, often one that is cooled with liquid nitrogen. The extreme temperature differences between the vapor source and the collection surface lead to very low thermodynamic efficiencies. Nonetheless, the simplicity of the method makes it quite useful for laboratory applications, and even for some large scale synthesis systems, i.e., a number of carbon black synthesis systems employ a cold surface to quench the soot (carbon black) formation and oxidation and to collect the product.

The mechanism by which the particles are collected in these systems is *thermophoresis*, a thermal diffusion process in which the particles migrate from regions of high temperature to lower temperatures. The thermophoretic migration velocty can be expressed as

$$v_T = -\frac{\text{Th}\mu}{\rho T}\frac{dT}{dx} \qquad (16)$$

where the dimensionless thermophoretic parameter, Th, is essentially independent of particle size in the free molecular regime and is predicted to lie between 0.42 and 1.5, with experimental measurements in the range of 0.5[6]. Thermophoretic velocities are low unless the temperature gradients are quite high. One could readily design thermophoretic collection systems that achieve higher temperature gradients than produced by natural convection over a cold surface, but the energy losses are inherently large in these systems.

Particles could be efficiently collected by filtration, although the possibility of filter fibers contaminating the sample must be considered. Electrostatic precipitation can also be made quite efficient for nanoparticle collection, with a lower risk of contamination. The design basis for these separation systems are thoroughly described in texts on air pollution control, e.g., Flagan and Seinfeld[6].

The aerosol techniques described in this paper can readily be used to produce materials that exhibit size-dependent optical and electronic properties. To verify this size dependence and to take advantage of them in electronic and photonic devices, the sizes of the product particles must be controlled to within very narrow intervals. Since aerosols grown by coagulation inherently have broad size distributions, high resolution size classification is needed to achieve this control. The differential mobility analyzer can achieve the necessary precision, but efforts to process large quantities of aerosol lead to disortions of the classifier response due to space charge effects[1, 3]. New classifier designs will be required to work at the higher concentrations required to classify nanoparticles at high rates.

References

1. Alonso, M., & Kousaka, Y. 1996. Mobility shift in the differential mobility analyzer due to Brownian diffusion and space charge effects. *J. Aerosol Sci.*, **27**, 1201–1225.
2. Birringer, R., Gleiter, H., Klein, H. P., & Marquardt, P. 1984. *Phys. Lett.*, **102A**, 365–369.
3. Camata, R. 1997. *Aerosol Synthesis and Characterization of Silicon Nanocrystals.* Ph.D. thesis, California Institute of Technology.
4. Ellerby, H. M., Weakliem, C. L., & Reiss, H. 1991. Toward a molecular theory of vapor phase nucleation. 1. Identification of the average embryo. *J. Chem. Phys.*, **95**, 9209–9218.
5. Flagan, R. C., & Lunden, M. M. 1995. Particle structure control in nanoparticle synthesis from the vapor phase. *Mater. Sci. Eng. A.*, **204**, 113–124.
6. Flagan, R. C., & Seinfeld, J. H. 1988. *Fundamentals of Air Pollution Engineering.* Engelwood Cliffs, NJ: Prentice-Hall.
7. Forrest, S. R., & Witten, T. A. 1979. *J. Phys. A.: Math. Gen.*, **12**, L109–L117.
8. Frenkel, J. 1945. *J. Phys.*, **9**, 385.
9. Granqvist, C. G., & Buhrman, R. A. 1976. Ultrafine metal particles. *J. Appl. Phys.*, **47**, 2200–2219.
10. Hiram, Y., & Nir, A. 1983. *J. Colloid Interface Sci.*, **95**, 462.
11. Johnson, D.L. 1968. New Method of Obtaining Volume, Grain-Boundary, and Surface Diffusion Coefficients from Sintering Data. *J. Appl. Phys.*, **40**, 192–200.
12. Katz, J. L. 1992. Homogeneous nucleation theory and experiment - A survey. *Pure Appl. Chem.*, **64**, 1661.
13. Katz, J. L., & Donohue, M. D. 1982. Nucleation with simultaneous chemical reaction. *J. Colloid Interface Sci.*, **85**, 267–277.
14. Kingery, W.D., & Berg, M. 1955. Study of the Initial Stages of Sintering Solids by Viscous Flow. Evaporation-Condensation, and Self-Diffusion. *J. Appl. Phys.*, **26**(10), 1205–1212.
15. Knutson, E. O., & Whitby, K. T. 1975. Aerosol classification by electric mobility: apparatus, theory, and applications. *J. Aerosol Sci.*, **6**, 443–451.
16. Kobata, A., Kusakabe, K., & Morooka, S. 1991. Growth and transformation of TiO_2 crystallites in aerosol reactor. *AIChE J.*, **37**, 347–359.
17. Koch, W., & Friedlander, S. K. 1990. The effect of particle coalescence on the surface area of a coagulating aerosol. *J. Colloid Interface Sci.*, **140**, 419–427.
18. Koch, W., & Friedlander, S. K. 1991. Particle growth by coalescence and agglomeration. *Part. Part. Syst.*, **8**, 86–89.
19. Kruis, F. E., Kusters, K. A., Pratsinis, S. E., & Scarlett, B. 1993. A simple model for the evolution of the characteristics of aggregate particles undergoing coagulation and sintering. *Aerosol Sci. Technol.*, **19**, 514–526.
20. Kuczinski, G.C. 1949. Self-Diffusion in Sintering of Metallic Particles. *Trans. Am. Inst. Mater.. Eng.*, **185**, 169–178.
21. Lai, F. S., Friedlander, S. K., Pich, J., & Hidy, G. M. 1972. The self-preserving particle size distribution for Brownian coagulation in the free-molecule regime. *J. Colloid Interface Sci.*, **39**, 395.
22. Matsoukas, T., & Friedlander, S. K. 1991. *J. Colloid Interface Sci.*, **146**, 495.
23. McClurg, R. B. 1997. *Homogeneous Nucleation Theory.*
24. McClurg, R. B., C., R., & Goddard, W. A. 1997. Influences of binding transitions on the homogeneous nucleation of mercury. *Nanostructured Mater.*, **9**, 53–61.
25. Meakin, P. 1986. *Pages 111-135 of:* Stanley, H., & Ostrowsky, N. (eds), *On Growth and Form.* Boston, MA: Martinus Nijhoff.
26. Mountain, R. D., Mulholland, G. W., & Baum, H. 1986. *J. Colloid Interface Sci.*, **114**, 67.
27. Nguyen, H. V., & Flagan, R. C. 1991. Particle formation and growth in single stage aerosol reactors. *Langmuir*, **7**, 1807–1814.
28. Pluym, T. C., Lyons, S. W., Powell, Q. H., Gurav, A. S., Kodas, T. T., Wang, L. M., & Glocksman, H. D. 1993. Palladium metal and palladium oxide particle production by spray pyrolysis. *Mater. Res. Bull.*, **28**, 369–376.
29. Rao, N. P., & McMurry, P. H. 1990. Effect of the Tolman surface-tension correction on nucleation in chemically reacting systems. *Aerosol Sci. Technol.*, **13**, 183–195.
30. Rossell-Llompart, J., Loscertales, I. G., Bingham, D., & d. l. Mora, J. F. 1996. Sizing nanoparticles and ions with a short differential mobility analyzer. *J. Aerosol Sci.*, **27**, 695–719.
31. Shen, Y. C., & Oxtoby, D. W. 1996. Nucleation of Lennard-Jones fluids - A density-functional approach. *J. Chem. Phys*, **105**, 6517–6524.
32. Ulrich, G. D. 1971. Theory of Particle Formation and Growth in Oxide Synthesis Flames. *Combust. Sci Technol.*, **4**, 47–57.
33. Ulrich, G. D. 1984. Flame synthesis of fine particles. *Chem. Engr. News*, **62**, 22–29.
34. Ulrich, G. D., & Riehl, J. W. 1982. Aggregation of Growth of Submicron Oxide Particles in Flames. *J. Colloid Interface Sci.*, **87**, 257.
35. Ulrich, G.D., Milnes, B.A., & Subramanian, N.S. 1976. Particle Growth in Flames II. Experimental Results for Silica Particles. *Combustion Science and Technology*, **14**, 243.

36. Ulrich, G.D., and Subramanian, N.S. 1977. "Particle growth in flames III. coalescence as a rate controlling process". *Combust. Sci Technol.*, **17**, 119–126.

37. Wu, M. K., Windler, R. S., Steiner, C. K. R., Bors, T., & Friedlander, S. K. 1993. Controlled synthesi of nanosized particles by aerosol processes. *Aerosol Sci. Technol.*, **19**, 527–548.

38. Xiong, Y., & Pratsinis, S. E. 1993a. Formation of agglomerate particles by coagulation and sintering 1. A 2-dimensional solution of the population balance equation. *J. Aerosol Sci.*, **24**, 283–300.

39. Xiong, Y., & Pratsinis, S. E. 1993b. Formation of agglomerate particles by coagulation and sintering 2. The evolution of the morphology of aerosol-made titania, silica, and silica-doped titania powders. *J Aerosol Sci.*, **24**, 301–313.

40. Zhang, S. H., Akutsu, Y., Russell, L. M., & Flagan, R. C. 1995. Radial differential mobility analyzer. *Aerosol Sci. Technol.*, **23**, 357–372.

CHEMICAL SYNTHESIS AND PROCESSING
OF NANOSTRUCTURED PARTICLES AND COATINGS

G. M. CHOW
Material Science and Technology Division, code 6323
Naval Research Laboratory
Washington, DC 20375, USA
email: gmchow@anvil.nrl.navy.mil

1. Abstract

An overview of the synthesis and processing of nanostructured particles and coatings using chemical routes is presented. Solution chemistry approaches offer advantages of the design of materials at the molecular level that can result in better homogeneity for multiphase materials, and cost-efficient bulk quantity production in many cases. Of particular importance, solution chemistry allows for the control of particle size and particle size distribution, and the control of agglomerate size and agglomerate size distribution, through effective manipulation of the parameters determining nucleation, growth and agglomeration at the molecular level. In this paper, selected examples of metallic, ceramic and hybrid materials prepared by aqueous, nonaqueous, sol-gel and surfactant mediated methods are given. The effects of the synthesis and processing conditions on the phases, microstructures, control of particle size and agglomeration, impurities incorporation, defects formation and properties are addressed.

2. Introduction

Nanostructured particles and coatings with grain size ≤ 100 nm can be synthesized by many methods [1]. The three major approaches to the fabrication of these materials are, namely, physical and chemical vapor deposition, solution chemistry and mechanical attrition. In both the vapor phase and solution phase methods, materials are assembled from atoms (or molecules) to nanoparticles in a bottom-up approach. In mechanical attrition, coarse grained materials are broken down and/or reacted to obtain nanostructures. Both powders and coatings can be synthesized using vapor phase and solution phase approaches, whereas only powders can be made using mechanical attrition.

Chemical synthesis of materials is based on the manipulation of atoms and molecules to assemble materials using the bottom-up approach [2]. It has a long history in such nanostructured materials as colloids and supported catalysts. The recently popularized term "nanostructured materials" can indeed be used to

G.M. Chow and N.I. Noskova (eds.), Nanostructured Materials, 31–46.

characterize many chemically synthesized materials. Chemical synthesis offers its unique strength in the field of nanostructured material science and technology by manipulating the assembly of matter at the nanometer regime to achieve novel material properties. Often chemical methods are used to prepare precursors in the form of a fine, coprecipitated solid mixture or a gel by allowing the reactants to be mixed at the atomic level. The precursors can be subsequently converted to the final product at a lower temperature and a shorter time of reaction due to the smaller diffusion distances, when compared to the conventional mixing of coarse-grained materials.

Chemical synthesis and processing of nanostructured materials is a rapidly growing field. Many traditional methods have been used or revised, with many new ones being continually developed, to provide a powerful arsenal of means to synthesize organic, inorganic and hybrid materials. In chemical methods, caution should always be exercised when handling complex and hazardous chemistry. The handling of chemicals, reaction byproducts and post-reaction wastes may vary in different reactions. Special steps are often needed to remove the entrapped impurities from the products and to avoid post-synthesis contamination. Many (but not all) chemical methods are suitable for economical scale-up production of bulk quantity. It should be noted that the parameters used in a laboratory scale experiment may not always be linearly scaleable when it comes to large-quantity production. The so-called "synthesis parameters" such as temperature, pH, reactant concentration and time should be ideally correlated with the factors such as supersaturation, nucleation and growth rates, surface energy and diffusion coefficient in order to ensure the reproducibility of reactions.

Because of its wide breadth, a review of every aspect of this field is beyond the scope of this lecture paper. Interested readers are encouraged to consult the cited literature in the reference section for a more comprehensive review and current progress in this subject [1-9]. In this paper, only selected examples of our work and collaborations are discussed. These include aqueous and nonaqueous chemistry for elemental, alloy and composite metallic particles; nonaqueous chemistry for metal coatings; sol-gel and colloidal chemistry for ceramic powders and coatings; and self-assembled membrane mediated chemistry for dispersed particles.

3. Particle Synthesis

A common technique of synthesis of fine particles is precipitation of solids from a solution [10-11]. A multicomponent solid can be synthesized by coprecipitation of ions in a batch solution. Since different species may not precipitate at the same conditions, the reactions need to be carefully controlled in order to ensure the chemical homogeneity and stoichiometry of the final product. The precursors, in the form of solids or liquids such as salts or organometallics, may be dissolved or mixed in a suitable solvent for a specific reaction. When the solution becomes supersaturated with the product concentration, condensation will occur to form *nuclei* of the particles. The nucleation can be either heterogeneous or homogeneous, which refers to the formation of stable nuclei with or without the aid of foreign species, respectively.

In a growth process, the thermodynamics competes with the kinetics[12]. In particle growth, kinetics factors such as reaction and transport rates, accommodation, removal and redistribution of matter compete with the influence of thermodynamics. The reaction and transport rates are affected by factors including concentration of reactants, reaction temperature, pH, and methods of adding and mixing the reagents. The morphology is determined by factors such as supersaturation, nucleation and growth rates, colloidal stability, recrystallization and aging processes, and different habit modifications due to the impurities in the solvent or the solvent itself [11]. The supersaturation has the predominant role on the morphology of precipitates. At low supersaturation, the particles are small, compact and well formed; and the shape depends on the crystal structure and surface energies. At large supersaturation, large and dendritic particles form. At even larger supersaturation, smaller but compacted and agglomerated particles are formed [11]. The growth in solution is interface-controlled when the particle is small and becomes diffusion-controlled after reaching a certain critical size [13]. The formation of monodispersed particles, i.e. unagglomerated particles with a very narrow size distribution, requires all the nuclei to form at nearly the same time, and subsequent growth must occur without secondary nucleation or agglomeration of the particles [14].

3.1 METALS

Fine metal particles have many applications including electronic and magnetic materials, explosives, catalysts, pharmaceuticals, and in powder metallurgy. Precious metal powders for electronic applications can be prepared by adding soluble reducing agents to the aqueous solutions of suitable salts at adjusted pH [15]. Many reducing agents such as sodium formate, formic acid, borohydride, sodium hydrosulfite and hydrazines can be used. Care must be taken during washing, filtering and drying of nanostructured powders to avoid hydrolysis or oxidation.

3.1.1 Aqueous borohydride methods

For example, borohydride reduction can be used for the synthesis of amorphous and crystalline metals and borides. Aqueous potassium borohydride reduction was used to make ultrafine amorphous Fe-Co-B alloy powders for applications in ferrofluids and magnetic memory systems [16]. The amorphous phase was formed at reaction temperatures below the glass transition temperature and was stabilized by a high concentration of boron atoms. Aqueous sodium borohydride reduction was used to prepare nanoscale Co_2B particles as the primary product [17], whereas non-aqueous reduction of Co ions in diglyme produced primarily nanostructured Co particles [18].

Nanocrystalline equilibrium and metastable solid solutions can be formed by chemical techniques. For example, iron and copper are immiscible at the equilibrium state. Metastable alloys of Fe-Cu can be formed by far-from-equilibrium processing techniques such as vapor phase quenching or mechanical alloying. Nanocrystalline Fe_xCu_{100-x} (x in atomic %) alloys and composite powders were synthesized by reducing aqueous solutions of ferrous chloride and cupric chloride (in various molar ratio) using sodium borohydride solution [19].

The reaction was carried out at room temperature with stirring for 5 min. The atomic ratio between Fe and Cu in the product was very close to that in the original aqueous solutions. When $x \leq 40$, only fcc metastable alloys were formed. At higher Fe concentration such as $x \approx 70$, phase separation occurred and both fcc Cu and bcc Fe were synthesized. The formation of Cu_2O was also observed and its concentration scaled with that of Cu. The powders were agglomerates of nanocrystallites. The crystallites were between 30 to 45 nm for alloys, and between 10 to 15 for Fe and 30 to 40 nm for Cu in the composites. As-synthesized alloys were magnetically soft with coercivity ranging from 10 to 40 Oe, due to the lack of nearest neighbor interaction in solid solutions. As-synthesized composite powders had coercivity as high as 400 Oe. The formation of crystalline phases was controlled by decreasing the boron concentration in the powders when a higher molarity of borohydride was used in the reaction.

Another example is Co-Cu systems that only form terminal solid solutions at equilibrium. Nanostructured Co_xCu_{100-x} powders were synthesized by sodium borohydride reduction of aqueous cobalt chloride and cupric chloride solutions [20]. As-synthesized powders were a mixture of fcc and amorphous phases. The concentration of amorphous phase increased with the ratio of Co/Cu. The annealed powders phase separated to fcc Co and fcc Cu at about 500 °C. The annealing led to significant surface sintering and some grain growth (grain size $\approx$ 30 to 40 nm), and boron impurity was found to segregate at the sintered surfaces. Coercivity increased with annealing temperature to a maximum of 620 Oe (Fig. 1), similar to that of the nanostructured Co-Cu films prepared by annealing the sputtered alloy films.

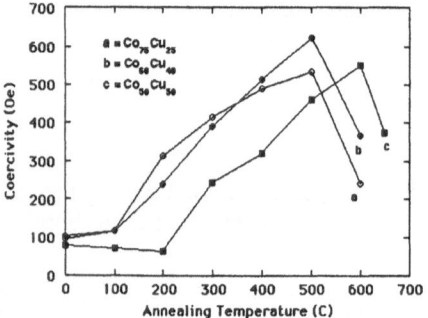

Figure 1. Realtionship of coercivity and annealing temperature of Co-Cu powders synthesized by borohydride reduction [20].

3.1.2 Nonaqueous polyol methods

A non-aqueous synthesis method known as the polyol process has been used to make fine, dispersed single element metallic particles such as Cu, Ni and Co in the micron and submicron size range [21]. In this method, precursor compounds are either dissolved or suspended in ethylene glycol. The mixture is heated to the refluxing temperature (about 194 °C) for 1-3 h. During the reaction, metal ions are reduced and metal particles precipitate out of solution. Submicron size particles can be made by increasing the reaction temperature or by inducing heterogeneous nucleation by either adding foreign nuclei or forming foreign nuclei in-situ. Nanocrystalline powders such as Fe, Co, Ni, Cu, Ru, Rh, Pd, Ag, Sn, Pt, Fe-Cu, Co-Cu and Ni-Cu were also prepared using this method [22-25]. Oxidation of powders was found to be less severe than in the aqueous borohydride method.

For example, nanostructured powders of Co_xCu_{100-x} ($4 \leq x \leq 49$ at. %) [24-25] were synthesized by reacting cobalt acetate tetrahydrate and copper acetate hydrate in various proportions in ethylene glycol. The mixtures were refluxed at about 190°C for 2 h. The powders precipitated out of the solution, and were subsequently collected and dried. The reaction rate was slower and the reaction time (i.e. 2 hr) was much longer than that of the aqueous borohydride reduction for Co-Cu synthesis [20]. In this case, the overall kinetics did not favor the formation of metastable solid solution. The nucleation of the more easily reducible Cu (compared to Co) occurred first. Cobalt was found to subsequently nucleate on Cu crystallites. X-ray diffraction, commonly used to study the formation of metastable alloys, was used to investigate the structure of powders. The XRD results seemed to suggest that metastable alloy could have been formed. However, detailed studies of the powders using complementary techniques such as extended x-ray absorption fine structure (EXAFS) spectroscopy and solid state nuclear magnetic resonance (NMR) (both for studying local atomic environment), high resolution transmission electron microscopy (HRTEM) and vibrating sample magnetometry (VSM) confirmed that the nanocomposites of Co-Cu were synthesized by the polyol method. Since no attempt was made during the synthesis to control the dispersion of particles, the powders formed by this method and the borohydride methods were very agglomerated (Fig. 2).

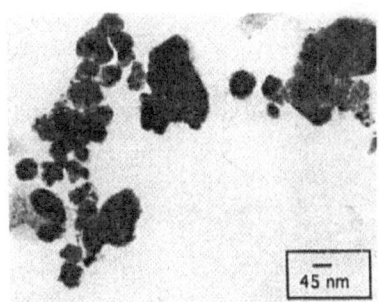 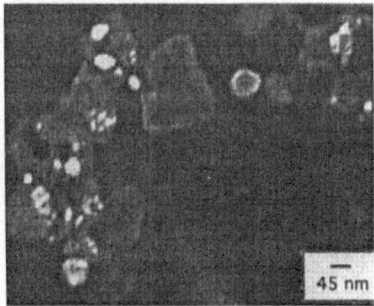

Figure 2. TEM bright field image (left) and corresponding dark field image (right) of $Co_{49}Cu_{51}$ powder synthesized by the polyol method [24].

3.1.3 Powders consolidation for structural applications

Bulk nanostructured materials for structural applications can be fabricated by consolidation of the starting nanoscale powders. Nanostructured precursor powders of M50 type steel (with a typical composition of 4.5% Mo, 4.0% Cr, 1.0% V, 0.8% C, with balance of Fe, in weight %) were chemically synthesized using thermal decomposition of metal carbonyls or co-reduction of metal halides [26]. The thermal decomposition method was easier and more cost-efficient to scale up for the production of large quantity of M50 type powder. It also did not produce residual impurities that would require removal at higher temperature prior to powder consolidation. The structural and microstructural development of the powders was controlled by subsequent consolidation such as hot pressing or hot-isostatic pressing [27-28]. During these consolidation processes, precursor powders transformed to a nanocrystalline M50 type structure with precipitation of carbides, and simultaneously, pressure assisted sintering of the powders occurred to produce a dense sample.

Porosity (about 5 % and with a wide size distribution) was found in the sample prepared by hot pressing the powders between 700°C and 850 °C, at 275 MPa for 0.5 to 2 h. Figure 3 shows a TEM micrograph of a consolidated sample. The matrix of the consolidated bulk was nanostructured α-Fe with a grain size between 5 to 70 nm. Clusters of 10 nm Mo_2C precipitates were observed. There also existed a size distribution of carbide precipitates. Large precipitates ($\approx$ 100 nm) were located at the triple point grain boundaries of smaller matrix grains. Smaller carbide precipitates were also observed within the large matrix grains, and they had little effect on preventing grain growth. The consolidation results using hot press and hot isostatic press indicated that increasing pressure and temperature did not have a significant effect on reducing the degree of porosity in all the samples. As expected, both normal and abnormal grain growth of the matrix and precipitates occurred with increasing temperature, due to the wide size distribution of starting crystallites. The retention of nanostructures in powder consolidation of multicomponent engineering materials such as M50 type steel remains a challenge.

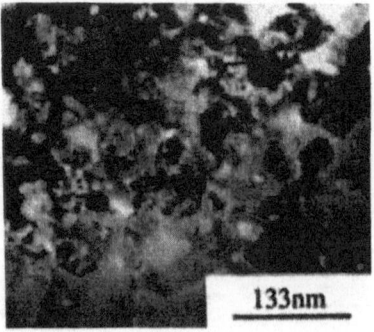

133nm

Figure 3. TEM bright field micrograph of a consolidated sample obtained by hot pressing M50 precursor powders (from thermal decomposition) at 700 °C and 275 MPa for 2 h [27].

3.2 CERAMICS

Ceramic oxides and hydroxide nanoparticles can be prepared by aqueous precipitation [3, 29] and sol-gel type methods [5, 30]. Depending on reaction kinetics and subsequent post-synthesis aging of particles, the oxide particles can be amorphous or crystalline. Nanoscale oxide powders may be converted to carbides or nitrides by subsequent thermochemical reactions.

3.2.1 Sol-gel or colloidal gel methods

Multicomponent gels can be chemically converted to nanocomposite powders. For example, a precomposite gel, prepared by reacting iron chloride hexahydrate, urea, and boric acid in a strongly basic aqueous solution at 150 °C, was reacted in ammonia at 500 °C to obtain Fe_xN/BN (x =3 or 4) nanocomposite powder [31].

Similarly, AlN_xBN_{100-x} ($0 \leq x \leq 100$) nanocomposite powders were synthesized from pyrolysis of the precomposite gels (with aluminum chloride hexahydrate as Al source) in ammonia [32]. The compositional and thermal effects on these AlN-BN nanocomposite powders were studied [33]. It was found that BN was amorphous when its concentration was below 35 mole %. For higher BN concentration, a mixture of amorphous and turbostratic-BN was observed. A more ordered hexagonal BN phase was formed by annealing the powders. Independent of compositions, both normal and abnormal grain growth of AlN and BN occurred. Both crystallites retained their nanometer size up to 1600 °C. The normal grain growth of AlN phase was impeded by the second phase immobile inclusion particles of BN. When normal grain growth of AlN was inhibited by inclusions, only large AlN grains with boundary curvatures larger than average could move past the inclusions. As a result, larger grains experienced exaggerated grain growth. These results indicated that initial crystallization of AlN in the composite powders produced grains with a mixed size distribution.

Nanostructured AlN powders for thermal management in electronics applications were synthesized by nitridation of oxide precursor powders [34-36]. The oxide powders were prepared by hydrolysis of aluminum tri-sec-butoxide at room temperature to favor precipitate formation instead of sol formation. The dry precursor (AlOOH) powders were calcined and subsequently nitrided in ammonia at temperatures up to 1100 °C for up to 10 h. Since small particles favored the diffusion-controlled nitridation kinetics, nanostructured AlN powders were synthesized at temperatures 400 to 600 °C lower than that used in the conventional carbothermal nitridation or direct nitridation of coarse-grained oxide or aluminum powders. It was shown that the formation of AlN was favored when the oxide precursor powders were either amorphous or highly disordered. Commercially available nanocrystalline α-alumina could not be fully nitrided under the same nitridation conditions. The powders were consolidated by pressureless sintering. Significant densification was observed for these nanostructured AlN powders when compared to the sintering of commercial coarse-grained AlN powder under the same conditions. As-synthesized AlN powders were highly agglomerated, and the agglomerates contained pores with a random size distribution. The largest of these pores hindered the sintering of powders to full density.

4. Surfactant-Mediated Synthesis [37]

Many materials such as paints, pigments, electronic inks, and ferrofluids are only useful if the particles are dispersed in an appropriate solvent or medium. Nanostructured particles have a very high surface area as the specific surface area is inversely proportional to the particle size. Because of the attractive van der Waals interactions between particles and the thermodynamics of minimizing the total surface energy of a system, nanoscale particles tend to agglomerate to form either lumps or secondary particles. Agglomeration of nanostructured particles may occur during any of the following stages: synthesis, drying, handling and processing. The examples shown above in section 3 were agglomerated powders. since no special attention was paid to the control the dispersion during synthesis.

When the particles are suspended in a liquid, they will move around by Brownian motion. If the potential energy barrier to agglomeration is larger than the attractive forces in question for a given system, the particles will approach one another and then separate without agglomeration. However, random agglomeration may still occur for some particles which have high enough energy to cross this repulsion barrier. In aqueous solvents, the surface of particles may become charged if the surface is hydrated or covered with an oxide. An electric double layer forms around each particle. The electrostatic repulsion (inversely proportional to the second power of separation distance between two charges) prevents the particles from agglomeration. On the other hand, the attractive van der Waals forces (inversely proportional to the third to sixth powder of separation distance) and other attractive forces promote agglomeration of particles. A stable dispersion will exist provided that the potential energy of repulsion is high enough to be a barrier to the interparticle approach that will lead to agglomeration. The stable distance of particle separation depends on the charges on the particles and the concentration of other ions in the diffuse region of the double layer. The collapse of the double layer, which can be caused by other ions in solution, leads to particle contacts and agglomeration.

To produce unagglomerated particles, surfactants can be used to control the dispersion during the synthesis stage or to redisperse as-synthesized agglomerated fine particles. A surfactant is any substance that lowers the surface or interfacial tension of the medium in which it is dissolved. A surfactant is a surface-active agent that need not be completely soluble and may decrease surface or interfacial tension by spreading over the surface. It has an amphipathic structure in that solvent, i.e. a lyophobic (solvent repulsive) and lyophilic group (solvent attractive). Surfactants are classified according to the charge configuration as either anionic, cationic, zwitterionic (bearing both positive and negative charges) or nonionic (no charges).

In aqueous media, particles can be prevented from agglomeration by electrostatic repulsion, which results from interactions between the electric double layers surrounding the particles. This can be achieved by adjusting the pH of the solution or adsorbing charged surfactant molecules on particle surfaces. Electrostatic stabilization is generally effective in dilute systems of aqueous or polar organic media, and it is very sensitive to the pH and the effects of other electrolytes in the solution. At the isoelectric point where there is no net surface charge on the particles, agglomeration may occur.

In most nonaqueous, nonpolar media, electrostatic repulsion is less important because ionization is insignificant. A different stabilization approach is necessary. This involves the steric forces that are produced by the adsorbed surfactant on the particle surfaces. The lyophilic, non-polar tails of the surfactant molecules extend into the solvent and interact with each other. Since these nonpolar tails do not have large van der Waals attraction, they provide the steric hindrance to interparticle approach. In order for steric stabilization to be effective, the size of surfactant molecules should be large enough that they serve as a stable adsorbed barrier on the surface without entangling each other. When the interparticle approach occurs, the extended lyophilic tails of the adsorbed surfactant will be forced into a restricted space. This interaction is thermodynamically unfavorable since it will decrease the entropy of the system. Thus the entropic repulsion takes place and prevents the particle agglomeration from occurring. Steric stabilization is effective in both aqueous and nonaqueous media, and it does not depend on the existence of the electric double layer. It is less sensitive to impurities, trace additives or other ions than electrostatic stabilization, and is also particularly effective in dispersing high concentrations of particles.

Colloidal synthesis in the presence of stabilizers has a long history. For example, stable colloidal dispersions of 5-15 nm Fe particles were synthesized by thermolysis of iron pentacarbonyl in dilute solutions of polymers [38]. The functional polymer was catalytic for the decomposition of the carbonyl and it induced particle nucleation in its domain. Monodispersed 3-5 nm particles of Pd-Cu were stabilized by poly(vinylpyrrolidone) (PVP) in refluxing the mixture of metal acetate precursors and PVP in 2-ethoxyethanol [39]. Monodispersed Ni powders of 14 nm diameter were made by reducing nickel hydroxide in the polyol with PVP [40]. A ferrofluid of 8 nm particles of metallic glass was prepared by refluxing the mixture of a surfactant (n-oleyol sarcosine), iron pentacarbonyl and Decalin (decahydronaphthalene) [41]. Deagglomeration of agglomerated powders can be achieved by breaking the agglomerates using milling or ultrasonication in a suitable solvent with a surface-active dispersant [42]. For example, agglomerated nanoscale AlN particles were dispersed using N-methylpyrrolidone and stabilized in a polyimide matrix [43].

4.1. SURFACTANT MEMBRANE STRUCTURES

Nanoparticles can be synthesized inside self-assembled structures [44]. Membrane structures are assemblies of molecules in which each possesses a polar head group and a nonpolar hydrocarbon tail. These molecules, held together by van der Waals forces, orient themselves so that the contact of nonpolar tails with the aqueous medium is minimized. The polar parts of the molecules are attracted to water by electrostatic and hydrogen bond interactions. Thus these molecules self-assemble into membranes with the minimum energy configuration. Self-assembly can result in monolayer films, Langmuir-Blodgett films, micelles, reversed micelles, vesicles and tubules, etc.

For example, glycerol monooleate, a lipid with a single fatty acid tail and a glycerol headgroup, can form bicontinuous cubic phases, in which size-controlled aqueous channels are connected in a three dimensionally periodic network. Size-controlled, about 4 nm Pd particles were synthesized by a polyol type reaction of a solution of tetrachloropalladate and glycerol monooleate [45]. The hydroxyl groups in the glycerol headgroup reduced the metal salt to metal at room temperature, and the narrow aqueous channels of the bicontinuous phase constrained the size of Pd particles.

Reverse micelles, also known as water-in-oil (w/o) microemulsions, are a single layer of surfactant molecules entrapping solubilized water pools in a hydrocarbon solvent. The size of the water pool depends on the amount of entrapped water at a given surfactant concentration. Vesicles are generally closed bilayer membrane assemblies. The membrane assemblies serve as nanoreactors in which nanoparticles are synthesized. Compared to bulk precipitation, better chemical homogeneity is achieved since the reactions occur in a more controlled environment. The size and size distribution of these membrane structures can be manipulated to control the particle size and particle size distribution. In this approach, the nanoparticles synthesized inside the membrane are prevented from agglomeration with each other by the membrane barriers. Since it is possible to incorporate functional groups on the vesicle membrane surface, nanoparticles may be carried by the functionalized membranes for targeted applications. Nanosize metal colloids such as Au and Ag [46]and semiconductors such as CdSe [47] have been made using the reverse micelle method.

Multilamellar vesicles have diameters in the range of 100-800 nm. Single bilayer vesicles are 30-60 nm in diameter. Sonication or extrusion can be used to control the vesicle size. Nanoscale oxide particles can be synthesized using the vesicle-mediated approach. Metal ions and lipid mixtures form vesicles upon sonication. After removal of exogenous ions, anions are added and allowed to diffuse through the membrane layers and intravesicular precipitation occurs. Due to the preferential anion diffusion across the membrane, generally only oxides such as silver oxide [48], iron oxide [49] and aluminum oxide [50] were formed.

Dispersed nanocrystalline metal particles were prepared using polymerized phospholipid vesicles [37, 51]. The non-cross-linked polymerization of the vesicle resulted in many individual polymer chains in the membrane structure. It enhanced the structural integrity of the vesicle and provided breaks in the polymer network, through which both anions and cations could diffuse across the polymerized membrane. This allowed the synthesis of metal to take place. In this approach, vesicles made from mixed phospholipids (negatively charged and zwitterionic) were UV polymerized. Positively charged Pd ions were selectively and chemically attached to the negatively charged headgroup molecule on the membrane surface of the interior compartment. After adding a Au salt and a reducing agent to the solution, Au ions diffused across the membrane. The interior membrane bound Pd species served as catalysts for the initiation of the autocatalytic electroless metallization of Au, which subsequently led to the formation of nanocrystalline Au particles inside the polymerized vesicles (Fig. 4).

When unpolymerized vesicles were used to make nanoscale Au particles using this electroless method [52], the Pd catalysts were bound both to the internal and external membrane surfaces and electroless metallization occurred on the external surfaces of vesicles. The osmotic pressure (built up inside the unpolymerized vesicles due to the difference of salt concentrations across the membranes) and the external electroless reaction weakened the structural integrity of the vesicles, which were eventually ruptured. In this case, the lipid molecules served as a surfactant-dispersant during the synthesis of Au particles. The extent of dispersion of the nanocrystalline Au particles was dependent on the thermal disordering of the phospholipid mixture, which became more effective as a dispersant with increasing temperature. In this approach [52], nanostructured Au particles, synthesized under the conditions of reaction temperatures of 25 °C and 40 °C for 3 h, were faceted and many were multiply-twinned particles (MTP). It has been known that MTPs can be formed by rapid vapor quenching in films prepared by vapor deposition. Vapor deposition is generally a "far-from-equilibrium" process, in which atoms rapidly condense on a substrate. The maximum departure from equilibrium (ΔG), is approximately 160 kJ/N_a for condensation of supersaturated vapor into a solid, where N_a is Avogardro's number. The precipitation of a solid from a supersaturated liquid has a maximum departure from equilibrium ΔG of 8 kJ/N_a [53]. Though the deposition kinetics in solution chemistry differs significantly from that of vapor phase, MTPs of Au have been synthesized in both cases. The mechanisms of formation of multiple twins in nanoscale particles from various deposition processes remain interesting. Issues such as the transition from short range to long range crystal ordering during initial growth, the roles of twin boundary and surface energies of nanoparticles, and their relationship with the size effect (if any) need further investigation.

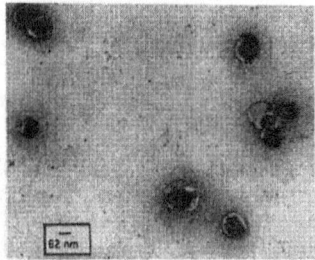

Figure 4. TEM micrograph showing gold particles synthesized inside polymerized vesicles [51].

5. Films and coatings

5.1. CERAMICS

Nanostructured ceramic oxide films and coatings can be deposited using sol-gel type methods[5, 54]. This approach is particularly useful to obtain homogeneous multicomponent coatings The problems of shrinkage and cracking and the limitation of coating thickness can be mitigated by increasing the particle loading in the sol-gel process [55]. The coatings are typically amorphous and may be converted to nanocrystalline structures by post-deposition heat treatment. The oxide coatings may also be carburized or nitrided to form nonoxide materials. Hybrid coatings can be fabricated by doping the sol with material of a different phase, followed by gelation and densification.

Ceramic nanoparticles can also be coated with organic materials for biomedical applications. For example, immunocompetent cells recognize antigens by interactions that are specific to the chemical sequence and conformation of the antigenic determinant. Current adjuvants to enhance immunity to antigens tend to either alter the antigen conformation through surface adsorption or shield critical functional groups. Diamond nanoparticles were surface-modified to provide conformational stabilization and a high degree of surface exposure to protein antigens, resulting in evoking a specific and stronger immune response than the conventional adjuvants in mussel adhesive protein [56].

5.2. METALS

Metal films and coatings can be deposited from aqueous solutions using chemical reactions in either an electrolytic or an electroless process, where the metal ions are reduced to metal by electrons. In the electrodeposition process, the deposition of a pure metal or alloy from the electrolyte solution occurs on the cathode when an external current is applied to the plating bath. A electrically conductive substrate is required. Nanostructured metal coatings such as Ni were deposited using electrodeposition [57-59]. Nanostructured grains were deposited when the plating variables such as bath composition, pH, temperature and current density are controlled to favor nucleation of new grains rather than sustaining the growth of existing grains.

In the electroless approach [60], electrons are generated by chemical reactions without the supply of external current. Unlike electrodeposition, electrical conductors are not required as substrates. Electroless deposition can occur by the following: deposition by ion or charge exchange; deposition by contacting the metal to be coated; and, autocatalytic deposition on catalytically active surfaces from solutions containing reducing agents. In the autocatalytic electroless process, a noncatalytic surface on which metal is to be deposited electrolessly is initially coated with catalytic particles. Reduction of a soluble metal ion or complex by a soluble reducing agent present in the plating bath leads to the deposition of metal atoms at the surface. Metal ions are reduced by electrons provided by the reducing agents. Subsequently each layer of freshly deposited metal atoms becomes the catalyst for deposition of the next layer. Conventionally Pd colloids with diameter ≥ 2 nm are used as catalysts to initiate the metallization.

Electroless deposition was used to deposit nanostructured Ni-P or Ni-B coating (the phosphorous or boron was in the composition of the plating solution) [60]. Depending on factors such as post deposition heat treatment and compositions, amorphous or crystalline structure and a range of grain size from 2 to 100 nm could be produced. The control of the size of the bound catalysts is the principal determining factor in controlling the particle size of the nanostructured electroless deposit [61]. Chemical modification of the substrate surface to allow for binding smaller catalysts led to a three- to four-fold reduction in the particle size of electroless deposits.

Electroless metallization was used to deposit nanostructured metal coatings on self-assembled biomolecular structures. Nanoscale protein tubules called rhapidosomes, approximately 17 nm in diameter and 400 nm in length, were electroless metallized with 10 nm Ni particles using molecular catalysts, instead of the traditional colloidal catalysts [62]. The surface amino acids initiated the catalyzation process of molecular catalyst particles. Self-assembled phospholipid hollow tubules, 0.5 μm average diameter and 50 to 80 μm long, are interesting materials due to the large shape anisotropy. These tubules were electroless metallized, magnetically aligned and cast into a polymer matrix composite. The magnetic anisotropy of electroless metallized Ni or permalloy tubules (with metal grain size about 2-4 nm, depending on metallization conditions), and the alignment of metallized tubules in the magnetic tubule-polymer composite was studied [63]. To fabricate an ungated vacuum field emission cathode structures for generation of a macroscopic electron beam current , the local electric field enhancement was achieved by exploiting the large aspect ratio of the metallized tubules, the radius of curvature and thickness of metal coatings at the edge of metallized tubules [64-65]. The selective removal of polymer matrix resulted in a composite base template of oriented exposed metallized tubules, and subsequent surface electrical contact was provided by a thin sputtered gold film. The resultant microstructures demonstrated vacuum field emission of current I > 10 μA at relatively low applied macroscopic electric fields of ~ 60 - 150 kV / cm.

When the substrate to be metallized or the coating deposits are susceptible to adverse hydrolysis and oxidation (thus leading to deterioration of the substrate-coating interface and the deposits), aqueous electrodeposition and electroless deposition are not attractive approaches. Recently, a nonaqueous coating process capable of producing fine-grained deposits has been developed using the polyol method. For example, Cu coatings were deposited on AlN substrates [66] and Co on WC substrates [67]. Unlike the traditional aqueous electroless metallization, this process does not require the use of catalysts on electrically insulative substrates. A combined surface study of grazing incidence asymmetric Bragg scattering and small angle X-ray scattering at glancing incidence revealed that the surface of Cu metal film consisted of 4 nm particles [66].

6. Summary

Chemical synthesis is a very versatile, powerful technological approach to the fabrication of advanced materials. Its role in the synthesis and processing of nanostructured materials, particularly hybrid materials, continues to grow rapidly. In this paper, selected examples of chemical synthesis and processing of nanoscale metallic and ceramic particles, films and coatings have been discussed.

ACKNOWLEDGMENT

The support of the NRL core programs on Nanocomposites, Membrane Structures, Nanoscale Hybrid Coatings, NRL TEW program on Nanostructured Ceramics, and the ONR Affordability Initiative program on Nanostructured Coatings is acknowledged. The author would like to thank his many colleagues for their valuable contributions and collaborations, particularly, L. K. Kurihara, M. A. Markowitz, A. Singh, K. Gonsalves, and L. Martinez-Miranda.

44

References

1. Edelstein, A.S. and Cammarata R.C. (eds.) (1996), *Nanomaterials: Synthesis, Properties and Applications*, Institute of Physics Publishing, Bristol and Philadelphia (author's comment: this is a reasonably updated comprehensive text book).
2. Ellis, A.B., Geselbracht, M.J., Johnson, B.J., Lisensky, G.C., and Robinson, W.R. (1993), *Teaching General Chemistry: A Materials Science Companion*, American Chemical Society, Washington, D.C.
3. Chow, G.M. and Gonsalves, K.E. (1996), Particle synthesis by chemical routes, in Edelstein, A.S. and Cammarata R.C. (eds.), *Nanomaterials: Synthesis, Properties and Applications*, Institute of Physics Publishing, Bristol and Philadelphia, pp. 55-71.
4. Herron, N., and Wang, Y. (1996), Synthesis of semiconductor nanoclusters, in Edelstein, A.S. and Cammarata R.C. (eds.), *Nanomaterials: Synthesis, Properties and Applications*, Institute of Physics Publishing, Bristol and Philadelphia, pp. 73-88.
5. Klein, L.C. (1996), Processing of nanostructured sol-gel materials, in Edelstein, A.S. and Cammarata R.C. (eds.), *Nanomaterials: Synthesis, Properties and Applications*, Institute of Physics Publishing, Bristol and Philadelphia, pp. 147-164.
6. Rolison, D.R. (1996), Chemical properties, in Edelstein, A.S. and Cammarata R.C. (eds.), *Nanomaterials: Synthesis, Properties and Applications*, Institute of Physics Publishing, Bristol and Philadelphia, pp. 305-321.
7. Chow, G.M. and Gonsalves, K.E. (eds.) (1996), *Nanotechnology: Molecularly Designed Materials*, American Chemical Society Symposium Series **622**, Washington, DC.
8. Special issue: Nanostructured Materials (1996), Chemistry of Materials **8**, Washington, DC.
9. Gonsalves, K.E., Chow, G.M., Xiao, T.D., and Cammarata, R.C. (eds.) (1994), *Molecularly Designed Ultrafine/ Nanostructured Materials*, Materials Research Society Symposium Proceedings **351**, Pittsburgh, Pennsylvania.
10. Nielsen, A.E. (1964), *Kinetics of Precipitation*, Pergamon Press, London, New York.
11. Walton, A.G. (1979), *The Formation and Properties of Precipitates*, Robert Krieger Publishing Company, Huntington, New York (reprint edition).
12. Lagally, M.G. (1993), An atomic-level view of kinetic and thermodynamic influences in the growth of thin films: a review, Japanese Journal of Applied Physics **32**, pp. 1493-1501.
13. Turnbull, D. (1953), The kinetics of precipitation of barium sulfate from aqueous solutions, Acta Metallurgica **1**, pp. 684 -691.
14. LaMer, V. K. and Dinegar , R. H. (1950), Theory, production and mechanism of formation of monodispersed hydrosols, J. American Chemical Society **72** , pp. 4847-4854.
15. Yang, K. C. and Rowan, B. D. (1984), Production of gold, platinum, and palladium powders, in *Metals Handbook Ninth Edition* **7** American Society for Metals, Metals Park, Ohio, pp. 148-151.
16. van Wonterghem, J. , Morup, S. , Koch, C.J.W. , Charles, S.W. and Wells, S., (1986), Formation of ultra-fine amorphous alloy particles by reduction in aqueous solution, Nature **322**, pp. 622-623.
17. Glavee, G.N., Klabunde, K.J., Sorensen, C.M. and Hadjipanayis, G.C. (1993), Borohydride reduction of cobalt ions in water. Chemistry leading to nanoscale metal, boride, or borate particles, Langmuir **9**, pp. 162-169.
18. Glavee, G.N., Klabunde, K.J., Sorensen, C.M. and Hadjipanayis, G.C. (1993), Sodium borohydride reduction of cobalt ions in nonaqueous media: Formation of ultrafine particle(nanoscale) of cobalt metal, Inorg. Chem. **32**, pp. 474-477.
19. Chow, G.M., Ambrose, T., Xiao, J.Q., Twigg, M.E., Baral, S., Ervin, A.M., Qadri, S.B. and Feng, C.R. (1992), Chemical precipitation and properties of nanocrystalline Fe-Cu alloy and composite powders, Nanostructured Materials 1, pp. 361-368.
20. Chow, G.M., Ambrose, T., Xiao, J., Kaatz, F., and Ervin, A. (1993), Nanostructured Co-Cu powders via a chemical route, Nanostructured Materials 2, pp. 131-138.
21. Fievet, F., Lagier, J.P., and Figlarz, M. (1989) Preparing monodisperse metal powders in micrometer and submicrometer size by the polyol process, Materials Research Society Bulletin **14**, 29-34.
22. Kurihara, L.K., Chow, G.M., and Schoen, P.E. (1995), Nanocrystalline metallic powders and films produced by the polyol method, Nanostructured Materials **5**, pp. 607-613.
23. Chow, G.M., Schoen, P.E., and Kurihara, L.K. (1995), Nanostructured metallic powders and films via an alcoholic solvent process, US Navy Case No. 76,572, US patent application pending.
24. Chow, G.M., Kurihara, L.K., Kemner, K.M., Schoen, P.E., Elam, W.T., Ervin, A., Keller, S., Zhang, Y.D., Budnick, J., and Ambrose, T. (1995), Structural, morphological and magnetic study of

nanocrystalline cobalt-copper powders synthesized by the polyol process, J. Mater. Res. **10**, pp. 1546-1554.

25. Chow, G.M., Kurihara, L.K., and Schoen, P.E.(1996), Synthesis of nanostructured composite particles using a polyol process, Navy Case No. 77,467, US patent application pending.

26. Gonsalves, K.E., Xiao, T.D., Chow, G.M., and Law, C.C. (1994), Synthesis and processing of nanostructured M50 type steel, Nanostructured Materials **4**, pp. 139-147.

27. Feng, C.R., Chow, G.M., Rangarajan, S.P., Chen, X, Gonsalves, K.E., and Law, C. (1997), TEM and HRTEM characterization of nanostructured M50 type steel, Nanostructured Materials **8**, pp. 45-54.

28. Chow, G.M., Feng, C.R., Rangarajan, S.P., Chen, X, Gonsalves, K.E., and Law, C. (1997), Microstructural study of nanostructured M50 type steel, in Ma, E., Fultz, B., Shull, R., Morral, J. and Nash, P. (eds.), *Chemistry and Physics of Nanostructures and Related Non-Equilibrium Materials*, the Minerals, Metals & Materials Society, pp. 157-162.

29. Gallagher, P.K. (1991), Chemical synthesis, in *Engineered Materials Handbook, Volume 4: Ceramics and Glasses*, ASM International, USA, pp. 52-64.

30. Shoup, R.D. (1991), Sol-gel processes, in *Engineered Materials Handbook, Volume 4: Ceramics and Glasses*, ASM International, USA, pp. 445-452.

31. Gonsalves, K.E., Chow, G.M., Zhang, Y., Budnick, J.I. and Xiao, T. D. (1994), Iron nitride/boron nitride magnetic nanocomposite powders, Advanced Materials **6**, pp. 291-292.

32. Xiao, T.D., Gonsalves, K.E. and Strutt, P.R. (1993), Synthesis of aluminum nitride/boron nitride materials, J. Am. Ceram. Soc. **76**, pp. 987-992; Xiao, T.D., Gonsalves, K.E., Strutt, P.R., Chow, G.M., and Chen, X. (1993), Synthesis of AlN/BN composite materials via chemical processing, Ceramic Science and Engineering Proceedings **14**, pp. 1107-1114.

33. Chow, G.M., Xiao, T.D., Chen, X and Gonsalves, K.E. (1994), Compositional and thermal effects on chemically processed AlN-BN nanocomposite powders, J. Mater. Res. **9**, pp. 168-175.

34. Kurihara, L.K., Chow, G.M., Choi, L.S., and Schoen, P.E. (1997), Chemical synthesis and processing of nanostructured aluminum nitride, in Battle, T.P. and Henein, H. (eds.), *Processing and Handling of Powders and Dusts*, the Minerals, Metals, and Materials Society, Warrendale, PA, pp. 3-12.

35. Kurihara, L.K., Chow, G.M., and Schoen, P.E. (1997), Nanostructured ceramic nitride powders and a method of making the same, Navy Case No. 77,219, US patent application pending.

36. Kurihara, L.K., Chow, G.M., Baraton, M.I., Schoen, P.E., Rayne, R., Bender, B., Lewis, D., and Choi, L.S. (1997), Synthesis and pressureless sintering of nanostructured AlN Powders derived from solution chemistry precursors, submitted to J. Am. Ceram. Soc.

37. Chow, G.M., Markowitz, M.A., and Singh, A. (1993), Synthesizing submicrometer and nanoscale particles via self-assembled molecular membranes, Journal of the Minerals, Metals and Materials Society **45**, pp. 62-65.

38. Smith, T.W., and Wychick, D. (1980), Colloidal iron dispersions prepared via the polymer-catalyzed decomposition of iron pentacarbonyl, J. Phys. Chem. **84**, pp. 1621-1629.

39. Bradley, J.S., Hill, E.W., Klein, C., Chaudret, B., and Duteil, A. (1993), Synthesis of monodispersed bimetallic palladium-copper nanoscale colloids, Chem. Mater. **5**, pp. 254-256.

40. Hedge, M.S., Larcher, D., Dupont, L., Beaudoin, B., Tekaia-Elhsissen, K. and Tarascon, J.M. (1997), Synthesis and chemical reactivity of polyol prepared monodisperse nickel powders, Solid State Ionics **93**, pp. 33-50.

41. van Wonterghem, J., Morup, S., Charles, S.W., Wells, S., and Villadsen, J. (1985), Formation of a metallic glass by thermal decomposition of $Fe(CO)_5$, Physical Review Letters **55**, pp. 410-413.

42. Shanefield, D.J. (1995), *Organic Additives and Ceramic Processing, with Applications in Powder Metallurgy, Ink and Paint*, Kluwer Academic Publishers, Boston, Dordrecht, London.

43. Chen, X, Gonsalves, K.E., Chow, G.M., and Xiao, T.D. (1994), Homogeneous dispersion of nanostructured aluminum nitride in a polyimide matrix, Advanced Materials **6**, pp. 481-484.

44. Fendler, J.H. (1987), Atomic and molecular clusters in membrane mimetic chemistry, Chem. Rev. **87**, pp. 877-899.

45. Puvvada, S., Baral, S, Chow, G.M., Qadri, S.B., and Ratna, B.R. (1994), Synthesis of palladium metal nanoparticles in the bicontinuous cubic phase of glycerol monooleate, J. Am. Chem. Soc. **116**, pp. 2135-2136.

46. Wilcoxon, J.P., Williamson, R.L. and Baughman, R. (1993), Optical properties of gold colloids formed in inverse micelles, J. Chem. Phys. **98**, pp. 9933 -.9950.

47. Kortan, A.R., Hull, R., Opila, R.L., Bawendi, M.G., Steigerwald, M.L., Carroll, P.J., and Brus, L.E. (1990), Nucleation and growth of CdSe on ZnS quantum crystallite seeds, and vice versa, in inverse micelle media, Journal of the American Chemical Society,J. Am. Chem. Soc., **112**, pp. 1327-.1332

46

48. Mann, S., and Williams, R.J.P. (1983), Precipitation within unilamellar vesicles. Part 1. studies oi silver oxide formation, J. Chem. Soc. Dalton Trans. pp. 311-316.

49. Mann, S., and Hannington, J.P. (1988), Formation of iron oxides in unilamellar vesicles, Journal oi Colloid and Interface Science **122** , pp. 326-335.

50. Bhandarkar, S., and Bose, A. (1990), Synthesis of submicrometer crystals of aluminum oxide by aqueous intravesicular precipitation, Journal of Colloid and Interface Science **135**, pp. 531-538.

51. Markowitz, M.A., Chow, G.M., and Singh, A. (1994), Polymerized phospholipid membrane mediated synthesis of metal nanoparticles, Langmuir **10**, pp. 4905-4102.

52. Chow, G.M., Markowitz, M.A., Rayne, R., Dunn, D.N., and Singh, A. (1996), Phospholipid mediated synthesis and characterization of gold nanoparticles, Journal of Colloid and Interface Science **183**, pp. 135-142.

53. Froes, F. H., Suryanarayana, C., Russell, K. C., and Ward-Close, C. M. (1995), Far from equilibrium processing of light metals, in Singh, J., and Copley, S.M. (eds.), *Novel Techniques in Synthesis and Processing of Advanced Materials*, the Minerals, Metals & Materials Society, pp. 1-21.

54. Brinker, C.J., Hurd, A.J., Schunk, P.R., Frye, G.C., and Ashley, C.S. (1992), Review of sol-gel film formation, Journal of Non-Crystalline Solids **147&148**, pp. 424-436.

55. Barrow, D.A., Petroff, T.E., and Sayer, M. (1995), Thick ceramic coatings using a sol gel based ceramic-ceramic 0-3 composite, Surface and Coatings Technology **76-77**, pp. 113-118.

56. Kossovsky, N, Gelman, A., Hnatyszyn, H.J., Rajguru, S., Garrell, R.L., Torbati, S., Freitas, S. S.F., and Chow, G.M. (1995), Surface-modified diamond nanoparticles as antigen delivery vehicles, Bioconjugate Chem. **6**, pp. 507-511.

57. Ross, C.A. (1994), Electrodeposited multilayer thin films, Annu. Rev. Mater. Sci. **24**, pp. 159-88.

58. Erb, U. (1995), Electrodeposited nanocrystals: synthesis, structure, properties and future applications, Canadian Metallurgical Quarterly **34**, pp. 275-280.

59. Palumbo, G., Gonzalez, F., Brennenstuhl, A.M., Erb, U. Shmayda, W., and Lichtenberger, P.C. (1997), In-situ nuclear steam generator repair using electrodeposited nanocrystalline nickel, Nanostructured Materials **9**, pp. 737-746.

60. Riedel, W. (1991), *Electroless Nickel Plating*, Finishing Publications Ltd., Stevenage, Hertfordshire, England.

61. Brandow, S.L., Dressick, W.J., Marrian, C.R.K., Chow, G.M., and Calvert, J.M. (1995), The morphology of electroless Ni deposition on a colloidal Pd (II) catalyst, Journal of the Electrochemical Society **142**, pp. 2233-2243.

62. Chow, G.M., Pazirandeh, M., Baral S., and Campbell, J.R. (1993), TEM & HRTEM characterization of metallized nanotubules derived from bacteria, Nanostructured Materials **2**, pp. 495-503.

63. Krebs, J.J., Rubenstein, M., Lubitz, P., Harford, M.Z., Baral, S., Shashidhar, S., Ho, Y.S., Chow, G.M., and Qadri, S. (1991), Magnetic properties of permalloy-coated organic tubules, J. Appl. Phys. **70**, pp. 6404-6406.

64. Chow, G.M., Stockton, W.B., Price, R., Baral, S., Ting, A.C., Ratna, B.R., Schoen, P.E., Schnur, J.M., Bergeron, G.L., Czarnaski, M.A., Hickman, J.J., and Kirkpatrick, D.A. (1992), Fabrication of biologically based microstructure composites for vacuum field emission, Materials Science and Engineering **A158**, pp. 1-6.

65. Kirkpatrick, D.A., Bergeron, G.L., Czarnaski, M.A., Hickman, J.J., Chow, G.M., Price, R., Ratna, B.R., Schoen, P.E., Stockton, W.B., Baral, S., Ting, A.C. and Schnur, J.M. (1992), Demonstration of vacuum field emission from a self-assembling biomolecular microstructure composite, Appl. Phys. Lett. **60**, pp. 1556-1558.

66. Chow, G.M., Kurihara, L.K., Feng, C.R., Schoen, P.E. and Martinez-Miranda, L.J. (1997), Alternative approach to electroless Cu metallization of AlN by a nonaqueous polyol process, Appl. Phys. Lett. **70**, pp. 2315-2317.

67. Eriksson, G., Siegbahn, H., Andersson, S., Turkki, T. and Muhammed, M. (1997), The reduction of Co^{2+} by polyalcohols in the presence of WC surfaces studied by XPS, Materials Research Bulletin **32**, pp. 491-499.

NANOSTRUCTURED MATERIALS PRODUCED BY HIGH-ENERGY MECHANICAL MILLING AND ELECTRODEPOSITION

MICHEL L. TRUDEAU
Emerging Technology
Institut de recherche d'Hydro-Québec (IREQ)
1800 boul. Lionel-Boulet
Varennes, Québec, Canada J3X 1S1

ABSTRACT. The field of nanostructured materials has gained worldwide prominence in recent years as an area with great potential for new technological advances. As the field develops, the need for large quantities of materials with complex nanostructures will become more and more pronounced. Probably one of the most efficient synthesis techniques for obtaining large quantities of these materials is high-energy mechanical milling. One of the goals of this paper is to review some of the concepts related to nanostructure design and processing using the milling process. Some physical considerations will be presented as examples of various nanostructured systems are discussed. The examples will also serve as a basis for examining some technological applications based on mechanically processed nanostructured materials.

If mechanical milling is considered as the method of choice for producing large quantities of nanostructured powders, electrodeposition is probably the most efficient synthesis technique to obtain dense, nanostructured end products for a variety of applications. Recent advances in controlling particle nucleation and growth during electrodeposition have resulted in a renewed interest in this processing method. Because of its enormous potential, the second part of the paper is devoted to recent advances in this area.

1. Introduction

There have been numerous very good reviews on high-energy mechanical milling [1-4]. Likewise a number of books have been written on electrodeposition [5-7]. This paper is designed to present these material synthesis processes in a slightly different way, i.e. from an applied-research point of view. The approach chosen for this review is to look back at ten years of research and development in the field of nanostructured materials using mainly high-energy mechanical milling and, recently, electrodeposition as synthesis methods. The focus of this review will be materials development. The presentation will partly follow a chronological line and will describe some of the nanostructured materials synthesized and subsequently studied at Hydro-Québec's research institute during the past ten years together with the reasoning behind those studies. Some of the concepts and recent results related to nanostructure synthesis using high-energy milling and electrodeposition will also be described. This approach should hopefully allow the reader not only to get a feel of the exciting world of materials synthesis, some of the reasoning behind particular experiments, and the

47

G.M. Chow and N.I. Noskova (eds.), Nanostructured Materials, 47–70.

problems as well as the surprises that can happen, but also to realize the enormous possibilities of these two synthesis techniques.

2. High Mechanical Energy

Mechanical energy has been around since the early ages of humanity, being the simplest method of processing materials. Early man used it to crush bones, grains, shells and to produce food, pigments, weapons... Over the centuries, the use of mechanical energy has evolved as well as sophistication of the end-products, whereas its simplicity of operation has remained fairly constant. A recent use of mechanical energy came about with the development of two new synthesis techniques: high-energy mechanical milling (MM) and mechanical alloying (MA). MA can be defined as the use of mechanical energy to produce new alloyed materials using a number of individual components. This technique was developed on a large scale in the 1960s by the International Nickel Company (INCO) for the development of new oxide dispersion-strengthen (ODS) alloys [8]. MM can be viewed as a subset of the MA method, where no alloying take place, the milling being performed on single-phase materials (e.g. pure elements or intermetallics).

When the author of this reviewed joined the Hydro-Québec's research institute, IREQ, in 1987, one of his first projects was devoted to develop more energy-efficient electrode materials for hydrogen evolution. At that time, the main focus was on amorphous Fe-based alloys prepared by rapid quenching. In order to be electroactive, these alloys needed some form of activation prior to hydrogen evolution, namely a cycle of oxidation-reduction which was found to produce fine Fe crystals on the surface of the amorphous ribbons. Concurrently, researchers around the world were developing and studying the formation of amorphous materials using high-energy mechanical milling. This interest in the technological potential of amorphous alloys and the desire to obtain a basic understanding of the structural transformation produced by high-energy milling led us to buy our first SPEX-8000 mill.

In 1988, word came that a new class of materials based on crystallites with an average size in the nanometer range (5-30 nm) had been produced and that they exhibited a number of unique properties. The first studies on these nanostructured materials, produced mainly by the gas-phase evaporation-condensation method, indicated that, when the grain size is below 20 nm, the number of atoms present at the grain boundaries increases dramatically and that the properties of solids are more and more influenced by the properties of the intercrystalline grain boundary regions. For example, for a solid with an average crystallite size of 5 nm and an intercrystalline thickness of 5 Å, about 50% of the atoms are located at the grain boundaries [9].

The discovery of this new class of solids with their very active surface coincided with the time we were considering the synthesis of materials by high-energy mechanical milling and, in particular, attempting to produce amorphous Ni-Mo alloys. Ni-Mo alloys were of particular interest since it was known that Ni-Mo coatings, prepared by means of a complex cycle of heat treatments, had one of the best efficiencies for hydrogen evolution. However, when we tried to prepare amorphous powders using high-energy milling, we inavertently produced nanocrystalline NiMo compounds, which are among the best cathode materials for hydrogen production [10].

2.1   MECHANICAL ALLOYING

Mechanical alloying is both a simple and very complex process. Experimentally, it is very simple: the amount of elemental powders required for a desired alloy composition is placed in a hard material vial (steel or WC) with a selected number of balls. In order to minimize surface and/or oxidation reactions, the vial is normally sealed in an inert atmosphere such as argon. The assembly is then milled for the desired length of time. To follow the material reactions, the milling is stopped at different time intervals and small amounts of powder are taken to monitor the structural evolution as a function of the milling process. Because of this simplicity of operation, people tend to forget that the structural changes occurring during high-energy milling are substantial and that a number of parameters may be playing a role, such as: powder to ball weight ratio, the number and size of the balls, the energy intensity, the milling temperature and atmosphere, the movement of the vial...

Macroscopically, a three-stage process for the nanostructure by high-energy milling was described by Fecht [3]:

→Deformation localization occurs in shear bands containing a high dislocation density; the atomic-level strain can increase up to 3%.

→The dislocations are annihilated and recombined to small-angle grain boundaries, creating a subgrain structure of nanoscale dimensions.

→The orientation of the grains with respect to their neighboring grains becomes completely random.

On a microscopic level, a good amount of work has been done to understand the structural changes occurring during high-energy mechanical alloying and milling process. Thermodynamic modeling of the structural changes produced by mechanical alloying is based on the model developed by Miedema on the formation enthalpies of various competing phases using the equation [11]:

$$\Delta H = \Delta H_{chem} + \Delta H_{elas} + \Delta h_{struc} \qquad (1)$$

where ΔH_{chem} is the chemical contribution to the mixing of atoms of two different metals, ΔH_{elast} is the elastic contribution due to the atomic-size mismatch and ΔH_{struc} is the variation in the lattice stability as a function of the average number of valence electrons per atom and calculation of Gibb's free energy for the different phases. A number of authors have looked into the possibility of predicting the formation of solid solutions, intermetallics or other microstructures. Calculation using the CALPHAD method has been successful in several systems [12-14].

Because of its simplicity, X-ray analysis is the chosen characterization tool for observing the different structural transformations resulting from high-energy milling. Figure 1 presents the X-ray spectrum of Ni-Mo powders taken before and after 20 h of milling for $Ni_{85}Mo_{15}$ and $Ni_{60}Mo_{40}$. With increased mechanical alloying time, the Ni diffraction lines broaden and shift to lower angles while at the same time the Mo lines slowly diminish in intensity [15].

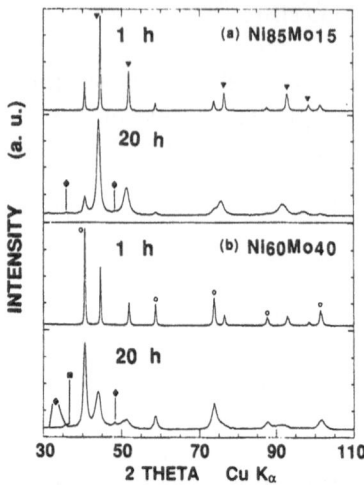

Figure 1. X-ray spectra for $Ni_{85}Mo_{15}$ and $Ni_{60}Mo_{40}$ before and after 20 h of milling (▲) Ni, (○) Mo and (♦) WC impurity [15]

⇒ Some important points should be clarified with regard to the analysis of X-ray line broadening. The broadening of the X-ray peaks is due to two main factors: the reduction in the crystallite size and the incorporation of inhomogeneous strain in the crystallites, $<e^2>^{1/2}$. It can be shown that, in reciprocal space given by: $s=2\sin\Theta/\lambda$, the broadening due to the size effect is independent of s, while that due to the strain is proportional to s [16]. Having said this, one could think that it is easy to obtain both values using a number of peaks. Unfortunately, such is not necessarily the case. A first difficulty is that an amount of broadening due to the instrument itself is added to the sample broadening. To subtract this component from the X-ray spectra, we must make some hypotheses on the functionality of broadening. If we assume that broadening is purely Gaussian, then $\beta_i^2 = \beta_m^2 - \beta_s^2$, where β_m is the sample peak breadth and β_s is the instrumental breadth. On the other hand, if Cauchy (Lorentz) distributions are assumed, then $\beta_i = \beta_m - \beta_s$. In most cases, the distributions are never purely Cauchy or Gaussian but more a convolution between these two distributions (Voigt function); a number of uncertainties will result from this correction, especially for large crystals (> 50 nm). A second difficulty concerns the functionality of the size and strain distribution: some people have argued that both distributions are Gaussian, others claim that they are Lorentzian and, finally, some say the distribution associated with the size is Lorentzian whereas a Gaussian distribution should be associated with the strain, the final distribution being a convolution of both. Depending on the approach taken, different relationships exist between the volume-average grain size and the strain, which can give similar or very different results, depending on the nanostructure of the samples. Lastly, another problem associated with X-ray analysis is the tendency to look for an average crystallite size using all reflecting planes. This can lead to large errors if the crystallites are not sperical or cubic in shape or if the strain is nonuniform. For example, it is well known that the amount of strain that can accumulate in the <111> direction will normally be less than in the <100> direction. Zhang et al. [17] found that the strain for nanostructured Cu prepared by severe plastic deformation was three times larger in the <200> direction than in the <111> direction. It should be mentioned that more elaborate techniques for X-ray spectra analysis have been developed, such as the Rietvel analysis [18], which models the crystalline structure, the experimental array and the background and which, combined with a least-squares minimization routine, allows the lattice parameters, average crystallite size, atom position, temperature factors associated with atom vibration, average microstrain, phase concentrations and atom occupancy to be determined [19]. Finally, one of the most accurate methods is that developed by Warren and Averbach based on the Fourrier analysis of the peak profile, which yields simultaneously the area-weighted and volume-weighted average grain size, both quantities allowing an estimation of the grain size distribution [20].

Going back to the analysis of the NiMo spectrum, it was found that the average Ni crystallite size decreased continuously with milling time down to values of about 5 nm. For the Mo, the smallest size obtained was about 20 nm. At the same time, the Mo peaks were found to decrease in intensity and the Ni peaks to shift to a lower angle. The lattice parameter for Ni increased from its tabulated value of 3.52 Å to about 3.62 Å for the $Ni_{60}Mo_{40}$ materials after 40 h of milling. All this experimental evidence indicated that, as the crystallite decreased in size, the Mo atoms were being incorporated into the Ni lattice. Since the Mo concentration in the Ni is directly proportional to the lattice parameter, it was possible to estimate the solubility of Mo in the Ni crystals as being about 27 at.% for the milling parameters used [21]. At the same time, the strain in the Ni and the Mo crystals increased to values of about 1.0%. This maximum strain was observed for Ni and Mo crystallites with an average size of about 10 and 30 nm, respectively. With further milling, the strain was observed to decrease for both Ni and Mo as the crystallites were further reduced. This maximum strain value was also observed in a number of systems including pure Ru and AlRu alloys by Hellstern et al. [22] and in fcc and bcc pure metals by Oleszak and Shingu [23].

The difference in the final grain size for the Ni and Mo in the ball-milled materials can be explained by the following factors. First, as recently discussed by Koch, there seems to be a relation between the crystalline structure and the minimum grain size that can be obtained; d_{min} following the relation fcc<bcc<hcp [2]. Another factor is the existence of a minimum length for the dislocation separation in a pileup, L, that originates from the equilibrium between the repulsive force between dislocations and the externally applied force; where L is given by the relation:

$$L = \frac{3Gb}{\pi(1-v)H_v} \qquad (2)$$

where G is the shear modulus, b the Burgers vector, v the Poisson ratio and H_v the hardness. Since the addition of a second phase increases the strength and hardness of a material, this results in a smaller value of L for a solid solution than for a pure metal and, thus, a smaller crystallite size, as found by Eckert et al. in Fe-Cu [24, 25]. Figure 2a presents the minimum grain size found in Fe-Cu as a function of alloy concentration

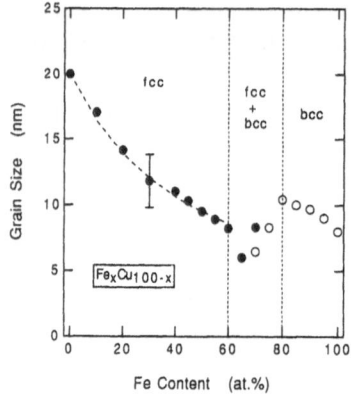 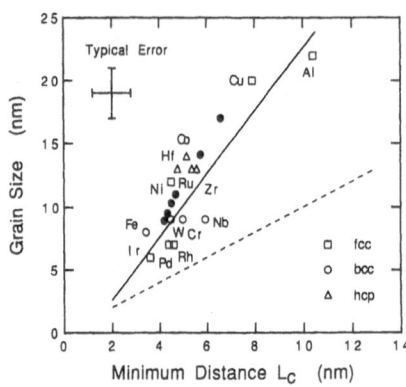

Figure 2. a) Minimum grain size in Fe-Cu as a function of the concentration [24];
b) same size as a function of L [25]

and structure, while in Figure 2b the crystallite size is plotted as a function of L. The reduced grain size for the fcc solid solution seems to be in good agreement with a hardening process due to the increased dissolution of Fe. For the bcc phase, a reverse effect is observed: the crystallite size increases as the amount of Cu in solution increases. It is known, however, that Fe(Cu) undergoes a softening effect due to the reduced grain boundary hardening, as found also in Ni(Cu) and Cr(Cu) solid solutions [26]. The very small grain size for the Ni(Mo) solid solution compared to the Mo crystals and even to pure Ni (30 nm [21]) would therefore be due mainly to solution hardening as a result of Mo insertion into the Ni lattice.

An interesting discovery in this study was the fact that the amount of Mo that could be put in solution was about 27 at.%, twice the value accepted in normal thermodynamic equilibrium. This deviation is observed for many systems prepared by mechanical alloying and, also, by other synthesis techniques used to produce nanostructured materials. High solubilities were obtained in systems that have very little or no solid solubility in normal equilibrium (positive heat of mixing).

As shown previously, Eckert et al. [24] produced single-phase fcc Cu(Fe) with 60 at.% of Fe and single-phase bcc Fe(Cu) with up to 20% Cu. Similarly, Xu et al. [27] succeeded in placing 4.3 at.% of Ni and Ag and 6.6 at.% of Ag in Ni, two elements that are normally totally immiscible with a ΔHmix of about +15 kJ/mol. The latter authors also found that, if the milling was done at higher temperatures, the solubility was reduced, while it increased for the Ag-rich composition if the vial was cooled with liquid nitrogen. They interpreted their results as a competition between mechanically driven alloying and the decomposition of thermodynamically unstable alloys. The difference in the solubility of the two elements was explained by the diffusional properties of Ag and Ni, resulting in a lower thermal stability of the Ag-rich phase.

Mechanical alloying is not a metallic exclusivity. The structural transformations produced by this simple technique were such that we milled nearly everything (however, we did not try plutonium!). We studied mechanical milling and mechanical alloying on high T_c superconductors and their oxide precursors and found that some solid solutions could also be formed in media such as $CaO\text{-}Bi_2O_3$ systems

[28]. Similarly, Chen and Yang [29] used high-energy milling to produce solid solution in ZrO_2-CeO_2 systems. The solid-solution limit of ZrO_2 in CeO_2 of 6 mole % in equilibrium at room temperature increased to 70 mole % after 90 h of milling.

2.2   MECHANICAL MILLING

High-energy mechanical milling is similar to mechanical alloying but is used for a single element or for an intermetallic compound. Normally, mechanical milling is used to transform the structure: crystalline size refinement or a change in the crystalline structure (e.g. ordered to disordered transformation).

After Ni-Mo, our research interest shifted from hydrogen evolution to hydrogen storage. Initial studies on the properties of nanostructured materials indicated that, because of their numerous grain boundaries, hydrogen diffusion could be considerably increased. Moreover, theoretical studies of metallic clusters suggested that hydrogen absorption could possibly be greatly enhanced by reducing the crystallite size. These two factors were at the basis of our investigation into the hydrogen absorption of nanostructured powders. The first system studied was FeTi, since it is a well known hydrogen storage compound. Nanostructured FeTi was produced by mechanically milling elemental Fe and Ti and, also, by milling the intermetallic compound FeTi, a CsCl ordered cubic structure [30]. High-energy milling was found to reduce the grain size down to about 7 nm, a value that is comparable to that obtained by mechanical alloying. However, high-energy milling was not able to produce a completely disorder edcrystalline structure. The intensity of the super-lattice peak after 40 h of milling was reduced by about 45% compared to the value found in a fully ordered sample. Bakker and Di [31] were able to cause chemical disorder in A15 compounds V_3Ga and Nb_3Au, but Di et al. [32] were not successful in producing B2 structures such as Co-Ga. Bansal et al. [33] studied a number of Fe_3X alloys (X=Al, As, Ge, In Sb, Si, Sn and Zn) and found that the trend for the as-milled phases largely follows the metallic radius of the solute atoms; as the radius increasingly differs from that of Fe, the structure of the final powders goes from a bcc solid solution, to a B2 ordered phase, to a partially B2 phase with the presence of intermetallic phases and, finally, to no alloying at all.

Numerous authors have looked into the formation of nanostructured pure elements using high-energy milling. Recently, Koch [2] reviewed a number of studies on the formation of nanostructured materials by high-energy milling and found an interesting correlation between the minimum grain size and the melting temperature, as shown in Figure 3. In his review, Koch suggests that the total strain, rather than the milling energy or ball-powder-ball collision frequency, was responsible for determining the minimum nanocrystalline size achievable by high-energy mechanical milling.

Mechanical milling was found to be just as effective in refining the microstructure of ceramics [34]. One of the disadvantages of high-energy milled metallic powders is that their effective surface is too small for direct utilization as catalytic materials. In order to increase this surface, catalytic materials are often dispersed on a catalytic support such as Al_2O_3. However a few high-energy milling experiments were performed on pure γ-Al_2O_3 before incorporating some new NiRu catalytic nanostructured powders, in order to try to increase its surface area. The results

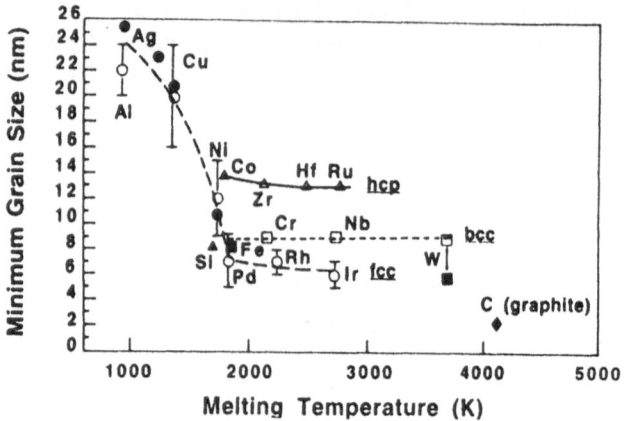

Figure 3. Minimum grain size as a function of the melting temperature for a number of pure elements [2]

were quite unexpected. High-energy milling was found to produce a structural change of the alumina, transforming various transition aluminas (γ, κ and χ) into the stable α form, a process normally achieved at temperatures of the order of 1100-1200°C. Moreover, the α-Al$_2$O$_3$ produced had a nanocrystaline structure, with a crystalline size of about 20 nm, a much lower value than that obtained with high-temperature processes. It was also found that this value was lower if the milling was performed in air rather than in argon.

2.3   MILLING PARAMETERS

It is worth briefly discussing the importance of milling equipment and milling parameters. A variety of milling equipment has been developed. Many studies have used a type of shaker mill, such as the SPEX-8000, which is considered a high-energy mill. The ball velocity in this mill is reported to be about 3 m/s [35]. Other studies used attrition mills, such as the Fritch mill, which offers the possibility of varying the milling intensity [37]. Figure 4 presents the ball movement in these two mills as well as for a less energetic tumbler mill. Low-energy ball mills such as the horizontal mill used by Oleszak and Shingu [23] have also been used. With this kind of milling equipment, it takes much more time to obtain a structural transformation similar to that produced by the SPEX. A milling time of the order of about 600 h was required to reduce the size of Fe below 10 nm, with a mass to powder ratio of 100:1. Normally in such a mill, the contamination produced by the equipment is very low: after 600 h, the total contamination was found to be < 0.1 wt.%.

Apart from the type of mill, a number of additional parameters, such as the milling energy and the temperature of the milling, can affect the structure of the final powders. An example of the impact of these two parameters can be seen in the work of Yamada and Koch [37] who compared the milling of a TiNi intermetallic in a SPEX at different temperatures and in a lower-energy vibratory mill. They found the energy of the SPEX mill was sufficient to reduce the grain size down to 5 nm, a critical value for

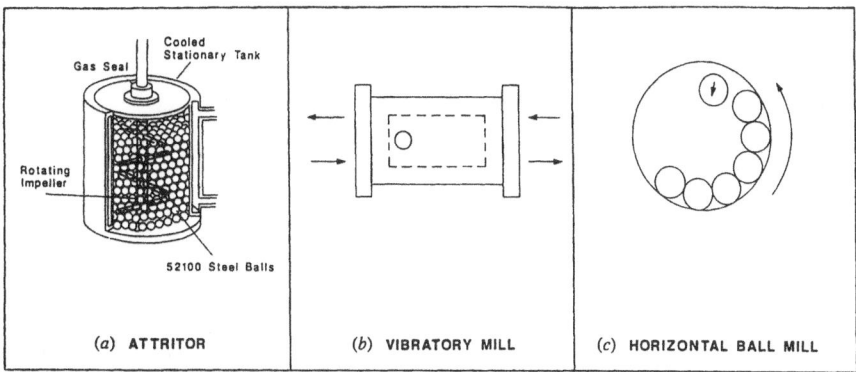

Figure 4. Ball movement for different types of mill [36]

effecting a structural transformation to an amorphous state, whereas the grain size stabilized at about 15 nm in the vibratory mill. The lattice strain was significantly higher for the powders milled in the SPEX. They also found, as seen in Figure 5, that the crystallite size decreases faster at lower temperatures, indicating that defect recovery is reduced at low temperatures.

The main point to remember is that the results of a milling experiment should be analyzed with great care, always considering the potential intercorrelation between the milling system and milling parameters used and the resulting structure.

2.4   MECHANICAL CRYSTALLIZATION

After working on the NiMo and FeTi systems, we had the chance to investigate an interesting new phenomenon. Some new insights into the process of mechanically induced structural transformation and the importance of shear bands were offered by the process of mechanical crystallization. This process is interesting since it offers a new perspective for examining the relationship between mechanical energy and structural transformation. Moreover, mechanical crystallization can be used to obtain some very unique nanostructured materials that are difficult to produce by other means.

As often the case, the discovery of this effect was accidental. The study originated from an interest in producing amorphous powder of Metglas-type ($Fe_{78}Si_9B_{13}$) by milling amorphous ribbons. In doing so, we found that the powders were slowly crystallizing. To understand this effect, we conducted many experiments by varying the composition and the atmosphere in the vial and by adding different elemental powders during the milling process [38, 39]. One of the major questions regarding high-energy milling at that time was the effective temperature during collision. It was argued that the temperature could be high enough to produce thermal crystallization of an alloy (400 to 500°C). The question then was, could the observed crystallization be due to a thermal effect, even if the crystallization temperature of this alloy was above 500°C? The addition of different elements proved that this concept could not be sustained. The first experiments showed that Co addition to $Fe_{78}Si_9B_{13}$ powder speeded up the crystallization process, while the addition of Ni drastically reduced crystallization. X-ray data showed rapid incorporation into the amorphous powders for both elements. What was more interesting was that the addition of Ni reduced the crystallization temperature

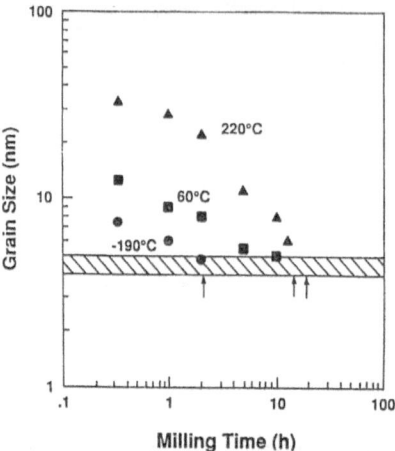

Figure 5. Variation of the crystallite size for TiNi as a function of milling time for different milling temperatures [37]

by more than 100°C. Clearly, these results indicated that simple thermal effects could not be at the basis of this process. Similar results were also found in amorphous $Fe_{64.5}Co_{18}Si_1B_{16}C_{0.5}$ [40] as well as in $Fe_{80}B_{20}$ [41].

Complete amorphization of Fe-B type alloys has always been difficult if not impossible by high-energy milling. These studies indicate that thermodynamic or kinetic factors prevent such amorphization. As discussed in the following section, there is also the chance that impurities may play a role.

To pursue the work on mechanical crystallization and to discount some of the possible effects related to the presence of metalloids, similar studies were conducted on Fe-Zr alloys [42]. This system is interesting because amorphization through rapid quenching is possible for $20 < x < 42$ and for $88 < x < 93$, while amorphous powders are obtained for $30 < x < 78$ using high-energy mechanical milling [43, 44]. Because of this difference in composition, it seemed important to explore further the effect of high-energy milling on amorphous Fe-riched samples. Figure 6 shows that, as anticipated, mechanical milling can partly crystallize an amorphous $Fe_{90}Zr_{10}$ alloy: the final structure of the powder was identical to that of mechanically alloyed powders. It was pointed out at that time that possible explanations for mechanical crystallization were the unstable nature of the materials and a change in the local atomic structure due to an enhancement of the diffusion by the mechanical process. On the other hand, Bansal et al. [45] argue that mechanical crystallization was due to the presence of contaminants such as oxygen. They found that an amorphous alloy subjected to mechanical crystallization could not be re-amorphized by rapid quenching due to the presence of too many Fe-oxide clusters or changes in composition.

It is worth pointing out that similar surface mechanical crystallization had already been observed by other researchers. For example, Miyoshi and Buckley [46] studied the wear properties of amorphous $Fe_{67}Co_{18}B_{14}Si_1$ and observed multiple slip bands due to shear deformation on the side of the wear tracks and two distinct types of wear debris: alloy debris and whiskey oxide debris. They suggested that strain could control the extent of the surface crystallization. Recently, the study of mechanical

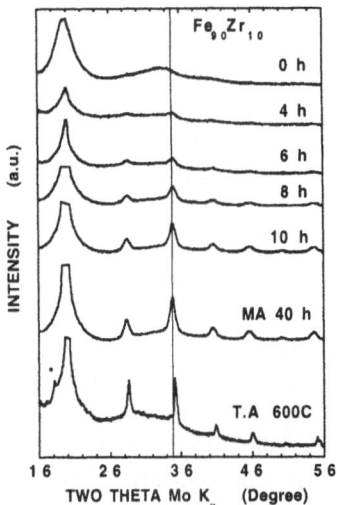

Figure 6. X-ray spectrum as a function of milling time for amorphous $Fe_{90}Zr_{10}$, for a MA 40-h powder and a partially thermally annealed amorphous ribbon. The major peaks are α-Fe, ($\bullet$) Fe_3Zr [42]

crystallization of various Al-based amorphous alloys such as $Al_{90}Fe_5Gd_5$ and $Al_{90}Fe_5Ce_5$ clearly indicated that mechanical energy and shear band formation are responsible for this phenomenon [47, 48]. For these alloys, ball milling for 1 min induced the formation of nanocrystalline Al particles. Interestingly, in other Al-riched alloys such as $Al_{85}Ni_5Y_{10}$, milling did not successfully produce any crystallization, even if the thermal stability of the materials changed. Studies of these alloys also showed that bending was sufficient to precipitate some very fine nanocrystals in the shear band. High-resolution elemental analysis of these precipitates indicates that a diffusion mechanism occurred, as suggested for FeZr mechanical crystallization [42]. This diffusion could be due to shear strain, which can be as high as 100 to 1000%, and can produce local changes in the topological or chemical short-range order leading to the formation of more stable Al nanocrystals or to some thermal mechanism [49].

Recently, a similar mechanically driven crystallization was observed in amorphous Si_3N_4 and Si-N-C ceramics. Fan et al. [50] found that a very fine nanocomposite formed of α-Si_3N_4 and β-Si_3N_4 with a remaining amorphous phase was formed after only 4 s of milling at high milling intensity. The crystallization products were the same as those observed after thermal annealing. No variation of the impurity level was found in the powders before and after milling. On the other hand, if a lower intensity of milling was used, no crystallization was observed.

The major point to be made about this process is that mechanical crystallization makes it possible to obtain nanostructured materials with unique nanostructural features. For example, the first α-Fe nanocrystals to precipitate in $Fe_{90}Zr_{10}$ have a crystallite size of about 3 nm and a composition that cannot be obtained by mechanical alloying. In the case of Al-based materials, precipitation of these nanocrystals was found to increase the mechanical strength of amorphous materials.

2.5 IMPURITIES

The foregoing discussion of the potential effects of impurities in high-energy milling provides a good starting point for this section, which is probably the most fundamental in this paper.

High-energy milling is plagued by one major inconvenience and one which cannot be easily disregarded, as many researchers (the present author included) tend to do at times: the incorporation of impurities coming from the wear of the milling tools, from the vial and balls, and from the reaction between the fresh surface of powders with trace amounts of reactive gas present in the atmosphere of the vial.

The first kind of contamination is usually easy to detect. Among the most common are Fe and the other elemental constituents of the vial and balls, such as Cr and Ni, since hardened steel is the material most widely used. For example, the first attempts to mill NiMo powders were made with steel vials. After 20 h milling, the composition of the materials was approximately $Fe_{40}Ni_{37}Mo_{14}Si_1P_4Cr_4$. Similarly, Beke found Fe contamination of the order of 15 at.% after 300 h milling of pure Ni in a vibratory mill [51]. He also found 1.5 to 2 at.% of Cr after milling pure Fe for a similar period of time. Knowing the amount of Cr in steel, we can follow the possible changes in composition of the starting materials. Because of its higher hardness, WC is often employed to make the tools commonly used for mechanical milling. However, with hardness comes increased brittleness, which can produce a large amount of contamination. For example, in the case of the mechanical alloying of $Ni_{60}Mo_{40}$ materials, typically 2.5 at.% tungsten is found after 20 h of milling.

⇒It is important to mention that WC particulates are bonded together by Co, which also tends to be forgotten. So, although WC is not too reactive, Co is easily soluble in many systems and is also ferromagnetic. This must be borne in mind when evaluating some properties, magnetic properties especially, wherever relatively large amounts of WC contaminants are found in the milled powders.

The amount of contamination from the milling tools depends on a number of factors [2]: the energy of the mill (as expected, the higher the mill energy, the greater the amount of contamination); the possible reactivity between the milled powders and the milling equipment; the mechanical properties of the powders, principally their hardness vs that of the vials and balls. Courtney and Wang [52] looked into contamination during the mechanical alloying of Ni-W alloys using steel tools and found very high Fe contamination which increased nonlinearly in time but seems to depend on the composition of the powder or its structure. One final factor is the quality of the tools, since a) the hardened steels of which they are made are not all equal and b) this type of equipment is not normally used for such precise experiments. This is even truer of WC-Co tools, since the final composition of the alloys is a tradeoff between high hardness and good mechanical integrity. Moreover, the quality and mechanical stability of the tools are a function of the powder processing and sintering used. For a given hardness, a large variability in the mechanical stability is often observed.

⇒Another point of concern is the method used to measure the amount of metallic contamination in powders. Most people use energy dispersive x-ray analysis (EDAX) but, as mentioned by Yvon and Schwartz [53], large errors can occur due to matrix effects. Standards, with a composition close to the estimated powder composition, should always be used if possible.

Contamination from the wear of the milling equipment is not, in the opinion of the present author, the biggest problem, however. Gaseous impurities that partly react with the milled powders or go into interstitial sites can present greater contamination problems, since they are not easily detected. Thus, while researchers are at least often measuring wear contamination and trying to explain their studies taking this factor into account, people tend not to even measure gaseous impurities, despite the many experiments that show them to be serious contaminants.

One highly illustrative example is related to the work on NiMo discussed earlier. Because this alloy was causing drastic wear of steel tools, we decided to do mechanical alloying in a small WC vial which was not hermetically sealed. Because of this, the oxygen content in the powder increased steadily with the milling time up to a value of about 2.6 wt.% after 20 h in the $Ni_{60}Mo_{40}$ alloy. At that time, it was thought that this amount could be regarded as too small to worry about and not significant for the electrocatalytic activity. Interestingly, however, when we tried to repeat these experiments and prepare new materials in a more controlled environment (with no air contamination), we found it impossible to reproduce the electrocatalytic activity. After a number of experiments, we found that only 1 wt% of O was able and, in fact, needed to drastically change the electrocatalytic properties of these materials [21].

Even more extreme results were obtained by Dussault et al. [54, 55] in a study of gaseous impurities on the mechanical crystallization of Metglas amorphous alloys. Figure 7 presents the variation in the crystallization phase as a function of the milling environment for experiments done in oxygen (x), air (■) and argon (●). These experiments underscore the importance of gaseous impurities in the mechanical crystallization process. Under a continuous flow of oxygen, the amorphous alloy crystallized in less than 15 h whereas it took more than 150 h to crystallize it in argon. The amount of O present in the powder after 15 h milling was 1.7 wt.%, compared to 0.1 wt.% for the Ar case. Careful x-ray analysis revealed no trace of oxide peaks for these crystalline powders.

The above examples show very clearly that greater care is needed when analyzing the results of mechanical milling where a liquid organic phase (such as ethyl or methyl alcohol) is used to avoid excessive sticking. Faudot et al. [56] examined the contamination effect of Fe-Cu powders milled with ethanol. They found large differences in the final metallic composition (reduced contamination from the milling tools) as well a significant difference in structure between powders milled with and without ethanol.

More insidious to the presence of gases such as O, N and C is the presence of hydrogen, since it is more difficult to detect. However, H can play a major role in the structural transformation that takes place during the milling as well as in the determination of the various properties of the final materials. Ivison et al. [57], for

60

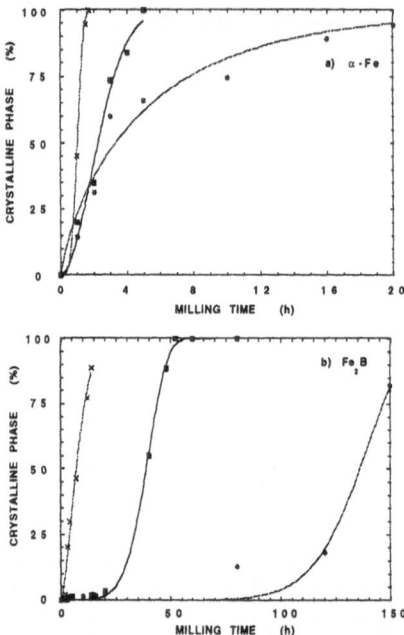

Figure 7. Amount of α-Fe and Fe₂B as a function of milling time for experiments
performed in (x) oxygen, (■) in air and (●) in argon [55]

example, found that the presence of H was necessary for complete amorphization of Ti-riched Cu-Ti alloys during high-energy mechanical alloying.

Finally, even if care is taken to seal the vial under an inert atmosphere, it should be remembered that the surface of most powders is often covered by a number of atomic layers of adsorbed gas and that, depending on the powder size used, large amounts of gaseous impurities can diffuse inside the crystallites or be trapped at grain boundaries during milling.

2.6   NANOCOMPOSITE FORMATION

One major deficiency of nanostructured materials prepared by high-energy mechanical milling for catalytic applications is their low surface area (around 1 to 10 m²/gr). In order to use such powders for catalytic reaction, we tried to increase their effective surface through the use of an inert support. High-energy milling was thus used to produce nanostructured heterogeneous nanocomposite catalysts. To accomplish that, very finely dispersed metallic nanocrystals on α-Al₂O₃ were easily obtained by milling NiRu nanocrystals and α-Al₂O₃ powders.

More interestingly, studies suggest that the hydrogen storage process could be greatly improved by the presence of palladium on the surface of the storage materials [58, 59]. The presence of small Pd nanocrystals in compounds such as FeTi or LaNi₅ allows the storage compounds to absorb hydrogen readily without the need for an activation stage. The materials also showed an improved absorption kinetics, at a lower temperature than for pure nanocrystalline compounds. Lastly, Pd-storage compound composites are less affected by poisoning effects.

In the field of catalysis and electrocatalysis, sub-oxide systems tend to be the material of choice, since they offer surface vacancies that can be used as reactive sites [60]. As shown in the Ni-Mo-O systems, high-energy mechanical milling is a very efficient technique for producing such nonstochiometric materials. In recent years, research efforts have been devoted to develop a new electroactive cathode for the production of chlorate. A similar process to that for the Ni-Mo-O was found, but this time in the Fe-Ti-Ru-O system [61, 62].

2.7   REACTIVE MILLING

Finally, an even more complex field deserves to be discussed here: reactive milling. This technique is gaining more and more attention these days because the basic principle is very attractive: high-temperature chemical reactions can be obtained at low temperature. An example of this reaction was presented by Suzuki and Nagumo [63] who milled Ti and Al powders with n-heptane to form a Ti-Al carbide as well as Ti hydride. Ding et al. [64], for their part, have shown that nanocrystalline Cu powders can be produced by milling $CuCl_2$ with Na via the solid-state reaction

$$CuCl_2 + 2Na \Rightarrow Cu + 2NaCl$$

and by washing the resulting powder with water. Crystallite sizes of the order of 20-30 nm were obtained using this mechano-chemical process. The latter authors also observed traces of oxide, probably due to a reaction with water. Similarly, nanocrystalline iron nitrides (Fe_3N and $\gamma'-Fe_4N$) and supersaturated solid solutions of N in Fe have been obtained by milling pure iron in liquid nitrogen and in ammonia [65]. Pure(?) Fe was produced by milling $FeCl_3$ with Na, Ca and, also, Al. The crystallite size was minimum for milling in Na (15 nm) and maximum in the case of Al (50 nm), while the saturation magnetization was about 165 emu/g when Na was the reductant and 205 when Al was used [66]. TiN could also be produced by simply milling Ti in air. Zhang et al. [67] concluded that the formation of this nitride could not be explained by a conventional gas-metal reaction mechanism but that TiN was formed by a solid-solution pumping mechanism. Furthermore, they found that the presence of Fe, as part of the milling medium, was necessary for the formation of the nitride phase, the Fe acting as an O getter or as a catalytic site for the dissociation of N_2 prior to adsorption.

In a combined approach of reactive milling and nanocomposite formation, metallic-ceramic nanocomposites were obtained by high-energy milling. Pardavi-Horvath and Takacs [68] produced iron-alumina nanocomposite by milling Fe_3O_4 and Al according to the following reaction:

$$3 \, Fe_3O_4 + 8 \, Al \Rightarrow 9 \, Fe + 4Al_2O_3$$

They added 10% excess Al to the mixture to compensate for the surface oxidation of the powder particles. They also produced small Fe_3O_4 clusters embedded in a Cu matrix by milling Cu and Fe_3O_4 and, also, CuO and Fe [69]. Similarly, Matteazzi and Le Caër [70] produced a number of metallic-alumina nanocomposites by ball-milling metallic oxides with aluminum. The size of the $\alpha-Al_2O_3$ was ~ 10 nm while sizes as low as 7 nm were measured for Fe. Other studies showed that it is possible to produce iron nitrides, fluorides and carbide by reactive milling [71].

2.8.   APPLICATIONS

There is an enormous potential for large-scale applications of nanostructured powders prepared by high-energy milling. Moreover, this potential would be greatly increased if suitable densification methods were found to overcome the difficulty of producing dense products while at the same time keeping the crystalline nanostructure of the powder. However, discussion of densification is beyond the scope of this work.

On the other hand, the high-energy milling technique offers enormous possibilities for further development of the field of nanostructured materials. The main advantage is the possibility of scale-up to produce tonnage quantities of nanostructured powders. Another advantage is the possibility to produce a wide range of materials from ceramics through metals to nanocomposites. For example, one atractive application is the development of new hard magnetic materials such as α-Fe/Nd$_2$Fe14B or α-Fe/Sm$_2$F$_{17}$N$_{2.5}$ which were prepared by high-energy milling and showed an enhanced remanence compared to a randomly oriented single-phase magnet. This improvement was found to be related to the very small grain size and the exchange coupling between the hard and soft magnetic phases [72].

Another field that lends itself to the application of nanostructured mill powders is catalysis or electrocatalysis, as illustrated by some of the examples discussed in this paper. The ease with which non-stochiometric oxides or active nanocomposites are produced make this technique one of the best methods for the future development of new active materials.

The interest in high-energy mechanical milling is not restricted to powder metallurgy alone. Some very concrete applications of the principle developed in the study of this process and the formation of nanostructures have recently been presented. One striking example is the work of Fecht and colleagues reported in a number of publications on the structural transformation of a high-speed railway track [73]. These authors found that, due to the train movement, the surface of the steel track was undergoing a solid-state transformation caused by local pressure on the track which can exceed 1 GPa. From this structural transformation, nanostructured regions with grain sizes between 15 and 20 nm were found to develop with a wavelength of 3-4 cm. These nanostructured regions had twice the wear resistance and a fivefold increase in hardness compared to the original material. The development of different surface regions with large differences in mechanical properties was found to cause noise and hamper further development of high-speed trains. Since the process that leads to the development of such nanostructures on the tracks was found to be very similar to high-energy milling, it is possible that the development of a better railway track can be based on the results obtained with small-scale milling equipment [74].

3. Electrodeposition

The current need to obtain fully dense nanostructured materials has focussed the research interest focused on electrodeposition. This well known process has been around for more than 100 years and is being used worldwide for electroplating components or electroforming pieces that are difficult to produce by conventional mechanical means. Our understanding of the electrodeposition process is still not perfect. A large number of process parameters can have a significant influence on the final microstructure: pH, temperature, concentration and chemistry of the solution, temperature, agitation,... with the result that most of the work being done is based on a trial-and-error approach rather than a search for a more fundamental understanding of the deposition process itself.

Electroplating is another field where nanostructured materials have been produced for a long time, unknown to those working in the field. Way back in the 1940s, for example, Schaffer and Gonser [75] studied the electroplating of Fe using a sulphate-chloride solution and found hardness values as high as ~ 700 (Knoop hardness), compared to ~ 80 for annealed samples. Using a similar bath, it was possible to show that the average crystallite size was about 7 nm, a value that could have explained the observed hardness [76]. At the present time, electrodeposition is probably one of the most promising techniques for producing fully dense bulk nanostructured materials, especially for large-scale applications. Only a few research groups have studied the relationship between the microstructure of plated materials and the electrodeposition parameters. Most of the recent work on this subject has been done by the group of Prof. U. Erb and by researchers at Ontario Hydro who have successfully shown that large-scale application was possible when they used electroformed nanocrystalline Ni to repair the inside of nuclear reactor steam generator tubing. In the latter case, the advantage of using nanocrystalline Ni is its superior mechanical properties and increased corrosion resistance compared to conventionally produced coarse-grain materials [77].

Electrodeposition is an extremely complex process. Recent work has shown that some of the major factors have a more direct effect on the production of nanostructured materials: pulse plating, for instance, and the control of some other electrodeposition parameters, such as the temperature, current density and pH as well as the electrolyte composition. When the electrodeposition conditions are such that massive nucleation and reduced grain growth occur, it is very easy to produce pure elements or alloys with an average crystallite size ranging from 5 to 50 nm [78]. Among the different parameters that play a role during electrodeposition, one that dominates for nanocrystal formation is the pulsation of the DC current. During pulse plating, a DC current is applied for a period of time t_{on} and is subsequently turned off or reversed for normally a longer period of time t_{off} . It is then possible to define a duty cycle coefficient γ defined as [79]:

$$\gamma = \frac{t_{on}}{(t_{on} + t_{off})} \tag{3}$$

or

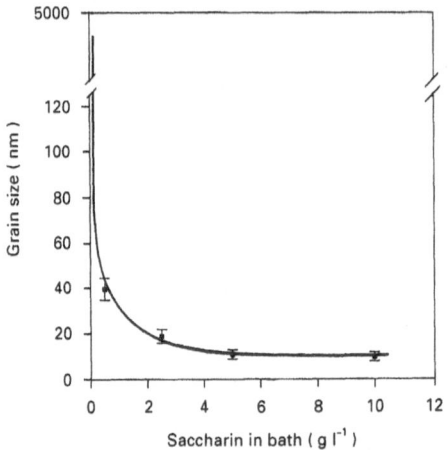

Figure 8. Grain size variation of pulse electroplated Ni with the saccharine concentration in the bath [81]

$$\gamma = \frac{t_c}{(t_c + t_a)} \qquad (4)$$

where t_c denotes the cathodic pulse duration and t_a the anodic pulse duration. For pulse plating, the average current will then be equal to:

$$I_{av} = \gamma \, I_c \qquad (5)$$

where I_c is the applied cathodic current, while for pulse reverse plating:

$$I_{av} = \frac{(I_c t_c - I_a t_a)}{t_c + t_a} \qquad (6)$$

Since a high overpotential and low surface diffusion rates promote the formation of nuclei, pulse plating with a high current density of the order of 50 A/dm^2 are used with a relatively short cathodic duty cycle. Typical values for t_{on} and t_{off} are about 2.5 and 45 ms, respectively. Another factor that helps to decrease the grain size, since it reduces the surface mobility of the adions, is the use of stress relievers or grain growth inhibitors such as saccharine that can be added to the solution [78, 80]. For example, Figure 8 presents the average grain size for electroplated nanocrystalline nickel as a function of the saccharine content in the electrolyte, the other plating parameters being constant. It should be stressed however that, as the grain size decreases, the amount of impurities, mainly C and S in this case, was found to increase [81].

Electroformed nanostructured materials have already proven that they exhibit unique properties and microstructure [82, 83]. They have also demonstrated the effects of porosity or contamination by impurities on certain properties. Mechanical studies have revealed that, as in the case of other nanostructured materials, the   hardness of electroplated nanomaterials is greatly enhanced when compared to their coarse-grain counterparts. However, these studies indicate that, contrary to some of the results on gas-phase-produced samples, electroplated nanostructures exhibit some softening at very small grain sizes, i.e. an inverse Hall-Petch relation. On the other hand, the Young's modulus was nearly the same as that of coarse-grain samples. Moreover, contrary to what was observed in a number of magnetic materials produced by gas-condensation, which showed a large reduction in the magnetization, Bs was only slightly lower in electroformed films. This difference has been explained in terms of large oxygen contamination of the gas-condensed powders.

Because of the absence of porosity, electroformed materials offer an enormous potential, for not only mechanical but also chemical applications. Recently Bryden and Ying, for example, looked at nanostructured Pd and Pd-Fe electroplated films for a hydrogen separation membrane reactor. They found that the Pd-Fe nanoalloys had a more stable structure and did not experience the $\alpha-\beta$ phase transition when exposed to hydrogen. They also observed fast hydrogen absorption kinetics with no need for a prior activation cycle [84].

Lastly, it should be noted that, contrary to what is widely believed, electrodeposition can be a very fast and efficient process. For example, common techniques with deposition rates of about 100 μm/min already exist for high-speed deposition of conventional materials [6]. It is interesting that fast electroplated materials were found to have a finer microstructure than normally plated solids.

4.   Conclusion

It is the author's hope that this paper has succeeded in illustrating the immense potential of high-energy mechanical milling and electrodeposition. As mentioned in the introduction, the main purpose was to give the reader a feel for materials development and, for that reason, a large chunk of very good work was unfortunately not discussed. As one researcher said during a seminar a number of years ago, materials development is key to the discovery of new physical properties. Both synthesis methods offer very cost-effective possibilities of producing a wide range of new nanostructured materials: from pure nanocrystalline elements to very complex nanocomposites, which will certainly give rise to new properties.

Finally, the other and, perhaps, the most important point of this paper is that, in order to fully understand the properties of these new nanostructured materials, an extremely careful characterization of the materials under investigation is indispensable, since the observed size effects could easily be the result of a number of other parameters, such as variation in the concentration of some impurities or in other macrostructures.

ACKNOWLEDGMENTS: I would like to thank the Organizing Committee for their invitation to participate in this International Workshop. Also, I am very grateful to all the people with whom I have collaborated on this very exciting subject over the years. Lastly, I would like to thank Lesley Kelley-Régnier for greatly improving the quality of the text.

66

References

1. Weeber A.W. and Bakker H. (1988) "Amorphization by ball milling. A review", Physica B **153**, 93-135.
2. Koch C.C. (1997) " Synthesis of nanostructured by mechanical milling: problems and opportunities", NanoStruc. Mater. **9**, 13-22
3. Fecht H.J. (1994) "Nanophase materials by mechanical attrition: synthesis and characterization" NATO ASI series E - Vol **260**, ed: G.C. Hadjipanayis and R.W. Siegel, 125-144.
4. Koch C.C. (1991) "Mechanical milling and alloying", Materials Sci. & Tech, Vol. 15, ed: R.W. Cahn, VCH.
5 . Lowenheim F.A. (1974) **"Modern Electroplating"** 3rd ed., Wiley & Sons, New-York.
6. Durney L.J. (1984) **"Electroplating Engineering Handbook"**, Fourth Ed., VNR, New York.
7. Dini J.W. (1993), **"Electrodeposition, The Materials Science of Coatings and Substrates"**, Noyes Publication.
8. Benjamin J.S., Sci. Amer. **234**, 40 (1976).
9. Siegel R.W. and Thomas G.J. (1992) "Grain boundaries in nanophase materials", Ultramiscroscopy **40**, 376-384.
10. Huot J.Y., Trudeau M.L. and Schulz R. (1991) "Low hydrogen overpotential nanocrystalline Ni-Mo cathodes for alkaline water electrolysis", J. Electrochem. Soc. **138**, 1316-1321.
11. Suryanarayana C., Chen G.-H., Frefer A. and Froes F.H. (1992) "Structural evolution of mechanically alloyed Ti-Al alloys", Mater. Sci. & Eng. **A158**, 93-101.
12. Oehring M., Klassen T. and Bormann R. (1993) "The formation of metastable Ti-Al solid solutions by mechanical alloying and ball milling", J. Mater. Res. **8**, 2819-2829.
13. Ma E., Atzmon M. and Pinkerton F.E. (1993) "Thermodynamic and magnetic properties of metastable $FexCu100-x$ solid solutions formed by mechanical alloying", J. Appl. Phys. **74**, 955- 962.
14. Gente C., Oehring M. and Bormann R. (1993) "Formation of thermodynamically unstable solid solutions in the Cu-Co system by mechanical alloying", Phys. Rev. B **48**, 13244-13252.
15. Trudeau M.L., Huot J.Y and Schulz R. (1991) "Mechanically alloyed nanocrystalline Ni-Mo powders: a new technique for producing active electrodes for catalysis", Appl. Phys. Lett. **58**, 2764, (1991).
16. Klug H.P. and Alexander L.E. (1974) **"X-ray Diffraction Procedures for Polycrystalline and Anorphous Materials"**, 2nd ed., Wiley & Sons, New York, pp. 618-686.
17. Zhangm K., Alexandrov I.V., Valiev R.Z., and Lu K. (1996) "Structural characterization of nanocrystalline copper by means of x-ray diffraction", J. Appl. Phys. **80**, 5617-5624.

18. Young R.A. (1993), "**The Rietveld Method**", IUCr Monographs on Crystallography Vol. 5, Oxford Science Pub..

19. Bokhimi, Morales A., Lucatero M.A. and Ramirez R. (1997) "Rietveld refinement of nanocrystalline phases", NanoStruc. Mater. **9**, 315-318.

20. Krill C.E. and Birringer R. (1997) "Estimating grain-size distribution in nanocrystalline materials from x-ray diffraction profile analysis", Phil. Mag. A, in press.

21. Schulz R., Huot J.Y., Trudeau M.L., Dignard-Bailey L., Yan Z.H., Jin S., Lamarre A., Ghali E. and Van Neste A. (1994) "Nanocrystalline Ni-Mo Alloys and their Application in Electrolysis", J. Mater. Res. **9**, 2998-3008.

22. Hellstern E., Fecht H.J., Garland C. and Johnson W.L. (1989) "", J Appl. Phys. **65**, 305-

23. Oleszak D. and Shingu P.H. (1996) "Nanocrystalline metals prepared by low energy ball milling", J. Appl. Phys. **79**, 2975-2980.

24. Eckert J., Holzer J.C., Krill III C.E. and Johnson W.L. (1993) "Mechanically driven alloying and grain size changes in nanocrystalline Fe-Cu powders", J. Appl. Phys. **73**, 2794-2802.

25. Eckert J., Holzer J.C., Krill III C.E. and Johnson W.L. (1992) "Reversible Grain Size Changes in Ball-Milled Nanocrystalline Fe-Cu Alloys", J. Mater. Res 7, 1980-1983.

26. Shen T.D and Koch C.C. (1996) "Formation, solid solution hardening and softening of nanocrystalline solid solutions prepared by mechanical attrition", Acta Mater. **44**, 753-761.

27. Xu J., Herr U., Klassen T. and Averback R.S. (1996) "Formation of supersaturated solid solution in the immiscible Ni-Ag system by mechanical alloying", J. Appl. Phys. 79, 3935-3945.

28. Tessier P., Trudeau M.L., Strom-Olsen J.O. and Schulz R. (1993) "Structural transformations and metastable phases produced by mechanical deformations in the Bi-Sr-Ca-Cu-O superconducting system", J. Mater. Res. **8**, 1258-1267.

29. Chen Y.L. and Yang A.Z (1993) "Formation of supersaturated solution in ZrO2-CeO2 system induced by mechanical alloying", Scripta Metal. et Mater. **29**, 1349-1354.

30. Zaluski L., Tessier P., Ryan D.H., Donner C.B., Zaluska A., Strom-Olsen J.O., Trudeau M.L. and Schulz R. (1993), "Amorphous and nanocrystalline Fe-Ti prepared by ball milling", J. Mater. Res. **8**, 3059-3068.

31. Bakker H. and Di L.M. (1992) "Atomic disorder and phase transition in intermetallic compounds by high-energy ball milling", Mater. Sci. For. **88-90**, 27-34.

32. Di L.M., Bakker H., Tamminga Y. and de Boer F.R. (1991) "Mechanical attrition and magnetic properties of CsCl-structure Co-Ga", Phys. Rev. **44**, 2444-2451.

33. Bansal C., Gao Z.Q., Hong L.B. and Fultz B. (1994) "Phases and phase stabilities of Fe3X alloys (X=Al, As, Ge, In, Sb, Si, Sn, Zn) prepared by mechanical alloying", J. Appl. Phys. **76**, 5961-5966.

68

34. Zielinski P.A., Schulz R., Kaliaguine S. and Van Neste A. (1993) "Structural transformation of alumina by high-energy ball milling", J. Mater. Res. **8**, 2985-2992.
35. Hong L.B. and Fultz B. (1996) "Two-phase coexistence in Fe-Ni alloys synthesized by ball milling", J. Appl. Phys **79**, 3946-3955.
36. Maurice D.R. and Courtney T.H. (1990) "The physics of mechanical alloying: a first report", Metall. Trans. A **21**, 289-303.
37. Yamada K. and Koch C.C. (1993) "The influence of mill energy and temperature on the structure of the TiNi intermetallic after mechanical attrition", J. Mater. Res. **8**, 1317-1326.
38. Trudeau M.L., Schulz R., Dussault D. and Van Neste A. (1990) "Structural changes during high-energy ball milling of iron based amorphous alloys: is high-energy ball milling equivalent to a thermal process?", Phys. Rev. Lett. **64**, 99-102.
39. Trudeau M.L., Van Neste A. and Schulz R. (1991) "High-resolution electron microscopy study of iron nanocrystals prepared by high-energy mechanical crystallization", Mat. Res. Soc. Symp. Proc. **206**, 487-492.
40. Giri A.K., Gonzalez J.M. and Gonzalez J. (1995) "Crystallization by ball milling: a way to produce soft magnetic materials in powdered form", IEEE Trans. Magne. **31**, 3904-3906.
41. Fan G.J., Quan M.X. And Hu Z.Q. (1996) "Induced magnetic anisotropy in $Fe_{80}B_{20}$ metallic glass by mechanical milling", Appl. Phys. Lett. **68**, 1159-1161.
42. Trudeau M.L. (1994) "Deformation-induced crystallization due to instability in amorphous FeZr alloys", Appl. Phys. Lett. **64**, 3661-3663.
43. Hellstern and Schultz L. (1986) "Glass-forming ability in mechanically alloyed Fe-Zr", Appl. Phys. Let. **49**, 1163-1165
44. Hellstern E., Schultz L. and Eckert J. (1988) "Glass-forming ranges of mechanically alloyed powders" J. Less-Comm Met. **140**, 93-98.
45. Bansal C., Fultz B. and Johnson W.L. (1994) "Crystallization of Fe-B-Si metallic glass during ball milling", NanoStruc. Mater. **4**, 919-925.
46. Miyoshi K. and Buckley D.H. (1984) "Mechanical-contact-induced transformation from the amorphous to the partially crystalline state in metallic glass", Thin Sol. Films **118**, 363-373.
47. Chen H., He Y., Shiflet G.J. and Poon S.J. (1994) "Deformation-induced nanocrystal formation in shear bands of amorphous alloys", Nature **367**, 541-543
48. He Y., Shiflet G.J. and Poon S.J. (1994) "Ball milling-induced nanocrystal formation in aluminium-based metallic glasses", Acta Metall. Mater. **43**, 83-91.
49. Csontos A.A. and Shiflet G.J. (1997) "Formation and chemistry of nanocrystalline phases formed during deformation in aluminium-rich metallic glasses", NanoStruc. Mater. **9**, 281-289.
50. Fan G.J., Quan M.X. and Hu Z.Q., Li Y.L. and Liang Y. (1996) "On the mechanically driven rapid crystallization of amorphous Si3N4 ceramics", Appl. Phys. Lett **68**, 915-916

51. Beke D.L. (1996) "Magnetic properties of nanocrystalline Fe, Ni(Fe) and Fe(Si)", Mater. Sci. Forum **225-227**, 701-706.

52. Courtney T.H. and Wang Z. (1992) "Grinding media wear during mechanical alloying of Ni-W alloys in a SPEX mill", Scripta Metal. et Mater. **27**, 777-782.

53. Yvon P.J. and Schwarz R.B. (1993) "Effects of iron impurities in mechanical alloying using steel media", J. Mater. Res. **8**, 239-241.

54. Dussault D., Trudeau M.L., Van Neste A. and Schulz R. (1995) "The influence of Oxygen on the Crystallization Process of Amorphous Powders by Mechanical Deformation", Advances Powder Particulate Mater. **8**, 13-21.

55. Trudeau M.L., Dignard-Bailey L., Schulz, R., Dussault D. and Van Neste V. (1993) "Fabrication of nanocrystalline iron-based lloys by the mechanical crystallization of amorphous materials", NanoStruc. Mater. **2**, 361-368.

56. Faudot F., Gaffet E. and Harmelin M. (1993) "Identification by DSC and DTA of the oxygen and carbon contamination due to the use of ethanol during mechanical alloying of Cu-Fe powders", J. Mater. Sci. **28**, 2669-2676.

57. Ivison P.K., Soletta I., Cowlam N., Cocco G., Enzo S. and Battezzati L. (1992) "The effect of absorbed hydrogen on the amorphization of CuTi alloys", J. Phys: Cond. Matter **4**, 5239-5248.

58. Zaluski L., Zaluska A., Tessier P., Ström-Olsen J.O. and Sculz R. (1996) "Nanocrystalline Hydrogen Absorbing Alloys", Mater. Sci. For. **225-227**, 853-858.

59. Zaluski L., Zaluska A., Tessier P., Ström-Olsen and R. Schulz (1996) "Hydrogen absorption by nanocrystalline and amorphous Fe-Ti with palladium catalyst, produced by ball milling", J. Mater. Sci. **31**, 695-698.

60. Trudeau M.L. and Ying J.J. (1996) "Nanocrystalline materials in catalysis and electrocatalysis: structure tailoring and surface reactivity", NanoStruc. Mater. 7, 245-258.

61. Van Neste A., Yip S.H., Jin S., Boily S., Ghali E., Guay D. and Schulz R. (1996) "Low overpotential nanocrystalline Ti-Fe-Ru-O cathodes for the production of sodium chlorate", Mater. Sci. For. **225-227**, 795-800.

62. Blouin M., Guay D., Boily S., Van Neste A. and Sculz R. (1996) "Electrocatalytic properties of nanocrystalline alloys: effect of the oxygen concentration in Ti_2RuFeO_x alloy on the structural and electrochemical properties", Mater. Sci. For. **225-227**, 801-806.

63. Suzuki T. and Nagumo M. (1992) "Mechanochemical reaction of Ti-Al with hydrocarbon during mechanical alloying", Scripta Metal. et Mater. **27**, 1413-1418.

64. Ding J., Tsuzuki T., McCormick P.G. and Street. R. (1996) "Ultrafine Cu particles prepared by mechanochemical process", J. Alloys & Comp. **234**, L1-L3.

65. Chen Y., Halstead T. and Williams J.S. (1996) "Influence of milling temperature and atmosphere on the synthesis of iron nitrides by ball milling", Mater. Sci. & Eng. **A206**, 24-29.

66. Ding J., Tsuzuki T., McCormick P.G. and Street R. (1996) "Structure and magnetic properties of ultrafine Fe powders by mechanochemical processing", J. Magne Magne. Mater. **162**, 271-276.

67. Zhang H., Kisi E.H. and Myhra S. (1996) "A solid solution pumping mechanism for the nitrogenation of titanium during mechanical deformation in air", J. Phys. D **29**, 1367-1372.

68. Pardavi-Horvath M. and Takacs L. (1992) "Iron-alumina nanocomposite prepared by ball-milling", IEEE trans Magne. **28**, 3186-3188.

69. Pardavi-Horvath M. and Takacs L. (1993) "Magnetic properties of copper-magnetite nanocomposite prepared by ball milling", J. Appl. Phys. **73**, 6958-6960.

70. Matteazzi P. and Le Caër G. (1992) "Synthesis of nanocrystalline alumina-metal composites by room-temperature ball-milling of metal oxides and aluminum", J. Am. Ceram. Soc. **75,** 2749-2755.

71. Matteazzi P. and Le Caër G. (1992) "Exchange reaction milling in iron nitrides, fluorides and carbides", J. All. & Comp. **187**, 305-315.

72. McCormick P.G., Ding J., Feutrill E.H. and Street R. (1996) "Mechanically alloyed hard magnetic materials", J. Magne. Magne. Mater. **157/158**, 7-10.

73. Baumann G., Knothe K. and Fecht H.J. (1997) "Surface modification and nanostructured formation of high speed railway tracks", NanoStruc. Mater. **9**, 751-754.

74. G. Baumann G., Y. Zhong Y. and Fecht H.-J. (1996) "Comparison between nanophase formation during friction induced surface wear and mechanical attrition of a pearlitic steel", NanoStruc. Mater. **7**, 237-244.

75. Schaffer R.M. and Gonser B.W. (1943) "A sulphate-chloride solution for iron electroplating and electroforming", Trans. Electrochem. Soc. **84**, 319-334.

76. Trudeau M.L., unpublished

77. Gonzalez F., Brennenstuhl A.M., Palumbo G., Erb U. and Lichtenberger P.C. (1996) "Electrodeposited nanostructured nickel for in-situ nuclear steam generator repair", Mater. Sci. For. **225-227**, 831-836.

78. Choo R.T.C., Toguri J.M., El-Sherik A.M. and Erb (1995) "Mass transfer and electrocrystallization analyses of nanocrystalline nickel production by pulse plating", J. Appl. Electrochem. **25**, 384-403.

79. Grimmett D.L., Schwartz M. and Nobe K. (1990) "A comparison of DC and pulse Fe-Ni alloy deposits", J. Electrochem Soc. **140**, 973-978.

80. Clark D., Wood D. and Erb E. (1997) "Industrial applications of electrodeposited nanocrystals", NanoStruc. Mater. **9**, 755-758.

81. El-Sherik A.M. and Erb U. (1995) "Synthesis of bulk nanocrystalline nickel by pulsed electrodeposition", J. Mater. Sci. **30**, 5743-5749.

82. Erb U., Palumbo G., Szpunar B. and Aust K.T. (1997) "Electrodeposited vs. consolidated nanocrystals: difference and similarities", NanoStruc. Mater. **9**, 261-270.

83. Erb U. (1995) "Electrodeposited nanocrystals: synthesis, properties and industrial applications", Nanostruc. Mater. **6**, 533-538.

84. Bryden K.J. and Ying J.Y. (1997) "Electrodeposition synthesis and hydrogen absorption properties of nanostructured palladium-iron alloys", NanoStruc. Mater. **9**, 485-488.

PERSPECTIVE ON
NANOPARTICLE MANUFACTURING RESEARCH

M.C. Roco
Directorate for Engineering, National Science Foundation
4201 Wilson Blvd., Suite 525, Arlington, VA 22230, USA

Abstract
Nanoparticles are seen either as building blocks of tailored properties for nanostructured materials, nanocomponents and nanodevices, or as enhancement agents of various physical, chemical or biological processes. Corresponding manufacturing processes may be arranged into four groups: synthesis of nanoparticles; processing/conversion of nanoparticles into functional nanostructures; utilization of nanoparticles for specific effects or phenomena; and process control. Particle synthesis via aerosols, colloids, combustion, plasmas, self-assembly techniques and other processes at high production rates has been a major research objective in the last few years. Future research is focused on particle processing into functional nanostructures and devices, hierarchical modeling, and particle utilization. The multiphase transport aspects (including particle-fluid and interparticle forces, rheological properties of powders, sorting, mixing, filtration, sintering, assembling, and interaction with external fields) and transient aspects have received less attention than the analysis of physical, chemical and biological aspects. A brief overview of research programs on nanoparticles and nanotechnology existing in several countries are presented.

1. Introduction

Nanotechnology is the development and utilization of structures and devices with a size range from 1 nm (molecular scale) to 100 nm where new physical, chemical and biological properties occur as compared to bulk materials. It is recognized as an emerging technology of the 21st century, besides the already established computer, information and biotechnology. The new properties are generally realized at the nanoparticle or nanocrystal level, and assembling of precursor nanoparticles is the most generic route to nanostructured materials.

71

G.M. Chow and N.I. Noskova (eds.), Nanostructured Materials, 71–92.
© 1998 *Kluwer Academic Publishers.*

A main issue is how the scientific paradigm changes will translate into novel concepts in manufacturing processes. Nanoparticles are used either as precursors to generate more complex structures and devices, or modifiers (enhancing or changing) of physical phenomena, chemical or biological processes. Nanoparticle manufacturing processes aim to take advantage of two kinds of effects:

a.   New physical, chemical or biological properties and laws that are caused by the reduced size of the particle, tube, layer, or other nanostructure. Smaller particle size determines larger surface area quantum-confinement and transport size effects. New properties are due to size reduction to the point where interaction length scales of physical and chemical phenomena (for instance, the magnetic, laser, photonic, and heat radiation wavelengths) become comparable to or larger than the size of the particle or respective microstructure.

b.   Generation of new atomic, molecular and macromolecular structures of materials by using either chemistry (ex: macromolecular structures and self-assembling techniques), nanofabrication (ex: creating nanostructures on surfaces), or biotechnology (ex: evolutionary approach) routes.

Few nanoparticle synthesis processes have developed their scientific base decades ago, long time before other nanotechnology areas have emerged. We find in this category the pyrolysis process for carbon blacks and flame reactors for pigments (Williams and Loyalka [1]), particle polymerization techniques (Sperling [2], Poehlein et al. [3], El-Aasser and Fitch [4]), self-assembly of micelles (Eiche et al. [5]), and chemistry self-assembly (Lehn [6]). Several kinds of nanoparticles are routinely produced for commercial use via aerosol and colloids reactors in the U.S., Japan and Europe. The powder price is in a wide range as a function of market size, from $0.5/kg for Carbon Black and Alumina (each with over 5M t/year worldwide market) to about $1,000/kg for superconductors (about 2t/year) and Silicon Carbide whiskers (about 1t/year). The challenge is to control the nanoparticle size, morphology and properties, assemble them for a given purpose, and to do this with a variety of materials. Nanoparticle processing into functional structures has received more attention only in the last few years. Main areas of relevance are advanced materials and manufacturing, electronics, biotechnology, pharmaceutics and sensors.

2. Manufacturing Processes

Nanoparticle systems, including nano-clusters, -tubes, nanostructured particles, and three-dimensional structures in the size range between 1 and 100 nm are seen as tailored precursors for nanostructures materials and devices. The improved properties are obtained in this dimension range as a function of material. Powder processing (sintering, extrusion, plasma activation, etc.) is the most general method of preparation of nanostructured materials and devices.

Nanoparticle manufacturing processes may be separated into four groups:

- *Nanoparticle synthesis.* This includes precipitation from solutions (colloids), gas condensation (aerosols), chemical, plasma, combustion, spray pyrolysis, laser ablation, expansion of supercritical fluids, polymerization, mechanical attrition, molecular self-assembly, hydrodynamic cavitation, and other processes (see, for instance, Trudeau et al. [7], Roco et al. [8], Shaw et al. [9]).

- *Processing nanoparticles into nanostructured materials (such as advanced ceramics), nanocomponents (such as thin layers), and nanodevices (such as sensors).* Examples of processing methods include sintering (German [10]), creation of nanostructures on surfaces, evolutionary biotechnology (Schuster [11]), and self-assembly techniques (Lehn [6]).

- *Utilization of nanoparticles to create or enhance an effect or a phenomenon.* It includes processes of mechanical (erosion resistance, friction reduction, sintering aids, etc.), chemical (catalysts, sensors, filtration agents, etc.), optical (pigmentation, filters, waveguides, etc.), electrical (quantum dots, superconductors, insulation, electroceramics with nonlinear electrical response, etc.), magnetic (giant magnetoresistive, superparamagnetic effects) and biological (separation and filtration, active agents, etc.) nature. Examples of more frequently used manufacturing processes of this kind are: (a) nanoparticle deposition and removal from surfaces, with relevance to microcontamination and chemical vapor deposition; (b) use of particles for surface modification, such as in chemical-mechanical polishing (CMP); (c) separation processes, including filtration, mass spectroscopy, and bioseparation; (d) nanoparticle emission and control.

including combustion pollution control; (e) biomedical applications, including drug delivery and health diagnostics; (f) and use of nanoparticles as catalysts in chemical plants.

Process control and instrumentation aspects. This includes off- and on-line measuring techniques for fine particles and their structures. Besides the better known characterization methods for particle size, shape and composition, new instruments are developed to measure particle interaction forces, their roughness, electrical, magnetic and thermal properties. A special attention is given to scanning probes, optical and laser-based diagnostic techniques. For example, the previously developed Raman spectroscopic analysis applicable to larger particles has been extended to nanoparticles (Vehring et al. [12]). An increase of the measurement sensitivity has been achieved by electrodynamically trapping and illuminating the assembly of trapped particles. In another example, scanning thermal microscopy has been developed recently (Majumdar et al. [13]). By using thermal sensors mounted on AFM tips (Figure 1), a spatial resolution of 25 nm has been obtained for temperature measurements on a solid surface. The scanning Joule expansion microscopy is an improvement of the technique that simplifies the fabrication of the sensor and enhances the resolution to about 10 nm (Varesi and Majumdar [14]).

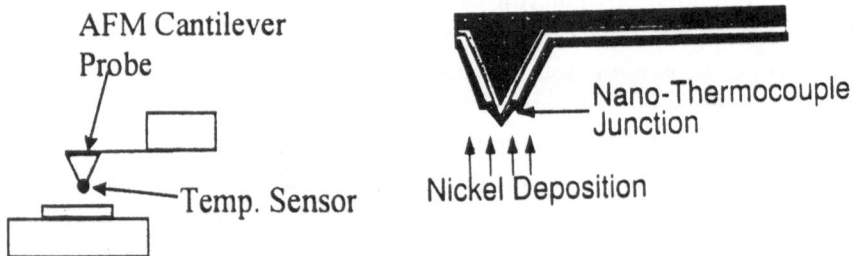

Figure 1. Scanning thermal microscope (after [13])

3. Nanoparticle Synthesis

At the beginning of any investigation, one is confronted with the selection of the synthesis method, the experimental and simulation techniques to be used, and the choice of materials (metals, ceramics, polymers, organics or carbon-based, composites). The main challenge is relating the final product properties and production rates to the material properties of the precursors and process conditions. The product may be either homogeneous or composite nanostructured particles, with one or multi chemical species, aerogels, and includes particle coating and particle doping.

After the nature of the process, the particle generation methods may be either physical (such as mechanical size reduction, gas-phase condensation), chemical, biological or a combination of these (such as spray and flame pyrolysis).

After formation, the particles may be either in dispersion in a volume, on surfaces, in tubes or even in matrix lattice spaces (zeolites).

The precursors may be gaseous, liquid or solid. The available synthesis methods may be separated into five groups after the medium from which particle are generated:

- *Liquid-phase methods*, including bulk precipitation in solutions, sol-gel processing, inverse micelles (using surfactants), spray conversion, electrospray, using super-critical fluids, infiltration in tubular templates, and various aqueous or non-aqueous chemical routes. These methods have some restrictions concerning the choice of materials and material contamination. Solution chemistry has the advantages of chemical homogeneity and accessible control parameters for nucleation and particle growth. The control parameters are the temperature and concentrations of reactants, additives, and solvents.

- *Gas-phase methods,* including gas evaporation, condensation and deposition, laser vaporization and pyrolysis, plasma chemical reaction and expansion, combustion, vapor infiltration in a porous matrix (template) and a variety of chemical routes in aerosol reactors.

- *Vacuum synthesis techniques*, including sputtering, laser ablation, exploding wires, and ionized-beam deposition.

- *From bulk solids by size reduction*, including mechanical attrition, mechanical alloying, spark erosion, severe plastic deformation, and combustion synthesis.

- *From molecules or finer nanoparticles by self-assembly techniques*, including physical, chemistry and biology methods.

The selection of the most suitable method is generally based on several criteria related to (a) *the desired product*: size distribution (particle size, monodispersity), morphology (single particles, agglomerates, etc.), composition (material, single component versus multicomponent, purity, stoichiometry), particle generation flowrate, and to (b) *the particle production facility*: its complexity, operation conditions (temperature, pressure, toxicity, etc.), and cost.

An illustration of the selection process of an aerosol process is given by Pratsinis and Kodas [15]. A subset of initial conditions and selection criteria (particle size, size distribution, morphology, maximum temperature, material, and degree of complexity of the production process) is used to identify spray pyrolysis as advantageous in comparison to flame, evaporation/condensation/reaction, laser, plasma and hot wall aerosol processes. Another group of initial conditions and criteria is used, for instance, by Brock [16] to identify laser ablation of fine powders as the method of choice.

Synthesis of nanoparticles/nanostructures with engineered properties and their processing into microstructured materials with tailored technological functions require new knowledge in a variety of areas (such as aerosols, colloids, thermal, plasma and combustion processing), and is at the intersection of a broad spectrum of disciplines (physics, chemistry, mathematics, biology and others). Particle nucleation and growth mechanisms are important scientific challenges. Use of hierarchical modeling and simulation is necessary in order to complement and generalize the increasing body of experimental data. Simulations aim to incorporate phenomena at different scales: molecular (1nm), nanoscale macromolecules (10 nm), mesoscale molecular assemblies (100 nm), microscale (1000 nm-1micron), and macroscale. Study of multibody interactions combined with simplified molecular dynamics techniques are possible avenues for an approach for mesoscale.

Uyeda [17] and Sheka (WTEC [18]) published two review papers describing activities less known to the community at large in Japan and Russia, respectively.

4. Multiphase Transport Processes in Nanostructured Systems

Most studies in nanoparticle manufacturing have focused on steady-state chemical, material and biological aspects. However, the basic transport mechanisms for mass, momentum and heat exhibit particularities that must be clarified in order to understand particle interactions, nanophase formation, heat and mass transfer at interfaces, and many other basic processes. We note that the term "ultradispersed system" has been used in Russia (Petrunin, in WTEC [18]) for nanostructured systems, that implies the study of the system rather than individual grains. The analysis should include basic aspects of multiphase transport processes, such as interaction between phases, rheology, particle settling and fluidization, mixing and segregation, separation, and then move to aerosol or colloidal reactors.

Let's consider the fluid-particle interactions, i.e. the drag, interparticle and particle-wall forces caused by the interstitial fluid. The continuity models in fluid mechanics breakdown for motion of very fine particles in fluids when particle size or interparticle distance is comparable to those of the fluid molecules. Molecular dynamics simulation becomes a useful tool, and the Knudsen number (the ratio of the mean free path of the molecules to the particle diameter) is an important parameter to identify that threshold. Modeling of nanoparticle interactions has similarities with that for rarefied and compressible gases for which there are previous studies (Crowe et al. [19], and Henderson [20]). Typically, particle velocity has continuous fluctuations as a result of the colliding molecules. Other molecular level phenomena, such as van der Wall attraction and electrical charges, become important. Multibody interaction must be considered. Similar issues are for flow in microchannels and nanotubes. Momentum, mass and heat transfer, chemical reactions and biological interactions should be analyzed starting from this background. The most challenging issues are in the intermediate mesoscopic regime, between the regimes where are clearly suitable the molecular or continuum approach. The advantages of molecular dynamics simulations are the lack of restrictions concerning the number of bodies and their arrangement, the nature of the fluid, and gas molecule-particle surface interaction. The disadvantage is that the simulations are case specific and computer intensive. Figure 2 illustrates the differences obtained for the drag force of a spherical particle trailing another sphere by using continuum mechanics and molecular dynamics (Peters [21]). Alternative methods for bridging molecular

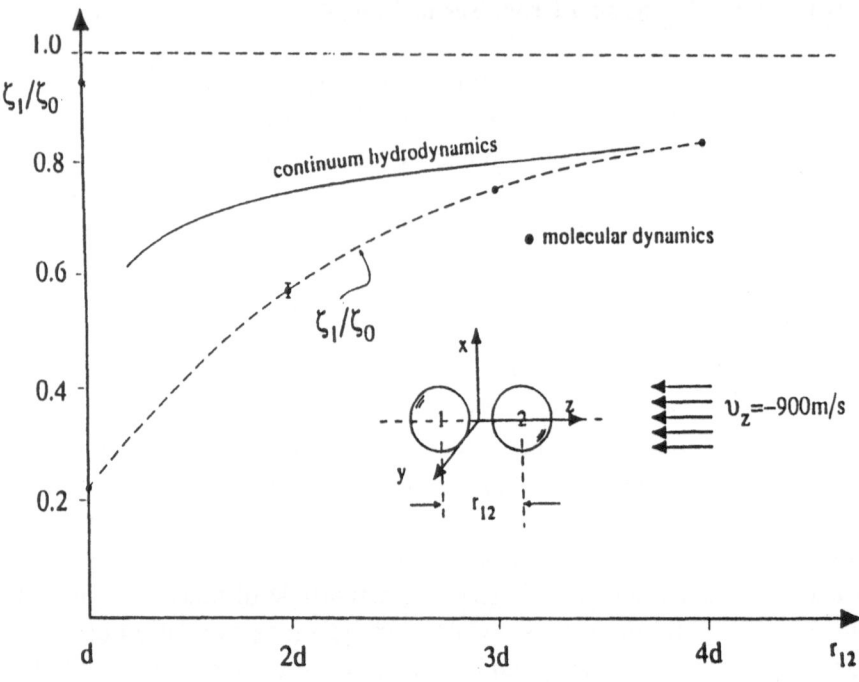

Figure 2. Comparison between molecular dynamics simulation and continuum hydrodynamics for the friction (drag) tensor of the downstream particle (1) (after [21])

dynamics studies (with time scales in the range 10 exp(-14) seconds) and
Brownian dynamics studies (with time scales in the range 10 exp(-5)
seconds) have been proposed (see for example, Peters [21]).

Understanding of particle-wall interactions has been the subject of a
considerable research effort (Israelachivili [22]). Besides the Atomic Force
Microscope and other more frequently used microscopes, the Total Internal
Reflection Microscopy (TIRM) was recently develop to measure the
colloidal particle-wall interaction force at distances as small as few
nanometers (Prieve and Walz [23]). A schematic of the basic experimental
device is shown in Figure 3. A laser beam is reflected from a solid-fluid
interface at sufficiently high incident angle that an evanescent wave is
formed in the fluid, and its intensity is decreasing exponentially with
distance from the wall. A particle located close to the interface will scatter

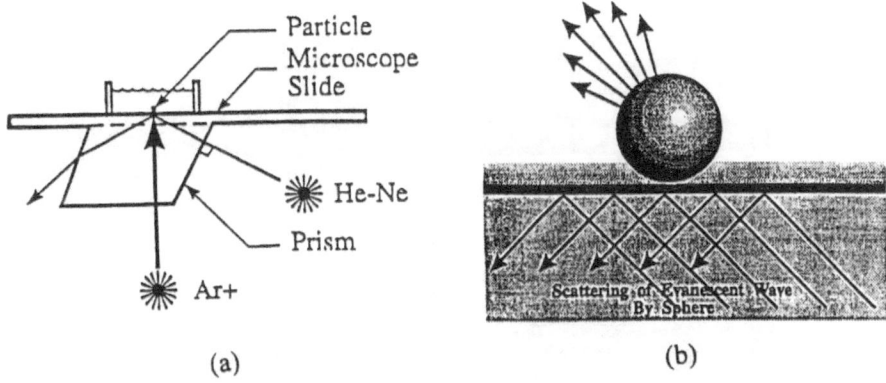

Figure 3. TIRM to measure dynamic particle-wall interactions in the nanometer range

the evanescent wave with an intensity that varies with the particle-wall separation distance. A main difference from other techniques is that the particle is not attached to a fixed frame, and dynamic characteristics can be obtained as a function of the separation distance in a liquid. The particle double layer and particle shape and roughness can also be investigated

Other basic phenomena and processes related to particle assembly that need to be better understood include nanoparticle rheology, particle settling and fluidization, particle mixing and segregation, separation from fluids and other solids. The knowledge from large particle investigations does not extrapolate directly to nanoparticles. A further step would be the control the multiphase processes in "ultradispersed" systems, by using for instance particle coating and manipulation with external fields. Hierarchical and adaptive modeling to include molecular dynamics and grain boundary methods into engineering solutions are in development.

5. Paths to New Nanoparticle Manufacturing Processes

We illustrate three examples of nanoparticle manufacturing processes that suggest three development paths to a new process:

(a) One originates from curiosity-driven research and not yet applied,

(b) Another originates from applications and is not yet well understood,

(c) A non-traditional field of application (biomedicine) of particle technology, at the intersection of disciplines.

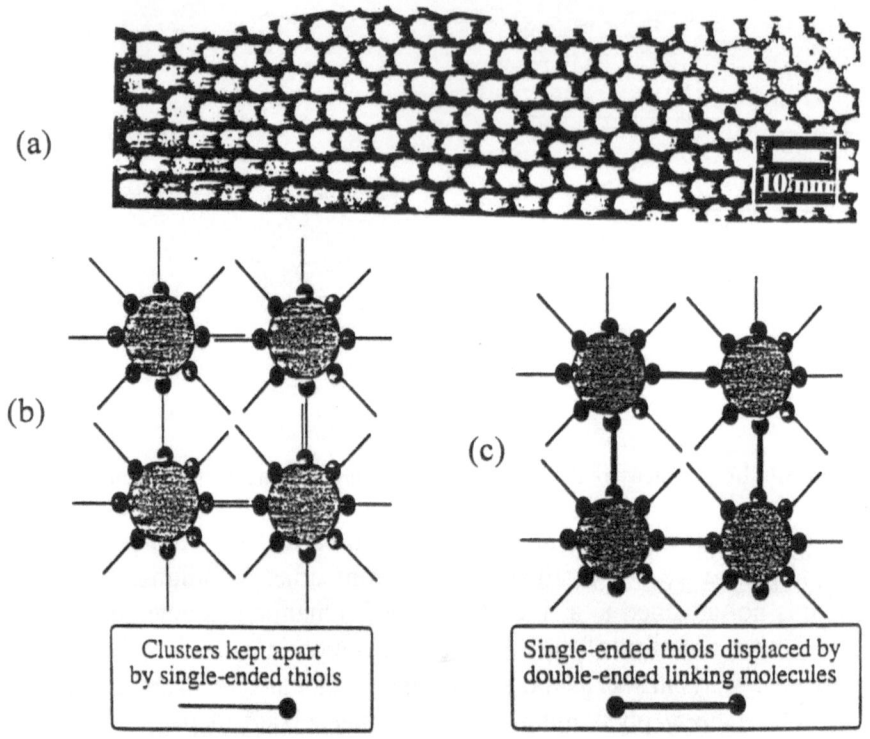

Figure 4. Linked cluster networks (after Andrews, 1996):
a. TEM micrograph of monolayer film of 3.7 nm gold clusters supported on a thin
flake of MoS2; b. The clusters are kept from coalescing by encapsulating each cluster in
deodecanethiol; c. Double-ended linking molecules are used to displace the single-ended
alkyl thiols and provide electronic coupling between clusters

a.   *Electronically linked cluster networks* (LCN) (Andres et al. [24]). It
combines metallic nanoclusters and molecular wires to fabricate a two-
dimensional lattice of nanometer sized gold clusters electronically linked by
dithiol organic molecules. The first step is to form a dense array of clusters
that are kept from coalescing by encapsulating each cluster in dodecanethiol
(single-ended thiols); the second step is to use double-ended linking
molecules to displace the single-ended thiols and provide electronic
coupling between the clusters. The authors are currently investigating the
use of LCN for device applications such as: interconnector networks for
high density integrated circuits, that may replace the need for complex
lithographic steps; opto-electronic devices by "programmable" LCN, where

the conductivity in selected regions can be increased by orders of magnitude by "doping" or exposure to an external optical signal; and chemical sensors, where the connectors' conductivity can be changed by orders of magnitude in the presence of particular chemical species.

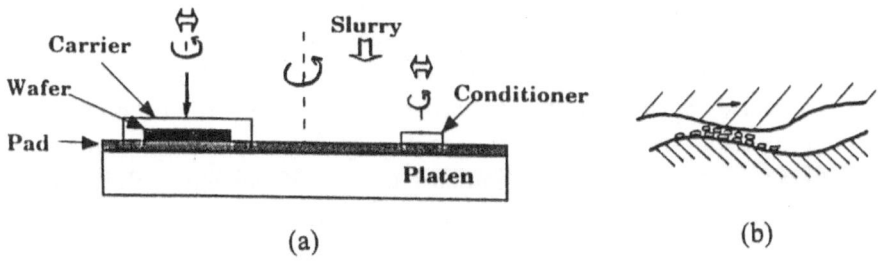

Figure 5. Schematic of a CMP unit

b.   *Chemical-Mechanical Polishing (CMP)* is an example where the particles play a useful (erosive/corrosive) role without being part of the final product.   A schematic of a CMP unit is shown in Figure 5.   Global planarization of the wafers between successive deposition and processing steps is a key objective in chip production.   Submicron tolerances for thin layers require continuously smaller particles in the tens of nanometer range or even smaller.   This is a manufacturing process widely used in electronic industry, and for which basic understanding of the influence of particles is lacking.   Silica, alumina or zircoria colloidal slurries are used.   The slurry effect is a function of particle size, shape, chemistry, concentration, stability, pH, and other parameters.   Basic understanding of three-body-interaction, erosion and corrosion mechanisms at the molecular level, effect of slurry chemistry, role of surfactant additives, and near surface chemical and mechanical changes are critical research issues.

c.   *Drug delivery systems in biomedicine.*   Colloidal and aerosol particles can be used as drug delivery systems, as contrast agents in diagnostics and as carriers of vaccines (Coombs and Robinson [25]).   Drug delivery systems are of particular interest because the size, composition and surface properties of nanoparticles may be changed, and their size allows them to penetrate capillaries. The surface properties of the particles play a critical role. The adsorption of plasma proteins, that enhances or diminishes the recognition events and rates of drug deliveries, is determined by the selective hydrophilicity of the particle surface and surface steric stabilization with a macromolecular layer. By attaching for instance polyoxyethylene residue to the particle surface, the particles have a much slower retention rate by the liver and remain longer in the circulatory system. By attaching homing ligands to these particles, site-specific delivery (such as speen, tumors, bone marrow, and sites of inflammation) can be provided. More recently, genetically fused proteins have been adopted because of the easy stoichiometric control (Kotabake et al. [26]). Biocompatible, biodegradable and non-toxic materials for the carrier particles have been obtained from natural substances (starch, albumin, polymeric materials, and surfactants). Loading and release of drugs or diagnostics agents is the second process. Loading and activation of coatings to achieve site-specific or condition-specific delivery is the third typical process. A schematic diagram illustrating a particle prepared for a medical application, including doping with monoclonal antibody, lipid-tagged antibody, steric barrier, and drug components, is shown in Figure 6. Besides active targeting as described passive targeting exploits natural mechanisms such as the filtration of particles by capillaries of the lung or the red pulp of the spleen.

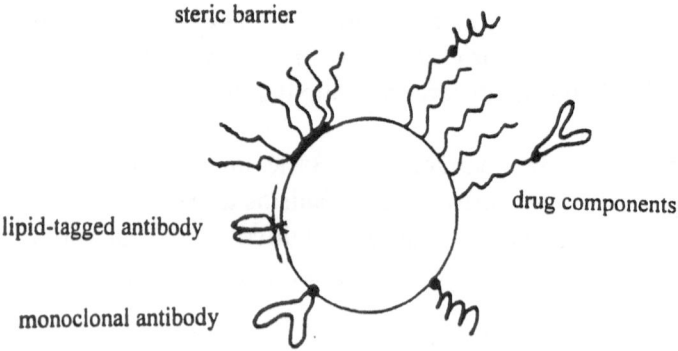

Figure 6. Schematic diagram of an engineered particle for biomedical applications

The particles may be delivered as microspheres, microcapsules, phopholipid vesicles or emulsions by a variety of routes (in either intravenous system, intra-arterially, intramuscularly, subcoutaneously, mucosal surfaces of the nose, lung, gastrointestinal track, etc.). Particle displaying organ or cell targeting characteristics can be loaded with drugs, antigens, or contrast agents for medical applications. It is important that the particles reach the correct cell type within the targeting organ. It is often advantageous to deliver the drug to cytoplasmor nucleus of the cell in situations such as vaccines and gene therapy.

Particle manufacturing processes are based either on assembling nanoparticles from molecules into patterned ensembles with complex functionality, or on modifying bulk materials and miniaturization. At the beginning of the next century, it is predicted that both the miniaturization approach and building-up approach will join forces to construct complex functional structures at scales in the range 1 to 100 nm. Rohrer [27] noted that the corresponding combination of methods will bring nanofabrication closer to the nature's way of developing complex, multifunctional structures (see for illustration the schematic in Figure 7)

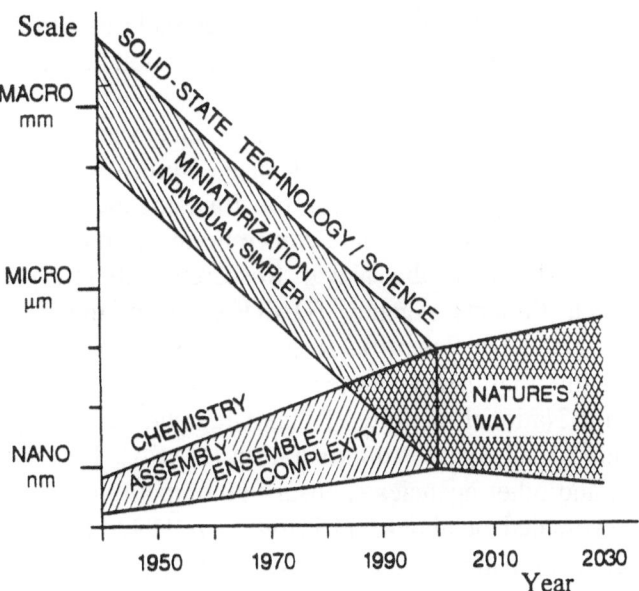

Figure 7. Schematic suggesting the convergence of solid state science and technology and supramolecular assembling chemistry after year 2000 (after [27])

Most successful nanoparticle manufacturing areas have found applications in cosmetics and colorants, catalysts, abrasives systems, coatings, magnetic recording, pharmaceutical and delivery medical systems, reversible and more powerful batteries, thermal insulation and hard materials for aeronautical and aerospace industry, selfassembled polymeric materials, special ceramics and metals with structural functions, membranes, and miniaturized electronic devices.

6. Relevant Programs in the World

The international interest is highlighted by the existence of comprehensive programs in USA, Japan, and several European countries, and more recent developments in Australia, Canada, China, Taiwan, Korea and Singapore. Since research on nanoparticle manufacturing is part of nanotechnology research, we will refer below to relevant research programs on nanotechnology and their components on nanoparticle research. Nanotechnology is qualitatively different from size reduction to micron dimensions as defined by MEMS (micro-electro-mechanical systems, as called in USA) or MST (microsystems technology, as called in Europe). The specific phenomena encountered at nanometer scale as compared to bulk materials add a new dimension to ultraminiaturization under 100 nm.

In USA, the federal government, large electronic, chemical and pharmaceutical companies, small and middle size enterprises, as well as state and private foundations provide support for fundamental, precompetitive nanotechnology research. The National Science Foundation (NSF), Office of Naval Research (ONR, with the Naval Research Laboratory), Defense Advanced Research Projects Agency (DARPA), Department of Energy (DOE), Air Force Office of Scientific Research (AFOSR), Army Research Office (ARO, with the Army Research Laboratory), Department of Commerce with the National Institute of Standards and Technology (NIST), National Aeronautics and Space Administration (NASA), National Institute of Health (NIH), and other agencies sponsor precompetitive research in this area at a level estimated at about $115M/year in 1997/98 (WTEC [28]). Large multinational companies, such as Dow, DuPont, Motorola, Lucent, and Hewlett Packard, have nanotechnology groups in their long-term research laboratories where the total precompetitive research expenditure is estimated to be comparable with government funding. Small business, such as Aerochem Research Laboratory and Particle Technology Inc., have generated an innovative competitive environment in various technological areas including dispersion, filtration, synthesis, functional nanostructures, and various nanoparticle manufacturing processes. U.S. companies specialized in commercially producing a broader spectrum of nanoparticles are Nanodyne (currently at about 48 t/y and going to 500 t/y) and Nanostructured Materials (currently at about 50 t/y and expanding). Centers of excellence formed in the last years or currently developing at universities create a continuously growing public research and education infrastructure for this field. Examples are the University of Illinois at Urbana (including the Engineering Research Center on Microelectronics in collaboration with the Beckman Institute), University of North Carolina, RPI, Brown University (the Material Research Science and Engineering Center), and University of Washington.

Current interest in nanotechnology is broad-based, and there are several common themes among funding agencies, as well as particularities:

- A main goal has been the synthesis, processing, properties and characterization of nanostructured materials, including high rate production of nanoparticles for potential industrial use. Advanced generation techniques for nanostructures with controlled properties (including chemistry/bio-selfassembling techniques, and artificially structured materials), methods to simulate structure growth at molecular and mesoscale levels, instruments and sensors based on novel concepts and principles, tools for quantum control and manipulation, and interdisciplinary research including biology, are important components of the current research activities (NSF).

- Research on thermal spray processing and chemistry-based techniques for deposing multilayered nanostructured coatings, processing of nanoscale powders into bulk structures and coatings has been undertaken (ONR).

- Nanofabrication with particular focus on electronic industry is another major theme. It includes development of technologies seeking improved speed, density power and functionality beyond that achieved by simply scaling transistors, operation at room temperature, use of quantum well electronic devices, and computational nanotechnology addressing physics and chemistry related issues in nanofabrication (DARPA).

- Research on nanoscale materials for energy applications has a focus on synthesis and processing of materials with controlled structures, surface passivation and interface properties. The initially targeted energy-related applications are catalysis, optoelectronics and soft magnets (DOE).

- Miniaturization of spacecraft systems and theoretical modeling addressing physical and chemical of nanostructures is another area of focus (NASA).
- Biomimmetics, smart structures, microdevices for telemedicine, compact power sources, supperlattices and buckminsterfullerene are developed in an interdisciplinary environment (ARO).
- Neural communication and chip technologies have been investigated for biochemical applications and sensor development (NIH).
- Metrology activities for thermal, mechanical properties, magnetism, micromagnetic modeling, and thermodynamics of nanostructures have been initiated. Nanoprobes to study nanometer material structures and devices with nanometer length scale accuracy and picosecond time resolution have been developed and other are in development (NIST).

In longer term, the building-up approach from molecules and nanoparticles/ nanotubes/ nanolayers is the most promising from the scientific, expectation of innovations and industrial relevance points of view.

Previous NSF activities include:

- the Advanced Materials and Processing Program dealing with generation, properties and characterization of nanostructured materials;
- the Ultrafine Particle Engineering initiative covering synthesis and processing of nanoparticles with controlled properties, with a focus on high rate production processes; Nanoparticles, nano-clusters, layers and tubes are seen as precursors of tailored properties for nanostructures materials and devices. The initiative, started in 1991, has been focused on new concepts and fundamental research on methods to generate nanoparticles at high rates. The work has included contributions on fundamental physics and chemistry for nanoparticle generation with tailored properties via aerosols, colloids, plasma, combustion, sol-gel, chemical vapor deposition, molecular and cluster assembling, and other synthesis methods. Emphasis has been placed on awards for small groups, for multidisciplinary projects with a focus on fundamental research. The level of NSF support was approximately $6M/year in the last five years. Two review conferences have been held in 1994 [8] and 1997 [9] in collaboration with NIST. Novel ideas have been proposed

in laser ablation of materials to generate nanoparticles used in nanoelectronics, production of polymer semiconductor composites for development of non-linear optics for waveguides, molecular and nonostructure self-assembly techniques, high performance catalysts, control of nanoparticles resulted from combustion and plasma processes, and special sensors applied in chemical plants and environment.

- sponsoring various centers including the National Nanofabrication Users Network connecting researchers and facilities at five universities to fabricate nanometer-scale structures, with an initial focus on miniaturization in electronic industry; and

- an initiative on Instrument Development for Nano-Science and Engineering (NANO-95) for new instrumentation to advance atomic scale measurements of molecules, clusters, nanoparticles and nanostructured materials.

Most active programs on nanoparticle research include the Ultrafine Particle Engineering in fundamental research at NSF, research on magnetic nanoparticles at NIST, research on metallic and ceramic particles at ONR, and polymeric particles and self-assembling techniques at ARO.

In Europe, there is a combination of national programs, collaborative EC networks, and large corporations funding nanotechnology research. The ESPRIT Advanced Research Initiative in Microelectronics and BRITE/EURAM projects on materials science are partially dedicated to nanotechnology. The PHANTHOM (Physics and Technology Mesoscale Systems) is a network created in 1992 to stimulate nano-electronics, nanofabrication, opto-electronics, and electronic switching. Another related network for electronic applications with partial focus on nanotechnology is NEOME (Network for Excellence on Organic Materials for Electronics). Since 1995 the European Science Foundation is sponsoring a network for Vapor-phase Synthesis and Processing of Nanoparticle Materials (NANO) in order to promote bridges between the aerosol and materials science communities working on nanoparticles. In 1996, the European Consortium on Nanomaterials was created with the center in Switzerland, and the Joint Research Center Nanostructured Materials Network was created with the center in Ispra, Italy. The largest national support for nanotechnology is coming from the Federal Ministry of Research and Technology (BMFT) in Germany (the total support in 1997 is estimated at $50M/year). The

Fraunhofer Institutes, Max Plank Institutes and several universities have formed centers of excellence in the field. In UK, the National Physical Laboratory established a forum (NION, National Initiative on Nanotechnology) for promoting nanotechnology in universities, industry and government funding. A network type of program (LINK Nanotechnology Programme) has been launched in 1988 with an annual budget of about $2M/year. The Engineering and Physical Sciences Research Council (EPSRC) is funding materials science projects related to nanotechnology with a total value of about $7M for a five year interval (1994-1999). About $1M is specifically used for nanoparticle research. Other national nanotechnology programs have been established in France, Sweden, Switzerland, Spain, and Finland. Large companies with research activities include IBM (Zurich), Philips, Siemens, and Hitachi. Degussa Co. is a commercial supplier of micro- and lately nanoparticles since 1940.

In Russia and other FSU countries, there are relatively comprehensive research activities on nanoparticle generation and processing into nanostructured materials. The research institutes of the Academy of Sciences and universities have contributed with innovative particle formation processes (plasma chemical reaction, exploding metallic wires, chemical precipitation, and laser evaporation), compacting technologies (by severe plastic deformation, pulsed magnetic compressive waves), particle characterization (vibration spectra of nanoparticles, quantum-chemical characterization at atomic level, computational vibrational spectroscopy) and applications of nanoparticles (for purification, oil improvements, flame arresters and as biologically active systems) (Sheka, in WTEC [18]).

In Asia, there are national programs on nanotechnology in Japan, China, Australia, Taiwan, Korea and Singapore.

The largest programs are in Japan. The first five-year program on ultrafine particles started in 1981 under the Exploratory Research for Advanced Technologies Program (ERATO), and an overview of the results was published in 1991 (Uyeda [17]). The Japanese government organizations and very large corporations currently are the main source of funding for nanotechnology, including nanoparticles. Small and medium size companies have a minor role. Funding for nanotechnology research should be viewed in the context of the overall increase of basic research support in Japan since 1995 as a result of the Japanese Science and

Technology Basic Plan (Government of Japan [29]). The main government organizations sponsoring nanotechnology in Japan are MITI (Ministry of Industry, Technology and International affairs), STA (Science and Technology Agency) and Monbousho (The Ministry of Science, Education, Culture and Sports).

- It is estimated that MITI/AIST (The Agency of Industrial Science and Technology) has a budget of approximately $50M/y for nanotechnology in 1996 considering the current exchange rate. For example, the Joint Research Center for Atom Technology (JRCAT) has a ten year budget of about $220M (1992-2001), with $25M/y in 1996. Research on Cluster Science has been funded at a level of about $10M for the interval 1992-1997.

- The STA has its main nanotechnology facilities at the Institute of Physical and Chemical Research (RIKEN) (nanotechnology is included in the Frontier Materials Research initiative); the National Research Institute for Metals (NRIM); the National Institute for Research in Inorganic Materials (NIRIM), ($0.8M/y for nanotechnology); and the Japan Research and Development Consortium. The ERATO Program (Exploratory Research for Advanced Technology) includes nanotechnology-related projects, each with total budgets of $13-18M for five years: Quantum Wave Project (1988-1993); Atomcraft Project (1989-1994); Electron Wavefront Project (1989-1994); and Quantum Fluctuation Project (1993-1998).

- The Monbousho supports nanotechnology programs at universities and national institutes, as well as via the Japan Society for Promotion of Science. The most active universities are the Tokyo University, Kyoto University, and Tokyo Institute of Technology, Tohoku University and Osaka University. New ideas are searched in the framework of the Institute of Molecular Science, and the Exploratory Research on Novel Artificial Materials and Substances for Next Generation Industries (5-year university-industry research projects). Sponsored by the *JSPS, the "Research for the Future" initiative* has programs on magnetic nanoparticles at the Tohoku University.

Large companies play a driving role for nanotechnology research in Japan. Important research efforts are at the Hitachi Central R&D Lab. (nanotechnology is about ¼ of the long-term research), NEC Fundamental Research Laboratories (nanotechnology is estimated to be about ½ of the precompetitive research), NTT, Fujitsu, SONY and Fuji Photo Film Co.

Nihon Shinku Gijutsu (ULVAC) is part of a conglomerate of 35 companies, and produces $10M particles sales per year for electronics, optics and arts.

In China, nanoscience and nanotechnology has received attention since mid-eighties. Approximately 3,000 researchers were contributing to this field in 1996. The ten-year "Climbing Project on Nanometer Science" (1990-1999) and a series of advanced materials research projects are core activities. The Chinese National Science Foundation and Academy of Sciences have funding programs on nanotechnology.

7. Concluding Remarks

Nanotechnology is qualitatively different from MEMS or "microsystems technology" because of the new phenomena encountered only at nanometer scale as compared to bulk materials. Since the specific properties are realized at the nanoparticle, nanolayer or nanocrystal level, assembling of precursor nanoparticles is the most generic route to generate nanostructured materials; thus nanoparticle manufacturing is an essential component of nanotechnology.

Several nanoparticle synthesis processes have developed their scientific base decades ago, long time before other nanotechnology areas have emerged. Most of them are currently developing the scientific base. There are production facilities at industrial scale for manufacturing processes such as deposition from aerosol reactors and colloidal chemistry. Many nanoparticle manufacturing processes are moving from the basic question that was asked one decade ago "what if?", to new questions "how to do it" and "at what cost?"

Research in USA, Japan and Europe is advancing towards developing a suitable research infrastructure, and promoting a series of industrial developments in dispersions, electronics, multimedia and bioengineering. The largest governmental funding opportunities for nanotechnology are provided in USA by NSF (approximately $65M/year for fundamental research) and DARPA (for product development), in Japan by MITI (approximately $50M/y for fundamental research and product development), in Germany by BMFG (approximately $50M/y for fundamental, applied research and product development) and in France by CNRS (approximately $50M/year for fundamental and applied research withinits own institutes). While large multinational companies are developing nanotechnology research activities in many developing countries, a particularity in the USA is the presence of an active group of small and medium size companies introducing new processes to the market.

While the current investigations have more focus on generation of nanoparticles and nanostructures, a shift towards nanoparticle processing and their utilization is noticeable. Future work aims to expand basic understanding of phenomena and mechanisms specific at nanoscale, combine synthesis and assembling into functional materials and devices into a continuous process, introduction of new principles of operation for devices, hierarchical simulation techniques at mesoscale, and development of new experimental tools.

Acknowledgments

Partial support received from NATO ASI is acknowledged. The opinions expressed in the paper are not necessarily those of NSF.

References

1. Williams, M.M.R. and Loyalka, S.K. (1991) *Aerosol Science: Theory and Practice*, Pergamon Press, London, UK.
2. Sperling, L.H. (1981) *Interpenetrating Polymer Networks and Related Materials*, Plenum, New York.
3. Poehlein, G.W., Ottewill, R.H. and Goodwin, J.W. (1983) Eds., *Science and Technology of Polymer Colloids*, 2 vols., NATO ASI Series, Martinus Nijhoff, Amsterdam.
4. El-Aasser and Fitch, R.M., eds., *Future Directions in Polymer Colloids*, NATO ASI Series, E (Applied Sciences 138), Kluger Academic Publishers, 1987.
5. Eicke, J.C., Shepherd W., and Steinemann (1976) A., J. *Colloid Interface Sci.*, 56, p. 168.
6. Lehn, J.-M. (1995) *Supramolecular Chemistry, Concepts and Perspectives*, VCH, Weinheim.
7. Trudeau, M.L., Provenzano, V., Shull, R.D. and Ying, J.Y., eds. (1997) Proc. of the Third International Conference on Nanostructured Materials, *Nanostructured Materials* 9, No. 1-8, Pergamon Press.
8. Roco, M.C., Shaw, D.T., and Shull, R.D. (1994) Ultrafine *Particle Engineering*, Proc. NSF-NIST Conference, Washington, D.C.
9. Shaw, D.T., Roco, M.C. and Shull, R.D., eds. (1997) *Nanoparticles: Synthesis, Processing into Nanostructures, Properties and Characterization*, Proc. NSF-NIST Conference, Arlington, Virginia.
10. German, R.M. (1994) *Small Particle Sintering*, Proc. Workshop, Penn State, University Park, PA.

11. Schuster, P. (1996) Evolutionary Biotechnology: Theory, Facts and Perspectives, *Acta Biotehnol*, Vol. 16, 3-17.
12. Vehring, R., Aardahl, C.L., Davis, E.J., Schweiger, G. and Covert, D.S. (1997) Raman Spectroscopy for Nanoparticles, *Rev. Sci. Instrum.* **68**, p. 70.
13. Majumdar, A., Lou, K., Shi, Z. and J. Varies (1996) Scanning Thermal Microscopy at Nanometer Scales: A New Frontier in Experimental Heat Transfer, *Experimental Heat Transfer*, Vol. 8, and p. 83.
14. Varies J. and Majumdar, A. (1997) Scanning Joule Expansion Microscopy at Nanometer Scales, *Applied Physics Letters* (in press).
15. Partisans, SE and Kodak, T.T. (1993) Manufacturing of Materials by Aerosol Processing, in Wallace, K. and Baron, P.A., Eds., *Aerosol Measurement*, Van Nostrand Reinold, New York.
16. J. Brock (1997) Nanoparticle Synthesis, in *Nanostructured Materials: Science and Technology*, Ed.: G.M. Chow, Proc. NATO ASI, Kluwer Academic Publ. Dordrecht
17. Uyeda, R. (1991) Studies of Ultrafine Particles in Japan, *Progress in Materials Science*, 35, p.1, Pergamon Press.
18. WTEC (1997b) *Review of Russia R&D Status in Nanoparticles and Nanostructured Materials*, Eds. Ovidko, I. and Roco M.C., Loyola University, Baltimore, Maryland (Includes E.F Sheka, "Some Aspects of Nanoparticle Technology in Russia"; and V.F. Petrunin, "History and Programs of Ultradispersed Materials in Russia").
19. Crowe, C.T., Babcock, W. and Willoughby, P.G. (1973) Drag Coefficient for Particle in Rarefied, low Mach number flows, *Prog. Heat and Mass Trans.* 6, p. 419.
20. Henderson, C.B. (1976) Drag Coefficients of Spheres in Continuum and Rarefied Flows, *AIAA J.* **14**, p. 707.
21. Peters, M.H. (1994), A Nonequilibrium Molecular Dynamics Simulation of Free Molecule Flows in Complex Geometries: Application to the Brownian Motion of Aggregate Aerosols, *Phys. Rev. E* **50**, 4609-4617.
22. Israelachivili, J.N. (1992) *Intermolecular and Surface Forces*, 2nd Ed., Academic Press, London.
23. Prieve, D.C. and Walz, J.Y. (1993) Scattering of an Evanescent Surface Wave by a Microscopic Dielectric Sphere, *Appl. Opt.* **32**, p. 1629.
24. Andres, R.P. et al. (1996) Self-Assembly of a Two-dimensional Supperlattice of Molecularly Linked Cluster Networks, *Science* **273**, 1690-1693.
25. Coombs, R.R.H. and Robinson, D.W. (1997) *Nanotechnology in Medicine and Biosciences*, Gordon and Breach Publ., UK.
26. Kobatake, E., et al. (1997) A Fluoroimmunoassay Based on Immunoliposomes Containing Genetically Engineered Lipid-Tagged Antibody, *Analytical Chemistry* **69**, No. 7, 1295-1298.
27. Rohrer, H., 1997, *The Nanometer Age, Challenges and Chances*, Lecture at NSF, Arlington, Virginia.
28. Rohrer, H., 1997, *The Nanometer Age, Challenges and Chances*, Lecture at NSF, Arlington, Virginia.
29. WTEC (1997a) Review *of U.S. R&D Status and Trends in Nanoparticles, Nanostructured Materials, and Nanodevices*, Eds. R. Siegel, E. Hu, M. Roco and G. Holdridge, Loyola University, Baltimore, Maryland.
30. Government of Japan (July 1996) *Science and Techn. Basic Plan*, Tokyo.

THE NANOCRYSTALLINE ALLOYS: THE STRUCTURE AND PROPERTIES

N.I.NOSKOVA

*Institute of Metal Physics, Ural Division of Russian Academy
of Sciences, 18 Kovalevskaya Str., GSP-170, Ekaterinburg,
620219,Russia,*

Abstract

We were interested in studying the strength and, plasticity , the structure of nanograins and their boudaries for nanocrystalline polyphases alloys. We have investigated nano-crystalline, $Fe_{73.5}Nb_3Cu_1Si_{13.5}B_9$, $Fe_{73}Ni_{0.5}Nb_3Cu_1Si_{13.5}B_9$, $Fe_5Co_{70}Si_{15}B_{10}$, $Pd_{77.5}Cu_6Si_{16.5}$ and $Pd_{81}Cu_7Si_{12}$ ribbons produced by superfast quenching from the melt followed by fast heating to 723-923 K, in vacuum. The annealing time was from 10 s to 1 h. The alloy $Pd_{81}Cu_7Si_{12}$ was producedby rapid quenching of melt and crystallized during creep tests in the temperature's range between 623 and 823 K and the stresses between 0.7 and 39 MPa. Under creep at 723 K at stress of 2.1 MPa , the resulting alloy had a nanocrystalline structure with a grain size of <10 nm. Under these conditi-ons the alloy exhibited an elevated plasticity. Also reported are the results of dynamic experiments on the compaction with temperature of nanocrystalline TiN and Al_2O_3 powder in die ranging 573-873 K. Phase composition and the microstructure of alloys were studi-ed using the transmission electron microscopy method. The microstructure of the nano-crystalline $Fe_{73.5}Nb_3Cu_1Si_{13.5}B_9$ alloy was studied in situ at different stages of crystalli-zation of the amorphous ribbons in the column of an electron microscopy. High-resolu-tion transmission electron mocroscopy (HRTEM) was used to study the structure of nanophase crystals and their interfaces in nanophase alloys. It was shown that the inter-faces between chemically similar nanophases may have different structure. The tensile strength of the alloys was determined by stretching ribbon specimens to failure at a rate of $1.6x10^{-3}$ - $7x10^{-5}$ s^{-1} at 293-723 K.

1. Introduction

One of the methods used for obtaining high-strength alloys with a grain size of the order of $D \cong 0,1$ μm consists in crystallization of an amorphous alloy at a temperature $\geq 0,6$ of the melting point . In recent years, alloys with a grain size $D \leq 10$ nm were obtained by specially selecting the composition of an amorphous matrix and the temperature of crystallization anneal [1,2]. Accordingly, the former materials ($D \cong 0,1$

G.M. Chow and N.I. Noskova (eds.), Nanostructured Materials, 93–119.

μm) are classified as the microcrystalline alloys, and the latter are referred to as the nanocrystalline alloys. Nanocrystalline metals and alloys can also be obtained by other methods, including sublimation followed by hot pressing of the disperse powder and techniques involving high plastic deformations. The results of structural investigations show that the state of grain boundaries and the degree of material perfection in the bulk of nanocrystalline samples depend on the method used for their production. The strength and plasticity of nanocrystalline metals and alloys have still not been sufficiently studied. Available publications contain data on microhardness [3] and the results of compression and bending tests [4,5]. The present paper reports on the structure and properties of $Fe_{73.5}Cu_1Nb_3Si_{13.5}B_9$, $Fe_{73}Ni_{0.5}Cu_1Nb_3Si_{13.5}B_9$, $Fe_5Co_{70}Si_{15}B_{10}$, $Pd_{77.5}Cu_6Si_{16.5}$ and $Pd_{81}Cu_7Si_{12}$ alloys tested under conditions of uniaxial tension and creep.

A recently rapidly quenched $Fe_{73.5}Cu_1Nb_3Si_{13.5}B_9$ alloy, which was first described in [1], has been the subject of intensive investigations. Amorphous in its initial state, the alloy crystallized into a nanocrystalline soft magnetic material with high magnetic characteristics (coercivity $H_c=0,4$ Am^{-1}). In the nanocrystalline state, the alloy exhibits a zero magnetostriction and shows no manifestations of magnetocrystalline anisotropy because of small grain size (10 nm).The alloy has two crystallization temperatures, 783 and 843K. At temperatures higher than 783K nanocrystalline grains form for 1 h. Above 843K the grains quickly grow in size while magnetic properties deteriorate [2,3]. To obtain high magnetic parameters, this amorphous alloy is generally annealed at 793-833K for 1 h.

The starting temperature for crystallization of amorphous alloys increases with a heating rate. Under rapid heating the two crystallization temperatures of the amorphous $Fe_{73.5}Cu_1Nb_3Si_{13.5}B_9$ alloy should shift towards higher temperatures, allowing the nanocrystalline grains to form at higher temperatures than usual and, probably, at a higher rate. Increasing heating rate and the temperature of short crystallization annealling could affect the phase composition, the size, and crystal structure of growing grains, and thus magnetic properties of the alloy [3]. We recently showed [3,6] that low-temperature annealing at 623 or 723K influences the structure and strength of the crystallized glass. The goal of this work is investigating the influence of temperature increases and the effect of similar preliminary treatments on the grain size of crystallized $Fe_{73.5}Cu_1Nb_3Si_{13.5}$ B_9 glass.

The coercive force H_c of amorphous soft magnetic alloys is well known to abruptly increase during their crystallization. For example, the coercive force of the alloy of an approximate composition of $Fe_5Co_{70}Si_{15}B_{10}$ rises from 0.5 to 3000 A/m i.e., by a factor of 6000 [3,7,8], during its crystallization at 723K for 1 h. This effect is due to the precipitation of fine-grained crystalline phases of different compositions. Cubic α-Co was found to be the most coercive phase in the alloy. It is also known [3,11] that the coercive force of ferromagnetic particles increases as their size decreases, within certain limits. This raises the question if there is a way to decrease the grain size of crystallized phases so that the coercive force of a crystallized alloy increases significantly. We attempted to solve this problem.

Earlier, when producing a nanocrystalline state in $Fe_{73.5}Cu_1Nb_3Si_{13.5}B_9$ alloy, we found that rapid crystallization at an elevated temperature (rapid heating and short soaking) produced grains of a smaller size and phases that slightly differed from those obtained upon long annealing at the crystallization point [3]. Similar data were obtained in (6) for the FeNiSiB-type amorphous alloys. We applied the same procedure to refine grains in an $Fe_5Co_{70}Si_{15}B_{10}$ alloy and to compare its magnetic properties with those of the alloy crystallized by a conventional procedure. The present work also include results of study of the structure and the tensile strength and microhardness of these materials. The problem which has yet to be solved is the production of a nanocrystalline material with the optimal structure (with the equilibrium grain boundaries between nanocrystalls, with the lowest internal elastic stresses in the space of the nanocrystals) and, in particular, with sufficiently good plasticity.Inoue et al. [12] suggested a method for producing a nanocrystalline alloy with an elevated plasticity, in which the amorphous alloy $Al_{88}Ni_{10}Ce_2$ is first annealed at a low temperature and subsequently subjected to tensile straining by 45% in the temperature range between 450 and 465 K. The low-temperature annealing causes precipitation of Al particles in the alloy, which during subsequent crystallization gives rise to nanocrystalline grains and simultaneously causes macrodeformation of the alloy. In the present study, the alloy $Pd_{81}Cu_7Si_{12}$ was crystallized under the creep conditions in order to enhance its plasticity. The plasticization effect could be produced by means of such treatment because the crystallization of the alloy had to proceed simultaneously with the dynamic recovery processes and even the dynamic recrystallization of the alloy. At the same time, the fine (nanocrystalline) grain structure had to be preserved owing to the presence of dispersed intermetallic phase particles that were precipitated during crystallization of the alloy and that inhibited grain growth in the alloy. Mechanical vibrations (elastic oscillation fields) are now widely applied in engineering to treat materials. The use of vibration techniques to relieve or redistribute residual stresses in products is promising and profitable. Changes in the residual stress level upon vibration treatment can be caused by the following factors: (1) Because of superposition of residual and external alternating stresses, yield stress can be reached in those sample (product) areas where the residual stress level is at a maximum. In these regions, a microplastic deformation occurs, which causes relaxation of internal stresses. (2) Under additional alternating loading, the stress $\sigma_{0.2}$ of the plastic flow onset. Up to now, only the effects of an alternating elastic field on crystals were studied. However, the advent of new materials (amorphous; quasi-,micro-, and nanocrystalline) requires knowledge about the effect of elastic oscillations on their structure and properties. Our work is devoted to studying this last problem. Recent works report the effect of elastic oscillations on the properties and the structure of metallic glasses [13], but to the best of our knowledge no works address the effect of ultrasonic oscillations on nanocrystallines structure and properties. That is why we were interested in studying the effect of a powerful ultrasonic treatment on the structural, and defects of nanocrystalline Cu and $Fe_{73.5}Cu_1Nb_3Si_{13.5}B_9$ alloy.

2.Experimental

Amorphous alloys were produced by rapidly quenching from the melt on a rotating copper cylinder. The components concentration in amorphous alloy ribbons was controled by chemical and spectral analysis[2 - 4,10].
We determined the nanocrystalline-grain sizes in the alloy from electron micrographs using the linear intercept metod [7]. We used 3000 grains when calculating the number

of large, medium, and small grains. In order to construct the curve of grain size distribution, the number of nanocrystallites of a certain size was normalized to the maximum number of the nanocrystallites. Phases in the nanocrystalline alloys were determined from electron diffraction patterns.

The samples were ribbons 1 - 1.5 mm wide, 20 - 25 μm thick, and 100 mm long. They were rapidly crystallized as follows. The ends were spot-welded to the thin nichrome wire, 0.1 mm in diameter, with a ferromagnetic ball ribbon end. The sample was placed into a horizontal quartz tube, pumped to obtain a rough vacuum. The ribbon is easy to position in any place in the tube using a permanent magnet outside the tube that attracts the ferromagnetic ball. A furnace heated to a required temperature was mounted around the central part of the tube. At the beginning, the sample was in the cold part of the tube to be then drawn into the heated furnace by the magnet. The heating rate under these conditions was as high as 10^3 K/s. After a required holding at a given temperature, as a rule from 1 to 20 s, the sample was pulled out into the other cold part of the tube. The reference sample was heat-treated accoding to the standard schedule: heating to a required temperature at a rate of 300 K/s, soaking for 1 h, and cooling in the furnace at the same rate. For magnetic measurements and electron-microscopic investigations, section 10 mm long were cut from the middle part of the ribbon. A vibrating sample

tests at T=573 K and that of the crystallized alloy in the tensile tests at T=293 K. The results on the deformation and failure of the Pd-Cu-Si alloy in the bending tests also indicated that the selected test stress ranges were significantly lower than the yield strength at the creep test temperatures.

3.Results and discussion

The Alloy FeCuNbSiB

The Structure . Ref.[3] shows the microstructure and microdiffraction patterns for a sample of the alloy $Fe_{73.5}Cu_1Nb_3Si_{13.5}B_9$ immediately after the rapid quenching. The structure appears as amorphous and homogeneous. The microdiffraction pattern shows the first and second diffuse halos and an inner halo (at small angles). The first (main) diffuse halo exhibits an internal structure represented by two rings. These features of microdiffraction suggest the presence of precipitation (the crystal phase nucleation of the FeSi α-solid solution) in the amorphous matrix. The diffraction rings of the sample annealed for 1 h at 813K [3] show multiple smaller reflections. In the main halo, the reflections are discrete and a diffuse background of the former halo is retained between them. This implies that the alloy structure obtained after such a treatment retains to some extent the amorphous matrix. The micrographs of the annealed sample to be composed of strongly misoriented crystal grains with a size of an order of 10 nm (Figure 1 a-I). By statistically treating the micrographs, we have constructed the function of grain size distribution [3]. The most frequently occurring grain size is 8-10 nm (Figure 2).There are several features that should be noted concerning the structure of nanocrystalline grains obtained upon crystallization of the amorphous alloy $Fe_{73.5}Cu_1Nb_3Si_{13.5}B_9$. Frequently observed was a nimbus surrounding the crystallites on the dark-field images of grains studied in the reflections corresponding to the diffuse halo (Figure 1 a,b-IV). This feature may be an indication that a phase of a different chemical composition (having a dissimilar lattice parameter) or an amorphous layer is formed at a crystallite boundary. Upon tilting the sample holder by even a considerable angle (up to 10°), part of a crystallite is still capable of participating in the reflection. This can be explained, apparently, by a large misorientation of the crystallite lattice, e.g., due to bending. The micrographs of the samples annealed at 933K for 1 s show that the grain size is smaller in the former samples. As can be seen from the size distribution of crystallites, the most common crystallite size in the alloy annealed at 933K is 6-8 nm (Figure 2).

Another feature of the crystal structure is that crystallites are better formed in the alloy crystallized at higher temperatures for 5-10 s, as can be judged from the electron micrographs (Figure 1a-II). In the micrograph of the samples annealed at 933K for 10 s the diffraction lines are more narrow and the point reflections on the rings are more closely spaced. This feature was confirmed by the fact that new crystallites gave reflections when the sample was tilted by 4° with respect to the electron beam. No reflections from any new crystallites were recorded from the samples annealed at 813 K even at the tilt of up to 10°. The dark field images. show a halo around many grains in both cases, which could be explained by the presence of a new phase on the edges of the grains. Figure 3 and Table 1 shows the microstructure and composition of phases for FeCuNbSiB alloy immediately after the rapid quenching and for sample of FeCuNbSiB

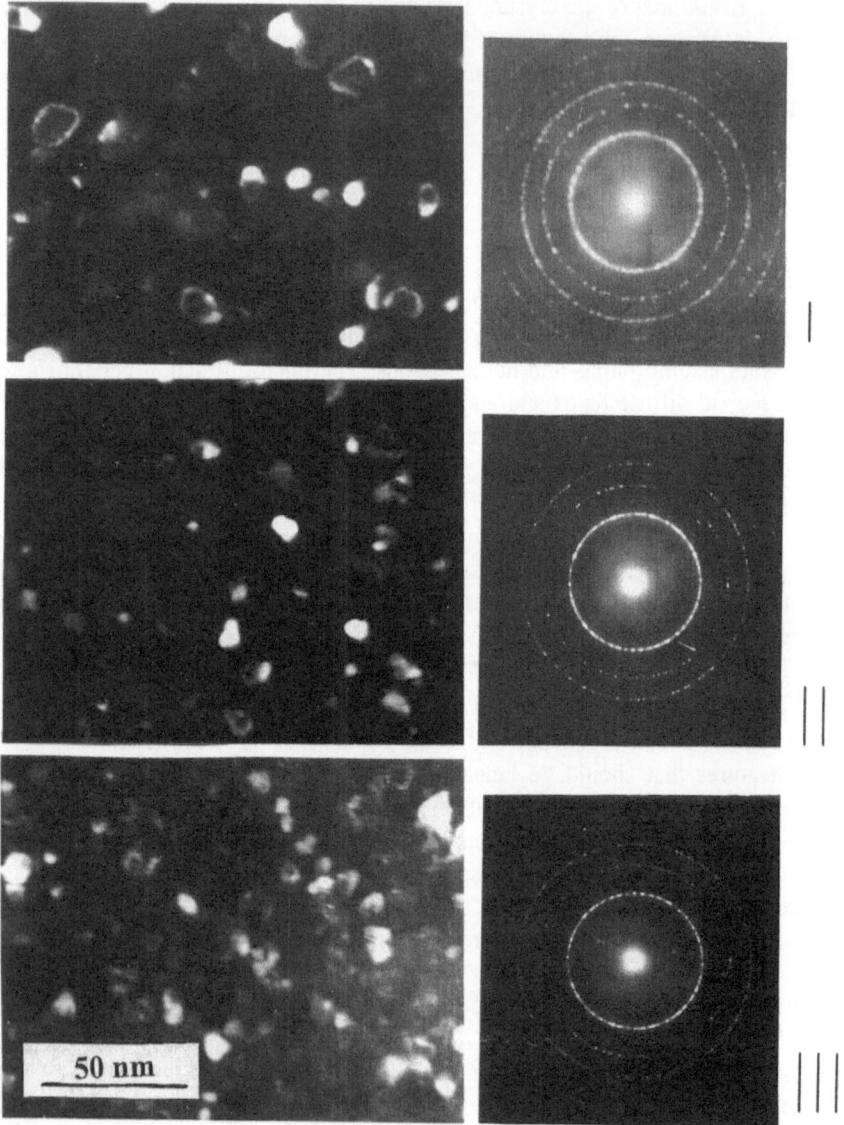

Figure.1. Electron-diffraction patters of $Fe_{73.5}Cu_1Nb_3Si_{13.5}B_9$ **(a)** ard $Fe_{73}Ni_{0.5}Cu_1Nb_3Si_{13.5}B_9$ **(b,c)** alloys: **a(I),b**-annealing at 813 K for 1 h, **a(II,III),,c**-annealing at 893 K for 0.001 h ;**a,b(IV)**- dark field micrographs and electron diffraction patters, **b(I,II,III),c**-bright field.

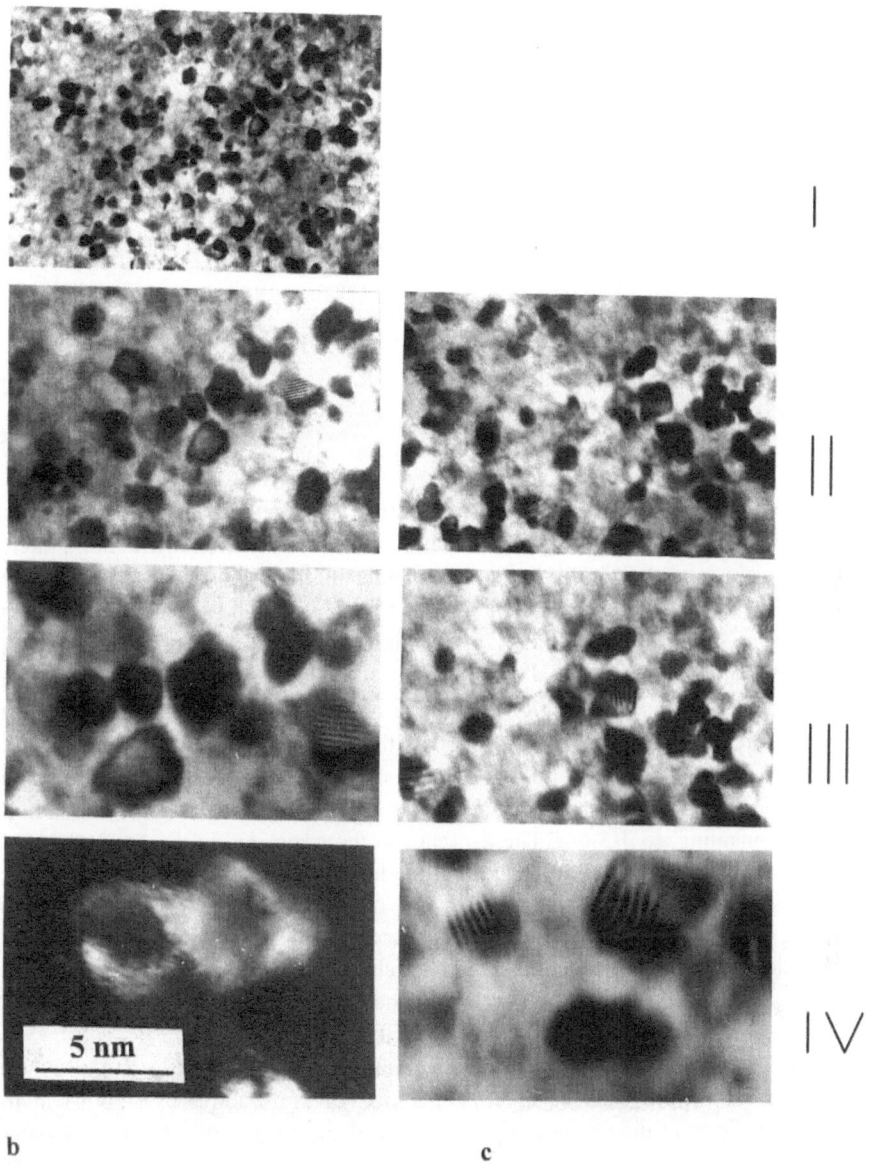

Figure 1.**b(I)**- x 100000; **b(II),c(II)**- x200000; **b(III),c(III)**- x400000; **b(IV),c(IV)**-
x 650000

alloy after the rapid quenching and deformation (tension from crack) by crystallization
in situ on a electron microscopy. Rapid crystallization at an elevated temperature and

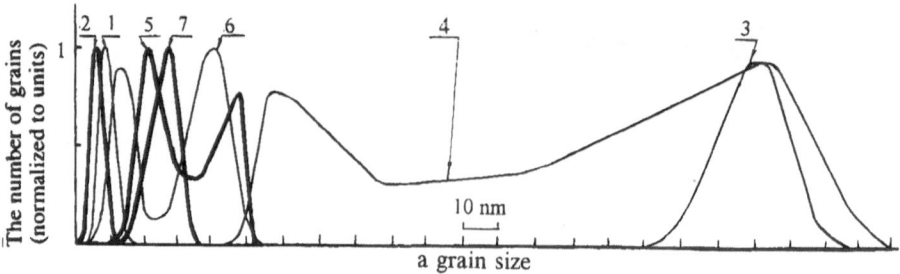

Figure 2. Grain size distribution in various nanocrystalline alloys: **1,2,3**-
Fe$_{73.5}$Cu$_1$Nb$_3$Si$_{13.5}$B$_9$,annealing at 813Kfor 1 h (**1**), at 923K for 0.003 h (**2**), at 1123K
for 1 h (**3**); **4,5**-Fe$_5$Co$_{70}$Si$_{15}$B$_{18}$, annealing at 873K for 1 h (**4**), at 923K for 0.003 h (**5**)
6,7- Pd$_{81}$Cu$_7$Si$_{12}$, annealing at 773K for 1 h (**6**),   at 823K for 0.003 h (**7**).

crystallization deformations alloy produced grains and nanophases whose size was
smaller than in the case of usual crystallized alloy. From the electron diffraction patterns
calculations were made of the phases in the nanocrystalline alloys crystallized at 933
and 813K [2,3]. Interplanar spacing characteristic of a Fe-Si solid solution tend to vary:
in the rapidly crystallized (933K for 1-5 s) sample, the values typical of a Fe-Si solid
solution with a smaller Si content, about 12-13 at. % Si, as compared to 15-18 at. % Si
in the sample crystallised at 813K for 1 h. Some reflections characteristic of Fe$_3$Si and
Fe$_2$B phases are not observed, indicating that these phases may have a preferential
orientation after rapid crystallization. It is seen in [6] that low-temperature annealing
results in a decrease of the grain size from 6-8 nm, which is characteristic of only a high-
temperature annealing, to 4-5 nm. The decrease in the grain size in the Fe-Cu-Nb-Si-B
alloy, which was initiated by step annealing, made the structure of the nanocrystalline
alloy resemble that of the pure metals with 2 to 3 nm grain size, that one can obtain, for
example, in nanocrystalline palladium.The decrease in the grain sizes due to the
preliminary treatment was not accompanied by a change of the phase composition in the
Fe$_{73.5}$Cu$_1$Nb$_3$Si$_{13.5}$B$_9$ alloy, which suggests that the phases are formed complete during
high-temperature annealings for short time (1 - 10 s).
The microstructure and phases of the nanocrystalline Fe$_{73.5}$Cu$_1$Nb$_3$Si$_{13.5}$B$_9$ alloy was
studied in situ at different stages of crystallization of the amorphous ribbons in the
column of an electron microscope. On the basis of the results of the study, we may
suggest the following scheme of crystallization. Upon heating to 773 K, the amorphoous
Fe$_{73.5}$Cu$_1$Nb$_3$Si$_{13.5}$B$_9$ alloy begins to crystallize with   the   precipitation   of   the
crystalline α-Fe-Si solid solution (Figures. 3a and 3b). High elastic distortion appears
at the boundaries of these precipitates.
As the heating temperature uncreases (to 853 K), the size and amount of the α-Fe-Si
solid solution perticles, and the Fe$_3$B phase appears in the alloy. The high elastic
distortions are still present at the phase boundaries. A further increase in temperature
(to 933 K) leads to the formation of a twinned (Fe,Nb)$_2$B phase. Supposedly, the
twinning of this phase is caused by high distortions of the uncrystallized amorphous

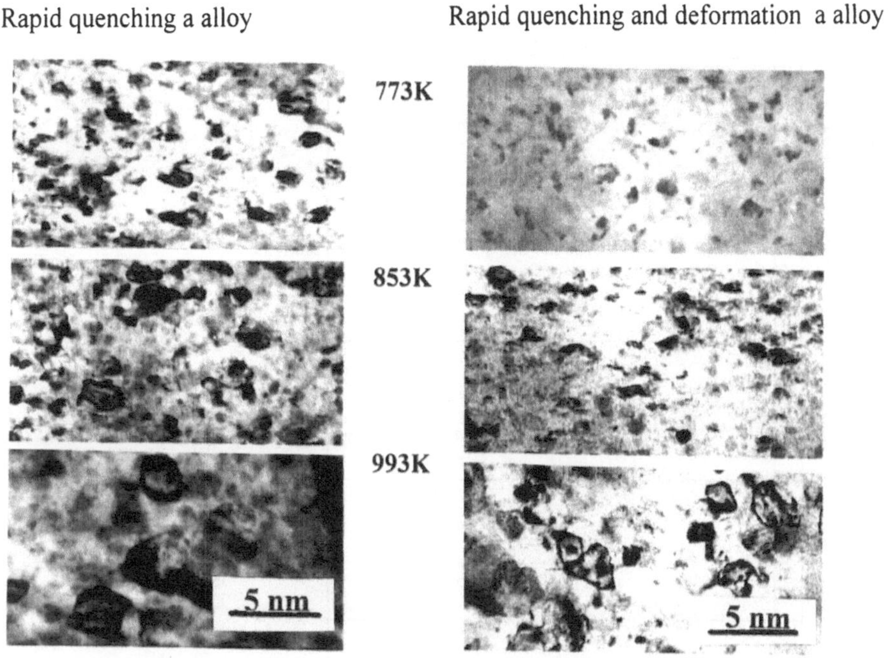

Rapid quenching a alloy   Rapid quenching and deformation a alloy

773K

853K

993K

5 nm   5 nm

a

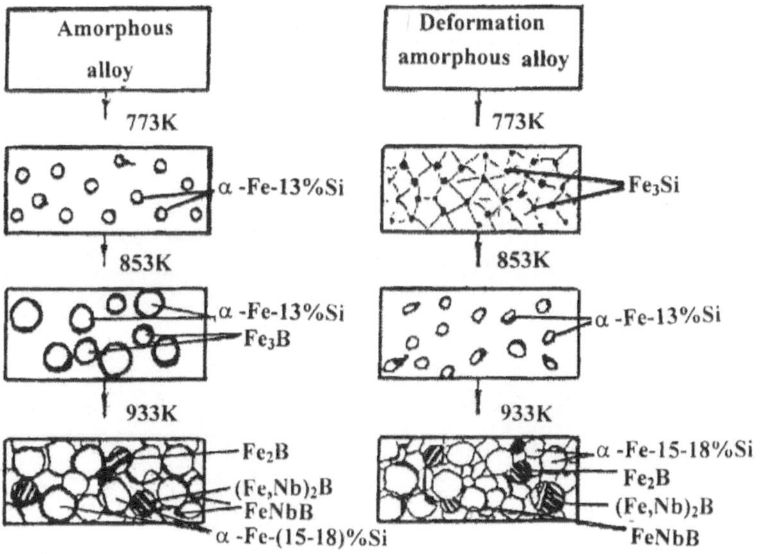

Amorphous   Deformation
alloy   amorphous alloy

773K   773K

α -Fe-13%Si   Fe₃Si

853K   853K

α -Fe-13%Si   α -Fe-13%Si
Fe₃B

933K   933K

Fe₂B   α -Fe-15-18%Si
(Fe,Nb)₂B   Fe₂B
FeNbB   (Fe,Nb)₂B
α -Fe-(15-18)%Si   FeNbB

b

Figure.3. Electron micrographs take after crystallizationfrom amorphous state in situ heating of the $Fe_{73.5}Cu_1Nb_3Si_{13.5}B_9$ alloy with-out deformation and after deformation (**a**) and scheme of crystallization alloy with-out deformation and after deformation (**b**)

Table 1

Phase compositions and interplanar spacings obtained from electron diffraction data in $Fe_{73,5}Cu_1Nb_3Si_{13,5}B_9$ alloy

Experimental (RQ ribbons)			α-Fe	Fe+13,9% Si	Fe+18,2% Si	Tabular				
T=773 K	T=853 K	T=933 K				FeNbB	FeNb	Fe₃ Si	Fe₃ B	Fe₂B
0.199 v.s. 0.139 s. 0.116 s.	0.195 v.s. 0.139 s. 0.115 s. 0.100 w.	0.209 w. 0.195 s. 0.185 w. 0.165 w. 0.156 w. 0.138 w. 0.126 w. 0.113 s. 0.109 w..	0.203 s. 0.143 m. 0.117 s.	0.200 v.s. 0.141 m. 0.116 m.	0.199 v.s. 0.140 m. 0.115 m.	0.213 w. 0.206 w. 0.190 w. 0.150 w. 0.148 s. 0.142 v.s. 0.135 m.	0.239 m. 0.221 s. 0.204 s. 0.201 s. 0.157 w. 0.138 w. 0.134 v.s. 0.130 v.s.. 0.125 v.s. 0.122 w. 0.119 v.s.	0.326 w. 0.282 w. 0.199 w. 0.163 s. 0.141 s. 0.129 v.s.	0.230 m. 0.209 s. 0.203 v.s. 0.193 m. 0.188 s. 0.177 w. 0.168 m. 0.136 w. 0.123 w. 0.122 m. 0.118 w. 0.117 m..	0.256 w 0.212 w 0.201 v.s. 0.181 m. 0.162 m. 0.144 w. 0.135 w.

Experimental (deformation ribbons)		
T=773 K	T=853 K	T=933 K
0.272 m. 0.199 v.s. 0.121 w.	0.197 v.s. 0.140 s. 0.115 s.	0.247 m. 0.211 m. 0.193 v.s. 0.177 m. 0.161 m. 0.153 m. 0.138 s. 0.127 m 0.119 m.

matrix and can, therefore, be considered as a possible relaxation process decreasing the elastic distortions. The FeNbB phase appears to complete the crystallization and is found as discrete particles at the boundaries of nanophases. Note that, according to [13] the niobium - rich phase surrounds the nanograins by continuous rings.

The deformed (by room-temperature tension to failure) amorphous $Fe_{73.5}Cu_1Nb_3Si_{13.5}B_9$ alloy also begins to crystallize upon heating to 773 K in the column of electron microscope. In this case, however ultrafine particles of the metastable phase Fe_3Si precipitate (Table 1). Simultaneously, the crystalline α-Fe-Si solid solution also appears in the alloy. As was noted above, the Fe_3Si disappear with increasing temperature (at 853 K), being replaced by the α-Fe-Si phase, but, judging from the distribution of the diffraction contrast, the fields of internal elastic stresses in this case are small.

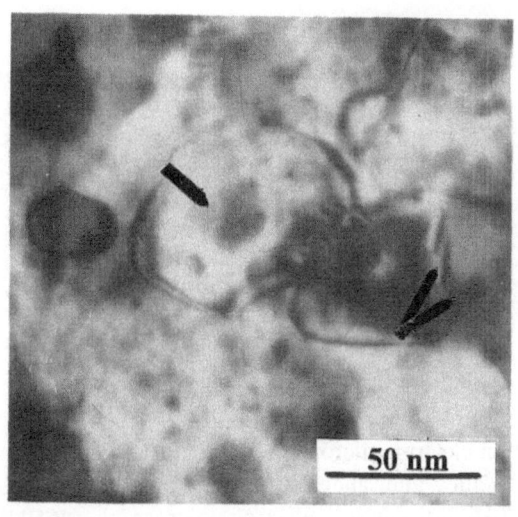

50 nm

Figure 4. Microstructure of $Fe_{73}Ni_{0.5}Cu_1Nb_3Si_{13.5}B_9$ alloy: after heating in an amorphous state to 1023 K in situ in column of an electron microscope:
→ - the nanophase, ⇗ - nanophase boundaries.

alloy upon heating in situ, we found that misfit dislocations from at grain junctions (Figure 4). However, we failed to observe diffraction contrast from dislocations by conventional TEM in multiphase nanocrystalline FeCuNbSiB alloy with an average grain size of 8-10 nm.

Figure 5 shows a HRTEM micrograph of the multiphase nanocrystalline $Fe_{73}Ni_{0.5}Cu_1Nb_3Si_{13.5}B_9$ alloy and a scheme of the structure observed. One feature of the HREM method used is the formation of a weak contrast in the presence of a background contrast, which is reduced to a minimum by the use of stigmators in the microscope; it can , therefore, be clearly seen from Figure 5 (in spite of the weak contrast) that atomic planes with interplanar spacings of 0.32 and 0.19 nm are well resolved. Figure 5b

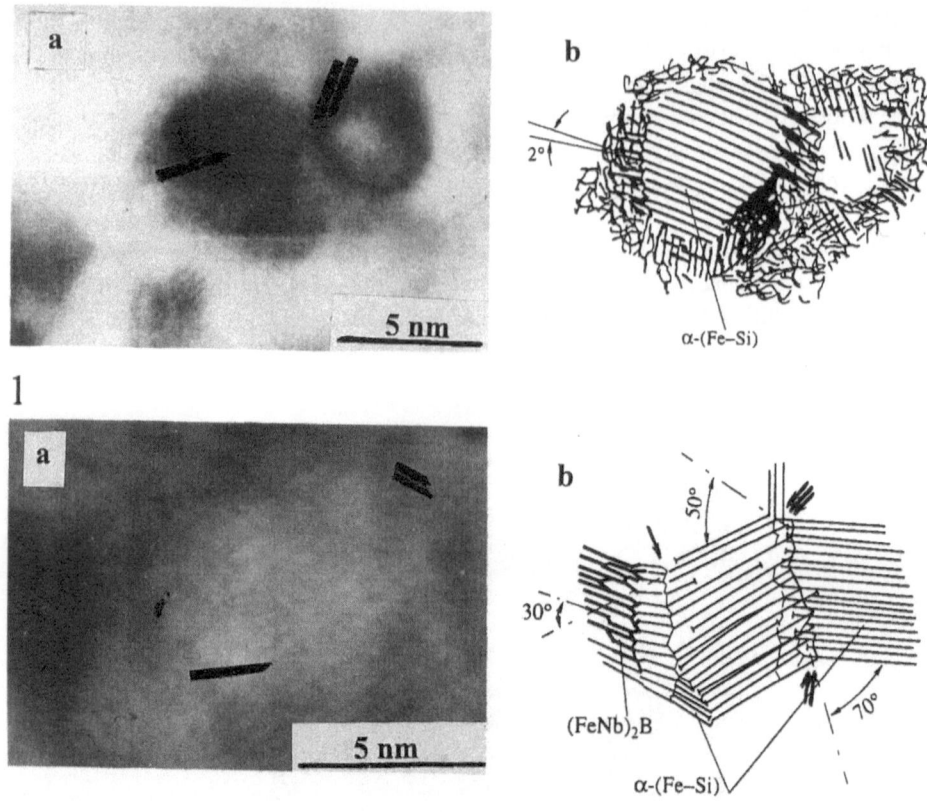

1

2

Figure.5. **1-** High-resolution electron micrograph (x400000+x12 upon reproduction) of the $Fe_{73}Ni_{0.5}Cu_1Nb_3Si_{13.5}B_9$ alloy after annealing at 923 for 0.003 h **(a); s**cheme of the structure observed **(b)**. **2-**High-resolution electron micrograph (x650000+x12 upon reproduction) alloy after annealing at 813 K for 1 h; scheme of the structure observed **(b)**. A nanocrystal and boundary are shown by a single arrow and adouble arrow, respenctively.

shows an approximate scheme of the arrangement of structural components in the micrograph in Figure 5a (1,2). Figure 5a (1) shows a HRTEM micrograph of a smaller nanocrystal (4 nm in diameter). As it is seen from Figures 5a (1) and 5b (1) , the bulk of the grain is virtually free from defects and only characterized by a slight elastic bending of atomic planes. In this case, the nanocrystal is surrounded by a matrix with a highly distorted (quasi-amorphous) structure and is separated by a crystal-like transition layer (interphase boundary) from the adjacent chemicaly distinct nanocrystal; this boundary only contains isolated dislocations and corresponds to a misorientation of $> 2^0$. The scheme displays angles of misorientation of some sets of atomic planes with respect to the same set of planes in the adjacent nanocrystals (Figure 5b (2)). The angles of misorientation were found to reach 70^0. Figures 5a,b (2) also show dislocation dipoles inside the nanocrystals; at nanocrystal boundaries, diffraction contrast is observed that may be identified as being due to dislocations (half-planes). Thus, it is seen from Figure 5, that the grain (phase) boundaries can have crystalline structure, or can be twin boundaries, or contain many dislocation; these types of the gran boundary structure are shown in Figure 5b(2) by single arrow, double arrow, and triple arrow, respectively.

The scheme displays angles of misorientation of some sets of atomic planes with respect to the same set of planes in the adjacent nanocrystals (Figure 5b (2)). In this case, the interface between chemically identical phases is no more than 0.2 nm in width, whereas the boundaries between chemically different nanocrystals are wider and reach 2 nm. We assume that the central nanocrystal in Figure 5a (2) and the left nanocrystal differ in the chemical composition). Because the intermediate layer between the α-(Fe-Si) and Fe_3Si nanocrystals has a different interplanar spacing (Figure5b(2)), we assume that this region of the interface contains a different phase, which may be FeNbB.

Properties. Magnetic properties of the samples isothermally annealed for 1 h are known to abruply deteriorate at 843 K because of a rapid growth in size of the crystalline grains. Therefore, elevated temperatures were chosen for the short crystallization period of the initially amorphous alloy. The minimum annealing temperature was taken to be 853 K. Figure 6 shows the values of coercivity H_c of the samples annealed at temperatures between 853 and 943 K for 5,10,20 and 60 s. Annealing at 853K for 10 - 30 s results in coercivity close to that obtained by conventional annealing at 813K for 1 h. A further increase in the annealing temperature for the same durations (1 - 30 s) leads to an increase of H_c, probably due to the onset of the second stage of crystallization. A 5-s annealing at 853K is not sufficient for a state

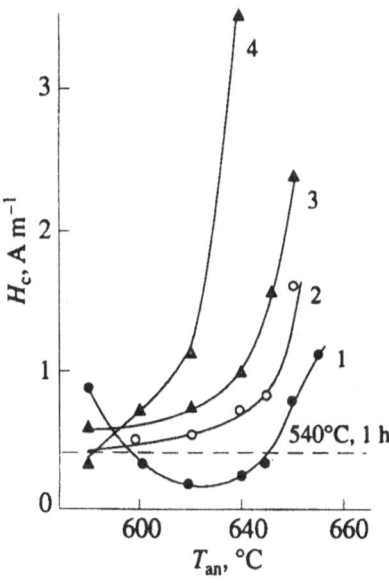

Figure 6. The dependence of coercivity H_c upon the annealing temperature T_{an} of the $Fe_{73.5}Cu_1Nb_3Si_{13.5}B_9$ alloy. The annealing time:5 s(**1**); 10 s(**2**);20 s(**3**); 60 s(**4**). Dashed line is the H_c value for the sample annealed at 540° C for 1 h.

Table 2

The microhardness (H) and Grain size (D) of nanocrystalline alloys and ceramic.

Alloy	Grain size, nm from..........to	Microhardness, GPa from.........to	Produced	Ref.
TiAlNb	25-10	5,5-3,5	Crystallized RQ alloy	9
FeCuSiB	100-25	8-10	-"-	9
FeSiB	100-25	6,2-11,8	-"-	9
FeSiB	100-25	8,8-13,5	-"-	4,15
FeSiB	25-8	13,5-6,0	-"-	4,15
FeCuNbSiB	200-6	6,0-15,0	-"-	4,15
FeMoSiB	200-10	6,1-10,0	-"-	17
CoFeSiB	100-25	10,5-13,0	-"-	4.15
CoFeSiB	25-8	13,0-9,0	-"-	4,15
PdCuSi	100-10	4,0-7,5	-"-	4,15
PdCuSi	10-4	7,5-2,8	-"-	4,15
TiAl	625-25	3,0-12,0	Mech.activation	9
TiAl	25-12	12,0-6,5	-"-	9
Nb_3Sn	100-6	8,0-9,0	-"-	9
$NbAl_3$	100-20	3,8-7,0	-"-	9
$NbAl_3$	20-9	7,0-5,7	-"-	9
Al-1,5%Mg	150-16	0,4-1,6	Severe plastic deformation	18
TiN	40	21,0	-«-	16,19
Al_2O_3	70	22,0	Synthesis	19
ZrO_2	150	6,4	-"-	19
TiO_2	10	17,0	-"-	19
SiN	400	18,0	-"-	19

with a low H_c to form, while annealing at 873-923K leads to an H_c value smaller than that obtained by a conventional technique. At 893K, for instance, it is twice as small.

Still shorter annealings (1 s) expand the range of lower H_c values to 943 K. The nanocrystalline state of the $Fe_{73.5}Cu_1Nb_3Si_{13.5}B_9$ alloy can be obtained within a very short time (a number of seconds) by crystallizing an initial amorphous alloy at the temperatures from 873 to 953 K. A nanocrystalline grain turns out smaller than that after annealing at 813 K 1 h. This decrease to be one of the factors leading to a decrease of coercivity. Other factors can be a more perfect crystal structure [14].

Rapid crystallization of a metallic glass at an elevated temperature under rapid heating and cooling was shown to results in significant gain in microhardness (Table 2) and greater phase homogeneity.

The Alloy FeCoSiB

The Structure. In order to account for the results obtained, the crystal structure of the samples of $Fe_5Co_{70}Si_{15}B_{10}$ alloy with a maximum H_c after rapid and slow crystallization was studied by transmission electron microscopy [3,7]. The results are presented in Figure 7 and in Table 3. It can be seen that the grain size in the sample rapidly heated to 923K and annealed for 10 s at this temperature is markedly smaller (15-50 nm) than that in the sample slowly heated to 873K and annealed for 1 h (50-200 nm). Twins are also clearly

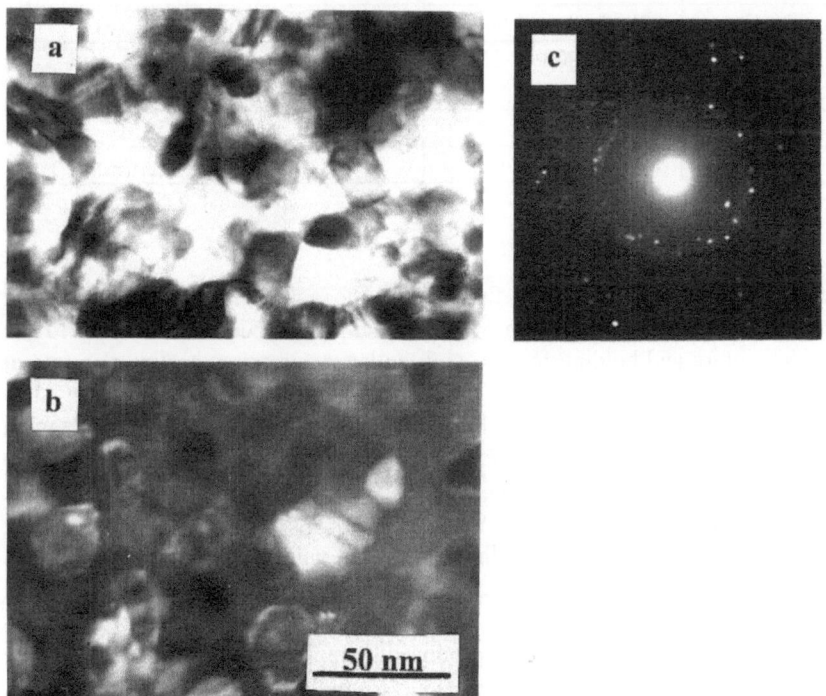

Figure 7. Bright-field (**a**) electron micrograph and dark-field electron micrograph(**b**) and electron diffraction pattern (**c**) of $Fe_5Co_{70}Si_{15}B_{10}$ alloy after rapid crystallizations.

visible in the structure of the rapidly crystallized alloy. Examination of a large number of electron micrographs showed that twins do exist in slowly crystallized samples, although much more seldom. The electron diffraction patterns also indicate that grains in the rapidly crystallized alloy are smaller than in the slowly crystallized one. Table 3 lists phases in nanocrystalline alloys derived from the electron diffraction patterns, and those for the phases that may be expected to crystallize in the alloy, accoding to [3,7]. An analysis of the data in the $Fe_5Co_{70}Si_{15}B_{10}$ alloy showed that α-Fe, α-Co, β-Co, Fe_3Si, Co_2Si and $(Fe,Co)2B$ phases would exist in the alloy after rapid crystallization, where as α-Co, β-Co, Co_2Si and CoB wold exist after slow crystallization.

TABLE 3
Phases in nanocrystalline alloys

Composition	T K, t (h)	D nm	Phases
$Fe_{73.5}Cu_1Nb_3Si_{13.5}B_9$	813 - 1	80-10	α-Fe-(15-18%)Si,$(FeNb)_2B$, FeNbB, Fe_2B
-"-	923 - 0.003	5-6	α-Fe-(12-13%)Si,$(FeNb)_2B$, FeNbB, Fe_3Si
$Fe_5Co_{70}Si_{15}B_{10}$	873 - 1	50-200	α-Co, β-Co, Co_2Si, CoB
-"-	923 - 0.003	15-50	α-Fe, α-Co,β-Co Fe_3Si, Co_2Si, $(FeCo)_2B$
$Pd_{81}Cu_7Si_{12}$	823 - 1	10, 50	Pd, Pd_5Si, γ- Pd-Cu
-"-	823 - 0.003	25	Pd, Pd_5Si, γ-Pd-Cu, i-Pd_xSi

Properties. Figure 8 shows H_c as a function of the annealing temperature during conventional and rapid crystallizations. The former consists in slow heating to the annealing temperature and holding for 1 h, and the latter comprises heating at a rate of 10^3 K/s and holding for 10 s. It can be seen that the maximum H_c upon slow crystallization is 300 Oe, in agreement with the value obtained in [3,7]. The maximum H_c upon rapid crystallization reaches 820 Oe, i.e., 2.7 times greater than in the former case. The temperature for H_c to reach its maximum during rapid annealing is 50-100K higher than upon slow annealing. Figure 9 shows M_r/M_s as a function of the annealing temperature for the same modes of annealing. The M_r/M_s value is rather small, as long as H_c is low and abruptly increases with H_c. The relative remanent magnetization for rapid crystallization reaches 60%, which is about 10% greater than that for slow crystallization . The high-coercivity state, produced by rapid crystallization, is thermally stable: the annealing, at 673K for 1 h, of the sample rapidly cooled from 923K does not change its magnetic properties. This means that the rapid annealing effect is not due to internal stresses produced during quenching, because they relax in metallic glasses at 673K. Coercive force ,Curie temperature, Junga's. modul and shear moduls of nanocrystalline polyphase alloys are presented in Table 4 .

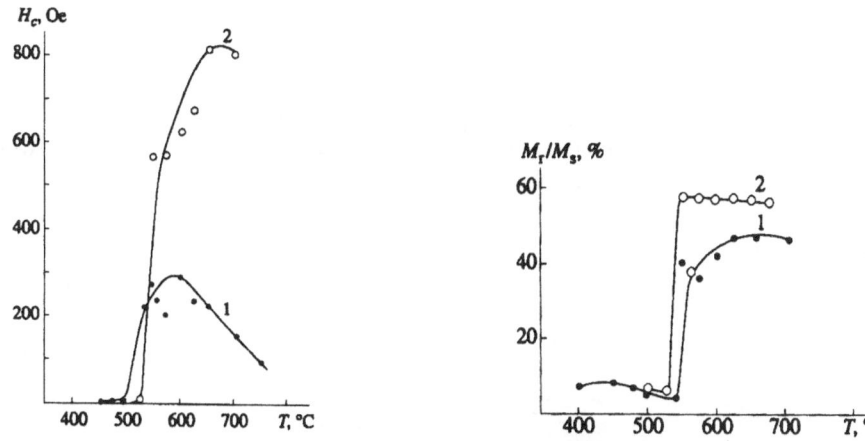

Figure 8. Coercive force H_c as a function of the crystallization temperature: (**1**) - slow crystallization; (**2**) - rapid crystallization

Figure 9. Relative remanent magnetization M_r/M_s as a function of the crystallization temperature: (**1**) - slow crystallization, (**2**) - rapid crystallization.

Table 4.

Coercive force(Hc), Max. magnetization (Ms), Curie temperature (Tc), specific heat (q), Junga's Modul (E) and shear Modul (G) of nanocrystalline polyphase alloys.

Alloy	d, nm	Hc, A/m	Ms, T	Tc, K	q, J/м.K	E, GPa	G, GPa
$Fe_{73.5}Cu_1Nb_3Si_{13.5}B_9$	10	0,4	1,25	593	-	-	-
-"-	6	0,2	1,30	813	-	-	-
$Fe_5Co_{70}Si_{15}B_{10}$	200	$0,02.10^6$	-	-	-	-	-
-"-	25	$0,07.10^6$	-	-	-	-	-
$Nd_{12-16}Fe_{100-x-y}B_{6-8}$	1500	$0,08.10^6$	-	-	-	-	-
-"-	10	$0,32.10^6$	-	-	-	-	-
$Fe_{78}Si_9B_{13}$	30	-	-	-	28,25	-	-
-"-	200	-	-	-	25,36	-	-
$(Fe_{80}Ni_{20})_{82}Si_6B_{12}$	50	-	-	-	-	151	55,5
-"-	5000	-	-	-	-	185	70

The Alloy PdCuSi

The Structure .Figure 10a displays micrographs showing the structure of the alloy after annealing for 10 s at 823 K (as described in [3]) and after the creep tests at a temperature below the crystallization temperature for the alloy (Figure 10b), and at a temperature above the crystallization temperature of the alloy (Figures 10c and 10d). In Ref. [10] it is shown the histograms of grain (phase particle) size vs. the number of grains (number of the phase particles), presents the statistical treatment results obtained from the micro-graphs where the number of fine, medium, and coarse crystal grains were counted for the specimens subjected to various creep conditions and after the crystallization annea-ling. According to the data shown in Figure 10 and in Ref. [10] , the sizes of the grains produ-ced in the Pd-Cu-Si alloy under the creep conditions (above the crystallization tempera-

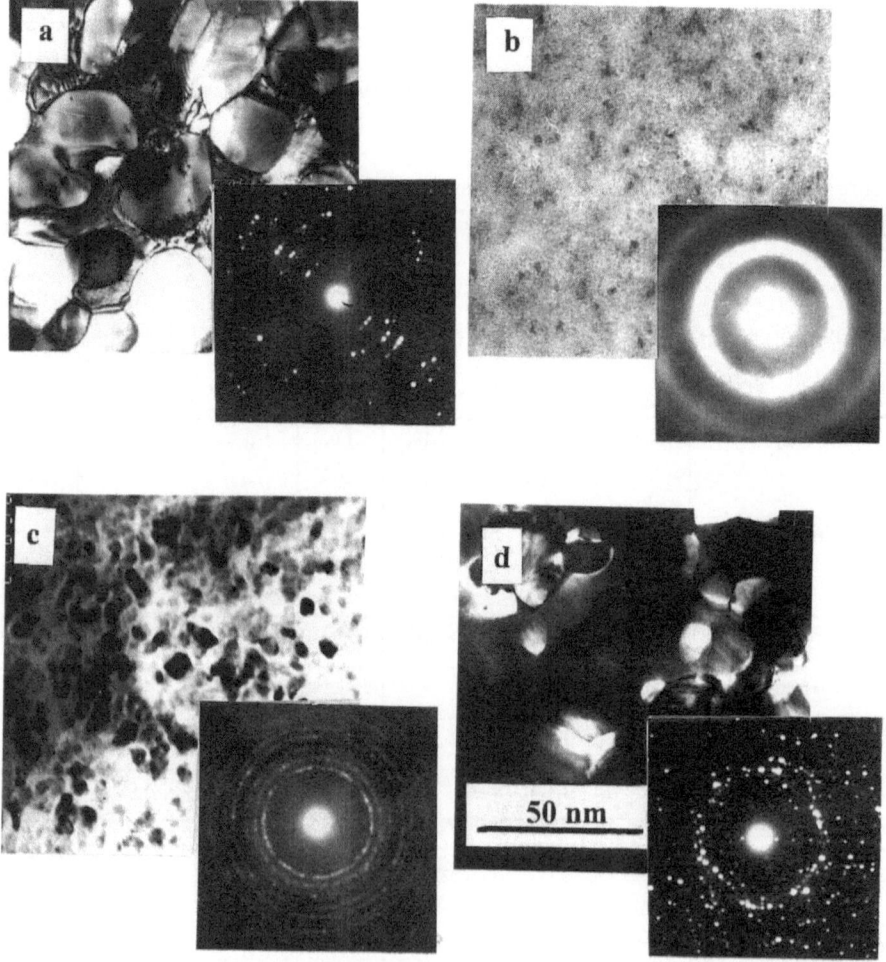

Figure 10. Electron micrographs and electron microdiffraction patterns of the $Pd_{81}Cu_7Si_{12}$ alloy after the following treatments: (**a**)-Annealing at 823 K for 10 s; (**b**)-Creep at 673 K; (**c**)-Creep at 773 K, test time=20 min.;(**d**)-Creep at 773 K,test time=1 h.

ture) differ significantly from the size of the grains in the alloy crystallized from the amorphous state under the conditions of conventional annealing. In addition, the alloy undergoes incomplete crystallization at the creep temperature of 728 K and under such conditions the alloy has the structure of an amorphous matrix containing particles of Pd and the crystalline phase of the γ-Pd-Cu solid solution (see Figure 10b). Conventional annealing at 723 K for 10 s causes complete crystallization of the alloy; the mean grain size in this cause is about 30 nm. During creep at a temperature of 773 K and stress of 2.1 MPa, short-term loading (20 min) of the specimens produces nanocrystalline grains in the alloy (the mean grain size is 10 nm) which give rise to diffraction rings (see Figure 10c) consisting of numerous reflections in the microdiffraction pattern (with a selector diaphragm of 0.25 μm). As in the case of the annealing producing the nanocrystalline state of the alloy, annealing of the alloy at 823 K for 10 s gives rise to crystal grains with a mean size about 40 nm (see Figure 10a). The grain size increases when the time of the creep test at 773 K is increased to 1 h (see Figure 10d). The resulting grain size depends on the applied stress. For instance, when the stress was decreased from 2.1 to 1.5 MPa, differences in grain sizes developed: some grains were as small as 4 nm and some were as large as 40 nm. The decrease of the stress in the creep tests from 1.5 to 0.7 MPa produced an increases in the grain size on the whole, but the differences in the grain size were preserved This was evidenced by the corresponding patterns of electron microdiffraction showing the first and second diffuse halos and no unusual features. In the crystalline state, the alloy samples exhibited a multiphase structure. Data on the phase composition of alloys tested by methods TEM and HRTEM in various regimes are presented in Table 3. The results of statistical treatment of nanocrystalline grains are given in Table 3 and Table 5.

Properties.Figures 11a-c shows the stress-strain diagrams of alloys tested at various temperatures. As is seen from data presented in Table 5 and Figure 11, the room temperature yield strength depends on the thermal treatment of alloy samples. The Table 5 also indicates that the temperature of annealing significantly affects the grain average size. Figure 12a shows the dependence of the yield strength σ_s on the grain size D (average up to everything nanophase particles). In order to provide a correct comparison of various alloys, these are characterized by the reduced value (σ_s/G), were G is the shear modules of the corresponding alloy. The plot of (σ_s/G versus the grain (nanophases particle) size shows that the Hall-Petch relation may be useful in the analysis of the dependence of yield strength on the grain size in nanophase materials with D $\geq$10 nm. In addition, we have measured the microhardness of all the alloys after various treatments leading to different dimensions of nanograins. Figure 12b and Table 2 show the plots of microhardness versus $D^{-1/2}$ for various nanophase alloys. The Hall-Petch relation ship for microhardnesses holds in all the nanophase alloys studied in the range of nanophase dimensions from 10 to 100 nm In addition to the grain size and phase-particle size, and strength, the data in Table 5 include the plastic properties of the crystalline alloys prodused in the experiments. It can be seen from these data that the alloy exhibits the highest plasticity when creep occurs at temperatures of 673 and 728 K and at relatively high stresses (39 and 3.9 MPa, respectively). In these tests, the time before the specimen failed did not exceed 1 or 2 min and the alloy had a mixed amorphous-crystalline structure. After the creep tests under such conditions, the microhardness of the alloy was 5.3 GPa. The alloys after crystallization under the creep

Table 5
The structure and properties of nanocrystalline polyphases alloys

Annealing	Condit. of tests	The structure and properties				
T(K), t(h)	T(K) - (MPa)	D(nm)	Phase	σ_s ,(MPa)	σ_B,(MPa)	δ, %
$Pd_{81}Cu_7Si_{12}$ -creep						
RQ	623 - 39.0	-	amorphous	-	-	80
RQ·	728 - 3.9	2	Pd,γ-Pd-Cu	-	-	53
RQ	773 - 2.1	10	Pd,γ-Pd-Cu, Pd_5Si, Pd_9S_{i2}	-	-	>15
RQ	773 - 1.5	4; 40	-«-	-	-	>10
823 - 0.003	823 2.1	30	-«-	-	-	10
$Pd_{77.5}Cu_6Si_{16.5}$ - uniaxial tension						
RQ	300	-	amorphous	-	820	0.0
573 - 1.0	300	4	Pd	700	710	0.3
RQ	573	4	Pd, Pd_5Si	310	550	4.3
573 - 1.0	573	4	Pd, Pd_5Si	350	640	1.6
RQ	773	6	Pd,Pd_5Si	140	160	1.6
573 - 1.0	773	6	Pd,Pd_5Si, Pd_9Si_2	60	140	1.0
$Fe_{73.5}Cu_1Nb_3Si_{13.5}B9$						
RQ	300	-	amorphous	2000	2105	1.8
623 - 1.0	300	4	Fe+13%Si	258	258	0.2
723 - 1.0	300	10	Fe+13%Si,	280	287	0.3
		12-18	$(FeNb)_2B$			
		-	FeNbB			
813 - 0.5	300	10	Fe+13%Si,	530	548	0.8
		10	$(FeNb)_2B$			
		-	FeNbB			
813 - 0.5	673	10	Fe+13%Si,	520	1620	1.2
		10	$(FeNb)_2B$			
		-	FeNbB			
$Fe_5Co_{70}Si_{15}B_{10}$						
RQ	300	-	amorphous	-	1180	0.0
873 - 1.0	300	50-200	α-Co, β-Co,	945	950	0.3
		90-200	Co_2Si, Co_2B	1880	2100	1.0
923 - 0.003	300	15-50	α-Co, α-Fe,			
		50	β-Co,Fe_3Si, Co_2Si			

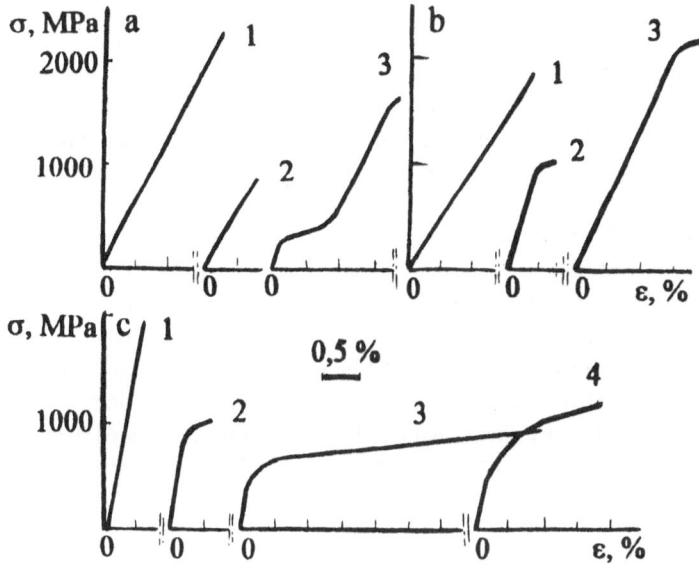

Figure 11. Stress-strain diagrams of alloys: **a** - $Fe_{73.5}Cu_1Nb_3Si_{13.5}B_9$ (**1**-amorphous state tested at 300 K, **2**- 813K-0.5 h tested at 300 K, **3**- 813 K-0.5 h tested at 673 K), **b**-$Fe_5Co_{70}Si_{15}B_{10}$(**1** -amorphous state tested at 300 K, **2** -873 K-1 h tested at 300 K, **3** - 923 K-0.003 h tested at 300 K) , **c**-$Pd_{77.5}Cu_6Si_{16.5}$(**1**-amorphous state tested at 300 K, **2** - 573 K-1 h tested at 300 K, **3** - amorphous state tested at 573 K, **4** - 573 K-1 h test at 573 K)

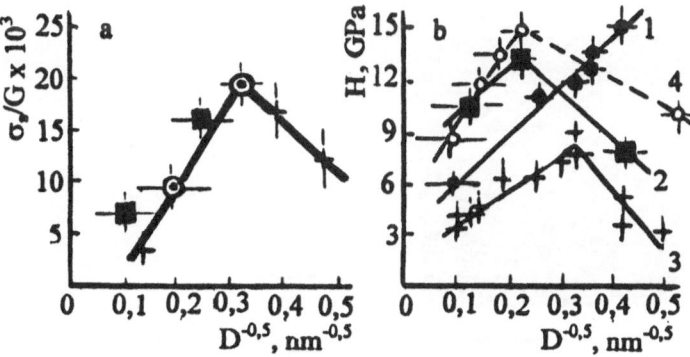

Figure 12. The plots of yield strength (**a**) and of microhardness (**b**) versus grain size D for nanocrystalline poliphases alloys: **1**- $Fe_{73.5}Cu_1Nb_3Si_{13.5}B_9$; **2**-$Fe_{81}Si_7B_{12}$, **3**- $Pd_{81}Cu_7Si_{12}$; **4**-$Fe_5Co_{70}Si_{15}B_{10}$.

conditions virtually always prove to be sufficiently longlived and have a fairly high plasticity. For these alloys, the table gives the strains after the tests rather than at failure.
The crystallized specimens never failed in the tests. It should be noted that the alloy
specimen with the 10-nm grain size proved to have both high plasticity and sufficiently
high strength. After being subjected to the stress of 2.1 MPa at 773 K for an hour, not
only did the specimen not fail but even its deformation did not increase. It can be supposed that the deformation of this specimen was easy until the grain size increased to
about 40 nm. The microhardness of this specimen was 7.3 GPa. We may suppouse, apparently, that the plasticity of the Pd-Cu-Si alloy depends on the structural, temperature
and loading conditions. For instance, the plasticity of thit alloy is high if the alloy has an
amorphous-crystalline structure and if the grain size dose not exceed 10 nm and, possibly, this is why the alloy exhibits high creep rates. When the test time is increased (>1h),
the alloy deformation increases with increasing grain size but the increase does not seem
to be very significant. Figures 13a-d show ,that the deformation relief extends over the
entire surface of nano-phase samples and is characterized by an insignificant step height

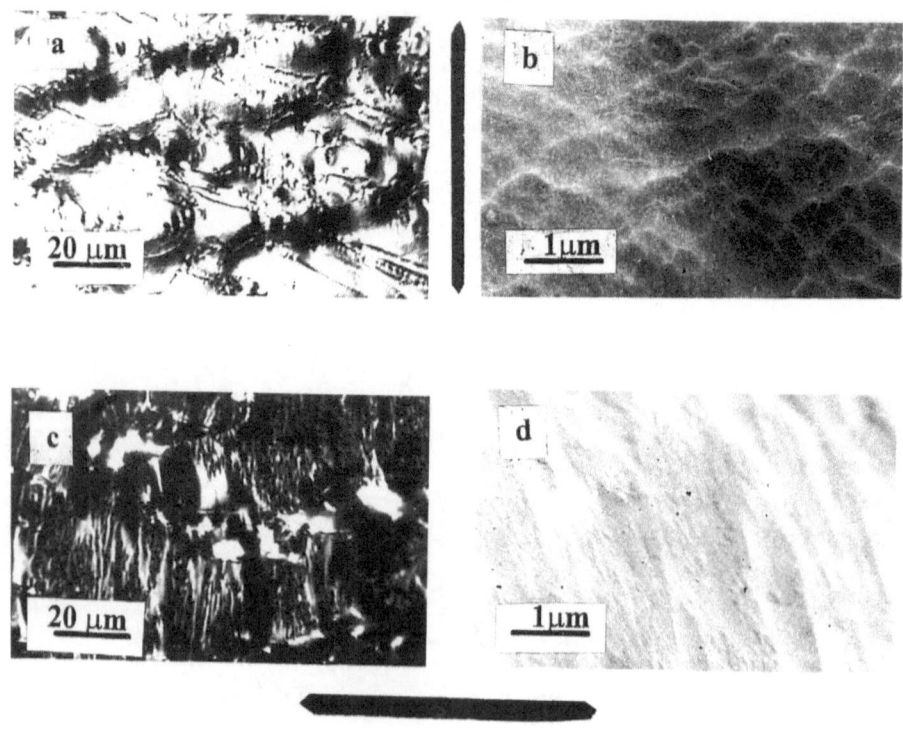

Figure 13. Micrographs of a deformation relief formed on the surface of an amorphous
alloy annealed at 573 K for 1 h and tested under conditions of (a,b) high- test rate
stretchng and (c,d) creep temperature 773 K .

The mechanism of plastic deformation in nanophase materials is characterized by the term deformation channels. These channels in nanophase materials are generally represented by interphase interfaces with high elastic stress fields in their vicinity. This conloy is shown in Figures 14a-c. Extended diffraction effects are observed as straight and cept is confirmed by the diffraction effects revealed. Microstructure of the deformed alcurved contrast lines (shown by arrows in Figure 14 a). In these sites, stitches of luminous points (shown by an arrow in Figure 14c) are observed in the dark-field image. The above diffraction contrast may be caused by the presence of deformation bands mediating plastic shear upon the deformation of the nanophase alloy, whereas the presrence of the stitched phase with variable concentration may be considered as an indication that deformation-induced flows leading to the transfer of matter appear in the deformation bands (deformation channels).

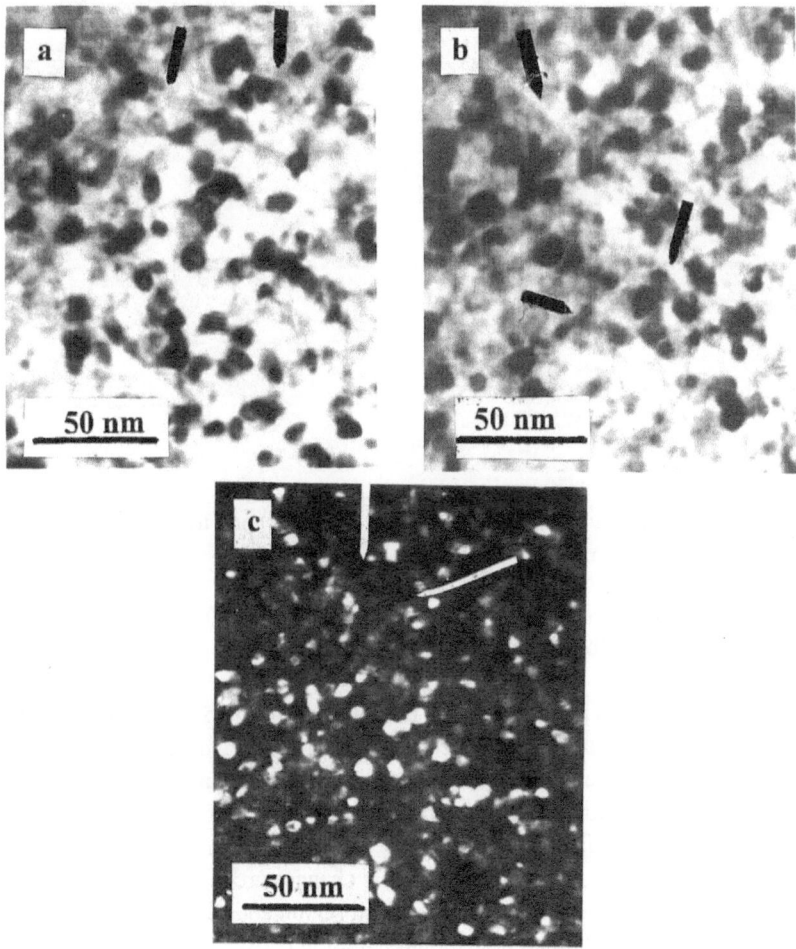

Figure 14. Electron micrographs of the deformed FeCuNbSiB alloy: (a) and (b) bright-field images and (c)dark-field image[20].

The nanocrystalline Cu and Mo. The Structure.

The structure of the nanocrystalline Cu before and after ultrasonic waves (UW) treatment (exposure to UW for 10 min) is shown in Figu-res 15a and 15b. The electron microscopic photographs were taken from different parts of the same specimen after irradia·tion. By investigating the regions in which the focus of the UW beam was situated we were able to obtain the parameters of the dislocation structure of the ultrasonically irradiated materials Figure 15b, and the structural parameters of the unirradiated material were obtained from the electron microscopic photographs of regions which were a long way from the focus Figure 15a. With this experimental procedure, the difference in the influence of UW radiation can only be due to the degree of localisation of the UW intensity, since the temperature distributions is almost uniform. Exposure to a UW beam focused on the surface of the deformation disk did not produce any particular rearrangement of the dislocation structure of the copper, although the cell boundaries did"break

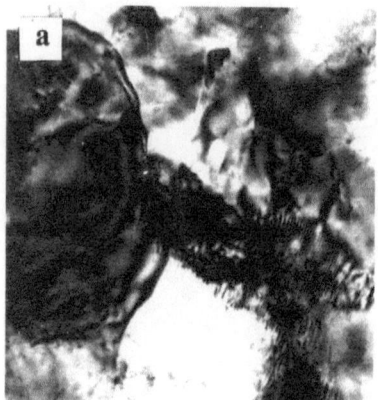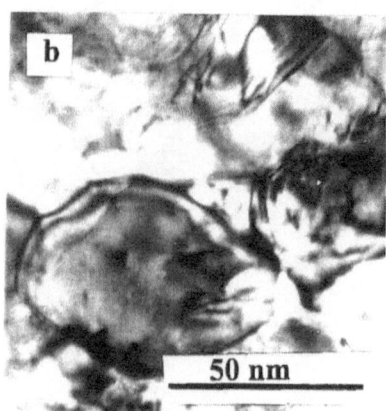

Figure 15.The structure of the nanocrystalline Cu before (**a**) and after (**b**) UW treatment

up" (Figure 15b). Electron diffraction from this regions indicates a reduction in misorientation of substructural formations, suggesting that relaxation processes have taken place.

Properties. In the copper it appears that the randomly distributed dislocations are partially annihilated and new boundaries formed for the structural elements under the effect of a UW beam. The microhardness of nanocrystalline Cu before and after UW treatment varied from 0.95 GPa to O.55 GPa. Our experimental results show conclusively that a focused UW beam does, in fact, have a local effect on the structure of the deformation disk when it is in highly stressed state. The visible changes are associated with an overall decrease of dislocation density in the irradiated region and rearrangement of the dislocation structure, with a reduction in the misorientation between substructural elements Note that no marked changes in the structure of the nanocrystalline Mo (D=40 nm) occur after the ultrasonic treatment. Within the accuracy of measurements, its microhardness measured and after the UW treatment was found to be (6.2 - 6.3) GPa.

4. Conclusions

1.The investigation of the structure and phase composition of the nanocrystalline multiphase $Fe_{73.5}Cu_1Nb_3Si_{13.5}B_9$ and $Fe_{73}Ni_{0.5}Cu_1Nb_3Si_{13.5}B_9$ alloys has shown that the alloys crystallized from the amorphous state at 773-933 K for certain holding times has a nanophase structure. The nanophases contained in the alloy are as follows: the α- Fe-Si (13-18 at.%) solid solution as the matrix phase; the twinned $(Fe,Nb)_2B$; and the FeNbB phase present as discrete precipitates at the boundaries of nanograins. The Fe_3Si phase behaves as a metastable phase and its appearance in the alloy depends upon the conditions of crystallization. The nanophase structure is characterized by the presence of internal elastic stresses, which attain the maximum values at the boundaries of nanophases and appear at early stages of crystallization. Another distinction of the nanophase structure is a variable composition of the α-Fe-Si solid solutition 2. Rapid crystallization of a metallic glass at an elevated temperature under rapid heating and cooling was shown to result in a significstion gain in the coercive force and remanent magnetization. This gain may be due to a grain refinement and changes in phase composition in comparison with the conventionally crystallized alloy. This effect offers a possibility for developing high-coercivity alloys of a new type. 3. An increased brittleness (low plasticity) of the nanophase $Fe_{73.5}Cu_1Nb_3Si_{13.5}B_9$ al-loy may be explained by a high level of internal stresses at nanophase boundaries and a low ability of the alloy to the relaxation of these stresses. The precipitation of primary metastable phases that do not cause high internal stresses is recommended to increase plasticity of the alloy. The crystallized alloy $Pd_{81}Cu_7Si_{12}$ may exhibit a sufficiently high plasticity under creep conditions if it has an amorphous-crystalline structure or if the grain size does not exceed 10nm. 4.The Hall-Petch relationship can be employed in analysis of the yield strength versus grain size plots of the nanophases alloys studied in this work, provided that the grain size is not less than 10 nm. Deformation of the nanocrystalline alloy (the grain size is less that 10 nm) occurs through the deformation channels represented by phase boundaries with high stress fields .

5. Acknowledgments

This work supported by the Russian Foundation for Basic Researches under No 96-02-17567. A author thank the colleaguaes , mentioned in the references for their useful discussions and kind cooperation.

6.References

1. Yoshizawa, Y., Oguma, S.V., Yamauchi, K. (1988) New Fe-Based Soft Magnetic Alloys Composed of an Ultrafine Grain Structure, *J. Appl. Phys.* **64**, 6044 -6046.
2. Noskova, N.I., Serikov, V.V., Glazer, A.A., Kleinerman, N.M., and Potapov A,P. (1992) An Electron-Microscopic and Mossbauer Stuty of the Composition and

Structure of Alloy $Fe_{73.5}Cu_1Nb_3S_{i13.5}B_9$ in a Nanocrystalline State, *Phys.Met. and Metallogr.***74**,52-57.

3. Glazer, A.A.,Lukchina, V.A.,Potapov, A.P., and Noskova, N.I.(1992) Nanocrystalline $Fe_{73.5}Cu_1Nb_3S_{i13.5}B_9$. Alloy Rapidly Crystallized from an Amorphous State at Elevated Temperatures, *Phys. Met. Metallogr.* **74**, 163-166; and(1993) Effect of Rapid Crystallization of $Fe_5Co_{70}Si_{15}B_{10}$ Glass on Its Magnetic Properties, *Phys. Met. and Metallogr.***76**, 222-224.

4. Noskova, N.I.,Ponomareva, E.G., Pereturina, I.A., and Kuznetsov, V.N. (1996) Strength and Plasticity of a Pd-Cu-Si Alloy in the Amorphous and Nanocrystalline States,*Phys. Met. and Metallogr.* **81**, 110-115.

5. Siegel, R.W. (1994) Nanostructured materials- mind over matter, *NanoStructured Materials* **4**, 121-138

6. Noskova, N.I., Ponomareva, E.G., Glazer, A.A., Lukshina, V.A., and Potapov, A.P. (1993) The Effect of Preliminary Deformation and Low-Temperature Annealing on Grain Size in Nanocrystalline $Fe_{73.5}Cu_1Nb_3Si_{13.5}B_9$ Alloy Produced by Crystallization of Amorphous Ribbons, *Phys. Met.and Metallogr.* **76**,171-173.

7. Noskova, N.I.,Ponomareva, E.G.,Lukshina, V.A., and Potapov, A.P. (1995) Effect of Rapid Crystallization of $Fe_5Co_{70}Si_{15}B_{10}$ Glass on Its Properties, *NanoStructered Materials* **6**, 969-972.

8. Noskova, N.I., Vildanova, N.F., Potapov, A.P., and Glazer, A.A. (1992) The Effect of Deformation and Annealing on the Structure and Properties of Amorphous Alloys, *Phys. Met. and Metallogr.* **73**, 181-187.

9. Siegel, R.W., and Fougere, G.E. (1995) Mechanical Properties of Nanophase Metals, *NanoStructured Materials* **6**, 205-216

10. Noskova, N.I., Ponomareva, E.G., Kuznetsov, V.N., Glazer, A.A., Lukshina, V.A.,and Potapov ,A.P. (1994) Crystallization of the Amorphous Pd-Cu-Si Alloy under Creep Conditions, *Phys. Met. and Metallogr.* **77**, 509-512.; and (1996) Nanocrystallization of amorphous Pd-Cu-Si alloy under creep conditions, *J. of Non-Crystalline Solids* **205-207**, 829-832

11. Herzer, G. (1990) Grain Size Dependence of Coercivity and Permeability in Nanocrystalline Ferromagnets, *IEEE Trans.Magn.* **26**, 1397-1402.

12. Inoue, A., Kim, Y.-H., and Masumoto, T. (1992) A Large Tensile Elongation Induced by Crystallization in Amorphous $Al_{88}Ni_{10}Ce$ Alloy, *Mater.Trans. JIM* **33**, 487-490.

13. Noskova, N.I., Klyachin, V.M., Ponomareva, E.G., and Boltachev, V.D. (1992) Structure and Properties of Metallic Glasses Subjected to Focused Ultrasonic Waves, *Phys. Met. and Metallogr.* **75**, 64-69.

14. Koster, U., Schunemann, U., Blank-Bewersdorff, M., Brauer, S., Sutton, M., and Stepherson, G.R. (1991) Nanocrystalline Materials by Crystallization of Metal-Metalloid Glasses, *Mater. Sci. Eng.* **A133**, 611-615

15. Noskova, N.I. (1997) The Structure and Properties of Nanocrystalline polyphases Alloys, *NanoStructured Materials* **9**,505-508.

16. Andrievskii, R.A., Vikhrev, A.N., Ivanov, V.V., Kuznetsov, R.I.,Noskova, N.I., and Sazonova, V.A. (1996) Magnetic-Pulse and High-Pressure Shear-Strain Compaction

of Nanocrystalline Titanium Nitride, *Phys. Met.and Metallogr.* **81,** *92-97*

17. Olofinjana, A.O., and Davies, H.A. (1995) Preparation and mechanical properties of Fe-Si-B-Nb-Cu Nanocrystalline alloy wire, *NanoSuctured materials,*.**6**, 465-468.

18. Valiev, R., Korznikov, A., and Mulykov, R. (1993) Structure and properties of ultrafine-grained materials produced by severe plastic deformation, *Materials Science and Engineering* **A168**, 141-148.

19. Noskova, N. (1997) The Structure and Properties of Nanocrystalline polyphase Alloys, *Proc.VII Inter. Seminary, The Structure, Defects, and Properties of Nanostructured Ultrafine-Grained and Multilayers Materials, Ekaterinburg, Russia,25-29 March 1996, UD RAS , Ekaterinburg,* 5-20.

20. Noskova, N.I.,Ponomareva,E.G. (1996) .Structure, Strength, and Plasticity of Nanophase $Fe_{73,5}Cu_1Nb_3S_{i13,5}B_9$ Alloy: I-Structure, *Phys. Met.and Metallogr.*,**82**, 542-548; and (1996) II-Strength and Plasticity, *Phys. Met.and Metallogr* **82**,630-633.

PROCESSING AND PROPERTIES OF NANOSTRUCTURED MATERIALS PREPARED BY SEVERE PLASTIC DEFORMATION

R.Z. VALIEV, I.V. ALEXANDROV, R.K. ISLAMGALIEV
Institute of Physics of Advanced Materials
Ufa State Aviation Technical University
12 K. Marksa, 450000 Ufa, Russia

ABSTRACT.
A review of recent works dealing with processing of ultrafine-grained, nanostructured materials by severe plastic deformation and investigation of their properties is given in the present paper. The paper considers details of processing methods, structural features of processed ultrafine-grained materials and discusses their unusual deformation behavior and properties.

1. Introduction

The procedure of processing nano- and submicrocrystalline metals and alloys termed severe plastic deformation (SPD) [1,2] implies large plastic deformations under high applied pressures at relatively low temperatures (usually less than 0.4 T_m). Significant refinement of structure during heavy deformation, in particular, extrusion and rolling, has been revealed rather long ago [3,4], however, these methods usually resulted in formation of cellular structure with low angle interfaces. Only recently it has been shown that severe plastic deformation can provide formation of ultrafine-grained, nanocrystalline structures having high angle grain boundaries [1,2]. Special methods of mechanical deformation were developed and used for this purpose such as torsion straining under high pressure, equal channel angular (ECA) pressing and others. The SPD procedure has a number of advantages as compared to other methods of nanostructured materials processing, namely, condensation in inert gas atmosphere [5], rapid quenching [6] or ball milling [7]. Though, SPD allows to obtain the ultrafine-grained materials with a mean grain size of about 100 nm and this value is a little more in comparison with samples processed by other methods, an important advantage of SPD is a possibility to fabricate massive specimens free of residual porosity and impurities. The given specimens can be successfully used for thorough investigations of achieved properties including mechanical testing and forming. Moreover, the SPD procedure can be applied for formation of ultrafine-grained structures in various metallic materials (metals, alloys and intermetallics) using both starting monolithic ingots and powders.

The recent studies showed that many of the properties of ultrafine-grained materials processed by SPD significantly differed from those of corresponding coarse-

G.M. Chow and N.I. Noskova (eds.), Nanostructured Materials, 121–142.

grained ones [1,2,8]. In particular, it was established that even fundamental structure insensitive parameters such as the Curie and Debye temperatures, elastic moduli, saturation magnetization and others, can be changed. It was observed that the features of these changes are similar to the ones revealed in nanocrystals processed by other methods. From this point of view, the ultrafine-grained materials processed by SPD can be classified as nanostructured. High engineering properties such as superstrength, superplasticity, elevated damping properties, attractive electric and magnetic properties being characteristic to the processed ultrafine-grained materials are of a great interest either.

All these unusual properties are attributed to a specific internal structure and this, in turn, determines the necessity in thorough structural investigations of ultrafine-grained materials. For investigation of their structure various mutually supplementing methods, first of all, transmission electron microscopy and X-ray analysis were applied. The results of these studies were used for developing a structural model of these nanostructured materials.

Currently ultrafine-grained materials fabricated by SPD are objects of comprehensive investigations aimed at developing and optimizing methods of their processing, conducting thorough structural examination and investigation of properties and mechanical behavior of these materials [2].

The present paper is a brief review in a field of development of processing methods and investigations of nanostructure formation during intense deformations, structural examination of fabricated ultrafine-grained materials, analysis of their deformation behavior and engineering properties attractive for their practical application.

2. Processing

Processing of nanostructures by SPD can be realized via special deformation methods, namely, torsion straining under high pressure, ECA pressing and multiple forging.

A method of torsion straining under high pressure (Fig. 1a) is used for fabrication of disk type samples. An ingot is held in a deformation cell (P) between a support and a plunger under a hydrostatic pressure (P) of several GPa, where it is strained in torsion to different true strains. In this procedure, the true logarithmic strain is given by

$$\varepsilon = ln(\vartheta r/l) \qquad (1)$$

where ϑ is the rotation angle in radians and r and l are the radius and thickness of the disk, respectively. According to this relationship, the logarithmic torsion strain of samples, 20 mm in diameter and 1 mm in thick, was 6, that of samples, 10 mm in diameter and 0.2 mm in thick, was 7 at the perimeter of the disk. The structural homogeneity of the resulting samples has been demonstrated by the uniform distribution of microhardness values measured across each sample.

During ECA pressing a polycrystalline ingot is machined and pressed through a special die to give a continuous repetitative passage (Fig. 1b). This procedure was

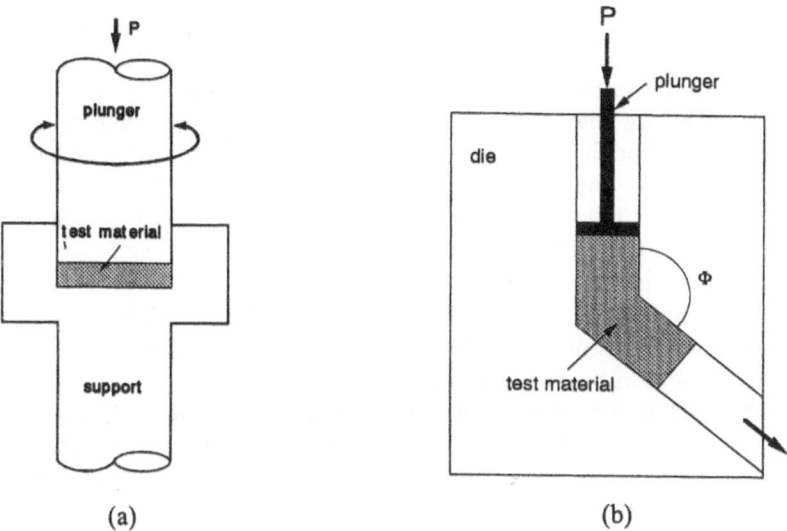

Fig. 1. Methods of severe plastic deformation:
(a) torsion straining ; (b) ECA pressing.

developed many years ago in order to introduce an intense plastic strain into materials
with no change in the cross-section area [9] and it is now established as a method of
attaining a submicrometer or nanometer grain size [1,10-12]. In these experiments
cylindrical samples or samples with the square cross section are cut from initial ingots.
The linear size of the cross section is usually of about 20 mm and the lengths of the
samples are in the range ~ 70 - 100 mm. These samples are pressed through the die
using an ECA facility in which the angle of intersection of two channels is usually 90°.
Following pressing, the samples were air cooled and pressed again through the same
die. A relationship is available to estimate the strain introduced into a sample on a
single passage through a die [2, 9, 12] and it follows from this relationship that a single
pass in the present experiments gives a strain of ~ 1. Usually, ECA pressing has been
conducted for 8-12 passes.

The experimental materials of recent investigations were pure metals, alloys
and steels as well as intermetallics where nanostructures were processed by using
different SPD methods [1, 2, 13].

Intense deformations lead to formation of nanostructures in different materials
where an attained grain size and a level of fields of internal elastic stresses depend on
an applied SPD methods, processing regimes and phase composition of a material.
Usually in pure metals the intense torsion strain results in formation of equi-axed

granular structure with a mean grain size of 50-150 nm while ECA pressing provides a grain size of 200-300 nm. At the same time a decrease in grain size can be attained by varying different parameters, applied pressure, temperature and strain rate being the most important among them. On the other hand, a stacking fault energy and a degree of alloying, especially by interstitial atoms, also significantly contribute to structure refinement. For example, it was established that a structure with a grain size of 15-20 nm can be processed in high carbon steel Fe-1.2%C [14], immisible Ag-50%Cu alloy [15] by severe torsion straining. Typical structures processed by severe torsion straining and ECA pressing in Cu, Al alloy and intermetallics Ni₃Al(Cr) doped B are shown in Fig. 2 and Fig. 3. It is seen that due to SPD sufficiently homogeneous and equi-axed structures can be formed. This was confirmed by a large number of spots arranged in circles in diffraction patterns from a foil area of about 1 μm², and data of the direct estimation of parameters of grain misorientations [16].

The recent studies showed that SPD could be used not only for structure refinement but also as a method of powder consolidation [17]. It was shown that high pressure of several GPa during torsion straining at room temperature can provide sufficiently high density almost equal to 100% and significant strength and ductility of fabricated massive samples. Massive nanocrystalline samples can be processed not only from traditional powders but also from powders subjected to ball milling.

There is an exciting question connected with the application of SPD for processing ultrafine-grained materials: What is the mechanism of ultrafine-grained structure formation during SPD? However, this question has not been studied adequately. Here we should note the recent investigation of structure evolution during severe torsion straining of Armco-iron [18]. It was established that during deformation the cellular structure having low angle boundaries subsequently transforms to a granular one having high angle boundaries. At the same time, the grain size decreases to a constant value of about 100 nm. The obtained experimental data confirm the following scheme of structure evolution during severe plastic deformation of iron (Fig.4).

The main idea of the given scheme is based on the fact that in the process of transformation of the cellular structure (Fig. 4a) to the granular one, when the dislocation density achivies some critical value, a partial annihilation of dislocations of different signs occurs at cell boundaries (Fig. 4b). As a result, excess dislocations of single signs remain (Fig. 4c). The excess dislocations play various roles: dislocations with Burgers vectors perpendicular to the boundary lead to an increase of misorientations and in the case when their density rises they affect the transformation to a granular structure; at the same time long range stress fields are connected with glide dislocations which can also lead to sliding of grains along grain boundaries.

3. Structural characterization

In recent years a number of structural studies aimed not only at measuring grain size and shape but also at analyzing defect structure, including grain boundaries, in ultrafine-grained materials fabricated by SPD were conducted. These studies testified the presence of strongly non-equilibrium (non-relaxed) grain boundaries in these materials.

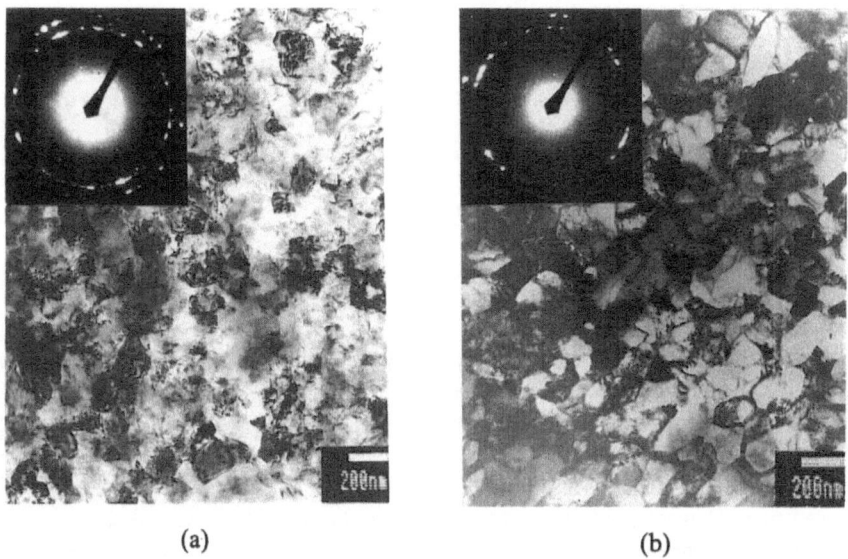

(a)   (b)

Fig. 2. Typical microstructures of ultrafine-grained materials processed by severe
plastic deformation: (a) Cu after torsion straining; (b) Cu after ECA pressing.

(a)   (b)

Fig. 3. Typical microstructures of ultrafine-grained materials processed by severe
plastic deformation : (a) Al alloy 1420 after torsion straining;
(b) $Ni_3Al(Cr)$ doped B after torsion straining.

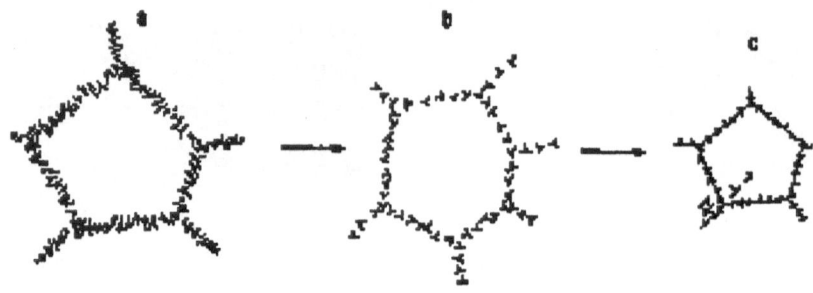

Fig. 4. Schematic model of dislocation structure evolution at different stages during severe plastic deformation in armko iron.

The first observations of non-equilibrium grain boundaries in materials imposed to SPD were made in some aluminum alloys [11]. A typical sample of these boundaries is given in Fig. 5, where we can see TEM photographs of the alloy Al-4%Cu-0.5%Zr with a grain size of about 200 nm after SPD ($e = 7$) (Fig. 5a) and after additional annealing at 160 °C for 1 h (Fig. 5b). In both cases, a granular type of structure and high-angle grain boundaries with random misorientations exist.

However, the structures of these states differ considerably. One can observe a specific "diffusive"contrast of inclined grain boundaries in the alloy after SPD (compare Figs. 5a and 5b). This contrast being a typical feature of non-equilibrium grain boundaries with a higher energy and long-range stresses [19] is observed in different ultrafine-grained materials [1,20,21]. Another evidence for the non-equilibrium state of grain boundaries is given by an occurrence of extinction contours inside of grains (Fig. 5a), which are linked with elastic stresses originated from grain boundaries. With increasing annealing temperature, one can observe a gradual decrease in the extinction contours in grain interiors and simultaneous appearance of usual grain boundaries with a typical banded contrast (Fig. 5b) associated with some grain growth as well.

HREM investigations performed for several pure metals and alloys, subjected to SPD, enabled to observe two important features of a grain boundary structure [25, 26]. First, the crystallographic width of grain boundaries, as in usual materials, is narrow and does not exceed 1 nm. It is known, that an analogous conclusion was made for nanocrystals processed by a gas condensation method [22,23]. Secondly, near a grain boundary there is an elastically distorted layer, about 10 nm in width, where a value of elastic strains is rather high and is as large as 3-5%. Such elastic distortions are obviously responsible for the appearance of "diffusive" contrast observed by TEM [24].

The lattice image of a region near a typical grain boundary in the as-strained ultrafine-grained Al-3%Mg alloy (Fig. 6) reveals that the boundary exhibits a periodic stepwise arrangement of facets parallel to (111) and each facet consists of about ten layers of (111) planes [25]. The facet density on the boundary is estimated to be ~5×10^8 m^{-1}. Comparative investigations of regions near grain boundaries and inside the grains have revealed a significant distortion or bending of lattice fringes near the grain boundary in the two-dimensional lattice image. The lattice dilatations up to a maximum

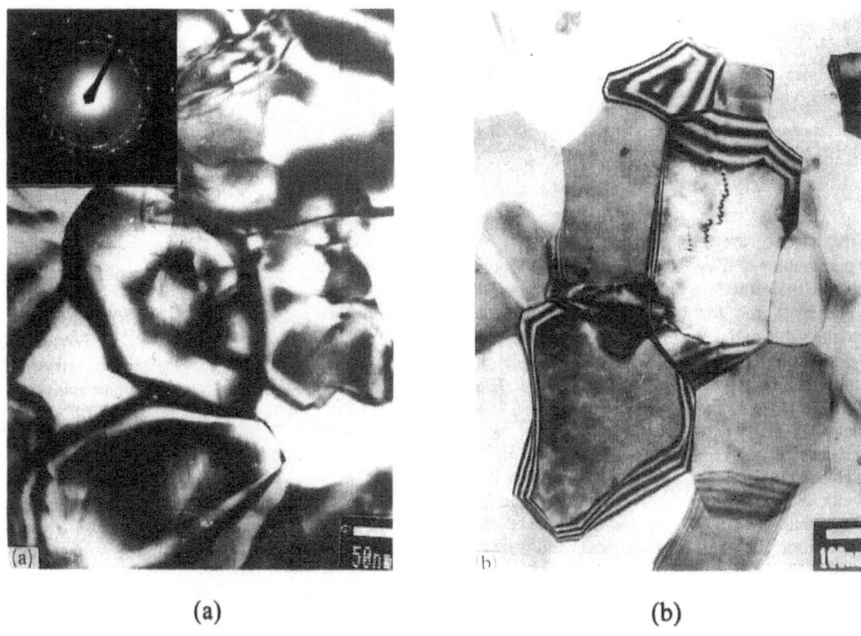

(a)   (b)

Fig. 5. TEM micrographs of the alloy Al-4%Cu-0.5%Zr : (a) after torsion straining; (b) after additional annealing at 160 °C for 1 h.

of 1.5% of interplane distances were observed within ~ 6 nm of grain boundaries in Ni and Ni_3Al [26]. Some lattice fringes are even terminated, thus suggesting the possible presence of dislocations. Though for detailed interpretation of HREM observation it requires the computer simulation, the conducted investigations have supply the direct evidence of the dislocation disturbed structure of grain boundaries in nanostructured materials and these extrinsic grain boundary dislocations create long range stress fields resulting in the observed distortions of the crystal lattice near grain boundaries.

The results of X-ray structural analysis show that X-ray patterns of the studied materials fabricated by SPD are characterized by a number of specific features as compared to corresponding coarse-grained materials. Firstly, relative maximum intensities of peaks on X-ray patterns of ultrafine-grained materials processed by SPD significantly differ from corresponding table data (Fig. 7). For example, torsion under high pressure leads to considerable relaxation of both maximum and integral intensities of all X-ray peaks as compared to peaks (111). This is specially prominent in the case of samples fabricated by consolidation of ultra disperse powders having the smallest grain size (15-20 nm) among ones considered [27]. The investigations showed that the revealed facts were connected not only with the structure refinement and the appearance of microdistortions of a crystal lattice but also with the intense processes of texture formation occurring during SPD.

The second typical feature of the X-ray pattern of ultrafine-grained materials fabricated by SPD is an increased fraction of the Lorentzian function achieving

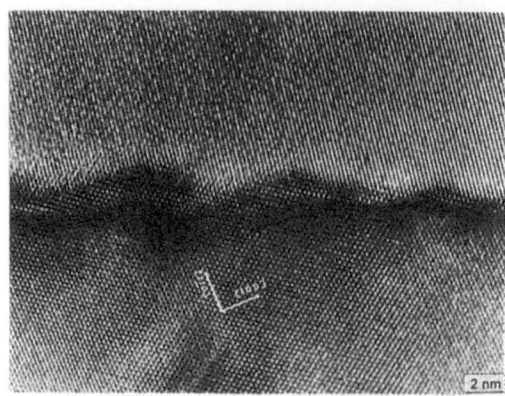

Fig. 6. TEM micrograph of a grain boundary in Al-3%Mg alloy
subjected to torsion straining.

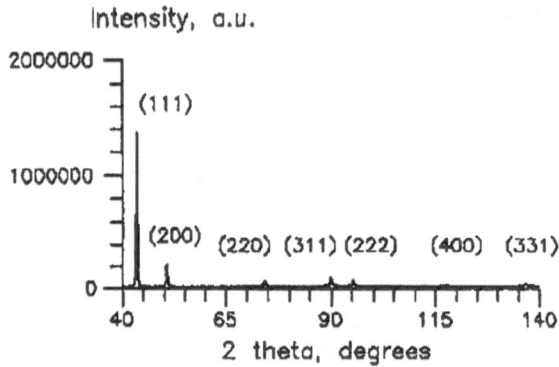

Fig. 7. X-ray diffraction pattern of Cu processed by torsion straining.

90-100% in the function describing a profile of X-ray peaks [27, 28]. At the same time
in the case of corresponding coarse-grained materials fractions of the Lorentzian and
Gaussian functions are close to each other. The performed analysis [29] showed that a
significant role of the Lorentzian function in description of X-ray peak profiles for
ultrafine-grained materials fabricated by SPD was evidently connected not only with
lognormal law of size distribution of grains or small size crystallites (areas of coherent
scattering), [30], but also with the specific character of a defect structure resulting in a
specific distribution of microdistortions in a crystal lattice. This distribution is
characterized by significant non-homogeneity. The given fact is confirmed by results of
modeling of X-ray patterns of nanocrystalline materials. The modeling was made taking
into account various grain size, grain boundary thickness, position of atoms in grains,

fields of elastic long-range stresses, extrinsic grain boundary dislocations and crystallographic texture [31].

Investigations of X-ray patterns of ultrafine-grained Cu showed that SPD led to some growth in intensity of X-rays scattering diffusive background achieving $6 \pm 3\%$ as compared to corresponding coarse-grained Cu. An increase of integral intensity of X-rays scattering diffusive background testifies that during SPD the density of crystal structure defects grows. In addition to point defects being a main reason for diffusive background growth one can mention extrinsic grain boundary dislocations. According to the HREM data [25] their density is very high and near core of these dislocations, atoms are significantly displaced from their equilibrium positions in a crystal lattice similar to point defects.

The method of X-ray structural analysis allows us to determine a mean crystallite size (coherent scattering domains). In the case of nanostructured materials their size can coincide with that of definite grains. The conducted investigations showed that a linear version of the Warren-Averbach method was more reliable as compared to other techniques of the X-ray analysis in the case of crystallite size determination for nanostructured materials processed by SPD [32]. A crystallite size of pure ultrafine-grained Cu determined by this technique was less than that determined by the TEM method. One can note the following possible reasons resulting in the mentioned difference: the presence of subgrains; specific features of the X-ray method allowing ones to determine only a size of a grain interior undistorted by fields of long range elastic stresses of extrinsic grain boundary dislocations; the difference in a character of grain size averaging (by a grain surface in a plane coincident with a sample surface in the TEM method, and by a surface or a volume fraction of grains in the direction normal to a sample surface in the X-ray structural method).

The results of X-ray structural analysis showed that a value of the Debye-Waller parameter of Cu subjected to SPD is higher than in starting Cu by 50%. The estimated values of atom displacements $<\mu^2>^{1/2}$ for nanostructured Cu was equal to 0.0126 ± 0.0003 nm, that was equal to 5.0% from the shortest distance between atoms. The Debye temperature was equal to 233 ± 6 K. For nanocrystalline Cu this value was less than a table value by 23% [28]. Such significant changes of the Debye temperature could not be explained by insignificant volume fraction of the grain boundary atoms in the investigated Cu with a mean grain size of about 100 nm. The analysis of the obtained data in terms of specific features of structure of ultrafine-grained materials allows us to assume that the influence of a large distorted regions near grain boundaries can be a main reason for the Debye temperature decrease in ultrafine-grained materials processed by SPD. At the same time the Debye temperature inside of grains can remain the same as in a coarse-grained sample. Following this assumption the estimates of the Debye temperature in a near boundary region will be $\Theta_{GB}=131\pm1$ K.

The results of conducted texture investigations testify [33] that usually these materials are textured. Thus, while analyzing their mechanical properties this fact should be taken into account.

The obtained results are important for development of a structural model of materials with ultrafine-grained structure processed by SPD [8]. The given model is based on the concept of non-equilibrium grain boundaries containing extrinsic grain boundary dislocations characterized by high density [34, 35], and triple junctions with non-compensated partial dislocations [36, 37]. Dislocations and disclinations create fields of long range elastic stresses. These fields concentrated near grain boundaries and triple junctions are responsible for elevated energy and dilatation of a crystal lattice. High internal stresses can lead to formation of elastically distorted regions of a crystal lattice near grain boundaries where strain can achieve 1-2% or even more and interatomic distances (spaces) change [8] (Fig. 8a). A decrease in a grain size to 10-20 nm results in elastic distortions of a crystal lattice of a whole grain (Fig. 8b). Such an unusual atomic structure of grain boundaries and near boundary areas is responsible for unusual physical and mechanical properties of ultrafine-grained materials.

The influence of the latter fact on properties of ultrafine-grained materials can be illustrated by the following. An increase of a lattice parameter in elastically distorted regions near grain boundaries can result in a change of an interaction exchange integral thus in turn leads to a decrease in the Curie temperature and saturation magnetization as well as to an increase in heat capacity. This conclusion agrees with the results of recent measurements of the given parameters [1, 38]. On the other hand elastic lattice strains near grain boundaries increases a limit of alloying elements dissolution that can be increased by an order of magnitude. For example, two-phase steel Fe-1.2%C transfers to oversaturated solid solution [14]. The given situation is similar to the one that takes place during formation of Kottrel atmosphere near dislocation cores. Another important fact is an increase of dynamic mobility of atoms in elastically distorted regions. The given fact is closely connected with a free volume and an increase of the latter leads to an increase of activation energy of diffusion-controlled processes such as grain boundary sliding and migration of grain boundaries. The results show [39], that in ultrafine-grained materials the activation energies of these processes can be decreased by 25% as compared to the corresponding coarse-grained materials. The given fact indicates an elevated mobility of atoms in grain boundaries and near them that provides such unusual features of deformation behavior of ultrafine-grained materials as absence of hardening and high ductility.

In the frame of the given model one can describe evolution of microstructure during annealing in ultrafine-grained materials processed by SPD. The generalization of results of various investigations [40] allowed us to make a conclusion about the processes taking place in the given materials during annealing: 1) redistribution and annihilation of dislocations existing within a crystal body in states after SPD; 2) redistribution of dislocations forming wide, fuzzy arrays that separate neighboring crystallites concurrent with formation of high angle grain boundaries having a thickness equal to an atom size; 3) structural recovery of these non-equilibrium grain boundaries with concurrent disappearance of fields of long range elastic stresses and formation of equilibrium grain boundaries leading to appearance of a classic polymorphic structure with a very small grain size without realizing of a stage of nuclei formation; 4) further a coarsening of grain size occurs in recrystallized structure.

A review of the results of the conducted structural investigations testifies that SPD leads to formation of ultrafine-grained structure in various investigated materials. The nanostructured materials processing is accompanied by formation of a specific defect structure characterized by high density of defects in grain boundaries. These defects create fields of long range elastic stresses resulting in distortions of a crystal lattice. The distortions are extremely strong near grain boundaries. In this connection one can assume that non-equilibrium grain boundaries containing extrinsic grain boundary dislocations of high density may be responsible for unusual mechanical and

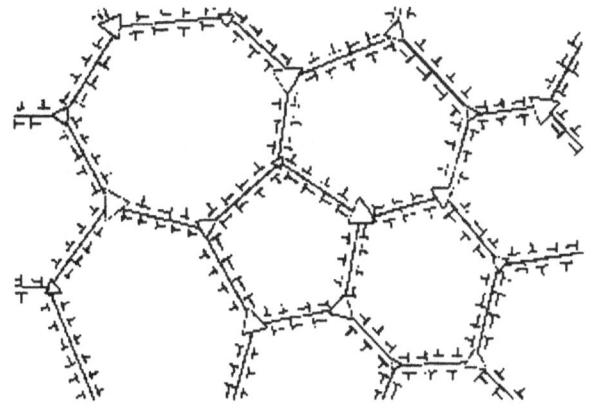

Fig. 8a. Schematic representation of ultrafine-grained structure having a mean grain size of about 100 nm. Triangles of different sizes and orientations designate disclinations of different powers and signs. Disclinations and grain boundary dislocations form elastically distorted layers (zones) near GBs.

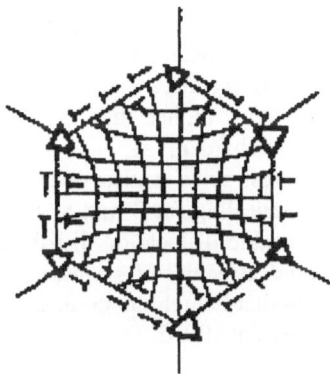

Fig. 8b. Schematic representation of nanostructured solid, i.e. a nanocrystal with dislocations disturbed grain boundaries.

physical properties of nanostructured materials processed by SPD. The developming approach may be useful for analysis of structure and properties of nanocrystalls processed by other method, as well. At the same time, this point of view requires the following studies.

4. Properties and Mechanical Behavior

Novel properties of nanostructured materials have aroused much interest among researchers in the materials science field. With respect to their mechanical properties one should note a hope to obtain extremely strength states that can be expected in the case of strong grain refinement according to the Hall-Petch relationship [5,41]. Another important aspect of these alloys is an effect of superplasticity which can be achieved at relatively low temperatures and/or high strain rates by reducing the grain size of the materials to the submicrometer or nanometer range [42, 43].

Fabrication of bulk samples by SPD provided an opportunity to start systematic investigations of mechanical properties of nanostructured metals and alloys [11, 44, 46].

4.1. TESTS AT LOW TEMPERATURES

A number of recent works were aimed at studying regularities of their deformation properties and investigating the relation between defect structure of grain boundaries and mechanical behavior in ultrafine-grained materials.

First let us consider the results of mechanical tests of ultrafine-grained Cu with a mean grain size of about 210 nm processed by ECA pressing [39, 45]. Fig. 9 present true stress-strain curves of these samples in tension and compression, respectively. For comparison the curves for annealed samples are also plotted in the figures. As compared with the well annealed coarse-grained state the ultrafine-grained samples are characterized by two significant features: firstly, by significantly high values of yield stress, several times higher than in conventional Cu samples, and, secondly, by less pronounced strain hardening.

Fig. 9b is of great interest since it also presents a 'stress-strain' curve for a sample subjected to additional annealing for 3 min at 473 K. Such annealing does not result in grain growth and noticeable dislocation distribution inside of grains but it is responsible for recovery of defect structure of grains, expressed in a sharp decrease of internal stresses [8,39]. It is seen that though the grain size in these two states is similar, their deformation behavior is essentially different. A curve for a state after a short period of annealing becomes similar to that for coarse-grained Cu and this fact is of great importance for the analysis of the Hall-Petch relationship in ultrafine-grained materials (see below).

The state of grain boundaries exerts a rather significant effect on fatigue behavior of ultrafine-grained Cu as well [46]. In Fig. 10 one can see cyclic hardening curves for four samples: two as-prepared samples, one sample subjected to annealing at 473 K for 3 min with the grain size remaining the same and finally the forth sample annealed at 773 K has a grain size of 50 μm. It is seen that, after a number of cycles, all samples reveal saturation but their saturation stresses are significantly different and depend on the preliminary heat treatment.

Significant differences in fatigue behavior of ultrafine-grained and coarse-grained Cu can be seen from Fig. 10b, showing changes in the Bauschinger energy parameter β_E with respect to the cumulative plastic strain.

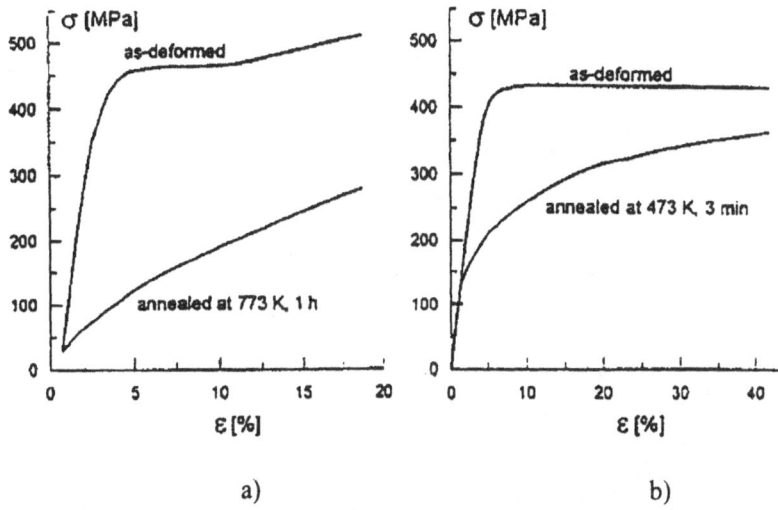

Fig. 9. True stress-strain curves for ultrafine grained Cu: (a) tensile tests; (b) compression tests at room temperature.

As seen from Fig. 10b a value of β_E in all states depends weakly on cumulative plastic strain energy and its maximum value is observed in ultrafine-grained Cu subjected to short time annealing. Thus, the considered data and other results of work [46] display unusual fatigue properties of nanostructured metals. First of all one should mention the revealed high values of saturation stress, and a reduced value of β_E in as-prepared samples as compared with the annealed ultrafine-grained state. The origin of these regularities is discussed below.

Another interesting example of deformation behavior of nanostructured materials is the investigation of mechanical properties of Ti having a grain size of about 100 nm [47]. These samples were processed by torsion severe straining and were 12 mm in diameter. That is why bending tests were used for their investigation. The obtained results allow us to determine the yield strength (σ_y), ultimate strength (σ_u) and a maximum value of flexure (Δ).

Similar to the nanostructured Cu the structure of Ti was of a granular type with high angle misorientations of grain boundaries (Fig. 11) [47]. Moreover, a crystal lattice within grains is strongly distorted. As indicated by TEM, during annealing of these samples noticeable structural changes started at a temperature of 250°C and higher. First of all this was seen in decreasing elastic distortions that was distinctly observed in the dark field image. In this case the average grain size measured by the TEM pictures becomes a bit larger though grain boundary migration has not occurred yet. Grain growth in Ti subjected to SPD starts at 350°C annealing for 2 h.

The results of microhardness measurements and tests by bending are given in Fig. 12. H_v values remain almost constant at annealing temperatures 250-350°C but at temperatures above 350°C they sharply decrease. At the same time a value of yield strength decreases significantly already after 250°C annealing . As for ductility

134

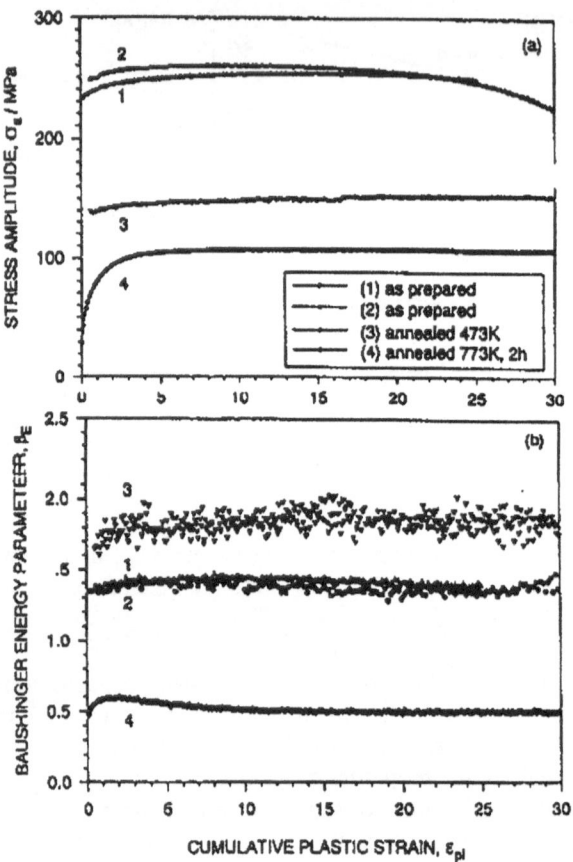

Fig. 10. Curve of cyclic hardening (a), Baushinger energy parameter versus cumulative
plastic strain (b) in UFG Cu in the as-prepared state and
after heat treatment.

measured by a value of a maximum flexure it is small in the starting state but starts to
increase after annealing and is equal to 0.35 mm at T=250°C where values of H_v and
ultimate strength are maximum [39,44].

It was assumed that a significant influence of defect structure of grain
boundaries on mechanical behavior of ultrafine-grained materials should be attributed
to the fact that non-equilibrium dislocation disturbed boundaries can accelerate
processes of structure recovery, and, consequently, can be responsible for the lack of
significant hardening. On the other hand, a high value of yield strength can be attributed
to the fact that generation of dislocations from non-equilibrium grain boundaries is
hampered in ultrafine-grained metals. From this point of view the revealed influence of
short time annealing at 473 K on a sharp reduction of the yield strength and the
appearance of strain hardening in ultrafine-grained Cu can be connected directly with
releasing the dislocation generation and decreasing the rate of recovery at grain
boundaries recovery having a more equilibrium structure.

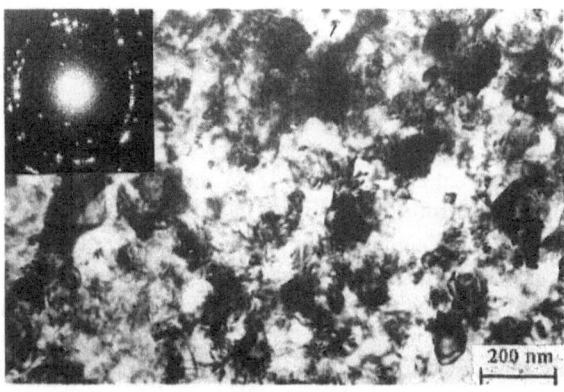

Fig. 11. Typical structure of Ti samples subjected to severe plastic deformation.

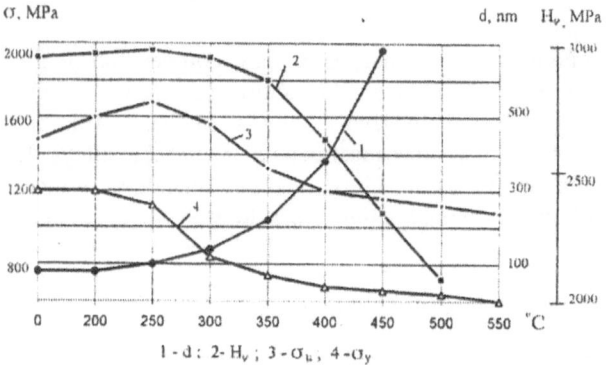

1 - d; 2- H$_v$; 3 - σ$_L$, 4 - σ$_y$

Fig.12. The dependence of grain size and mechanical properties on annealing temperature in Ti subjected to severe plastic deformation.

The developed hypothesis can be used to explain the unusual fatigue behavior of ultrafine-grained Cu. As shown, the short time annealing leading to a decrease in the yield strength provides an increase in the Bauschinger energy parameter (Fig. 10). It is known that an increase in the β$_E$ value is often connected with formation of dislocation pile-ups but their formation near equilibrium grain boundaries is preferable [19].

The considered statements are evidently true for the analysis of mechanical behavior of another material - ultrafine grained Ti [47]. In this case a strong dependence of yield strength on both the grain size and the grain boundary state is revealed (Fig. 12). After annealing to 300°C the yield strength value decreases considerably though the grain growth connected with the migration of grain boundaries is not observed, and the measured grain size slightly increases due to relaxation of internal elastic stresess [47]. This can be explained only by recovery of a grain boundary state, as confirmed by

TEM data. At the same time values of ultimate strength and microhardness preserve their level. On the basis of the obtained results one can make two important conclusions. Firstly, it becomes evident that the known empirical relationship $H_v/\sigma_y=3$ is not fulfilled in nanostructured materials if as-prepared and annealing states are studied. This can be explained in the following way. As known, the yield strength corresponds to the onset of plastic deformation but, while measuring the microhardness, an average value of strain is equal to 9-10% [48]. Consequently, one can expect that in the case of strong strain hardening a significant gap in the relationship between H_v and σ_y should be observed in the annealed samples as compared with the as-prepared nanostructured ones. As experiment has shown, strain hardening is not significant in the latter case. These results make us more careful in using the data on microhardness for the investigation of mechanical properties of ultrafine-grained materials. Secondly, the fact that in the case of a similar grain size the yield strength in ultrafine-grained materials can be essentially changed depending on a defect structure of grain boundaries indicates that investigations of the Hall-Petch relationship will be useless if one does not take into account the role of the defect structure of grain boundaries. That is why such strong controversy existing in the data available on the given question can be attributed to the lack of sufficient attention to the structural state of grain boundaries.

At the same time the fact that alongside with the ultrafine-grain size, the defect structure of grain boundaries and changes in phase composition significantly contribute to strength of materials subjected to SPD allows us to consider main principles for achieving extremely high strength states in metallic materials [49]. First of all such high strength states can be realized in nanostructured metastable alloys. It was shown that application of SPD in immisible alloys and intermetallics results in formation of metastable nanostructures whose aging leads to very high strength of these alloys (Table).

Table. Microhardness of aluminium alloys (MPa)

Alloy	1420	Al +11wt.%Fe
after quenching	540	–
after deformation	1750	1650
after ageing	2300	3020

4.2. TESTS AT ELEVATED TEMPERATURES: SUPERPLASTICITY

Superplasticity is defined as "the ability of a polycrystalline material to exhibit, in a generally isotropic manner, very high tensile elongations prior to failure" [42]. Superplastic deformation in metals requires a very small grain size (typically less than 10 μm) and a testing temperature above about 0.5 T_m, where T_m is the absolute melting temperature of the material. For metallic alloys, superplasticity occurs over an intermediate range of strain rates which tends to shift to larger strain rates with decreasing grain size and/or increasing temperature [42, 50, 51].

The high ductility associated with superplastic flow makes this process attractive for use in industrial forming operations. For example, superplastic forming has been used successfully to sporting applications [51]. However, these forming operations are usually limited because of the low speed and high temperature required for fabrication. It is very desirable from the scientific and technological viewpoint to explore the possibility of extending the superplastic forming capability to lower temperatures and/or higher strain rates. For industrial applications, lower temperatures are advantageous, because they lead to a minimization of tool wear. Faster strain rates are beneficial, since they decrease the total forming time and have the potential for extending the production run to larger numbers of units.

The recent investigations testify that formation of ultrafine-grained structure due to SPD can also significantly increase superplastic properties of metallic materials and an effect of superplasticity can be observed at rather lower temperatures and/or high strain rates. A sharp decrease in temperature of superplastic forming was first to be revealed in Al-4%Cu-0.5%Zr alloy having a grain size of 0.3 μm [11]. Later this effect was observed in a number of alloys with nano- and submicrocrystalline structure [43,52,53].

High strain rate superplasticity (HSR SP) has been defined formally as the ability to achieve high superplastic tensile elongations at strain rates faster than 10^{-2} s^{-1} [51]. This topic has received much attention in recent years, primarily because it is recognized that HSR SP has potential value in many practical applications. In superplastic forming, for example, an increase in the operating strain rate may lead to a significant reduction in the total time for processing and thereby provide the capability for the mass production of a large number of components.

As documented in detail elsewhere, observations of HSR SP to date have been confined exclusively to a limited range of Al matrix composites and mechanically alloyed materials and to some metallic alloys processed using powder metallurgy techniques [51, 54]. In practice, however, the ability to achieve HSR SP in more conventional commercial alloys is of considerable practical importance. The first demonstration of the potential for realizing HSR SP was shown recently in the work [55] for two Al alloys.

These experiments were conducted using two different alloys: 1) a Russian alloy designated 1420 with a chemical composition of Al-5.5%Mg-2.2%Li-0.12%Zr; (2) a British Al-2004 alloy known as Supral 100 and with a chemical composition of Al-6%Cu-0.5%Zr. Both alloys were subjected to ECA pressing to a total high strain of 12 at a temperature of 673 K followed by final passes at 473 K. It was revealed that these processed specimens exhibit exceptional ductility. For example, while the unpressed 1420 alloy exhibits only modest elongations even at relatively low strain rates, the ECA-pressed 1420 alloy gives elongations at 623 K (~0.7 T_m) as high as 1180% without failure at 1×10^{-2} s^{-1} and 910% at 1×10^{-1} s^{-1} (Fig. 13). These specimens represent a very clear demonstration of HSR SP, especially when it is noted that the highest elongation of >100% is achieved at 1×10^{-2} s^{-1} without the development of any visible necking within the gauge length.

The obtained results also provide some insight into the possible mechanism associated with HSR SP. In metal matrix composites exhibiting HSR SP the experiments are often performed at temperatures close to, or even slightly higher, than the solidus temperature for the metal matrix. This has led to the development of a reological model in which HSR SP takes place in a semi-solid material behaving in a

138

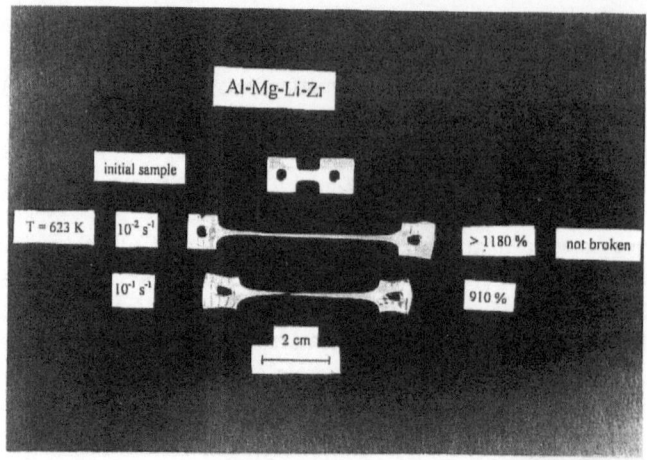

Fig. 13. The appearence of specimens of Al-5.5%Mg-2.2%Li-0.12%Zr (1420) alloy after superplastic tensile tests.

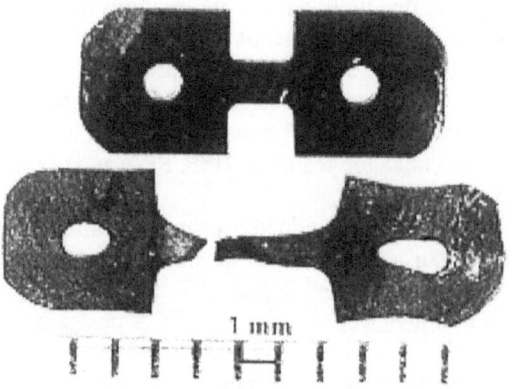

Fig. 14. View of nanocrystalline Ni$_3$Al(Cr) doped B: (a) before tensile test; (b) after tensile test at 650°C with a strain rate of 10^{-3} c^{-1} .

manner similar to a non-Newtonian fluid [56]. However, theories of this type cannot explain the present results where HSR SP was achieved in two conventional commercial alloys at relatively low testing temperature where there is no local melting at the grain boundaries. Thus, the results demonstrate that, contrary to some implications from experiments on metal matrix composites, partial melting is not a requirement for HSR SP.

Another exciting example of superplasticity is a ductile behavior of nanostructured intermetallics $Ni_3Al(Cr)$ doped B [49]. After severe torsion plastic straining a nanocrystalline structure with a mean grain size of 70 nm was processed in this alloy (Fig. 3b). Similar to the Ni_3Al intermetallics subjected to an analogous treatment [58], a complete atomic disordering was revealed. It was also shown that grain growth occurs at heating above 750°C. At the same time tensile tests showed that at room temperature this material had a low ductility less than 1-2%, but its ductile properties sharply increased at heating above 500°C and at the temperature 650°C the nanocrystalline intermetallic displayed a superplastic behavior, its elongation to failure being about 300% at a strain rate of 10^{-3} s^{-1} (Fig. 14).

This fact is rather extraordinary since the given alloy with a micrometer grain size usually displays superplastic properties at temperatures much higher than 1150°C .

The obtained results are very interesting for superplastic forming of intermetallics since these materials are usually very brittle.

5. Conclusions

The method of SPD which development was started in our laboratory in 1991-1992 has successfully been further developed and deepened in recent years. First of all this concerns fabrication of massive samples out of pure metals, various alloys and intermetallics with homogeneous ultrafine-grained structure, determination of main structural specific features and parameters of nanostructured materials fabricated by SPD, development of their structure model, investigation of relation between structure and attractive properties of these materials. A number of interesting results on this subject were published in the recent issue on ultrafine-grained materials fabricated by severe plastic deformation [2]. Moreover, the present review pays special attention on mechanical behavior and engineering properties of the processed materials. In addition to such attractive properties of ultrafine-grained metals and alloys as wear resistance, high damping [1] and unusual magnetic properties, high strength states and superplasticity observed at considerably low temperatures and/or high strain rates were revealed. The obtained rather interesting engineering properties allow one to expect their successful practical application in future.

6. Acknowledgments

This work was supported in part by the U.S. Civilian Research and Development Foundation under GRANT No. Re2-164, in part by the U.S. ARO under Grant No.68171-96-C-9006 and by the Russian Foundation for Basic Researches under No.96-02-16720. The authors thank the colleagues and friends, mentioned in the references for their useful discussions and kind cooperation.

140

7. References

1. Valiev, R.Z., Korznikov, A.V., Mulyukov, R.R. (1993) Structure and properties of ultrafine-grained materials, *Mat. Sci. Enginer.* **A168**, 141-148.
2. Ultrafine-grained materials prepared by severe plastic deformation, Valiev, R.Z. edit., 1996 *Annales de Chimie. Science des Materiaux* **21**, 369-520.
3. Langford, G., Cohen, M. (1969) Microstructure of Armco-iron subjected to severe plastic drawing, *Trans. ASM* **82**, 623-632.
4. Pavlov, V.A. (1989) Amorphisation during intense rolling. *Phys. Metal. Metallurg.* **67**, 924-932.
5. Gleiter, H. (1989) Nanocrystalline materials, *Progress Mater. Sci* **33**, 223-315.
6. Kim, Y.W., Griffith, W.M. eds. (1988) *Dispersion strengthened aluminum alloys*, TMS, Warrendale PA.
7. Koch, C.C., Cho Y.S. (1992) Nanocrystals by high energy ball milling, *Nanostructured Materials* **1**, 207-212.
8. Valiev, R.Z. (1995) Approach to nanostructured solids through the studies of submicron grained polycrystals, *Nanostructured Materials* **6**, 73-82.
9. Segal, V.M., Reznikov, V.I., Drobyshevskiy, A.E., Kopylov, V.I. (1981) Equal angular extrusion, *Russian Metally* **1**, 99-105.
10. Valiev, R.Z., Tsenev, N.K. (1991) Structure and superplasticity of Al-based submicton grained alloys. In Hot deformation of aluminum alloys. TMS, Warrendale PA, 319-329.
11. Valiev, R.Z., Krasilnikov, N.A., Tsenev, N.K. (1991) Plastic deformation of alloys with submicrograin structure, *Mater. Sci. Enginer.* **A137**, 35-40.
12. Iwahashi, Y., Wang, J., Horita, Z., Nemoto, M., Langdon T.G. (1996) Principles of equal-channel pressing, *Scripta Mater.* **35**, 143-148.
13. Salischev, G.A., Imaev, R.M., Imaev, V.M., Gabdulin, N.K. (1993) Formation of submicrocrystalline structures and superplasticity, *Mater. Sci. Forum* **113-115**, 613-619.
14. Korznikov, A.V., Ivanisenko, Yu.V., Laptionok, D.V., Safarov, I.M., Pilyugin, V.P., Valiev, R.Z. (1994) Influence of severe plastic deformation on structure and phase composition of carbon steel, *Nanostructured Materials* **4**, 159-167.
15. Shen, H., Li, Z., Guenther, B., Korznikov, A.V., Valiev, R.Z. (1995) Influence of powder consolidation methods on the structural and thermal properties of a nanophase Cu-50wt%Ag alloy, *Nanostructured Materials* **6**, 385-388.
16. Mishin, O.V., Gertsman, V. Yu., Valiev, R.Z., Gottstein, G. (1996) Grain boundary distribution and texture in ultrafine-grained copper produced by severe plastic deformation, *Scripta Mater.* **35**, 873-878.
17. Valiev, R.Z. (1996) Processing of nanocrystalline materials by severe plastic deformation consolidation. In: *Synthesis and processing of nanocrystalline powder (ed. by David L. Bowrell)*, The Minerals, Metals and Materials Society, 153-161.
18. Valiev, R.Z., Ivanisenko, Yu.V., Rauch, E.F., Baudelet, B. (1996) Structure and deformation behavior of armco iron subjected to severe plastic deformation. *Acta Metal.* **44**, 12, 4705-4712.
19. Valiev, R.Z., Gertsman, V.Yu., Kaibyshev, O.A. (1986) Structure of grain boundaries subjected to internal influences. *Phys. Stat. Solidi* **A97**, 11-32.

20. Musalimov, R.Sh., Valiev, R.Z. (1992) Dilatometric studies of aluminium alloy with submicro-grained structure. *Scr. Metall. Mater.* **9**, 95-100.
21. Valiev, R.Z. (1993) Grain boundary structure and mechanical properties of nanocrystalline materials, *Proc. NATO ASI* **233** (eds. Hadjipanayis, G.C., Siegel, R.W.), Kluwer Acad. Publ., Dordrecht-Boston-London, 275-282.
22. Thomas, G.J., Siegel, R.W., Eastman, J.A. (1990) HREM observations of nanocrystalline palladium, *Scr. Metall. Mater.* **24**, 202-207.
23. Wunderlich, W., Ishida, Y., Maurer, R. (1990) High resolution electron microscopy of nanocrystalline materials, *Scr. Metall. Mater.* **24**, 403-408.
24. Islamgaliev, R.K., Valiev, R.Z. (1995) Electron microscopy investigations of grain boundaries in ultrafine-grained germanium, *Solid State Physics* **37**, 3597-3606.
25. Horita, Z., Smith, D.J., Furukawa, M., Nemoto, M., Valiev, R.Z, Langdon, T.G. (1996) Effect of annealing on grain boundary structure in submicrometer-grained Al-3%Mg alloy observed by high-resolution electron microscopy, *Annales de Chimie. Science des Materiaux* **21**, 417-426.
26. Valiev, R.Z., Musalimov, R.Sh. (1994) High resolution electron microscopy of nanocrystalline materials, *Phys. Metals Metallogr.* **78**, 114-121.
27. Alexandrov, I.V., Zhang, K., Lu, K., Valiev, R.Z. (1997) Comparative X-ray analysis of nanocrystalline materials, processed by severe plastic deformation, *Nanostructured materials* **9**, 347-354.
28. Zhang, K., Alexandrov, I.V., Valiev, R.Z., Lu, K. (1996) The structural characterization of nanocrystalline Cu by means of the X-ray diffraction, *J. Appl. Phys.* **80**, 10, 5617-5624.
29. Valiev, R.Z., Alexandrov, I.V., Chiou, W.A., Mishra, R.S., Mukherjee, A.K. (1997) Comparative structural studies of nanocrystalline materials, processed by different techniques, *Mater. Sci. Forum* **235-238**, 497-506.
30. Islamgaliev, R.K., Pekala, K., Pekala, M., Valiev, R.Z. (1997) The determination of the grain boundary width of ultrafine-grained copper and nickel from electrical resistivity measuremenrs, *Phys. Stat. Solid.* (a), **162**, 559-566.
31. Alexandrov, I.V., Valiev, R.Z. (1996) Computer simulation of X-ray diffraction patterns of nanocrystalline materials, *Phil. Mag. B* **73**, 861-872.
32. Alexandrov, I.V., Zhang, K., Lu, K. (1996) X-ray studies of crystallite size and structure defects in ultrafine-grained copper, *Annales de Chimie. Science des Materiaux* **21**, 407-416.
33. Alexandrov, I.V., Wang, Y.D., Zhang K., Lu, K., Valiev, R.Z. (1996) X-ray analysis of textured nanocrystalline materials, *Proc. of the Eleventh Intern. Confer. On Textures in Mater. (ICOTOM-11).* Xi'an, China.
34. Nazarov, A.A., Romanov, A.E., Valiev, R.Z. (1993) On the structure, stress fields and energy of non-equilibrium grain boundaries, *Acta metall. Mater.* **41**, 1033-1040.
35. Nazarov, A.A., Romanov, A.E., Valiev, R.Z. (1994) On the nature of high internal stresses in ultrafine-grained materials, *Nanostructured Materials* **4**, 93-101.
36. Gertsman, V.Yu., Birringer, R., Valiev, R.Z, Gleiter, H. (1994) On the structure and strength of ultrafine-grained copper produced by severe plastic deformation, *Scripta metall. Mater.* **30**, 229-234.
37. Rybin, V.V., Zisman, A.A., Zolotarevsky, N.Yu. (1993) Triple junctions disclinations in plastically deformed metals, *Acta metall. Mater.* **41**, 2211-2219.

142

38. Valiev, R.Z., Korznikova, G.F., Mulyukov, Kh.Ya., Mishra, R.S., Mukherjee, A.K.. (1997) Saturation magnetization and Curie temperature of nanocrystalline nickel, *Phil. Mag. B,* **75**, 803-811.

39. Valiev, R.Z., Kozlov, E.V., Ivanov, Yu.F., Lian, J., Nazarov, A.A., Baudelet, B. (1994) Deformation behavior of ultrafine-grained copper, *Acta Metall. Mater.* **42**, 2467-2473.

40. Korznikov, A.V., Dimitrov, O., Korznikova, G. (1996) Kinetics of relaxation of disordered grain boundary dislocation arrays in ultrafine grained materials, *Annales de Chimie. Science des Materiaux* **21**, 443-460.

41. Weertman, J.R. (1993) Mechanical properties of nanocrystalline metals, *Mater. Sci. Eng.* **A116**, 16-26.

42. Chokshi, A.H., Mukherjee, A.K., Langdon, T.G. (1993) Superplasticity in advanced materials, *Mater. Sci. Eng.* **R10**, 237-274.

43. Valiev, R.Z. (1997) Superplastic behaviour of in nanocrystalline metallic materials, *Mater. Sci. Forum* 243-245, 207-216.

44. Valiev, R.Z. (1997) Structure and mechanical properties of ultrafine-grained metals, *Mater. Sci. Eng.,* in press.

45. Gertsman, V.Yu., Valiev, R.Z., Akhmadeev, N.A., Mishin, O.V. (1996) Mechanical properties of ultrafine-grained metals, *Mat. Sci. Forum* **233**, 80-90.

46. Vinogradov, A., Kanoko, Y., Kitagava, K., Hashimoto, S., Stolyarov, V.V., Valiev, R.Z. (1997) Fatigue behavior of ultrafine-grained copper, *Scripta Met.,* in press.

47. Popov, A.A., Pyshmintsev, I.Yu., Demakov, S.L., Larionov, A.G., Lowe, T., Sergeeva, A.V., Valiev, R.Z. (1997) Structure and strength of nanocrystalline titanium, *Scr. Mater.,* to be published.

48. Milman, V.Yu. (1994) Analysis of microhardness behavior, *Acta Metall. Mater.* **42**, 1349-1356.

49. Valiev, R.Z., Islamgaliev, R.K., Stolyarov, V.V., Mishra, R.S., Mukherjee, A.K. Processing and mechanical properties of nanocristalline alloys prepared by severe plastic deformation, *Proc. ISMANAM-11,* to be published.

50. Kaibyshev, O.A. (1992) *Superplasticity in metals, alloys and intermetallics.* Verlag.

51. Nieh, T.G., Wadsworth, J., Sherby, O.D. (1997) *Superplasticity in metals and ceramics.* Cambridge, Univer. Press.

52. Salischev, G.A., Valiakhmetov, O.R., Valitov, V.A., Mukhtarov, S.K. (1994) Submicrocrystalline and nanocrystalline structure formation in materials and search for outstanding superplastic properties, *Mater. Sci. Forum* **170-172**, 121-130.

53. Mabuchi, M., Iwasaki, H., Yanase, K. and Higashi, K. (1997) Low temperature superplasticity in Mg-alloy. *Scripta Mater.* 36, 681-685.

54. Mishra, R.S., Bieler, T.R., Mukherjee A.K. (1995) Superplasticity in Al-based alloys and composites, *Acta Metall. Mater.* 43, 877-891.

55. Valiev, R.Z., Salimonenko, A.D., Tsenev, N.K., Berbon, P.B., Langdon, T.G. Observations of high strain rate superplasticity in commercial aluminum alloys with ultrafine-grain sizes, *Scr. Mater.,* to be published.

56. Nieh, T.G., Wadsworth, J., and Imai., T. (1992) Superplastic Behaviour and Newtonian Flow, *Scripta Metall.* 26, 703-708

57. Languillanme, J., Chmelik, F., Kapelski, G., Bordeaux, F., Nazarov, A.A., Canova, G., Esling, C., Valiev, R.Z., Baudelet, B. (1993) Microstructures and Hardness of Ultrafine-grained Ni_3Al, *Acta Metall. Mater.* **41**, 2953-2962.

CHANGES IN STRUCTURE AND PROPERTIES ASSOCIATED WITH THE TRANSITION FROM THE AMORPHOUS TO THE NANOCRYSTALLINE STATE

A.L. GREER
University of Cambridge
Department of Materials Science & Metallurgy
Pembroke Street, Cambridge CB2 3QZ, U.K.

Abstract

The devitrification of amorphous alloys can produce nanometre-scale microstructures with excellent properties, particularly magnetic and mechanical. The origins of the microstructures and the corresponding properties are reviewed, and the advantages of the devitrification route assessed.

1. Introduction

This paper is concerned with obtaining nanometre-scale microstructures in alloys by crystallizing (devitrifying) an initial amorphous phase. It has been known for over thirty years that many alloy systems can be obtained in amorphous form by a wide variety of processes. The present work will focus on amorphous phases made by rapid liquid quenching or by thin-film deposition (mainly sputtering). The amorphous phase is always metastable, and is not normally obtained in alloys by conventional processing. Having had some difficulty in making an amorphous alloy, why then destroy the structure by partially or fully crystallizing it? What advantages are obtained by using the amorphous alloy precursor? This paper is largely about answering these questions. Crystallization of an amorphous phase occurs under conditions of high driving force and low atomic mobility; as noted in §2, these conditions favour nucleation rather than growth. Combined with the natural uniformity of an amorphous phase, the result is that devitrification has the potential to yield exceptionally fine and uniform microstructures. From the early days of studies on metallic glasses, it was recognised that such structures might be useful, but interest in them has increased enormously in recent years as it has been realised that the microstructures are often on a nanometre scale. Nanoscale microstructures are not only of fundamental interest, but have been shown to give attractive properties (magnetic and mechanical). The microstructures are of various kinds — primary phases (solid solution or compounds) in an amorphous matrix, or fully crystalline. The origins of these structures and properties will be explored.

143

G.M. Chow and N.I. Noskova (eds.), Nanostructured Materials, 143–162.

The principles of generating a fine structure by devitrification are well developed in conventional glass ceramics [1]. In many ways, the development of useful materials by alloy devitrification parallels existing practice in glass ceramics.

Nanoscale microstructures can also be formed directly by processing similar to that used to obtain an amorphous phase. For example, if direct liquid quenching is not quite rapid enough to give a fully amorphous product for the given composition, a nanocrystalline product may be obtained. The crystals form in the melt around the temperature of maximum crystallization rate, at a high undercooling similar to that encountered in devitrification. Although the microstructures formed by direct quenching and by devitrification may be similar, the latter process is likely to be more controllable and is emphasised in this review. In direct quenching the microstructure is very dependent on cooling rate, and this can vary from place to place, for example through the thickness of a melt-spun ribbon.

The metallic systems which can be made amorphous and are stable at room temperature are all alloys, and sufficiently concentrated that their equilibrium state is not a single phase. Thus devitrification leads to two-phase or multiphase materials, which with fine enough microstructures have been termed *nanophase composites* (although the phase arrangement is generated naturally in the alloys). For single-phase materials, devitrification is not a rival to the consolidation of condensed clusters, which was the technique which first brought nanocrystalline materials to prominence [2].

2. Advantages of Devitrification

It is first of interest to compare devitrification to conventional solidification as a way of generating fine grain structures in alloys. In casting, crystals grow inwards from the mould walls, the temperature gradients inhibiting nucleation ahead of the main growth front. Under these conditions, large columnar grains are expected. Finer, equiaxed and more uniform grain structures can be obtained by the addition of inoculants (grain refiners), as practised most notably in the DC casting of aluminium. A grain refiner is, however, regarded as effective if the average grain diameter falls to around 200 μm, a limit which is not improved by adding more refiner particles [3]. Nucleation and growth of crystals on added refiner particles in the undercooled liquid releases latent heat and this heating (recalescence) reduces the undercooling and stifles further nucleation events. In devitrification, in contrast, the crystal growth is so slow that temperature rises are negligible. Indeed, devitrification can be isothermal, and in this way all potential nucleation sites can be active. However, it is possible at higher devitrification temperatures for the growth to be sufficiently rapid to give significant temperature increases and for these to accelerate the growth; in this way the crystallization is autocatalytic or *explosive*. In such a case, non-uniform and not-so-fine microstructures may be obtained.

Figure 1 illustrates why high undercoolings, in the glass-formation range, are of interest for generating fine microstructures. The homogeneous nucleation frequency I and crystal growth rate U are shown as a function of temperature for polymorphic

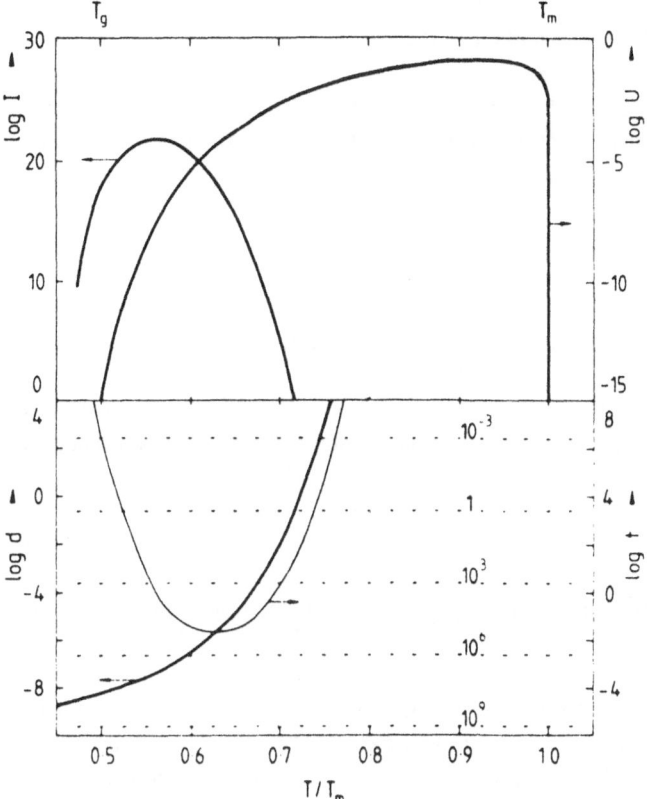

Figure 1. Calculated solidification kinetics (from the liquid or glassy state) for a typical glass-forming alloy, showing $\log_{10}$ of I (homogeneous nucleation rate, m^{-3} s^{-1}), U, crystal growth rate, m s^{-1}, d (grain size, $(U/I)^{1/4}$, m), t solidification time, $d/2U$, s). The dashed horizontal lines are marked with cooling rates in K s^{-1}. The solidification time t falls below these lines, recalescence will preclude grain refinement and the calculated value of d will not apply. (After [4])

crystallization of a hypothetical glass-forming alloy with a glass transition temperature T_g which is 50% of the absolute melting temperature. The maximum in the nucleation frequency is at high undercooling because of the interfacial energy barrier; except at high undercooling, heterogeneous nucleation would dominate. If the solidification were to proceed isothermally, the resulting grain size d would be related to I and U by [4]

$$d = (U/I)^{1/4} \tag{1}$$

As shown in Fig. 1, the expected d does decrease into the nanometre range in the region of T_g. Also shown are the temperature ranges in which recalescence would be significant for various imposed heat extraction rates. It is clear that recalescence will prevent a very fine structure being formed except for temperatures at or below T_g; this is the reason for the importance of the devitrification technique.

The major disadvantage of the devitrification route is the need for the amorphous precursor, normally obtainable only in thin sections (for example 10 to 100 μm thick melt-spun ribbon, or atomised powder) because of the need for rapid heat extraction. For some applications the devitrified materials can be applied (in ribbon or thin-film form) without any consolidation. Even if consolidation is required, the surface area of the material before consolidation is much less than for other routes (notably for the cluster-condensation route); consequently the requirements for negligible contamination are less severe, and the process can be expected to be lower cost. Consolidation may lead to some equilibration of the microstructure, but useful materials can still result, as described in §5.2.2

Mechanical alloying is well established as a method for mixing powders to obtain fine-scale composites, for example oxide-dispersion-strengthened superalloys [5]. However, the dispersoid is normally on the scale of micrometres rather than nanometres. In general with composites, there can be a problem with the chemical compatibility of the phases combined. In preparing nanophase composites by devitrification, in common with many metallurgical examples in which the microstructure is generated by phase transformations, the phases are automatically compatible.

Fine, even nanoscale, microstructures are not unknown in conventional metallurgy. Examples can be found in precipitation and in eutectoid decomposition. These are solid-state transformations with a possibility of a high density of heterogeneous nucleation sites (for example, crystal defects) together with low atomic mobility. The devitrification process is also effectively in the solid state. It provides access to nanoscale microstructures in systems for which suitable conventional solid-state transformations are not available.

3. Origins of Nanoscale Microstructure

3.1. HOMOGENEOUS NUCLEATION

For nanoscale microstructures a very high density of nucleation events is required, and given the high driving force for crystallization, homogeneous nucleation (otherwise very rare for metallic systems [6]) is a possibility. Typical population densities of heterogeneous nucleant particles in metallic melts are 10^{11} to 10^{12} m^{-3}. Even in amorphous alloys not showing nanometre-scale crystallites, the population density of crystallites greatly exceeds this range. For example, in $Fe_{80}B_{20}$ (all compositions in at.%) [7], the crystallite population density is ~10^{18} m^{-3}; the crystals appear to grow on homogeneous nuclei quenched-in during glass formation. In the nanophase composites of present interest, the population density can be as high as ~10^{25} m^{-3}, a number which seems most likely to arise by homogeneous nucleation during the devitrification annealing.

3.2 SPINODAL DECOMPOSITION

While often devitrified microstructures are very fine, and can be consistent with homogeneous nucleation during the quench into the amorphous state or during the devitrification anneal, they are not necessarily so. For example, a good glass-former such as $Pd_{40}Ni_{40}P_{20}$ (at.%) shows crystals sparsely distributed which can be sufficiently large to be visible with the naked eye [8]. In such a case, nucleation appears to be heterogeneous, and the homogeneous nucleation frequency is low. $Pd_{40}Ni_{40}P_{20}$ is a bulk glass-forming alloy, with a critical cooling rate for amorphous phase formation of less than 10 K s^{-1}. More recently developed Zr-based alloys are also bulk glass-formers, but show devitrification on a very fine scale. The new bulk metallic glasses — a typical composition is $Zr_{41.2}Ti_{13.8}Cu_{12.5}Ni_{10.0}Be_{22.5}$ — have been reviewed [9, 10]. It is suggested that the unexpected ultra-fine scale crystallization in such materials is associated with spinodal decomposition in the amorphous phase [9, 11, 12], analogous to the situation in some glass-ceramics [1]. In the bulk amorphous alloys the scale of the phase separation is of the order of 20 nm. It is becoming clear that impurities such as oxygen may play a significant role in the devitrification behaviour [13].

3.3. TYPES OF DEVITRIFICATION

Thermodynamically, three types of initial crystallization can be distinguished. In *polymorphic crystallization*, the amorphous alloy transforms to a single crystalline phase of the same composition. The crystalline phase is not normally a solid solution of the main element because such a phase, probably partitioning solute during its growth, would be an effective competitor to glass formation. The phase formed in polymorphic crystallization is normally a compound of some kind. In *primary crystallization*, a single phase is formed different in composition from the amorphous phase. There is partitioning of solute, and complete transformation to the primary phase may not be possible; the remaining amorphous matrix will then undergo a second, distinct crystallization. Finally, in *eutectic crystallization*, two crystalline phases grow co-operatively. There is no composition difference between the amorphous phase and the eutectic colony. Further details of the crystallization behaviour of amorphous alloys can be found in [14, 15].

Nanoscale microstructures are unlikely to arise, or be retained, in polymorphic crystallization. On the other hand, the solute partitioning (hindering growth) and the presence of more than one phase (stabilising a fine grain size), which are characteristic of primary and of eutectic crystallization, are conducive to the production and retention of fine-scale microstructures.

148

4. Magnetic Properties

4.1. HARD MAGNETIC MATERIALS

The first, and most important, industrial application of nanophase composites made by devitrification is in the production of permanent magnets. In general it is expected that finer microstructures would give greater pinning of domain boundaries and thereby higher coercivity. However, exceptional properties are obtained only with particular phases. The modern nanophase permanent magnets are based on the high-coercivity phase $Nd_2Fe_{14}B$. The materials can be made by sintering an alloy powder or by a melt quenching route. The quenching can be to a glass which is then annealed to devitrify, or directly to a nanophase structure, as illustrated schematically in Fig. 2 [16]. The melt-quenching method is now in use in full-scale production [17, 18]. The processing conditions must be carefully controlled to obtain the desired hard-magnetic phase, and to avoid completely the alternative soft magnetic phase. Grain sizes in optimised material appear to be in the range 14 to 50 nm.

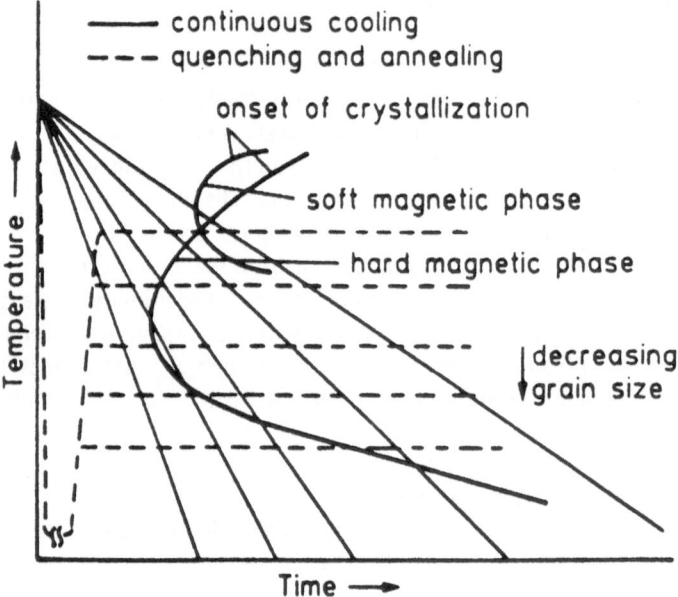

Figure 2. A schematic time-temperature-transformation diagram indicating conditions for producing Nd-Fe-B permanent magnets. (After [16])

4.2 SOFT MAGNETIC MATERIALS

4.2.1. *Soft-Magnetic Properties in Ultra-Fine Microstructures*

It is much more of a surprise that fine-grained structures could give useful soft-magnetic properties. As the grain size gets finer, the coercivity is expected to increase, and this behaviour is indeed observed down to about 0.1 μm (Fig. 3) [19]. However, at smaller grain sizes, the local magnetocrystalline anisotropies are averaged out, and soft-magnetic properties result; indeed, as shown in Fig. 3, the coercivity can be as low as in amorphous alloys. The origins of the soft-magnetic properties have been considered more fully [20].

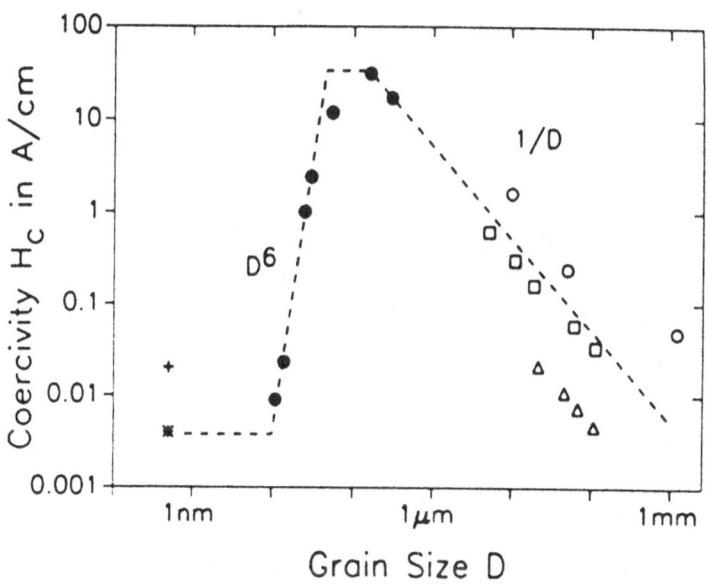

Figure 3. Grain size and coercivity H_c for various soft magnetic alloys:
✳ (+) amorphous Co (Fe) base; ● nanocrystalline Fe-Cu-Nb-Si-B;
○ Fe-Si 6.5 wt.%; ▫ 50 Ni-Fe; ▵ Permalloy. (From [19])

4.2.2. *Fe-Cu-Nb-Si-B alloys*

These alloys were developed in Japan [21] from the well known fully amorphous Fe-Si-B alloys used for soft-magnetic applications such as transformer cores [22], and are commercialised under the name Finemet™. A typical composition is $Fe_{73.5}Cu_1Nb_3Si_{13.5}B_9$. The Cu is added to promote nucleation of α-Fe(Si) crystallites, and the Nb is added to hinder their growth. Devitrification at ~800 K then leads to the desired microstructure of fine crystallites, 10 to 20 nm in diameter, embedded in a residual amorphous matrix. The matrix crystallises at higher temperature, typically ~920 K [23].

The nanophase composites are preferable to fully amorphous alloys because they have very low magnetostriction [24]. The magnetostrictions of the α-Fe crystallites and the residual amorphous phase are opposite in sign and substantially cancel [19, 20].

The devitrification mechanisms of the Finemet materials have been very widely studied; see, for example [23] and other papers in the same volume.

4.2.3. Fe-(Zr, Nb, Hf or Ta)-B alloys

These alloys show rather similar behaviour to those discussed in §5.2.3, but have the advantage of higher Fe content and consequently higher saturation magnetisation. A typical composition is $Fe_{91}Zr_7B_2$ [25]. On heating, devitrification occurs in two stages: at ~790 K there is a dispersion of ~15 nm α-Fe crystallites in a residual amorphous matrix; at 1020 K the matrix crystallizes to a mixture of α-Fe and $Fe_3(Zr, B)$. A further advantage of these materials is that they appear to be thermally more stable than those discussed in §4.2.2. The changes in properties brought about by the partial devitrification are indeed striking, as shown in Fig. 4. Both magnetisation and permeability improve markedly [10]. Further developments are discussed in [26].

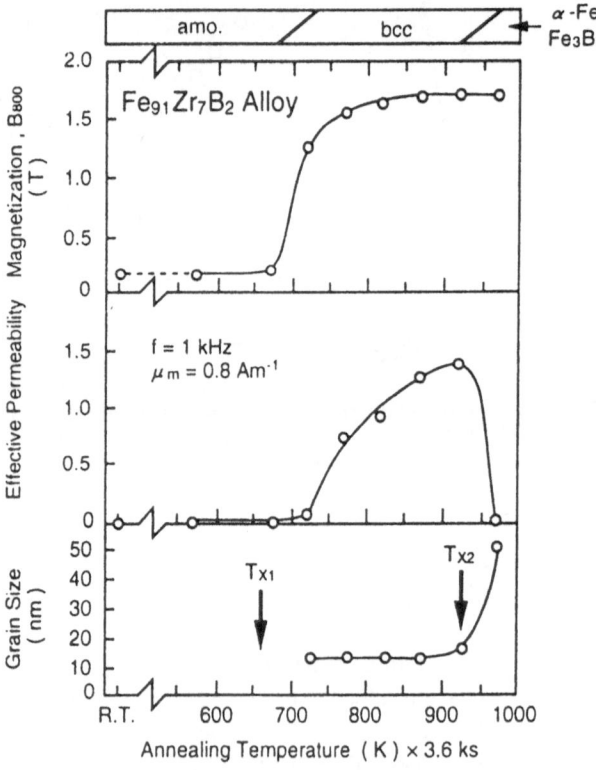

Figure 4. Changes in magnetization, permeability and grain size (α-Fe crystals) as a function of temperature of a 1-hour anneal of amorphous $Fe_{91}Zr_7B_2$. The magnetic properties are measured at 1 kHz and 0.8 A m^{-1}. (From [10])

4.2.4. Thermally Stable (Fe, Ni, Co)-M-C Alloys

The previous two categories of soft-magnetic nanocrystalline alloys are produced by rapid liquid quenching in ribbon form, and lack stability to high temperatures. Yet there are applications, for example in recording heads, which require thin-film soft magnetic materials with stability sufficient to withstand high-temperature bonding processes. Hasegawa and Saito [27-29] have developed Fe-M-C, Co-M-C and Ni-M-C nanophase composites (M = a group IVA-VIA metal, e.g., Ta) meeting these requirements. The alloys are sputter-deposited in an amorphous state, and then crystallized at ~700 K to obtain a nanocrystalline (~10 nm) metallic matrix, with a nanoscale dispersion of M-carbides (1 to 4 nm) mainly at the grain-boundary triple points [30]. The most investigated system has been Fe-Ta-C, marketed under the name Nanomax™. The key feature of these alloys is their thermal stability; with optimised composition, the nanocrystalline structure can be retained even at 1000 K. The stability appears to arise from Zener pinning of the Fe grain boundaries by the TaC particles. (For comparison, nanocrystalline Ni without carbides shows grain growth at ~350 K [31].) The grain size scales with the particle diameter as expected for such pinning, and the Ostwald ripening of the particles appears to be restricted by the low diffusivity of the Ta.

The devitrification of Fe-Ta-C involves primary crystallization of Fe, and appears to be preceded by nanoscale fluctuations of C-content in the amorphous phase [32-34]. Ta and C are rejected from the growing α-Fe crystallites, and when they are present in equiatomic proportions the alloy becomes fully crystalline through the formation of TaC precipitates. However, if there is an excess of Ta or C other compounds can form. Figure 5 maps the behaviour. It has been found that any residual amorphous phase, or other compounds degrade the magnetic properties, which are optimised for $Fe_{81.4}Ta_{8.3}C_{10.3}$ (at.%) [32].

Devitrified systems such as these appear to have much potential for further development [35, 36] for magnetic properties. As yet they have not been exploited for their mechanical properties, but in view of promising results on materials with nanoscale carbide dispersions made by other means (M50 steel [37, 38], and WC-Co cemented carbides [39-41]), this deserves to be investigated.

5. Mechanical Properties

5.1. MECHANICAL PROPERTIES IN ULTRA-FINE MICROSTRUCTURES

5.1.1. Approach to the Ideal Strength

It is well known that crystalline metals can flow plastically, as a result of dislocation motion, at stresses far below the *ideal strength*, calculable from the interatomic bonding potentials. The ideal strength is typically $\sim E/15$, where E is the Young's modulus of the material [42, 43]. Table 1 compares the tensile strengths for various types of steels with the ideal strength.

152

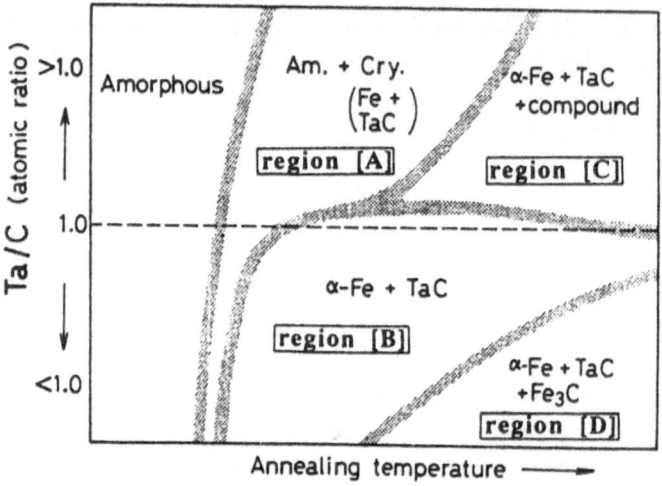

For aluminium, the best precipitation-hardened alloys have tensile strengths ~14% of ideal. Aluminium alloys offer fewer opportunities for microstructural manipulation than do steels, and conventional metallurgy does not appear to offer the prospect of exceeding one third of the ideal strength as seen in the Scifer wire.

Nanophase composites may offer route to ultra-high tensile strengths, but whether this is commercially worthwhile will depend on, besides many other factors, the property improvement over conventional (presumably cheaper) materials. Thus there appears to be a better prospect for Al-based nanophase materials than for Fe-based, simply because existing steels can have impressive properties.

5.1.2. Mechanical Properties of Amorphous Alloys

In considering the potential of devitrification as a processing route to obtain good mechanical properties, it seems appropriate to review the mechanical properties of the precursor amorphous alloys. Figure 6 is a deformation-mechanism map [45] showing the two regimes of behaviour. At ambient temperature and high strain rates, plastic flow is localised into shear bands, 10 to 20 nm thick [46]. These bands, shown schematically in Fig. 6, reflect the lack of work hardening, indeed work softening. The fracture surfaces show a characteristic vein pattern, indicating a liquid-like behaviour in the bands during deformation. As pointed out by Davis [47], the formation of shear bands is not unique to amorphous alloys, but can be found also in more conventional alloys which are at their hardening limit. Amorphous alloys can be brittle or not, depending on composition and heat treatment. Even when not brittle, however, there is no ductility in tension because of rapid failure along the shear bands. This is likely to be a problem for many applications.

At elevated temperatures and low strain rates there is the second regime, in which the plastic flow is homogeneous and can be considered as viscous flow of a highly undercooled liquid. The mechanical properties of amorphous alloys have been extensively reviewed [47-49]. Comparison with crystalline materials is not straightforward because the amorphous materials are concentrated alloys and often do not have simple crystalline counterparts at the same composition. However, amorphous alloys do generally show tensile strengths $\approx E/50$, i.e. about 30% of the ideal strength. Such high strengths are exceptional for alloys (Table 1), and reflect the lack of crystallographic slip planes and the lack of dislocation motion on them. While the yield stress is higher than for crystalline materials, the Young's and shear modulus values are generally lower.

Crystallization of amorphous alloys leads to the appearance of one or more new phases, and of course affects the mechanical properties. Usually annealing of amorphous alloys, inducing structural relaxation [50] and crystallization, leads to embrittlement of the alloys. However, when crystallization leads to a substantial volume fraction of a ductile phase, a reversal of embrittlement can occur [51].

In crystalline alloys it is standard practice to harden by the addition of dispersoids or precipitates to interfere with dislocation motion. It has been found that addition of WC particles (4 to 5 μm diameter) to amorphous alloys has a significant hardening effect [52, 53]. Since the amorphous alloys do not flow by dislocation motion, the hardening

154

mechanisms must be different, yet it seems clear that hard particles can interfere with the operation of the shear bands. This could be one of the effects underlying devitrification-induced hardening, although as pointed out in §5.2.1, there are other factors involved.

Devitrification gives ultra-fine microstructures with two or more phases. Such microstructures are classically what is needed for superplasticity at elevated temperatures, and indeed such properties are found (§5.2.2). In macroscopic terms, this is rather similar to the elevated-temperature viscous flow of amorphous alloys. Thus the main features of the mechanical properties of amorphous alloys and of the nanophase materials derived from them by devitrification are likely to be rather similar:

• unusually high yield stress at ambient temperature;
• plastic instability (work softening)
• low strength and superplastic or viscous flow at elevated temperature.

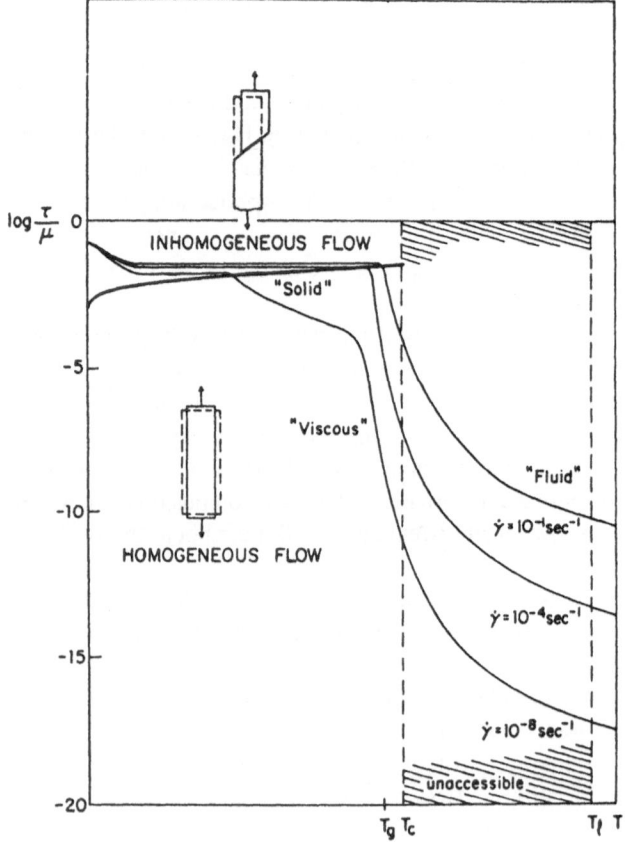

Figure 6. A deformation-mechanism map for an idealised metallic glass. T_g is the glass transition temperature, T_c the crystallization temperature, T_l the liquidus temperature, τ the shear stress, μ the shear modulus, and $\dot{\gamma}$ the shear rate. (After [45])

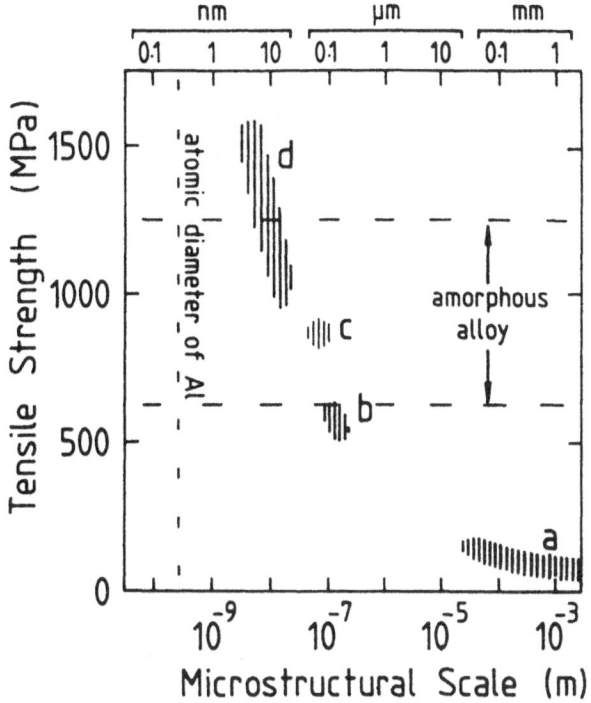

Figure 7. The tensile strengths of Al alloys compared as a function of microstructural scale: (a) commercial purity Al, (b) the strongest conventional precipitation-hardened Al alloys, (c) amorphous or partially amorphous Al-RE-TM alloys consolidated and crystallized [68, 70], (d) amorphous Al-RE-TM alloys partially devitrified (either during quenching or by annealing) to give nanophase composites of α-Al crystallites in an amorphous matrix [54-56]. The diameter of an Al atom, and the wide range of tensile strengths measured for fully amorphous Al-RE-TM alloys are shown for comparison. (RE...rare earth; TM...transition metal) (From [66])

5.2. LIGHT ALLOYS

5.2.1. Al-based Nanophase α-Al/Amorphous Composites
Interest in high-strength Al-based nanophase composites has focused on the systems Al-RE-TM (RE...rare earth; TM...transition metal). Containing 85 to 90 at.% Al, these alloys can be quenched into fully amorphous states, but they are of particular interest because the flow stress is increased by up to ~50% when there is a nanoscale dispersion of α-Al crystallites in the amorphous matrix [54, 55]. These nanophase composites have remarkable mechanical properties (Fig. 7): they are not brittle, and have tensile strengths as high as ~1.6 GPa at room temperature and as high as ~1 GPa at 573 K (these values being some 3× and 20× greater than for conventional high-strength Al

alloys) [56]. The room temperature strength is ~33% of the ideal strength for Al, a value comparable with that for the Scifer wire in Table 1. The nanoscale dispersion is obtained either by direct quenching on the edge of the glass formation [57], or by annealing [58] or deforming [59, 60] an initial fully amorphous state. Its development is most readily seen on devitrification, which has been extensively studied [58, 60-64]. The ccp α-Al crystallites, 3 to 10 nm in diameter, form on primary crystallization, and their dispersion in the residual amorphous matrix is remarkably stable. Optimum properties are obtained for crystalline volume fractions of ~25% and include hardness

system [62], that fully amorphous phases and nanophase composites with the same amorphous matrix compositions as those phases have essentially the same hardness; devitrification hardening in such cases can be attributed to solute enrichment in the amorphous matrix. The nanophase composite is, however, preferable to the fully amorphous phase because it has higher overall Al content, thereby having lower density and lower cost.

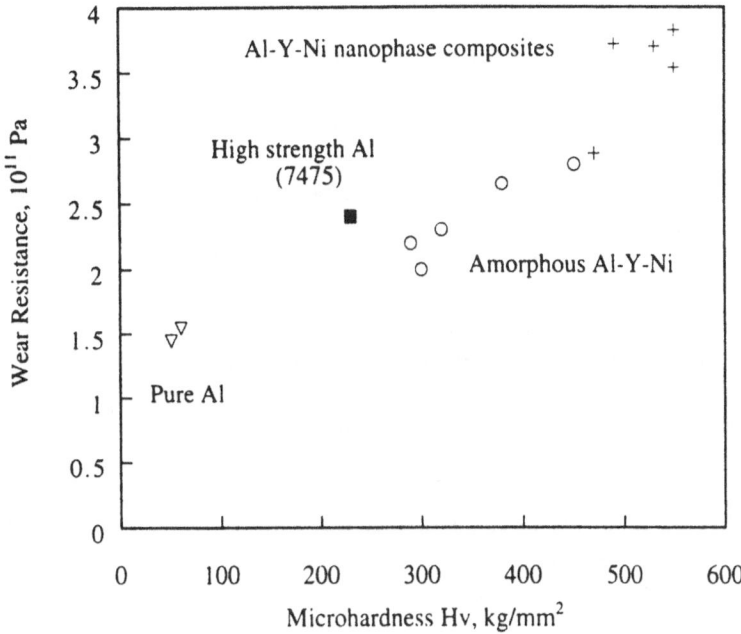

Figure 9. Comparison of abrasive wear resistance and microhardness for selected Al-based materials. The nanophase composites have clearly the highest values. (From [67])

5.2.2. *Consolidated Al-based Composites*

Materials of the type described in the previous section must be prepared using rapid quenching (for example, melt-spinning or atomisation), and are therefore available only in thin sections (10 to 100 μm in thickness or diameter) and not in bulk. However, consolidation of amorphous or nanophase α-Al/amorphous composites is possible. For example, warm extrusion of atomised amorphous powders (just below the normal crystallization temperature) gives a fully crystalline product with ~50 nm intermetallic particles dispersed in an Al matrix with grain size 100 to 200 nm [68, 69]. Although the tensile strength of this material is not as good as that of the partially amorphous nanophase composites (Fig. 7), the mechanical properties are still significantly better than those of conventional alloys. Al-RE-TM materials of this type show strengths of 800 to 1000 MPa, together with tensile ductility, a particularly good combination of strength and Young's modulus, and good fatigue strength [65]. In addition, consolidated material can show very impressive superplastic behaviour: elongations of >500% at strain rates as high as 1 s^{-1} [70]. Fabrication of small bulk components has been demonstrated [68], and the materials are now available commercially under the name Gigas™.

5.2.3. Other Al-based Nanophase Composites

In the Al-RE-TM nanophase materials discussed in §5.2.1, the primary phase is α-Al and it is the minority phase. However, by adjusting the types and amounts of solute in the Al-alloys many different microstructures, all on a nanometre scale, are possible by rapid quenching (reviewed in [65]). In melt-spun $Al_{94}V_4Fe_2$ [71], for example, a two-phase microstructure is obtained with ~7 nm grains of α-Al mixed with ~10 nm regions of amorphous phase. The predominance of α-Al is still greater in melt-spun $Al_{93}Ti_5Fe_2$ [72], which has α-Al as the majority phase in the form of 30 to 40 nm grains surrounded by a grain-boundary film of amorphous phase, ~7 nm thick.

It is also possible to have a compound rather than α-Al as the primary phase. Because of a low interfacial energy with the liquid [73], icosahedral phases are particularly easy to nucleate, and when the composition is chosen to give an icosahedral phase as the primary phase, nanoscale microstructures can result. In melt-spun $Al_{92}Mn_6Ce_2$ [74] and $Al_{94.5}Cr_3Ce_1Co_{1.5}$ [75] for example, the microstructure consists of primary icosahedral phase particles, 20 to 50 nm in diameter with a volume fraction of 60 to 70 %. These particles are separated by a thin film of α-Al. These materials have tensile strengths as high as 1350 MPa and good bending ductility. There is evidence from deformed microstructures that the particles of icosahedral phase (normally very brittle) can deform, and this is attributed to the defective nature of the phase in these materials [65]. It is significant that warm extrusion of atomised powders can give nearly fully dense bulk alloys consisting mainly of the icosahedral phase. The consolidated materials (not only of Al-Mn-Ce and Al-Cr-Ce-Co, but also Al-Mn-(Fe, Co, Ni, Cu) [65]) show impressive mechanical properties, including large impact fracture energy, and are certain to see further development.

5.2.4. Mg-based Nanophase Composites

Magnesium alloys are of particular interest for their low density, and there has been some work on the development on nanophase composites, though much less than for Al alloys. Nanophase composites of α-Mg crystallites in an amorphous matrix can be made by direct rapid quenching or by annealing fully amorphous precursors. This has been demonstrated for alloys in the systems Mg-Zn-Ce (e.g. $Mg_{85}Zn_{12}Ce_3$ [76]), Mg-Zn-La (e.g., $Mg_{80}Zn_{10}La_{10}$ [77]) and Mg-Al-Ca (e.g., $Mg_{66}Al_{19}Ca_{15}$ [78]). The hcp α-Mg crystallites are 2 to 20 nm in diameter and spaced 3 to 10 nm apart. They appear as a primary crystallization product, and in all the alloy systems studied they show evidence of substantial supersaturation in the solute elements. Devitrification hardening effects, analogous to those for Al nanophase composites (§5.2.1), are found. Indeed the flow stresses reached are ~$E/60$, similar to the nanophase Al alloys with similar volume fractions of crystallites [78].

Unlike Al alloys, Mg-based systems show examples of bulk glass formation — Mg-Cu-Y [79] and Mg-Ni-Nd [80]. Recently it has been shown that these bulk glasses can be the basis for nanophase composites produced on slow cooling. The system $Mg_{65-x}Li_xCu_{25}Y_{10}$ with 3 to 15 at.% Li shows dispersoids of bcc Mg_7Li_3 2 to 20 nm in diameter in the as-cooled state [81]. The samples, 1 mm thick, were prepared by casting into copper moulds. Continuous heating DSC shows two crystallization stages, the first being further primary crystallization of Mg_7Li_3, the second being the crystallization of the remaining amorphous phase to Mg_2Cu. There is evidence that the scale of crystallization is set by prior amorphous phase separation during cooling, into Mg-Li-rich and Mg-Li-poor regions. This work suggests a promising direction for future research in which elements can be added to known bulk glass-forming systems to induce phase separation and fine-scale crystallization. This would be analogous to the practice with some glass ceramics [1].

5.3. Fe- AND Ni-BASED ALLOYS

We turn finally to the category of alloy for which full devitrification was first used to obtain useful properties. These are Fe- and Ni-based, also containing Al, Ti, Cr, V, Mo, Co and W in many combinations, together with 5 to 12 at.% of B and 0 to 7.5 at.% of other metalloids (C, P, Si) to aid glass formation [82-85]. The materials are prepared by melt-spinning into amorphous ribbon, comminution of the ribbon, and consolidation by warm extrusion or hot isostatic pressing. During this processing full crystallization occurs. It is of particular interest that commercial materials have been developed for tool-steel applications, under the name Devitrium™ [86]. These commercial alloys, with compositions such as $Ni_{59}Mo_{29}B_{12}$ and $Ni_{56}Mo_{23}Fe_{10}B_{11}$, have a Ni-based matrix in which there is a very fine dispersion of intermetallic compounds (mainly Ni_2Mo and Ni_4Mo) together with coarser 0.5 to 2 μm boride particles ($NiMo_2B_2$). They show useful toughness (particularly when the processing is optimised), good hot workability, and are harder than conventional tool steels, particularly for temperatures >870 K. The processing conditions during consolidation have a large effect on the scale of the microstructure and the resulting properties. Subsequent ageing can also increase the hardness [86].

Related alloys can also show truly nanoscale microstructures, of the kind seen for Al alloys in §5.2.1. Melt-spun Ni-Si-B alloys (e.g. $Ni_{81}Si_{10}B_9$) can form nanophase composites of ccp Ni crystallites, 10 to 20 nm in diameter, in an amorphous matrix [87]. Analogous to the findings for the corresponding Al-based materials (§5.2.1), these composites have higher strengths than their fully amorphous counterparts. The optimum properties obtained are: tensile fracture strength 3.4 GPa, Young's modulus 130 GPa, Vickers hardness 960 kg mm^{-2}, and fracture elongation 2.7%. The tensile strength is ~25% of the ideal strength for Ni, a high fraction comparable with the values obtained for Al-based (§5.2.1) and Mg-based (§5.2.4) nanophase composites. This suggests that the principle of obtaining high strengths in this type of microstructure can be widely applied.

6. Conclusions

It is evident from the number of trade-marked names which have appeared in the review that there is considerable commercial interest in metallic nanometre-scale microstructures produced by devitrification of precursor amorphous alloys. At present, the main interests are in magnetic properties, both hard and soft, but there are strong prospects also for successful exploitation of the exceptional mechanical properties which can be achieved. The wide range of amorphous alloys, and the ability to tailor compositions, suggests that many possibilities remain to be explored.

7. Acknowledgements

The author is grateful for useful discussions with X.Y. Jiang, J.R. Wilde and Z.C. Zhong. His research in this area is supported by the Engineering and Physical Science Research Council (U.K.) and by the New Energy and Industrial Technology Development Organization (Japan).

161

23. Yavari, A.R. (1994) *Mater. Sci. Eng.* **A181/A182**, 1415.
24. Yoshizawa, Y., Bizen, Y. and Arakawa, S. (1994) *Mater. Sci. Eng.* **A181/A182**, 871.
25. Suzuki, K., Kataoka, N., Inoue, A. and Masumoto, T. (1991) *Mater. Trans. JIM*, **32**, 93.
26. Trudeau, M. (1995) *Mater. Sci. Eng.* **A204**, 233.
27. Hasegawa, N. and Saito, M. (1990) *J. Jpn. Inst. Met.* **54**, 1270.
28. Hasegawa, N. and Saito, M. (1991) *IEEE Trans. J. Magn. Jpn.* **6**, 91.
29. Hasegawa, N., Saito, M. Kataoka, N. and Fujimori, H (1993) *J. Mater. Eng. Perf.* **2**, 181.
30. Hasegawa, N., Kataoka, N., Hiraga, K. and Fujimori, H. (1992) *Mater. Trans. JIM*, **33**, 632.
31. Mehta, S.C., Smith, D.A. and Erb, U. (1995) *Mater. Sci. Eng.* **A204**, 227.
32. Hasegawa, N., Makino, A., Kataoka, N., Fujimori, H., Tsai, A.P., Inoue, A. and Masumoto, T. (1995) *Mater. Trans. JIM*, **36**, 952.
33. Hono, K., Hasegawa, N., Babu, S.S., Fujimori, H. and Sakurai, T. (1993) *Appl. Surf. Sci.* **67**, 391.
34. Hono, K., Zhang, Y., Inoue, A. and Sakurai, T. (1995) *Mater. Trans. JIM*, **36**, 909.
35. Hasegawa, N. and Saito, M. (1992) *J. Magn. Magn. Mater.* **103**, 274.
36. Hasegawa, N., Kataoka, N. and Fujimori, H. (1992) *J. Magn. Soc. Jpn.* **16**, 253.
37. Gonsalves, K.E., Law, C.C. and Chow, M.G. (1994) *Nanostruct. Mater.* **4**, 139.
38. Kear, B.H. and Strutt, P.R. (1995) *Nanostruct. Mater.* **6**, 227.
39. Kear, B.H. and McCandlish, L.E. (1993) *Nanostruct. Mater.* **3**, 19.
40. Chang, W., Skandan, G., Danforth, S.C., Kear, B.H. and Hahn, H. (1994) *Nanostruct. Mater.* **4**, 507.
41. McCandlish, L.E., Kear, B.H. and Kim, B.K. (1990) *Mater. Sci. Tech.* **6**, 953.
42. Ashby, M.F. and Jones, D.R.H. (1980) *Engineering Materials*, Pergamon, Oxford.
43. Cottrell, A.H. (1981) *The Mechanical Properties of Matter*, Krieger, New York.
44. Bhadeshia, H.K.D.H. and Harada, H. (1993) *Appl. Surf. Sci.* **67**, 328.
45. Spaepen, F. (1977), *Acta Metall.* **25**, 407.
46. Donovan, P.E. and Stobbs, W.M. (1981) *Acta Metall.* **29**, 1419.
47. Davis, L.A. (1978), in J.J. Gilman and H.J. Leamy (eds.), *Metallic Glasses*, ASM, Metals Park, pp. 190-223.
48. Li, J.C.M. (1978), in J.J. Gilman and H.J. Leamy (eds.), *Metallic Glasses*, ASM, Metals Park, pp. 224-246.
49. Li, J.C.M. (1993), in H.H. Liebermann (ed.), *Rapidly Solidified Alloys: Processes, Structures, Properties and Applications*, Marcel Dekker, New York, pp. 379-430.
50. Greer. A.L. (1993), in H.H. Liebermann (ed.), *Rapidly Solidified Alloys: Processes, Structures, Properties and Applications*, Marcel Dekker, New York, pp. 269-301.
51. Hillenbrand, H.G., Hornbogen, E., and Köster, U. (1982), in T. Masumoto and K. Suzuki (eds.) *Proceedings of the Fourth International Conference on Rapidly Quenched Metals*, JIM, Sendai, pp. 1369-1372.
52. Zielinski, P.G. and Ast, D.G. (1984) *Acta Metall.* **32**, 397.
53. Kimura, H. Masumoto, T. and Ast, D.G. (1987) *Acta Metall.* **35**, 1757.
54. Kim, Y.H., Inoue, A. and Masumoto, T. (1990) *Mater. Trans. JIM*, **31**, 747.
55. Chen. H., He, Y., Shiflet, G.J. and Poon S.J. (1991) *Scripta Metall. Mater.* **25**, 1421.
56. Inoue, A., Horio, Y., Kim, Y.H. and Masumoto, T. (1992) *Mater. Trans. JIM*, **33**, 669.
57. Kim, Y.H., Inoue A. and Masumoto, T. (1990) Mater. Trans. JIM, **32**, 747.
58. Inoue, A., Nakazato, K., Kawamura, Y. and Masumoto, T. (1994) *Mater. Sci. Eng.* **A179/A180**, 654.
59. Chen. H., He, Y., Shiflet, G.J. and Poon, S.J. (1994) *Nature*, **367**, 541.
60. Csontos, A.A. and Shiflet, G.J. (1997), in E. Ma, B. Fultz, J. Morral, P. Nash and R. Shull (eds.), *Chemistry and Physics of Nanostructures and Related Non-Equilibrium Materials*, TMS, Warrendale PA, pp. 13-22.
61. Foley, J.C., Allen, D.R. and Perepezko, J.H. (1996) *Scripta Mater.* **35**, 655.
62. Zhong, Z.C., Jiang, X.Y. and Greer, A.L. (1997) *Mater. Sci. Eng.* **A226-A228**, 531.

162

63. Jiang, X.Y., Zhong, Z.C. and Greer, A.L. (1997) *Mater. Sci. Eng.* **A226-A228**, 789.
64. Tsai, A.P., Kamiyama, T., Kawamura, Y., Inoue, A. and Masumoto, T. (1997) *Acta Mater.* **45**, 1477.
65. Inoue, A. and Kimura, H. (1997) *Curr. Opin. Sol. State Mater. Sci.* **2**, 305.
66. Greer, A.L. (1995) *Science,* **267**, 1947.
67. Greer, A.L., Zhong, Z.C., Jiang, X.Y., Rutherford, K.L. and Hutchings, I.M. (1997) in E. Ma, B. Fultz, J. Morral, P. Nash and R. Shull (eds.) *Chemistry and Physics of Nanostructures and Related Non-Equilibrium Materials*, TMS, Warrendale PA, pp. 3-12.
68. Ohtera, K, Inoue, A, Terabayashi, T, Nagahama, H. and Masumoto, T. (1992) *Mater. Trans. JIM*, **32**, 775.
69. Kawamura, Y., Inoue, A. and Masumoto, T. (1993) *Scripta Metall. Mater.* **29**, 25.
70. Higashi, K. Mukai, T., Tanimura, S., Inoue, A., Masumoto, T., Kita, K., Ohtera, K. and Nagahora, J. (1992) *Scripta Metall. Mater.* **26**, 191.
71. Inoue, A., Kimura H.M., Sasamori, K. and Masumoto, T. (1996) *Mater. Trans. JIM*, **37**, 1287.
72. Kimura, H.M., Sasamori, K and Inoue, A. (1996) *Mater. Trans. JIM*, **37**, 1722.
73. Holzer, J.C. and Kelton, K.F. (1991) *Acta Metall. Mater.* **39**, 1833.
74. Inoue, A., Watanabe, M., Kimura, H.M., Takahashi. F., Nagata, A. and Masumoto, T. (1992) *Mater. Trans. JIM*, **33**, 723.
75. Inoue, A., Kimura, H.M., Sasamori, K. and Masumoto, T. (1994) *Mater. Trans. JIM*, **35**, 85.
76. Kim, S.G., Inoue, A. and Masumoto, T. (1991) *Mater. Trans. JIM* **32**, 875.
77. Inoue, A., Nishiyama, N., Kim, S.G. and Masumoto, T. (1992) *Mater. Trans. JIM* **33**, 360.
78. Shaw, C., Rainforth, W.M. and Jones, H. (1997) *Scripta Mater.* **37**, 311.
79. Inoue, A., Kato, A, Zhang, T., Kim, S.G. and Masumoto, T. (1991) *Mater. Trans. JIM* **32**, 609.
80. Li, Y., Jones, H. and Davies, H.A. (1992) *Scripta Metall. Mater.* **26**, 1371.
81. Liu, W. and Johnson, W.L. (1996) *J. Mater. Res.* **11**, 2388.
82. Ray, R. (1981) *J. Mater. Sci.* **16**, 1924.
83. Ray, R. (1982) *Metals Prog.* **121**, 29.
84. Das, S.K., Okazaki, K. and Adam, C.M. (1985) in J.O. Stiegler (ed.) *High Temperature Alloys — Theory and Design*, TMS, Warrendale PA, p. 451.
85. Arnberg, L., Larsson, E., Savage, S., Inoue, A. Yamaguchi. S and Kikuchi, M. (1991) *Mater. Sci. Eng.* **A133**, 288.
86. Vineberg, E.J., Ohriner, E.K., Whelan, E.P. and Stapleford, G.E. (1985) in S.K. Das, B.H. Kear and C.M. Adam (eds.) *Rapidly Solidified Crystalline Alloys*, TMS, Warrendale PA, p. 301-306.
87. Inoue, A., Shibata, T. and Masumoto, T. (1992) *Mater. Trans. JIM*, **33**, 491.

MELT QUENCHED NANOCRYSTALS

A.GLEZER
Institute of Physical Metallurgy
107005 Moscow, Russia

1. Introduction

In this review we will try to present the latest information obtained in our group concerning the nanocrystalline structures formed directly during melt quenching and also after melt quenching.

It is known [1] that there is a variety of melt quenching techniques, and melt spinning is the most widespread among them. In this technique the melt is ejected under pressure onto the surface of a rapid rotating wheel. As a result, we obtain the ribbon with the thickness of 20-100 μm and with the structure depending upon the alloy composition and cooling velocity.

In general, three means of nanocrystal formation are possible:

1. Crystallization was fully completed during the melt quenching and we are dealing with the single- or multiphase submicrocrystalline structure (the first type of nanocrystals).

2. Crystallization was not fully completed during the melt quenching. As a result, amorphous-nanocrystalline structure was formed with very unusual properties (the second type of nanocrystals).

3. Melt quenching results in amorphous state formation. In this case nanocrystals can be obtained by annealing under appropriate conditions (the third type of nanocrystals).

In the following we will consider the features of structure and properties in each of the three means of nanocrystals formation.

2. The first type of nanocrystals

As an example of the first type of nanocrystals we shall consider below the structure and properties of binary and ternary melt quenched Fe-based alloys with desirable magnetic properties.

G.M. Chow and N.I. Noskova (eds.), Nanostructured Materials, 163–182.
© 1998 *Kluwer Academic Publishers.*

2.1. THE FEATURES OF CRYSTALLINE STRUCTURE

After melt quenching the crystallite (grain) size is varied from several tenth of micron

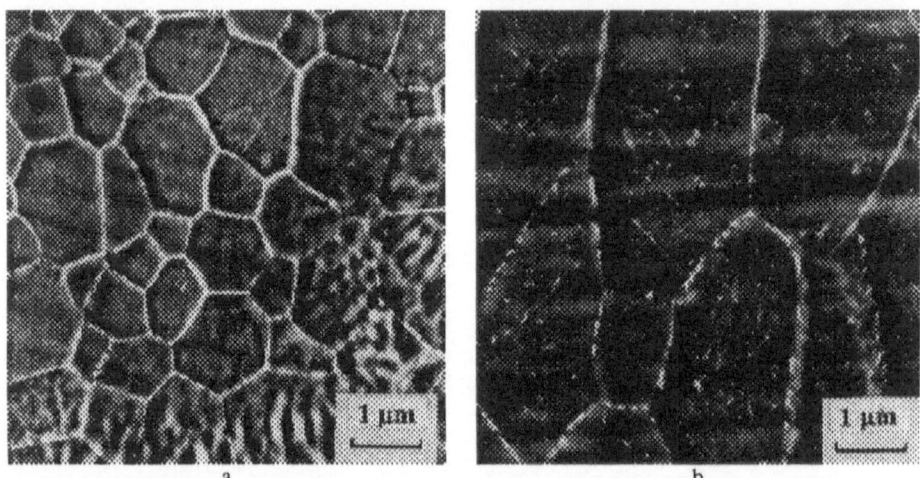

a   b
Figure 1. Equiaxied (a) and elongated (b) crystalline structure in the melt quenched Fe-Si alloy (SEM).

up to several microns. Grains are equiaxed or elongated in the direction normal to the ribbon plane depending on the ribbon thickness (Figure 1). The main peculiarities of the grain structure are [2]:

1. The existence of a well developed subgrain structure with different degrees of perfection.

2. Nonequilibrium character of the grain boundary structure which is evidenced by the well developed surface of grain boundaries and frequent variation of the grain boundary plane orientation. The most characteristic example is the screw-like grain boundary with the 180°-rotation of the boundary plane on the distances of several microns. Besides, this is also evidenced by the high density of defects (dislocations and submicroporosity) in the boundary plane.

3. High thermal stability of the grain structure. As was shown experimentally, the noticeable grain growth occurs at the temperatures of 800-900°C depending on the annealing time (1-10 h.) and the alloy composition (Figure 2). The possible nature of this phenomenon will be discussed below.

4. Sharp (on the order of magnitude) selective grain growth occurs at temperatures higher than 1000°C in the Fe-Si alloys. This process is accompanied by the sharpening of the crystallographic texture relative to the weak texture of the as-quenched state. The texture component {110} in the ribbon plane is the most enhanced. Certainly, this phenomenon resembles the secondary recrystallization process in cold worked crystals. Hence, we named this process as "quasi-secondary" recrystallization in the melt quenched alloys.

Melt quenching includes the stage of the rapid crystallization, giving grains with well developed internal cell substructure. Experiments have shown that in most cases solidification proceeds by the motion of a cellular crystallization front caused by the existence of the zone of constitutional undercooling of the melt. As a result, dendritic cells are formed in the interior of the grains with a size smaller than the grain size. There are two types of the dendritic cells (Figure 3).

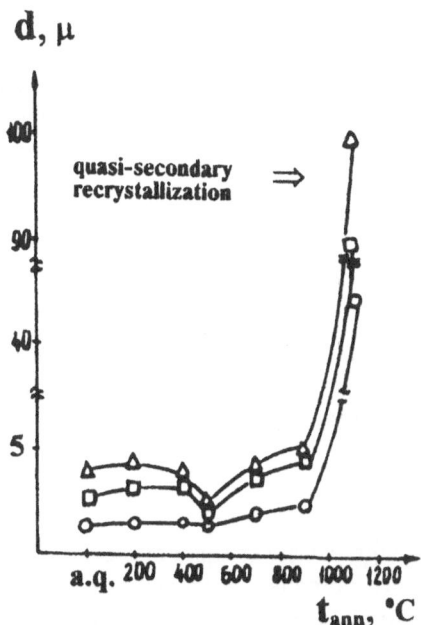

Figure 2. The grain size dependence on the annealing temperature for the melt quenched Fe-Si-Al alloy with ribbon thickness 25 μm (o), 30 μm ( ) and 35 μm (Δ); annealing time - 1 hour.

In the first case, the cells have a regular hexagonal form and are connected with the propagation of the crystallization front from the contact surface "wheel-ribbon". In the case of the sufficiently large ribbon thickness (more than 25-30 μm) the crystallization front is also formed on the free surface. The dendritic cells have in the last case the characteristic form of spherulites, i.e. fan-shaped cells meeting together on the free surface [3]. The dendritic-cellular mechanism of crystallization implies some difference in chemical composition between cell boundaries

Figure 3. Cellular structure on the contact (a) and on the free (b) ribbon surfaces of the melt quenched Fe-Si alloy (SEM).

and the interior of cells. The cell boundaries must be enriched by the solute atoms. The phase contrast on the cell boundaries seen with backscattered electrons in scanning electron microscopy can serve as an evidence of the chemical inhomogeneity during crystallization [4]. Cell boundaries and grain boundaries form two subsystems with different scales. The subsystems are not always in coincidence with each other. As it seems, the origin of the above discrepancy is in the capability of the grain boundaries for the conservative reconstruction under the effect of the quenching stresses, while at the same time the cell boundaries are incapable of such reconstruction. The diffusive dissolution of the dendritic cells and complete chemical homogenisation occur on annealing at 600-700°C.

2.2. QUENCHING DEFECTS

The imperfections formed during melt quenching can be conditionally divided into two groups [5]:
1. Defects are attributed to excessive vacancy concentration.
2. Defects are attributed to the quench stresses.

2.2.1. *The first group of defects.*
The most characteristic structural feature is the high density of small dislocation loops (Figure 4). Electron microscopy showed the intrinsic nature of these loops and the agglomeration mechanism of their formation. The mean size of the loops varied in the interval 20-40 nm in accordance with alloy composition and quenching conditions. The volume density of loops achieves a magnitude of order 10^{11} mm^{-3}, that is, it corresponds to the volume density of dislocation loops after hard radiation conditions. Calculation of the equilibrium vacancy concentration in the lattice at the temperature near melting point on the base of the mean size and the density of loops gives a value of $(1-2) \cdot 10^{-4}$ in good accordance with the theoretical estimate for bcc metals.

Several interesting features of the loop formation may be noted:
1. Dislocation prismatic loops show a clear tendency to segregate to the grain boundaries and cell boundaries [5] (Figure 4). The mean size of such loops is somewhat higher. The dislocation loop segregation inhibits grain boundary migration very much and may be the reason for the above mentioned high thermal stability of grain structure.
2. "Antiphase boundary disks" are formed in the loops, interior in the ordered alloys and because of this the loops, mobility becomes lower [6].
3. The loops possess high thermal stability. Their volume density initially undergoes some increase as the annealing proceeds. Probably, this fact evidences the retention of some concentration of excess vacancies in solid solution after melt quenching. This hypothesis is confirmed by positron annihilation experiments. A noticeable decrease of the dislocation loop concentration was observed at the annealing temperatures 500-600°C.
4. There exist alloys in which quenched-in prismatic loops are not formed at all [7]. As a rule, this was observed in alloys undergoing phase transformations (ordering, decomposition and so on), or in alloys inclined to the formation of complexes

"vacancy-solute atom". In these cases vacancies were retained in solid solution or were segregated at the interfaces or at the antiphase boundaries. Some examples of such segregations will be shown later.

5. Besides dislocation loops, the vacancies are forming submicropores with sizes up to

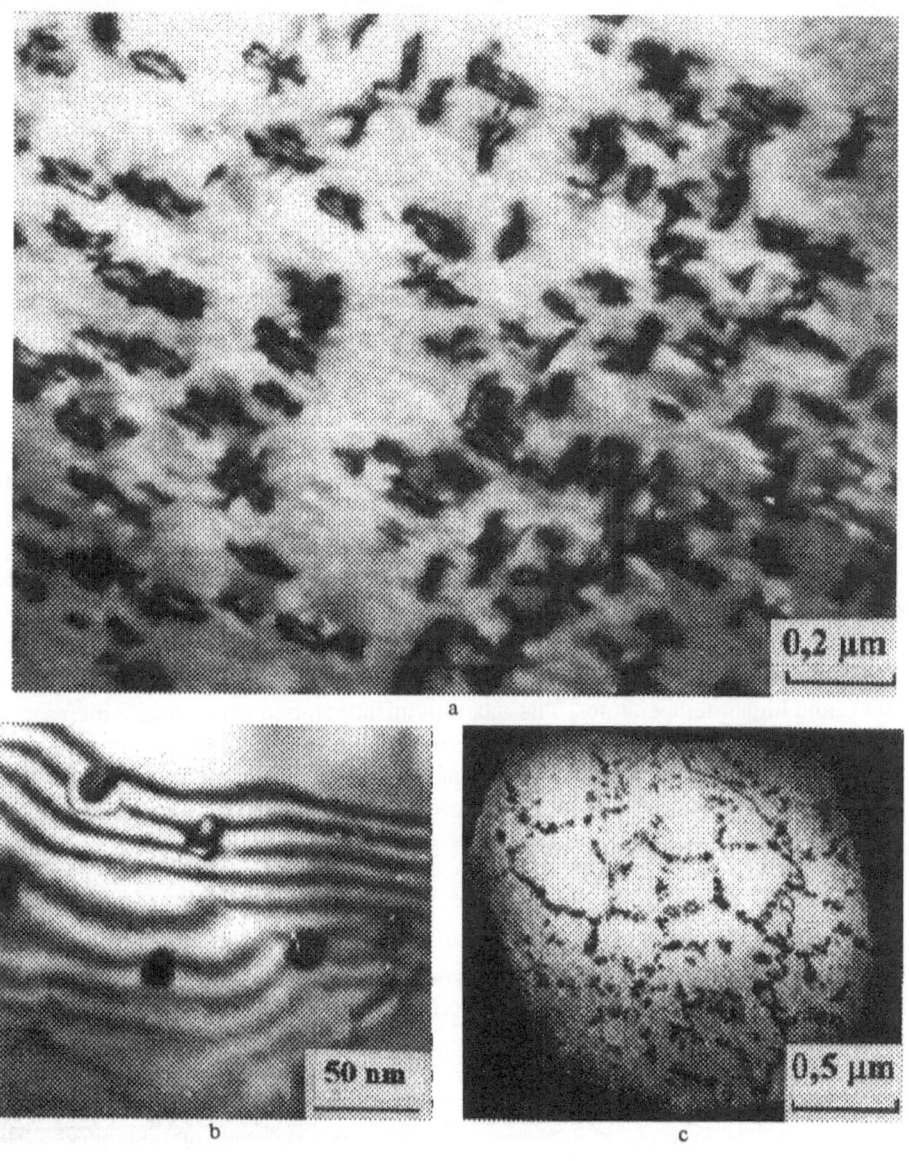

Figure 4. Dislocation loops inside grain (a), on the grain boundary (b) and on the cell boundaries (c) of melt quenched Fe-base alloys (TEM).

0.1μm (Figure 5) and even stacking fault tetrahedra (in fcc solid solutions). Interaction between vacancies and dislocations leads to observation of helicoidal dislocations.

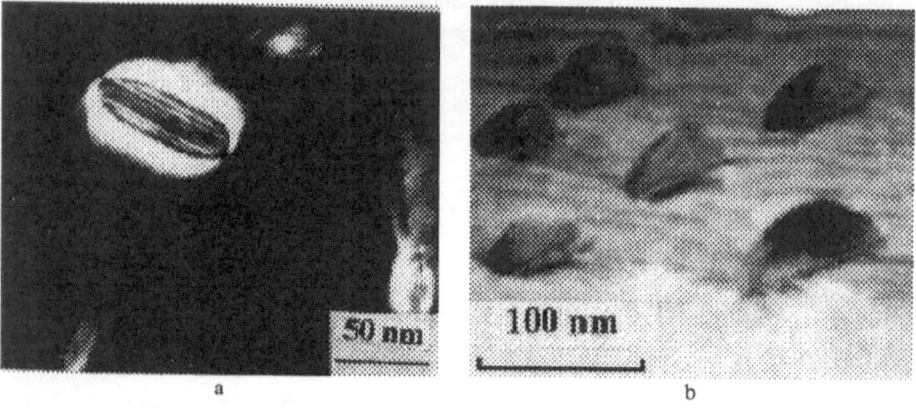

Figure 5. Micropores inside grain (a) and on the grain boundary (b) of the melt quenched Fe-Si alloy (TEM).

2.2.2. *The second group of defects.*

Quenched-in stresses cause local plastic flow and consequently high dislocation density [5]. The generation and interaction of dislocations occur in a wide temperature range, so there exists a wide spectrum of dislocation structures with various degrees of relaxation processes (from dislocation pile-up to subboundaries). Ragged subboundaries are frequently observed. These objects can be described as partial disclinations. Subgrain boundaries frequently coincide with dendritic cell boundaries since the latter are traps for all quenched-in defects. As the annealing proceeds, grain fragmentation initiates. The most intensive fragmentation process was observed at 500-600°C and higher temperatures. The subgrain misorientation continuously increases up to 5-7°, and becomes comparable with the misorientation angle for parent grains. The formation of low- and medium-angle boundaries (Figure 6) is the origin of the above mentioned anomalous decrease of grain size during annealing.

2.3. PHASE TRANSFORMATIONS

Melt quenching exerts some influence on the proceeding of various phase transformations. We will consider some examples of such effects on Fe-based alloys undergoing long-range atomic ordering, decomposition of supersaturated solid solution and martensitic transformation.

2.3.1. *Atomic ordering*

The character of long-range ordering processes and possibility of their suppression during melt quenching are a matter of keen interest. It is well known that superstructures of B2 and DO_3 type were formed in the investigated alloys in accordance with chemical composition. In this case the order-disorder transitions can be taken as first order or second order phase transformations. For example, the transition A2 (disordered bcc solid solution) $\Rightarrow$ B2 evolves as a second order transition in the Fe-Si alloys. The transition B2 $\Rightarrow$ DO_3 is realized then as a first order phase transition at lower temperatures for the same alloys [8]. The picture in the Fe-Al alloys was determined by the magnetic state of the A2-phase [9]. If the initial phase is in a ferromagnetic state, the long-range ordering proceeds as a first order transition. At the same time ordering is a second order transition in the case of A2 phase paramagnetic state.

The experiments showed the impossibility of suppressing the second order transformation by melt quenching. Only the partial lowering of the long-range order

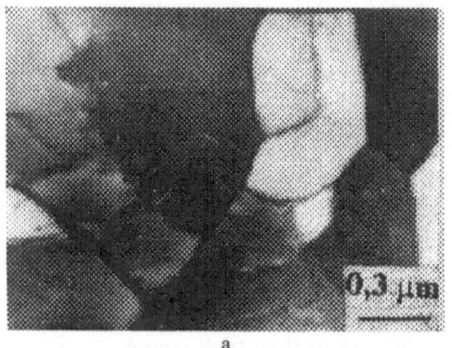 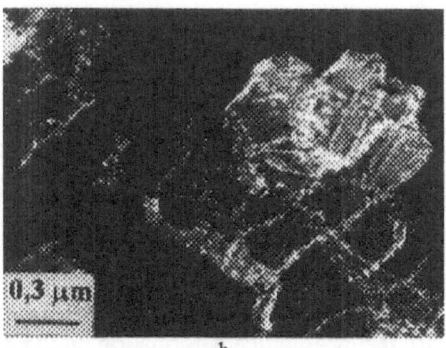

a   b

Figure 6. Dislocation structure of the Fe-Si-Al alloy after melt quenching (a) and after subsequent annealing (b) (bright and dark TEM images respectively).

degree is possible. For instance, the long-range order degree can be lowered from 0.8 to 0.4-0.5 by melt quenching of stoichiometric FeCo alloy where the superstructure B2 arised as the second order transition [10]. For the first order transitions long-range order can either be fully suppressed or its character can be qualitatively changed. In such a way melt quenching causes the suppression of the equilibrium first order phase transition A2 $\Rightarrow$ DO$_3$+A2 in Fe-22 at.%Al alloy and stimulates the A2 $\Rightarrow$ B2+A2 transition which is characteristic for higher aluminium content [4]. Moreover the two-phase state B2+DO$_3$ is realized after melt quenching of stoichiometric Fe-25 at.% Al alloy, though this two-phase field is absent on the equilibrium Fe-Al diagram.

Short-range order phenomena were observed after rapid quenching in some cases (Fe-Co and Fe-Cr-Al alloys) [10,11], though in usual conditions these phenomena are absent. There are diffuse scattering regions on the selected area electron diffraction patterns in the melt quenched state and mottled contrast with 5 nm dots is observed on the dark field images obtained (Figure 7). In this case the quenched-in prismatic dislocation loops were almost fully absent in the structure. As it may be suggested, the forming of the above mentioned short-range order is mainly determined by excess vacancies, which may play the role of an additional component of the solid solution. Agglomeration of vacancies into prismatic loops will be considerably hindered by formation of stable short-range order configurations with participation of vacancies.

Let us consider some interesting features of the antiphase boundary (APB) structure in the ordered alloys after melt quenching [4,12].

Creation and disappearance of the lattice deformation effects on the APBs. There is an abnormal fringe contrast on the electron microscope images of common Fe-Si alloys

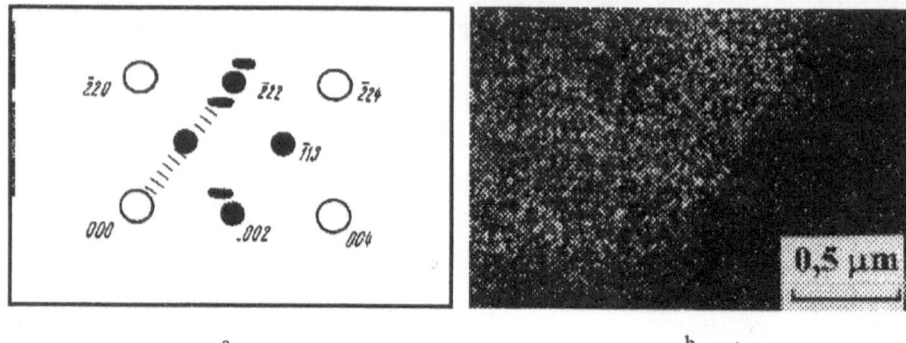

a   b

Figure 7. Short-range order effects on the electron diffraction pattern (a) and on the dark field 4/3<111> image (b) of the melt quenched Fe-Cr-Al alloy.

using of fundamental reflections (Figure 8a). This diffraction contrast is connected with the local lattice deformation on the APB. Because of this, such boundaries have a well-defined crystallographic orientation. After melt quenching of the same alloys, the APBs do not show the abnormal diffraction contrast and clear crystallographic orientation. On the other hand, that abnormal contrast was observed after melt quenching of the Fe-Al alloys, although this contrast was absent after ordinary heat treatment.

It may be definitely stated that the observed phenomena are directly connected with the realization of high cooling rates and with the segregation of excessive vacancies on the thermal APBs. This idea was supported by showing images the presence of quenched-in prismatic loops segregated at the plane of APB.

The calculation of the additional displacement vector at the APB in the common Fe-Si alloys with 11-17 at.%Si gives a value of 0.06-0.07 nm. That result allows us to estimate the excessive vacancy concentration needed for the compensation of this atomic displacement [12]. Calculations give the concentration of thermal vacancies about $0.5 \cdot 10^{-4}$ which is in correspondence with the magnitude estimated by the concentration of vacancy prismatic loops in disordered alloys. All that was said above allows us to judge the character of the local deformation on the APBs in common Fe-Si alloys because the vacancy segregation can compensate only local boundary lattice deformation of the compression type.

"Two-phase" character of APBs coalescence during the annealing. Annealing after melt quenching gives rise to unusual APB structure changes in the Fe-Al alloys. It is as if there are "two phases" in the interior of a single grain during the domain coalescence process: regions of the initial structure with small domain size and regions with enlarged domains [4] (Figure 8b). The "phase" with the enlarged domains grows at the expense of the other "phase" from the grain boundaries as the annealing proceeds.

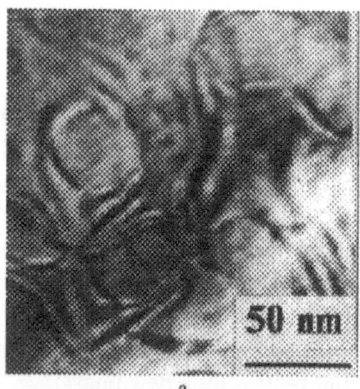

a

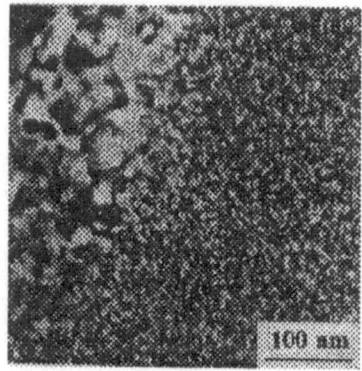

b

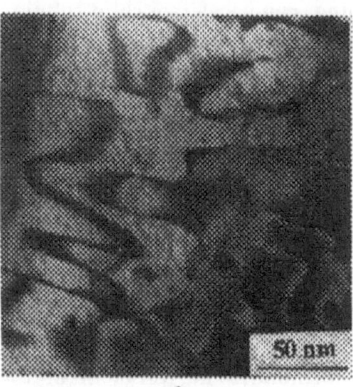

c

Figure 8. Antiphase boundary features of the melt quenched Fe-Al alloys: abnormal diffraction contrast (a), "two phase" structure (b) and elongated form of domains (c) (TEM).

The co-existence of domain structures with different scales as observed cannot be explained in the framework of the classic concepts of the "swiss cheese" type domain structure coalescence. As it may be supposed, near grain boundaries excess vacancies sink to the grain boundaries thus releasing the APBs from vacancy segregation and increasing their mobility. In due course, the "phase" without APB segregation consumes the "phase" with segregation which possesses the pinned APBs .

Elongated form of antiphase domains (APD). The APDs frequently have an elongated morphology in ordered alloys with a moderate critical point in contrast to the common equiaxed form (Figure 8c). This unusual APD morphology can be quickly eliminated during annealing as coalescence becomes appreciable. It may be supposed that the elongated APD morphology frequently observed in the vicinity of grain boundaries is connected with small displacements of the grain boundaries under the action of quenched-in stresses and also with the drag of some regions of the growing ordered phase in following the migrating boundary.

2.3.2. Decomposition of supersaturated solid solutions

As is widely accepted, melt quenching suppresses precipitation from solid solutions and assists their stabilization. Although this is true in most cases, we will consider paradoxical examples, where the melt quenching does not suppress the second phase precipitation, but on the contrary assists it.

The precipitation of a small quantity of second phase Cr_2Al takes place in the Fe-23 at.%Cr-10at.%Al alloy at triple junctions of grain boundaries without melt quenching only after annealing at 700-800°C with the duration of several hundred hours. A noticeable quantity of the same phase was presented in the same alloy directly after melt quenching (Figure 9). In this case the second phase creates a network of plate-like precipitates occupying the dendritic cell boundaries [11]. As stated above, the formation of the crystallization cells has its origin in the existence of the constitutional undercooling zone in the vicinity of the solid-liquid interface. This leads to chemical inhomogeneity between cell boundaries and cell volume depending on the distribution coefficient of the solute in the host metal. The segregation of chromium and aluminium at the cell boundaries occurs since their distribution coefficient in iron is greater than unity. As a result, the stimulus for the Cr_2Al phase formation increases considerably in the vicinity of cell boundaries. Second phase precipitation at cell boundaries accelerates also due to the existence of a high density of prismatic dislocation loops serving as nucleating centres for a new phase.

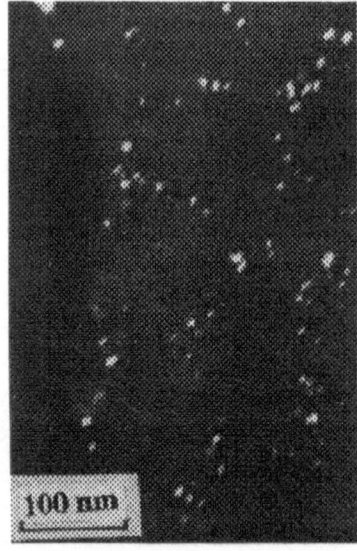

Figure 9. The precipitation of Cr_2Al nanophase on the cell boundaries of the melt quenched Fe-Cr-Al alloy (dark field TEM image).

A similar situation is found in melt quenched Al-Fe alloys, where nanocrystals of excess phase precipitated directly after melt quenching at the dendritic cell boundaries gives rise to a noticeable hardening. Besides, these results disprove the previously accepted view which connected the hardening of melt quenched Al-Fe alloys with the enriching of the fcc aluminium solid solution with iron atoms. The accelerated precipitation of the equilibrium and metastable phases on the identical origins was observed also in the melt quenched Ni-Fe-Nb alloys [7].

Thus, we revealed the interesting structural features of melt quenched decomposing alloys: despite the very high cooling rate, melt quenching can accelerate solid solution decomposition, but not suppress it. It may be stated that this phenomenon is due first of all to the cellular mechanism of crystallization, and also to the high density of the lattice defects related to vacancies.

2.3.3. *Martensitic transformation*

The grain size of the parent phase is of great importance for the progress of martensitic transformations. There are only a few references in which the influence of austenite grain size on the martensite start temperature M_s and the structure of the transformation product were estimated. One should also note the absence of the papers investigating the peculiarities of the martensitic transformation in melt quenched microcrystalline alloys. To fill in this gap we had formulated the problem of the production of superfine grains in melt quenched Fe-Ni alloys well-known from the point of view of the morphology of the thermally transformed martensite [13]. Investigations were made of the influence of the melt quenched state on the kinetics and morphological peculiarities of the martensitic transformation on cooling.

It has been detected that at the same average grain size, the martensitic transformation degree on the contact ribbon surface and on the free ribbon surface is different. Besides, the volume fraction of transformed initial phase does not depend on the grain size. But the initial temperatures of the martensitic transformation on contact and free ribbon surfaces are different. The experiments show that the M_s temperature at the contact side is higher than at the free one. As has been shown by Auger spectroscopy, the difference of matrensite points is connected with the difference of Ni concentration on the contact and on the free ribbon surface. This situation is typical of melt quenched ribbons. The fact of the composition difference through the cross section of ribbons is well defined for some melt quenched alloys. Thus we deal with a natural composite state greatly influencing the behaviour of melt quenched alloys.

We can conclude that the parent phase grain size influences the M_s temperature of the alloys with the athermal transformation kinetics but does not influence the volume fraction of martensite phase. Besides, melt quenching is shown to suppress the initiation of the so-called isothermal and surface types of martensite.

2.4. MECHANICAL PROPERTIES

We will consider only mechanical properties among the large variety of material properties. As a rule, all the Fe-based alloys considered in this report are brittle in the ordinary condition. Melt quenching increases their plasticity considerably. The ductile-brittle transition was observed in some instances (Figure 10). Thus, for example, the

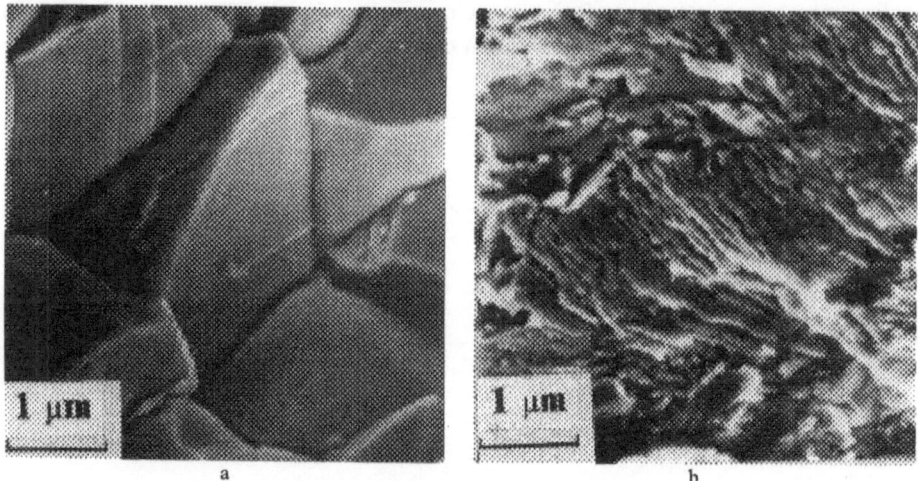

Figure 10. Fracture surfaces of the common (a) and melt quenched (b) FeCo alloy (SEM).

fracture surface shows the brittle character of fracture of the FeCo alloy in the common state and ductile fracture after melt quenching [10]. The reasons for the plasticity increase are: a small grain size, a well defined fragmented substructure, lower degree of long-range order and finally, presence of moving dislocations in the structure.

Typically, the strength of the alloys increases considerably after melt quenching simultaneously with the plasticity increasing. First of all, it results from the decreasing of the grain size in accordance with Hall-Petch relation. Also, it is attributed to the

174

high volume density of the vacancy-type defects and/or the second phase precipitation on the nanocrystalline scale.

3. The second type of nanocrystals

A completely different type of structure is expected when only of the initial stages of crystallization are completed under the conditions of rapid cooling at a critical rate. The critical rate corresponds to the minimum one for amorphisation under melt quenching. Our purpose was to analyze in detail the amorphous-crystalline structure of liquid-quenched alloys obtained by melt quenching with the rate close to critical one (when only crystals of nanometric size can develop during the rapid decrease of the melt temperature) and to measure their mechanical properties (uniaxial tension and bending tests, microhardness). Critical cooling experiments were made for the Fe-Cr-B and Co-B alloys by melt-quenching using the spinning technique with various cooling rates [14, 15].

3.1. MECHANICAL PROPERTIES

Figure 11 shows the change of microhardness HV with the decrease of the effective cooling rate during quenching from the melt near the critical cooling rate for Fe-Cr-B alloy. One can observe the sharp maximum corresponding to the transformation of the

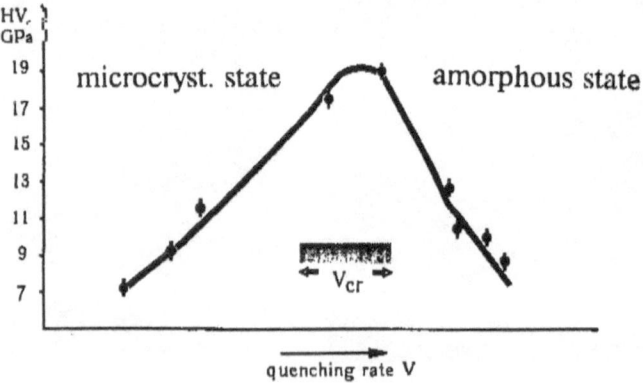

Figure 11. The change of microhardness *HV* depending on melt quenching rate *V* for Fe-Cr-B.

alloy to the crystalline state at the cooling rate close to V_{cr}. At $V>V_{cr}$ the alloy is in an amorphous state with the diffuse halo in the electron diffraction patterns peculiar to this state. At $V<V_{cr}$ we obtain the crystalline state formed by several phases. In the area of the HV(V) dependence corresponding to amorphous state one can observe the increase of electron diffraction effects connected with the existence of clusters with enhanced atomic correlation. This fact, in principle, determines the gradual HV increase with the decrease of V and its approach to V_{cr}. There exists together with

insignificant volume fraction of amorphous phase the crystalline nanophase in the area of the HV (V) curve maximum corresponding to the transition from the amorphous state to the crystalline one.

3.2. STRUCTURE

The attempts to interpret electron diffraction patterns and electron micrographs from these structures (Figure 12) led us to the assumption that there mainly exist in the

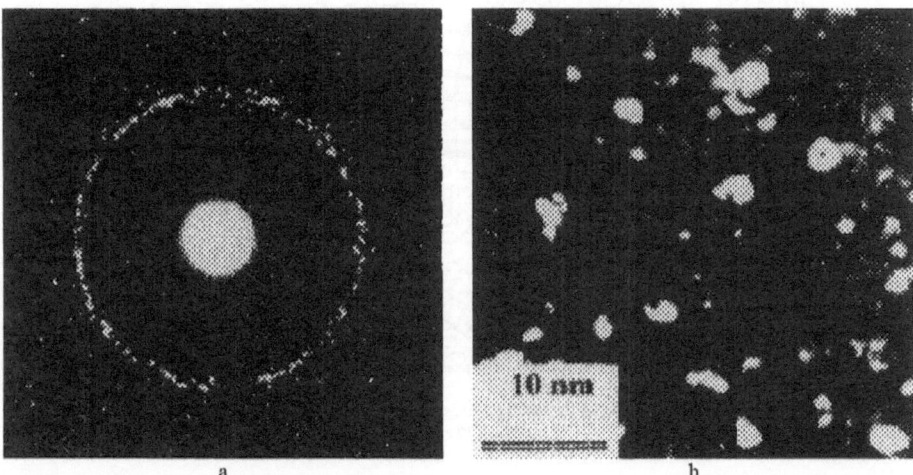

Figure 12. The diffraction pattern (a) and the dark field image (b) of the amorphous-nanocrystalline structure realized at the critical meltquench rate (TEM).

structure nanocrystals with b.c.c.crystal lattice (a≡0,285 nm) or with the crystal lattice close to it, and that there exists a remarkable spread of lattice parameters of the nanophase reaching several percent. The lattice parameter spread causes the appearance of the halo of reflections on the electron diffraction patterns with the obvious variation of the diffraction vector value for each of the reflection types from azimuthally misoriented nanoparticles. The average size of the nanocrystals is 8-10 nm (minimal size 1-2 nm). The morphology of nanocrystals is well visible in dark field images using of one or several reflections. The smearing of the reflections and the decrease of the intensity of primary diffraction contrast in the outlying parts of the dark field images of nanocrystals which were often observed allow us to assume that the lattice parameter in this case can be changed not only from one nanocrystal to another, but also from the central part of each nanocrystal to the periphery. Moreover, it is possible that at the earliest stages of the nanostructure formation ($V \approx V_{cr}$) the nanocrystal boundaries are partially amorphous. This phenomenon is confirmed by the fact that in the transitional amorphous-nanocrystalline state one does not observe the fringe α-type diffraction contrast from the grain boundaries.

On the basis of the aforementioned facts the following model for the amorphous-nanocrystalline state with the anomalous strength produced at melt-quenching with the critical rate can be suggested (Figure 13). Nanocrystals forming homogeneous conglomerates are characterized by the degree of crystalline order changing smoothly:

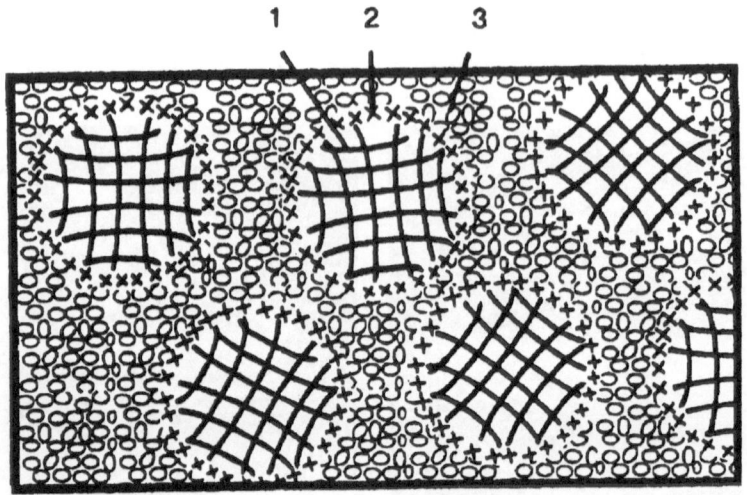

Figure 13. Structural model of the amorphous-nanocrystalline state realized at the critical melt quench rate: 1-nanocrystal, 2-interface, 3-amorphous state.

there is in the central area of every nanocrystal an ideal crystalline structure, which transforms gradually to an amorphous one while moving to the periphery. This is probably because the centre of the nanocrystal forms at a temperature significantly higher than the temperature at which the periphery forms. The amorphous phase layers being observed are enriched by metalloid (boron) atoms and they do not have clear interface boundaries with the crystalline phase. This type of structure can be considered as a common nanocrystalline one with diffused grain boundaries to such a degree that they represent rather extended areas of the amorphous phase.

3.3. PLASTIC DEFORMATION MODE

Static bending tests have allowed us to establish that the plasticity of the alloy in a transitional state is lower than the one in an amorphous state, but at the same time it is far from zero. Thus, we can ascertain that the state being investigated of Fe-Cr-B alloys possesses not only unique high strength, but also sufficient plasticity. Investigations of the plastic deformation structure of nanophase alloys have shown that the process of plastic yielding has the features inherent to the deformation of amorphous alloys. For example, the study of the patterns of shear bands was carried out by scanning-transmission electron microscopy. In this case sets of bands localized to a high degree were revealed (the height of the sliding steps reaches 0.3-0.4 μm), which correspond to the local deformation degree of several hundred percent. This fact is typical of amorphous materials. We could not observe by electron microscopy in the slip bands the signs of the existence of dislocations. Therefore the process of plastic deformation is localized in amorphous intergranular layers (Fig.4b) and is to a certain degree similar to the process of grain boundary sliding.

Besides, it should be noted that the described structural state of Fe-Cr-B alloys can be realized, possibly, in other amorphizing systems. Actually, we have recently observed the same state accompanied with very high strength (σ_f = 10 GPa) in the Co-B alloy.

4. The third type of nanocrystals

These nanocrystals have been investigated to a great degree [16]. Under the appropriate heat treating conditions single- and multiphase nanocrystalline structures were obtained for study. Very interesting nanophase materials in the Fe-Si-B system with small Cu and Nb additions have been discovered in Japan a few years ago; they were named "Finemet" [17]. A very large quantity of α-Fe-Si nanoparticles are formed in the melt quenched amorphous state. But these nanoparticles don't grow in size in rather wide interval of annealing temperatures. Moreover, as it turned out, the amorphous-nanocrystalline state manifests very good soft-magnetic properties exceeding the properties of corresponding amorphous alloys. These phenomena have been investigated in recent years very thoroughly.

We will consider in detail only some important problems of this nanocrystal type.

4.1. CORRELATION BETWEEN STRUCTURE AND PROPERTIES

The first problem consists of the possibility for obtaining a rather wide spectrum of physical and mechanical properties in the same alloy, depending on given nanocrystalline structure parameters. For example, let us consider the behaviour of magnetic and mechanical properties of Fe-Cr-B amorphous alloys under the influence of nanocrystalline structure formation during the annealing [18]. Here one can see the dependence of coercive force and saturation induction vs. short-time (a few seconds) annealing temperature for the amorphous alloy $Fe_{67}Cr_{18}B_{15}$. (Figure 14). The very sharp change of coercive force in the nanocrystalline state can be seen. It is typical that two sharp maxima of coercive force (up to 40 kA/m) coincide exactly with two maxima of microhardness (up to 20 GPa). Transmission electron microscopy showed the two-phase nanostructure in both maxima. The first maximum corresponds to the precipitation in the amorphous matrix of the α-Fe-Cr nanocrystals of about 5-10 nm in size. The second one corresponds to the precipitation of tetragonal boride $(Fe,Cr)_3B$ with the particle size 10-20 nm in the microcrystalline matrix α-FeCr (Figure 15).

4.2. DISAGREEMENT WITH THE HALL-PETCH RELATION

As in other cases, nanocrystals formed by melt quenching show abnormal mechanical properties. Nanocrystalline materials with a grain size less than 100 nm have unusual

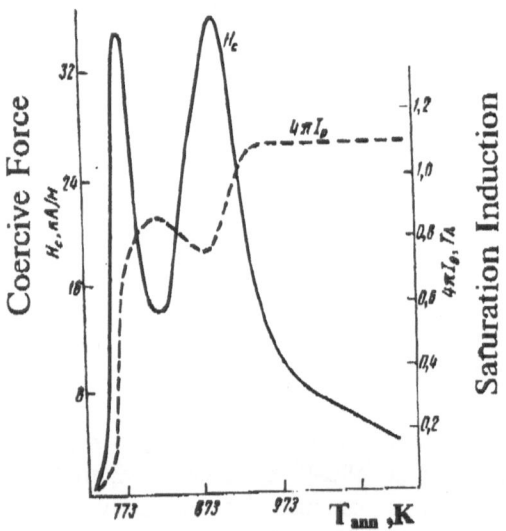

mechanical properties in comparison with common polycrystals. The difference between the mechanical behaviour of nanocrystalline materials obtained by various methods and having grains with identical average size has not only a quantitative, but also a qualitative character. The recently conducted researches on the yield stress dependence $\sigma_y=f(d)$ and the microhardness dependence H_v $=f(d))$ for the nanocrystalline materials have shown that the Hall-Petch relation

Figure 14. The magnetic properties dependence on the annealing temperature (annealing time - 10 sec.) for the amorphous Fe-Cr-B alloy.

$$\sigma_y = \sigma_0 + kd^{-1/2}, \qquad (1)$$

where σ_0 is determined by Peierls barrier and k depends on the resistance of grain boundaries to the dislocation movement, is not valid when the size of a grain d

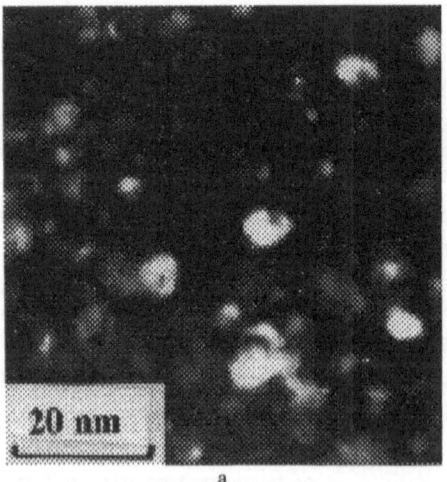
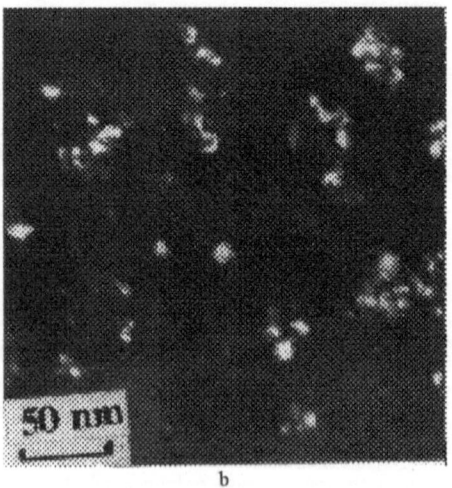

Figure 15. The dark field TEM images corresponding to the first (a) and to the second (b) maxima of the $H_c=f(T_{an})$ curve for the Fe-Cr-B alloy.

becomes less than 100 nm [19]. The degree of deviation is determined, as a rule, by the method of nanocrystal production. Some results even demonstrate a negative value of factor k. In nanocrystalline materials obtained by crystallization from amorphous precursor the dependence of σ_y on d has a point of inflection at $d = D_{cr}$ [16]. The nanocrystalline materials with a grain size $d < D_{cr}$ have a negative value of factor k. When $d > D_{cr}$ the dependence $\sigma_y(d)$ is described by the Hall-Petch relation (1).

The reason for such variety of contrary results for nanocrystals is not clear as yet. Concerning the melt quenched nanocrystals, a structural feature in which they differ from the other nanocrystals consists in the existence of amorphous interlayers between nanocrystals in certain stages of nanostructure evolution (Figure 16). Such an interlayer can be regarded as a certain specific "grain boundary" phase appreciably determining the mechanical behaviour of the two-phase system.

Suppose that in polycrystal grain boundaries there are separate amorphous ductile layers with the thickness Δ and the size L (Fig.5a). The amorphous material yield

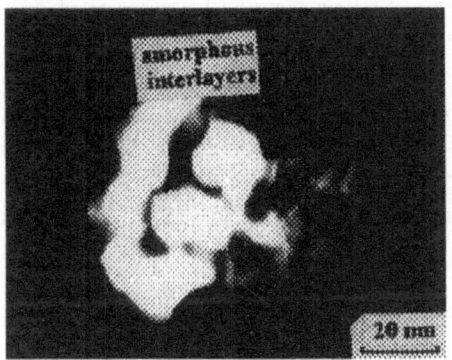

Figure 16. The dark field TEM image of annealed amorphous Fe-Cr-B alloy; bright contrast corresponds to nanocrystals, dark contrast corresponds to amorphous interlayers.

stress is $\sigma_0 = 2\tau_0$. For simplicity, let us consider that plastic deformation occurs by sliding along the layer plane. For $\tau_0 < \tau_s$, where τ_s is the average value of shear stress microsliding resistance, the first deformation stage ($\sigma_s < \sigma_A < \sigma_0$) will be defined by equation [20]:

$$\sigma_A = \sigma_s + AG\Delta\varepsilon/Lf, \qquad (2)$$

where f is the volume fraction of amorphous ellipsoidal inclusions. Here f < <1, because inclusion interaction effects are not taken into account. Another more realistic model for amorphous-crystalline material structure is blocks of N grains, surrounded by thin amorphous layers. The average block size is $L_N = N^{1/3}d$, where d is the average grain size. The yield stress of such a composite material at $\Delta << L_N$ is [20]:

$$\sigma_s = (\sigma_0/2)(1+2/3L_N/\Delta) \qquad (3)$$

To test experimentally the aforementioned theoretical estimations we have carried out the measurement of yield stress and microhardness of Fe-Si-B-Nb-Cu (Finemet) where gradual precipitation of Fe-Si nanocrystals takes place during amorphous state annealing [17]. Those dependences (Figure 17) show in accordance to equation (3) that the yield stress value grows linearly with the nanocrystal size and the reciprocal value of the average width of amorphous layers growth. Satisfactory coincidence of theory and experiment confirms the fact that plastic deformation of nanocrystals with amorphous layers is realized in the amorphous phase. In the limiting case the thickness of the amorphous intergranular layer becomes so small that we can go on to grain boundary structure for common nanocrystals. The results obtained show that the role of boundaries between separate crystals in nanocrystalline materials is very important.

Another model for the mechanical behaviour of nanocrystalline materials with amorphous interlayers considers deformation as the evolution of a space network of

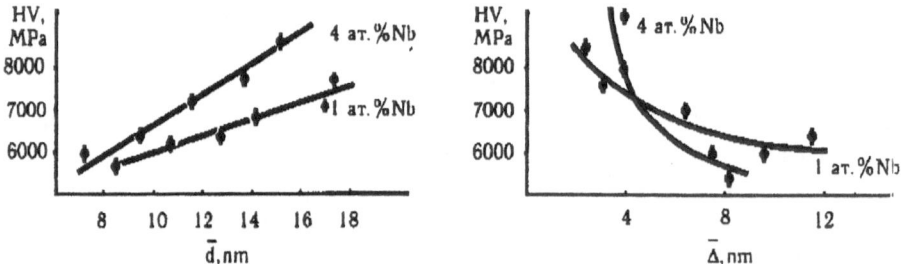

Fig. 17. The experimental dependence of microhardness HV on nanocrystal average size d and amorphous intergranular layer average thickness Δ for Fe-Si-B-Nb-Cu alloy (Finemet).

junction disclinations [21]. The defect structure of such nanocrystals is presented as a cubic network of junction disclinations the axes of which coincide with the triple junctions of intergranular boundaries, and for which the signs of Frank vectors of the same power are interleaved in staggered order. It was shown in [22] that junction disclinations are formed as a result of the mismatch of plastic rotations of grains forming triple junctions.

Moreover, this model takes account of the so-called "intergranular phase," that is of physical width of intergranular boundaries Δ, the existence of which is most clearly demonstrated by nanocrystals obtained by crystallization from amorphous precursors. Nanophase material, especially in the range of grains of small sizes, represents an aggregate of "hard" grains separated by "soft" intergranular phase, that is a medium with internal structure.

The mathematical analysis for the description of the mechanical behaviour of such media has been developed on the basis of micropolar elastic theory [21]. It should be noted that the micropolar elastic theory (for media with internal structure), contrary to the usual linear theory of elasticity, takes account of the dependence of stress concentration coefficient on the ratio between the grain size d and the width of intergranular boundaries (amorphous interlayers) Δ, as caused by the more hard phase.

The dependence of the stress concentration coefficient on the ratio between crystalline phase dimensions and the width of intergranular boundaries has been established in the framework of this model. It has been shown that nanocrystal plastic deformation begins when the internal stresses in the disclination quadrupole as an elementary network cell become fully compensated by the external loading. This plastic deformation occurs by the "go-ahead" mechanism of dislocation movement between network of disclinations.

It has been established that plastic incompatibility is accumulated in triple junctions only when the external loading increases more slowly than the internal stresses in the elementary cell. It follows from the suggested model that the dependence of the yield stress on the average grain size d has a point of inflection at $d = D_{cr} (\cong 25nm)$. When $d < D_{cr}$, σ_y increases with d growth, and when $d > D_{cr}$, σ_y obeys the Hall-Petch relation. The theoretical and experimental results are found to be in good agreement for a number of nanomaterials.

It should be noted from the analysis of the grain size dependence of the yield stress that the structural peculiarities of nanocrystals obtained by different methods play a very important role.

5. Summary

1. Melt quenching results in a very wide diversity of the nanoscale structures formed. Provisionally, nanocrystals obtained by means of melt quenching can be divided into three types: nano- and submicrocrystals formed directly during melt quenching, nanocrystals formed during melt quenching with the cooling velocity near to the critical amorphisation one and nanocrystals formed during the annealing of the amorphous state.

2. In the nanocrystals and submicrocrystals of the first type significant peculiarities of phase and structural transformations have been observed. They are connected with formation by melt quenching of microcrystallites, dendritic-cellular structure and high density of excess vacancies and quenched-in dislocations. As a consequence, the clear tendency to the strengthening and plasticity increase of the alloys is observed and in some instances a ductile-brittle transition takes place.

3. For second type of nanocrystals the melt quenching with the critical amorphization rate forms in the amorphous matrix nanocrystallites which have a degrees of crystallinity decreasing from the centre to the periphery. This nanocrystalline state leads to very high strength.

4. The presence of the amorphous interlayers at certain crystallization stages is typical of the third type of nanocrystals. Those interlayers can be regarded as a "grain boundary" phase. Such a peculiarity results in the abnormal dependence of yield stress on grain size with a deviation from the Hall-Petch relation.

6. References

1. Luborsky, F.E. (1985) *Amorphous Metallic Alloys*, Butterworths, London.
2. Glezer, A.M., Molotilov, B.V., and Sosnin, V.V. (1982) Melt quenched refined Fe-Si and Fe-Si-Al alloys, *Isvestiya Akademii Nauk SSSR, ser.fisicheskaya (in Russian)* **46**, 701-709.
3. Sosnin, V.V., Glezer, A.M., and Molotilov B.V. (1985) Structural features of the melt quenched sendust (Fe-Si-Al) alloy. III. Crystalline structure features, *Fizika metallov I metallovedenie (in Russian)* **59**, 509-516.
4. Glezer, A.M., and Sosnin, V.V. (1989) Structural and phase transfor mations in melt quenched Fe-Al alloys, *Isvestiya Akademii Nauk SSSR, ser. fisicheskaya (in Russian)* **53**, 671-681.

5. Glezer, A.M., Molotilov, B.V., and Sosnin, V.V. (1984) Structural features of the melt quenched sendust (Fe-Si-Al) alloy. II. Quenching defects, *Fizika metallov I metallovedenie (in Russian)* **58**, 370-376.

6. Glezer, A.M., Molotilov, B.V., Prokoshin, A.F., and Sosnin, V.V. (1983) Structural features of the melt quenched sendust (Fe-Si-Al) alloy. I. Long-range ordering study, *Fizika metallov I metallovedenie (in Russian)* **56**, 750-757.

7. Zhigalina, O.M., Sosnin, V.V., and Glezer, A.M. (1993) The effects of heat treatment on phase transformations in rapidly quenched Ni-Fe-Nb alloys, *The Physics of Metals and Metallography* **75**, 205-209.

8. Swann, P.R., Granas, L., and Lehtinen, B., (1975) The B2 and DO₃ ordering reactions in iron-silicon alloys in the vicinity of the Curie temperature, *Metal Sci.* **9**, 90-96.

9. Swann, P.R., Duff, W.R., and Fisher, R.M. (1972) the electron metallography of ordering reactions in Fe-Al alloys, *Met. Trans.* **3**, 409-419.

10. Glezer, A.M., and Maleeva, I.V. (1989) Structural features of the melt quenched FeCo alloy, *Fizika metallov I metallovedenie (in Russian)* **68**, 901-909.

11. Glezer, A.M., Maleeva, I.V., and Novoselova, N.G. (1990) The effects of melt quenching on the structure and properties of Fe-Cr-Al alloys, *Fizika metallov I metallovedenie (in Russian)* **69**, 122-130.

12. Glezer, A.M., Molotilov, B.V., and Sosnin V.V. (1985) The main regularities of structure formation in melt quenched Fe-Si alloys, *Isvestiya Akademii Nauk SSSR, ser.fisicheskaya (in Russian)* **49**, 1593-1602.

13. Glezer, A.M., Pankova, M.N. (1995) Martensitic transformation in microcrystalline melt-quenched Fe-Ni alloys, *J.Phys.IV* **5**, 299-303.

14. Glezer, A.M., Molotilov, B.V. and Ovcharov, V.P. (1987) Structure and mechanical properties of Fe-Cr-B alloys at the amorphous-crystallalline transition, *Fizika metallov I metallovedenie (in Russian)* **64**, 1106-1109.

15. Glezer, A.M., Pozdnyakov, V.A., Kirienko, V.I., and Zhigalina, O.M. (1996) Structure and mechanical properties od liquid-quenched nanocrystals, *Mater. Sci. Forum* **225-227**, 781-786.

16. Lu, K. (1996) Nanocrystalline metals crystallized from amorphous solids: nanocrystallization, structure and properties, *Mater. Sci. and Engin.* **R16**, 161-221.

17. Yamauchi, K., and Yoshihito, Y. Recent development of nanocrystalline soft magnetic alloys, *Nanostruct. Mater.* **6**, 247-254.

18. Drozdova, M.A., Glezer, A.M., Krasavin, J.I., and Savvin, A.A. (1989) Magnetic properties and structure changes of Fe-Cr-B amorphous alloys at the crystallization, *Fizika metallov I metallovedenie (in Russian)* **67**, 896-901.

19. Siegel, R.W., and Fougere, G.E. (1995) Mechanical properties of nanophase metals, *Nanostruct. Mater.* **6**, 205-216.

20. Glezer, A.M., and Pozdnyakov, V.A. (1995) Structural mechanism of plastic deformation of nanocrystals with amorphous intergranular layers, *Nanostruct. Mater.* **6**, 767-769.

21. Zaichenko, S.G., and Glezer, A.M. (in press) Disclination-dislocation model of the mechanical behaviour of nanocrystals prepared by crystallization from amorphous precursor, *Interface Sci.*

22. Zaichenko, S.G., Shalimova, A.V., Titov, A.O., and Glezer, A.M. (1996) Evolution of triple junction defect structure at plastic deformation of metallic polycrystals, *Interface Sci.* **3**, 203-207.

QUASIPERIODIC AND DISORDERED INTERFACES IN NANOSTRUCTURED MATERIALS

I.A.OVID'KO
Institute of Machine Science Problems,
Russian Academy of Sciences,
Bolshoj 61 , Vas.Ostrov, St.Petersburg 199178, Russia

1. Introduction

This lecture is concerned with quasiperiodic and disordered interfaces in nanostructured materials. The notion of quasiperiodic interfaces is rather new in both solid state physics and materials science. In particular, it is not widespread among experts in the area of nanostructured materials. Therefore, the most attention in this lecture will be paid to the structure and properties of quasiperiodic interfaces as well as to their contribution to macroscopic properties of nanostructured materials. In doing so, for brevity, we will concentrate our theoretical consideration on final results, while intermediate mathematical details will be only outlined.

It should be noted that studies of quasiperiodic interfaces in nanostructured materials are just at the starting point. In these circumstances, the lecture puts more new questions than answers to previously stated questions in this field. The main conclusions of this lecture are as follows: first, quasiperiodic interfaces are inherent elements of the nanostructured materials structure; second, quasiperiodic interfaces represent the special type of interfaces exhibiting the properties which, generally speaking, are different from those of periodic and disordered interfaces; third, quasiperiodic interfaces are capable of significantly contributing to macroscopic (at least, mechanical) properties of nanostructured materials; fourth, quasi-nanocrystalline materials (consisting of nanocrystallites and quasiperiodic grain boundaries) represent a new class of nanostructured materials.

In addition, here we will discuss the features of disordered interfaces in nanostructured materials. In doing so, the special attention will be paid to transformations of disordered interfaces into ordered (periodic and quasiperiodic) ones in nanocrystalline materials as well as to spreading of disordered interfaces in nanoamorphous materials (nanoglasses).

G.M. Chow and N.I. Noskova (eds.), Nanostructured Materials, 183–206.

The lecture is organized as follows. In section 2 definitions of periodic, quasiperiodic and disordered solid state structures are considered. Section 3 deals with description of quasiperiodic interfaces in film–substrate systems. In section 4 we examine quasiperiodic tilt boundaries in nanostructured polycrystals and discuss the notion of quasinanocrystalline materials. Section 5 is concerned with disordered interfaces in nanostructured materials. Section 6 contains concluding remarks.

2. Periodic, Quasiperiodic and Disordered Structures

Solid state structures are specified by their mass distributions in space. In this context, $\rho(\mathbf{r})$, a mass density at position $\mathbf{r}$ in a d–dimensional space, serves as the main characteristic of a d–dimensional solid. In general, there are the three basic categories of solid state structures: periodic, quasiperiodic, and disordered ones. Periodic and quasiperiodic structures exhibit a long–range translational ordering, in which case their diffraction patterns consist of sharp peaks. Disordered structures do not have any long–range translational order.

A periodic solid is by definition specified by mass density $\rho(\mathbf{r})$ that is periodic:

$$\rho(\mathbf{r}) = \rho(\mathbf{r} + \mathbf{R}) \tag{1}$$

for every vector $\mathbf{R}$ in some periodic lattice L. Due to periodicity in d–dimensional coordinate space, $\rho(\mathbf{r})$ can be expressed as a discrete Fourier series:

$$\rho(\mathbf{r}) = \sum_{\mathbf{q} \in Q} \rho_{\mathbf{q}}(\mathbf{r}) exp\{i\mathbf{q}\mathbf{r}\} \tag{2}$$

with $\mathbf{q}$ being vectors in a d–dimensional periodic lattice Q reciprocal to the d–dimensional periodic lattice L of vectors $\mathbf{R}$. The reciprocal lattice Q is generated by d basis vectors. Crystalline materials are examples of periodic solids.

A d–dimensional quasiperiodic solid is by definition specified by mass density $\rho(\mathbf{r})$ which can be expressed as a discrete Fourier series:

$$\rho(\mathbf{r}) = \sum_{\mathbf{q} \in \tilde{Q}} \rho_{\mathbf{q}}(\mathbf{r}) exp\{i\mathbf{q}\mathbf{r}\}, \tag{3}$$

with $\mathbf{q}$ being vectors in a reciprocal lattice $\tilde{Q}$ generated by $\tilde{d}$ basis vectors, where $\tilde{d} > d$, e.g. [1,2]. That is the number $\tilde{d}$ of basis vectors of the reciprocal lattice $\tilde{Q}$ is larger than dimension d of quasiperiodic solid. Quasicrystals serve as the most known example of quasiperiodic solids, e.g. [1-3].

Disordered solids do not exhibit any long–range translational order, in which case their diffraction patterns do not contain sharp peaks; such patterns consist of continuous extended regions. A disordered solid is specified by mass density $\rho(\mathbf{r})$ which can be expressed as a Fourier integral:

$$\rho(\mathbf{r}) = \int_{Q'} \rho_{\mathbf{q}}(\mathbf{r}) exp\{i\mathbf{q}\mathbf{r}\}d\mathbf{q}, \tag{4}$$

where $\rho_{\mathbf{q}}$ is a non–zero function in the continuous extended regions Q' of reciprocal space. Amorphous solids serve as well–known examples of disordered solid state structures.

3. Quasiperiodic Interfaces in Film–Substrate Systems

This lecture is concerned mostly with quasiperiodic interfaces. Let us start our discussion of quasiperiodic interfaces with description of such interfaces in film–substrate systems (in particular, in nanostructured–film–substrate systems). Such interfaces serve as the simplest example of quasiperiodic interfaces and, on the other hand, have the same basic features as more complicately structured, quasiperiodic grain boundaries in polycrystalline and nanocrystalline materials.

3.1. STRUCTURAL FEATURES OF QUASIPERIODIC INTERFACES

Let us consider a one–dimensional infinite interface between a two–dimensional film and substrate both having square lattices with different parameters. Such interfaces can be used as effective models of real interfaces in film–substrate systems, in which case analysis of model interfaces allows one to reveal the basic peculiarities of real interfaces, e.g. [4-7]. Let a and c be the square lattice parameters of film and substrate, respectively. For definiteness, we assume that $c > a$. Misfit parameter $f = 1 - a/c$ characterizes misfit between lattices of film and substrate.

Formation of ensemble of misfit dislocations at interface (Fig.1) is the effective way for relaxation of misfit stresses which is most often realized in real film–substrate systems [4-7]. Usually models of interfaces containing misfit dislocations deal with infinite interfaces between lattices with relatively rational parameters a and c, that is, $c/a = n/m$, where n and m are co–prime integers. In this situation, equilibrium spatial organization of misfit dislocation ensemble is periodic at an infinite interface with period $na = mc$ (Fig.1a).

However, in general, ratio c/a which characterizes infinite interface is either rational or irrational. In the second situation, interface with its misfit dislocations is quasiperiodically ordered (Fig.1b). As to details, let us determine and examine the quasiperiodic ordering of an interface with irrational a/c (and irrational $f = 1 - a/c$) by means of the geometric model.

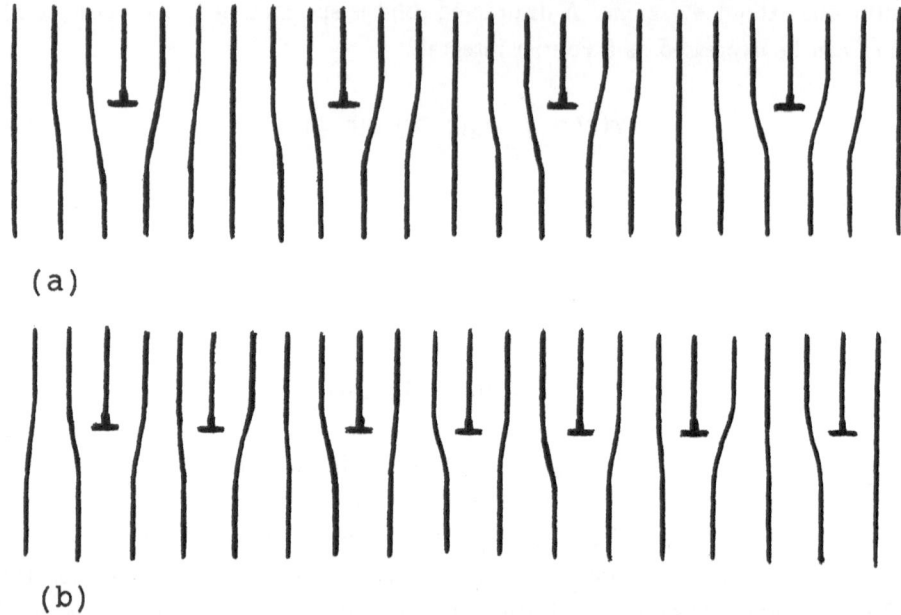

(a)

(b)

Fig.1. Misfit dislocations at interface. (a) Periodically and (b) quasiperiodically ordered ensembles of misfit dislocations.

Within the framework of the geometric model, an interface is treated as the two non–interacting atomic layers of film and substrate (Fig.2). This model directly deals with the symmetry properties of the interface, related to long–range ordering in adjacent film and substrate. In this model, atoms belonging to different layers do not interact, that is, each atom of the film (substrate) layer "feels" only the atoms of the film (substrate) layer.

When atoms of film and substrate are modeled as point balls with masses m_f and m_s, respectively, the atomic (mass) density function of the interface is

$$\rho(x) = \sum_{\alpha,\beta}[m_f\delta(x - U_f - \alpha a) + m_s\delta(x - U_s - \beta c)], \qquad (5)$$

where α and β are integers, U_f and U_s define the origins of the atomic chains of film and substrate, respectively, x denotes the coordinate along the interface line.

Since c/a is irrational, the basis one–dimensional vectors c and a of the model interface are relatively irrational. As a corollary, Fourier image of the interface (Fig.2) is a reciprocal lattice (determined by convolution of reciprocal lattices of film and substrate) having the two relatively irrational basic vectors, $2\pi/a$ and

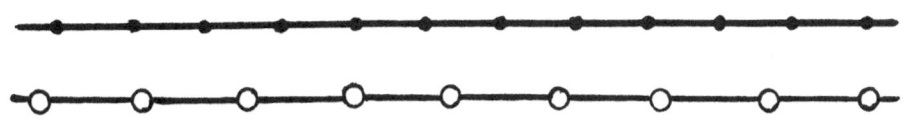

Fig.2. Quasiperiodic interface as a line with superimposed atomic layers of film and substrate. Atoms of film and substrate are shown as full and open circles, respectively.

$2\pi/c$. Then the atomic density function of the interface can be represented as:

$$\rho(x) = \sum_{p_1,p_2} \rho_{p_1 p_2} exp(\frac{i2\pi p_1 x}{a} + \frac{i2\pi p_2 x}{c} + i\varphi_1 + i\varphi_2) \tag{6}$$

where p_1 and p_2 are integers, $\varphi_1(= 2\pi U_f/a)$ and $\varphi_2(= 2\pi U_s/c)$ are the phases such that a displacement (along the interface line) of the atomic layer of the film (substrate, respectively) as a whole corresponds to a shift in the phase φ_1 (φ_2, respectively). The atomic density function (6) of the one–dimensional interface is quasiperiodic having the two relatively irrational basis vectors of its Fourier transform.

As to beyond the geometric model, analysis which takes into account both the symmetry properties of the interface and a short–range interaction between atoms belonging to film and substrate leads to the same conclusion: the interface is quasiperiodic having the reciprocal lattice generated by the two relatively irrational basic vectors (for details, see [1]).

The spatial positions of misfit dislocations are determined as follows. First, one connects each atom of the substrate with the nearest atom of the film (Fig.1). After this procedure, some atoms of the film are extracted which are not in the connection with atoms of the substrate. Such atoms of the film are treated to be associated with cores of misfit dislocations. As a result, we have a quasiperiodic interface as an interface with the smoothest (but non–periodic) arrangement of misfit dislocations separated by either distance $n'a$ or distance $(n'+1)a$, where n' is the positive integer such that $n'af < a < (n'+1)af$ (Fig.1b). It is related to the fact that misfits $-n'af$ and $-(n'+1)af$ accumulated at respective distances $n'a$ and $(n'+1)a$ can not be completely compensated by a misfit dislocation with Burgers vector a. Though n' is by definition the positive integer such that (irrational) quantities $n'f$ and $(n'+1)f$ are closest to 1, $n'af \neq a$ and $(n'+1)af \neq a$, because f is irrational. In this situation, the maximum that is possible in compensating misfit between film and substrate lattices with relatively irrational

expansion contains products of the form:

$$|\rho_1|^{s_1}|\rho_2|^{s_2} exp\{i(z_1\frac{2\pi x}{a} + z_2\frac{2\pi x}{c})\}exp\{i(z_1\varphi_1 + z_2\varphi_2)\} - (c.c.), \qquad (7)$$

where z_1 and z_2 are integers, $s_{1,2} = 0, ..., k$, $|z_{1,2}| \leq s_{1,2}$, and $s_1 + s_2 = k$.

To obtain an expression for the free energy, it is necessary to integrate the products of the form (7) with respect to the space coordinate x. For the case $z_1 2\pi/a + z_2 2\pi/c \neq 0$, all integrals of products of the form (7) with respect to x are equal to zero. The integrals of expressions (7) with $z_1 2\pi/a + z_2 2\pi/c = 0$, on the other hand, have non–zero values. However, since (one–dimensional) vectors $2\pi/a$ and $2\pi/c$ are relatively irrational, the linear combination of such vectors with the coefficients z_l is equal to zero only if $z_1 = z_2 = 0$. As a result, we find that only a product of the form (7) with parameters $z_1 = 0$ and $z_2 = 0$ are not transformed into zero by integration with respect to x. For such products, the factor $exp\, i(z_1\varphi_1 + z_2\varphi_2) = 1$ for any values of the phases φ_1 and φ_2. In this way, the free energy L is invariant with respect to changes in the two phases φ_1 and φ_2, which correspond to displacements of film and substrate, respectively. The phases φ_1 and φ_2 are hydrodynamic (Goldstone) modes or, in other words, the degrees of freedom of a quasiperiodic interface.

As noted above, hydrodynamic modes φ_1 and φ_2 parametrize respectively displacement of film $U_s = a\varphi_1/2\pi$ and displacement of substrate $U_f = c\varphi_2/2\pi$. Since the free energy of the quasiperiodic interface is invariant at any variations of the phases φ_1 and φ_2, it also is invariant for any changes of the joint displacement $U = \frac{1}{2}(U_f + U_s)$ and relative displacement $\Delta U = \frac{1}{2}(U_f - U_s)$ of film and substrate. So, the free energy is the same for any relative displacement ΔU of film and substrate. It is a very interesting feature that emphasizes the difference between quasiperiodic and periodic interfaces. In geometric terms, this feature is equivalent to the statement that the displacement–shift–complete (DSC) lattice of the one–dimensional quasiperiodic interface is the interface line. (One–dimensional periodic interfaces have rows of isolated points as their DSC lattices.)

Displacements of film relative to substrate. In spite of the fact that the free energy of a quasiperiodic interface is the same for any value of $\Delta U = U_f - U_s$, any displacement of the film relative to the substrate is related to overcoming an energetic barrier. As to details, let us consider the layer of film atoms placed in a periodic potential of substrate atoms, when the film and the substrate have relatively irrational periods (Fig.3a).

For a moment, let atoms of the film be assumed to do not interact with atoms of the substrate, that is, atoms of the film do not "feel" the periodic potential. In this situation (which corresponds to geometric model), as shown in the theory of incommensurate systems (e.g., [1,2]), the set of distances $\Delta_{\alpha\beta} = x_{f\alpha} - x_{s\beta}$ between film atoms (with coordinates $x_{f\alpha}$) and periodic potential maxima (with

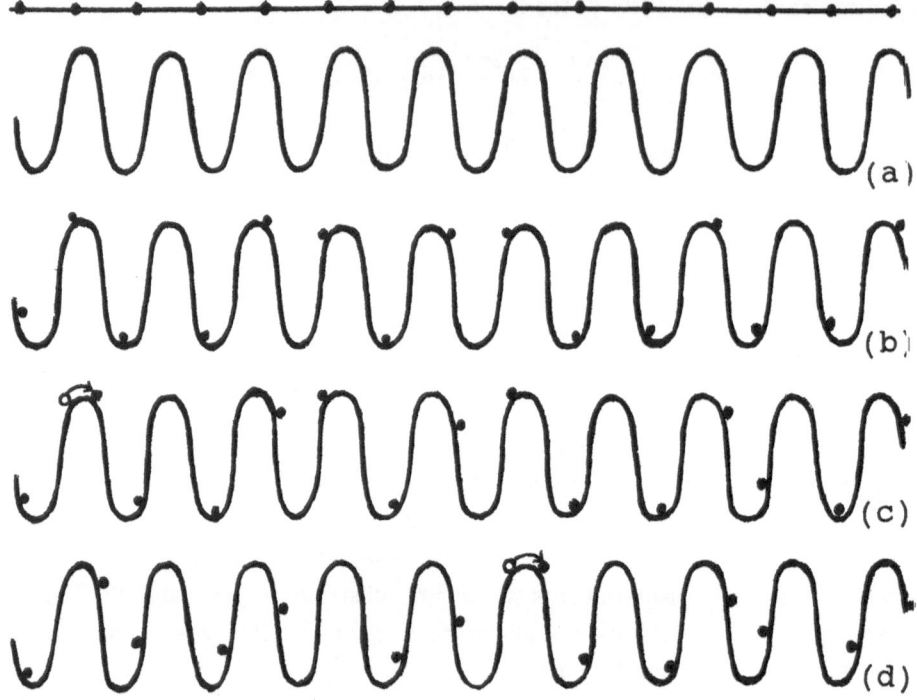

Fig.3. Atoms of film in potential created by atoms of substrate: (a) atoms of film do not "feel" the potential; (b) atoms of film "feel" the potential; (c) and (d) displacement of film occurs via jumps of film atoms.

coordinates $x_{s\beta}$) forms a DSC lattice being simply a line. As a corollary, for every distance $\Delta_{\alpha\beta}$ in the range from 0 to $c/2$, one can find a pair of atom α and its neighbouring maximum β of the periodic potential, which are separeted by distance $\Delta_{\alpha\beta}$.

Now let us return to the situation with the interaction between the atomic layers of the film and the substrate taken into account. This interaction causes atoms of the film and the substrate to be shifted to new positions, in which case the coordinates $x_{f\alpha}$ and $x_{s\beta}$ transform into $x'_{f\alpha}$ and $x'_{s\beta}$, respectively (Fig.3b). New coordinates $x'_{f\alpha}$ ($x'_{s\beta}$, resp.) depend on parameters of interaction between film (substrate, resp.) atoms, film–substrate interaction, and the coordinate $x_{f\alpha}$ ($x_{s\beta}$, resp.). For any realistic values of the above parameters, there is the following obvious tendency: the larger/smaller values of $\Delta_{\alpha\beta} = x_{f\alpha} - x_{s\beta}$ correspond to larger/smaller values of $\Delta'_{\alpha\beta} = x'_{f\alpha} - x'_{s\beta}$. In turn, a film atom characterized by a larger/smaller value of $\Delta'_{\alpha\beta}$ are placed at a lower/higher level of the potential (Fig.3b).

Let us consider a microscopic picture of a film displacement. Any such a displacement is realized via consequent-in-time jumps of film atoms associated with misfit dislocations (or, in other words, via consequent-in-time displacements of misfit dislocations) plus displacements of other atoms within "their" hollows of the potential (Fig.3c and d). In these circumstances, first, the atom jumps which is placed at the highest level of the potential, that is, the atom with the lowest value of $\Delta'_{\alpha\beta}$ (corresponding to $\Delta_{\alpha\beta} = 0$). This atom overcomes the lowest energetic barrier ΔU_{min}, when it jumps (Fig.3c). The jump of the atom and corresponding infinitesimal displacements of all other atoms within "their" potential hollows provide infinitesimal displacement of the film, resulting in a new (also quasiperiodic) configuration of the interface (Fig.3c). In this configuration a new atom of the film has the lowest value of $\Delta'_{\alpha\beta}$ (corresponding to $\Delta_{\alpha\beta} = 0$). When the film moves (by infinitesimal distance), this atom jumps and all other atoms displace within "their" potential hollows (Fig.3d). As a result, a new configuration of the atomic layers is formed with a new atom characterized by the lowest value of $\Delta x'_{\alpha\beta}$ (corresponding to $\Delta_{\alpha\beta} = 0$). Such a process repeatedly occurs providing a displacement of the film by any distance.

In the discussed scenario, consequent–in–time displacements of misfit dislocations (Fig.3c and d) occur in different places of the film–substrate interface, namely the places where the energetic barrier ΔU_q for a dislocation displacement is lowest: $\Delta U_q = \Delta U_{min}$. In this event, for any quasiperiodic configuration of the interface (or, in other terms, for any relative displacement $U_f - U_s$ of the film and the substrate), there is only one misfit dislocation characterized by the lowest barrier U_{min} for its displacement. Other misfit dislocations are characterized by values of $\Delta U_q > \Delta U_{min}$, where

$$\Delta U_q = \Delta U_{min} h(\Delta_{\alpha\beta}) \tag{8}$$

and function $h(\Delta_{\alpha\beta})$ monotonically increases as $\Delta_{\alpha\beta}$ increases. In the first approximation, we can write

$$h(\Delta_{\alpha\beta}) \approx 1 + \frac{\Delta U_{max}}{\Delta U_{min}} \frac{2\Delta_{\alpha\beta}}{(c - a)} \tag{9}$$

where ΔU_{max} is the maximum barrier for a misfit dislocation displacement, and $\Delta_{\alpha\beta}$ ranges from 0 to $(c - a)/2$.

When a film–substrate interface is periodic, displacements of a film are effectively realized via elementary displacements of misfit dislocations [8], which (locally) break the periodic translational order (Fig.4). Such breaks lead to an increase ΔU_{el} in the elastic energy of the initially periodic ensemble of misfit dislocations. In these circumstances, there is the following energetic barrier for a misfit dislocation displacement:

$$\Delta U_p \approx \Delta U_{min} h(a/2) + \Delta U_{el} \tag{10}$$

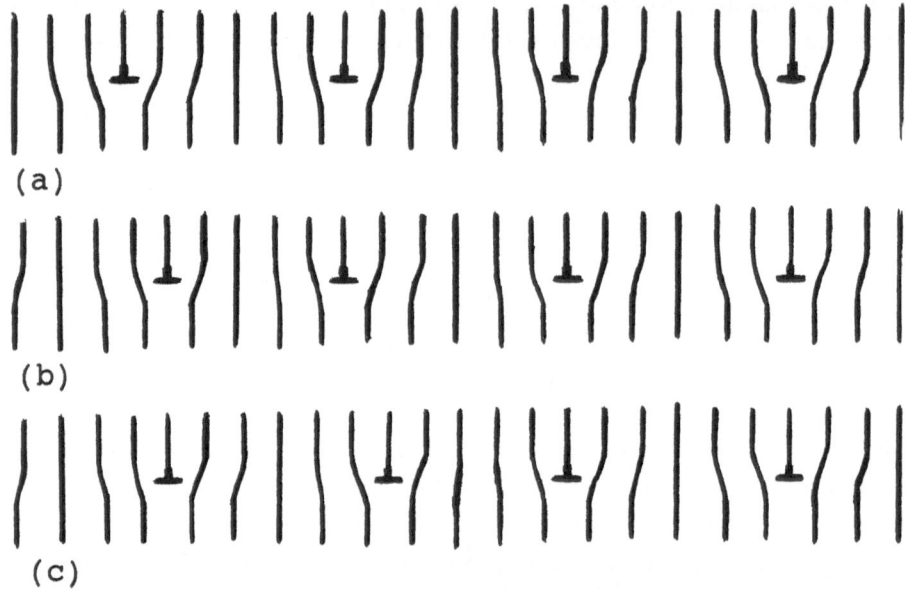

Fig.4. Displacement of film via transformations of initially periodic interface. (a) Periodically ordered ensemble of misfit dislocations (initial state). (b) and (c) Displacements of misfit dislocations, which locally break the initial periodicity.

where $\Delta_{\alpha\beta} = a/2$ corresponds to position of a film atom associated with a misfit dislocation core in the geometric model.

From (8)–(10) we find that $\Delta U_p > (or >>)\Delta U_q$. Hence film migration occuring via transformations of a quasiperiodic interface (Fig.3c and d) is more easier than that occuring via transformations of a periodic interface (Fig.4).

In paper [8] a simplified model of island film migration via displacements of misfit dislocations has been proposed. This model is simplified in the sense that it does not take into account any changes in the elastic energy of misfit dislocations, though it operates with displacements of either isolated misfit dislocations or their periodically ordered ensemble, which are undoubtedly related to changes in the elastic energy. At the same time, predictions of the model [8] are in a good agreement with experimental data (e.g., [9, 10]) on island film migration. This contradiction can be explained, if we accept that the model [8], in reality, describes island film migration occuring via transformations of quasiperiodic interfaces, in which case the contribution of ΔU_{el} to the energetic barrier for a dislocation displacement actually should be neglected. In this context, the agreement between theoretical estimates [8] and experimental data [9, 10] on island film migration can be treated as an indication of the specific properties of quasiperiodic interfaces.

3.3. QUASIPERIODIC INTERFACES OF FINITE EXTENT.
FORMATION OF QUASIPERIODIC INTERFACES

Up to now, we have examined infinite interfaces with infinite rows of misfit dislocations. Real film–substrate interfaces have finite lengths. In this subsection, we will demonstrate that interfaces of finite extent can exhibit quasiperiodic translational order even when the initial ratio a/c is rational. Let us consider an one-dimensional interface (consisting of the two atomic chains of a film and a substrate) with a finite length and a rational misfit parameter $f = 1 - a/c = 1 - m/n$. Let k_f and k_s be respectively the numbers of atoms in the chain of the film and the chain of the substrate, which belong to the interface. For given k_f, n and m (or, in other terms, k_f, a and c), the two following situations are possible: (i) ratio k_f/n is integer, and (ii) ratio k_s/n is non–integer. In the former situation, the interface is periodic, that is, misfit dislocations in the relaxed interface are periodically arranged with spacing $na(= mc)$ (Fig.1a). $k_f/n(= k_s/m)$ indicates the number of periods within the interface. In situation (ii), the smoothest arrangement of misfit dislocations is provided by quasiperiodic ordering of such dislocations. It minimizes the elastic energy density of the system [1] and, therefore, is realized when the film–substrate system reaches its equilibrium state.

In real processes of film deposition on substrates, film sizes (and, therefore, parameters k_f and k_s), in general, are arbitrary. In these circumstances, situation (i) with strict restrictions on values of n and m occurs very seldom as compared with situation (ii). In the first approximation, frequencies of appearance of situations (i) and (ii) are f and $1 - f$, respectively, where f denotes the misfit parameter. Usually, $f << 1$. As a corollary, within the framework of widely used model of film–substrate interfaces as one–dimensional interfaces with misfit dislocations [2], we come to conclusion that quasiperiodic interfaces are inherent elements of real film–substrate systems (at equilibrium).

3.4. CONCLUDING REMARKS

Thus quasiperiodic interfaces are theoretically revealed here to be inherent structural elements of film–substrate systems. In order to verify definitely this conclusion, the corresponding experimental investigations in future are of utmost interest, namely investigations which use the concept of quasiperiodic interfaces. (At

---

[1] In general, film edges influence on spatial positions of misfit dislocations, in which case this effect can violate the quasiperiodic arrangement of dislocations near the edges. Since the effect in question manifests itself near film edges only, hereafter, for simplicity, we will not take it into account.

[2] This model, being convenient for analysis, takes into account the basic features of real interfaces in film–substrate systems [4, 5].

present, in fact, most experimental studies of interfaces are based on the concept of periodic interfaces.) In addition, interpretation of available experimental data, which uses the notion of quasiperiodic interfaces in film–substrate systems, is also important.

4. Quasiperiodic Interfaces in Nanostructured Polycrystals. Quasinanocrystalline Materials

In this section we consider quasiperiodic grain boundaries in nanostructured poly-crystals and determine quasinanocrystalline solids (consisting of nanocrystallites and quasiperiodic grain boundaries) as a new type of nanostructured materials. In description of quasiperiodic grain boundaries, we will focus our attention to quasiperiodic tilt boundaries being a widespread type of interfaces.

4.1. STRUCTURAL-UNIT MODEL OF TILT BOUNDARIES. QUASIPERIODIC TILT BOUNDARIES

In terms of widely used model of structural units [11], a tilt boundary in a crystal is viewed as a packing of structural units of either one or (usually) two types, say, A and B. In general, two types, A and B, of structural units alternate with each other (in a certain consequence depending on misorientation of boundary) in boundary plane in direction perpendicular to tilt axis. Each row of structural units which is parallel with tilt axis consists of units of one type (either A or B). In these circumstances, only ordering of structural–units consequence in one direction (perpendicular to tilt axis) is significant for identification of a tilt boundary. Therefore, in context of structural–unit representation [11], hereafter we will treat tilt boundaries as one–dimensional consequences of structural units.

There are so-called favoured tilt boundaries each consisting of one type of structural units [11]. Say, boundaries described as consequences $...AAAAA...$ (Fig.5a) and $...BBBBB...$ (Fig.5b) are favoured. A tilt boundary in the misorientation range between two favoured boundaries is described as a consequence of deformed structural units of the favoured boundaries [11]. For example, tilt boundaries in the misorientation range between the favoured boundaries of $...AAAAA...$ and $...BBBBB...$ types can be represented as consequences of both A and B units (Fig.5c and d).

First, let us consider a periodic tilt boundary consisting of A and B units. It is effectively modeled as a periodic consequence of deformed units A and B, with r units A and s units B composing each one–period fragment of this boundary (Fig.5c). As with any periodic tilt boundary, value of r/s is rational, and period

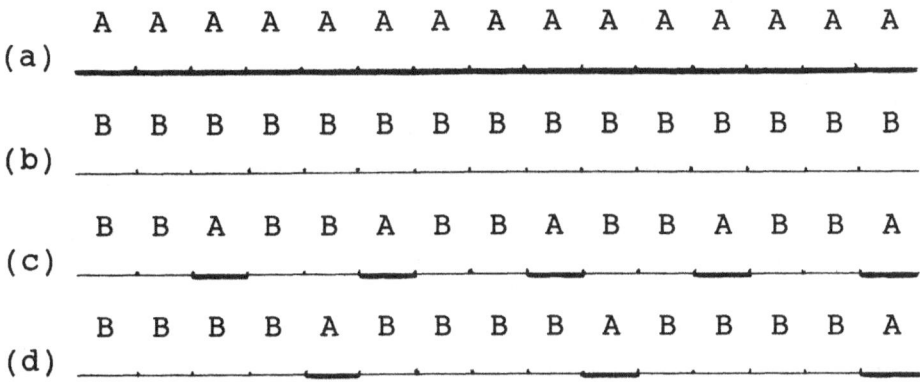

Fig.5. Structural–unit–representation of tilt boundaries. (a) and (b) Favoured tilt boundaries. (c) and (d) General tilt boundaries.

is finite.

Structural units (A or B) whose number is lower in a tilt boundary are called minority units. Hereafter, for definiteness, we will assume A units as the minority ones. With each minority unit a secondary grain boundary dislocation is associated [11]. Elastic fields of such dislocations determine the elastic energy of a tilt boundary.

Misorientation θ of a tilt boundary with the periodic structure is related to the characteristics of structural units and favoured boundaries as follows [12]:

$$\sin(\theta/2) = \frac{rd_A \sin(\theta_A/2) + sd_B \sin(\theta_B/2)}{H} \tag{11}$$

where d_A and d_B are the lengths of non–deformed A and B units , respectively; θ_A and θ_B are misorientations of the favoured boundaries consisting of only A units and only B units, respectively; H is the period of the tilt boundary. Since there are r deformed units A (each having length d'_A) and s deformed units B (each having the length d'_B) within one period of the tilt boundary, we find

$$H = rd'_A + sd'_B \tag{12}$$

The lengths d'_A and d'_B are the projections of respectively d_A and d_B on the boundary plane:

$$d'_A = d_A \cos(\theta + \kappa - \theta_A) \tag{13}$$

$$d'_B = d_B \cos(\theta + \kappa - \theta_B) \tag{14}$$

Here κ denotes the inclination angle characterizing assymetry of the boundary, in which case $\theta + \kappa - \theta_A$ ($\theta + \kappa - \theta_B$) is the tilt angle of the boundary plane relative

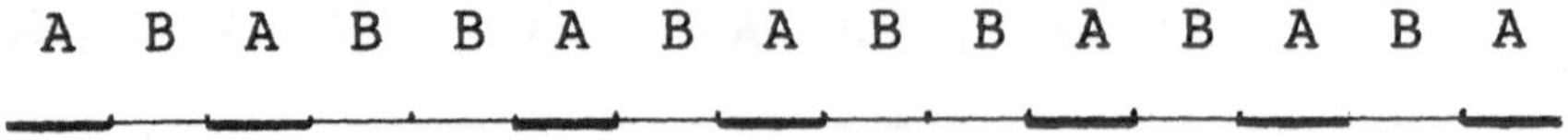

A B A B B A B A B B A B A B A

Fig.6. Quasiperiodic tilt boundary as a quasiperiodic consequence of structural
units A and B.

to the plane of the symmetric tilt boundary consisting of only non–deformed A
units (B units, respectively).

For periodic boundaries, the characteristic angle $\theta + \kappa$ is rational. When it is
irrational, d'_A and d''_B are relatively irrational (see formulae (13) and (14)). In this
case, a tilt boundary is quasiperiodic with its mass density expressed as:

$$\rho(x) = \sum_{p_1, p_2} \rho_{p_1 p_2} exp\{i2\pi(p_1 x/d'_A + p_2 x/d'_B) + i\varphi_1 + i\varphi_2\} \qquad (15)$$

where p_1 and p_2 are integers, φ_1 and φ_2 are the phases.

In terms of structural–unit model, a quasiperiodic tilt boundary with infinite
length can be modeled as a quasiperiodic consequence of $\tilde{r}$ units A and $\tilde{s}$ units
B with $\tilde{r}$ and $\tilde{s}$ being infinite and $\tilde{r}/\tilde{s}$ being irrational [13, 14]. Minority units
A in the quasiperiodic consequence are separated by either n' or $n' + 1$ units
B, where n' is the smallest positive integer such that $1/(n' + 1) < \tilde{r}/\tilde{s} < 1/n'$
(Fig.6). For given $\tilde{r}/\tilde{s}$, the quasiperiodic ordering in arrangement of $\tilde{r}A$ and $\tilde{s}B$
units provides the smoothest arrangement of minority units with corresponding
boundary dislocations. Since such dislocations repel each other, their smoothest
arrangement related to quasiperiodicity of tilt boundary provides minimization of
its elastic energy (associated with elastic fields of the dislocations) [13].

Let us consider a quasiperiodic tilt boundary with a finite length, which is
naturally defined as a finite–length fragment of an infinite quasiperiodic boundary
characterized by irrational values of $\tilde{r}/\tilde{s}$ and $\theta + \kappa$. For given finite length l
misorientation θ and inclination κ, one finds the characteristic finite numbers
r and s, of structural units as those satisfying relationships (11)–(14) with H
substituted by l. In this event, the rational value of r/s should be closest to
irrational value of $\tilde{r}/\tilde{s}$. Ordering in arrangement of A and B units in quasiperiodic
boundary with finite length is caused by the demand that elastic energy of the
boundary is minimal for given r/s or, in other words, that arrangement of minority
units with corresponding grain boundary dislocations) is smoothest for given r/s.

Let us consider briefly the properties (stress field, symmetry and grain boundary sliding) of infinite quasiperiodic tilt boundaries.

Stress field. In general, stress field components of quasiperiodic tilt boundaries are more homogeneously distributed along boundary planes as compared with periodic boundaries [15, 16].

Symmetry. Grain boundary dislocations. A DSC lattice of a quasiperiodic tilt boundary is simply a line lattice parallel to the tilt axsis [13, 17]. Since Burgers vectors of grain boundary dislocations are those of a DSC lattice, such dislocations can have any Burgers vector component normal to the boundary plane or perpendicular to the tilt axis in the boundary plane [13, 15].

Grain boundary sliding. Due to the symmetry reflected in a DSC lattice, a free energy of a quasiperiodic tilt boundary is invariant, in particular, at relative displacements of adjacent grains in the direction perpendicular to the tilt axis in the boundary plane [18, 19]. These relative displacements are realized via such rearrangements $(A \leftrightarrow B)$ of structural units that preserve the quasiperiodic ordering of the boundary [19]. Energetic barriers for such rearrangements are essentially lower than those for $A \leftrightarrow B$ rearrangements in a periodic boundary. Summing up, quasiperiodic tilt boundaries are capable of effectively carrying integrain sliding even without any involvement of grain boundary defects (dislocations). This special attribute distinguishes quasiperiodic from periodic boundaries.

4.2. QUASIPERIODIC TILT BOUNDARIES IN NANOSTRUCTURED POLYCRYSTALS

Let us consider conditions at which quasiperiodic boundaries with finite lengths are formed in polycrystalline and nanocrystalline materials. First, for example, we examine formation of grain boundaries in the highly nonequilibrium process of compacting of crystalline nanoparticles. Let a straight tilt boundary with length l be formed in cold welding of two crystalline particles. In general, inclination κ, misorientation θ and length l of this boundary can have arbitrary values as those determined by randomly varied geometric parameters of particles and their contact. Randomly-varied-in-specimen values of geometric parameters κ, θ and l of tilt boundaries, in turn, determine randomly-varied-in-specimen values of structural parameters r and s of tilt boundaries. In these circumstances, the two following basic situations can occur: (1) s/r is integer, and (2) s/r is non–integer. Since cold welding of particles is a highly nonequilibrium process, rA units and sB units in both situations (1) and (2) usually are disorderedly arranged in boundary sharing two just welded particles. More than that, the third following situation is also possible: a tilt boundary can contain structural units of other type(s), neither A or B. In all these events, a tilt boundary should be treated as a nonequilibrium (or defected) boundary.

However, relaxation processes driven by decrease in elastic energy of a tilt boundary, after some relaxation period, will result in equilibrium, ordered arrange-ment of structural units, A and B, in this boundary. In a relaxed state, ordering in arrangement of A and B units is either periodic or quasiperiodic, depending on the characteristic numbers r and s. If s/r is integer (situation (1)), a boundary is periodic. If s/r is non-integer (situation (2)), a periodic arrangement is impos-sible, and structural units of a tilt boundary of finite extent are quasiperiodically arranged.

Of utmost interest is the question: how many periodic and quasiperiodic boundaries do exist in compacted aggregates of particles, in particular, in nanos-tructured materials? Let us estimate ratio ν_p/ν_q of densities, ν_p and ν_q, of respec-tively periodic and quasiperiodic tilt boundaries [3] existing in a nanostructured material with the grain sizes in the range from 5 to 20 nm. For every tilt bound-ary in such a material, the quantity $r + s$ of structural units ranges tentatively

nanostructured polycrystals is determinitive. This allows us to view that nanostructured polycrystals with quasiperiodic grain boundaries as representing a new type of nanostructured materials with the structure and properties being different from those of conventional nanocrystalline materials with disordered and/or periodic boundaries (and of other nanostructured solids).

In order to emphasize the fact that the nanoscale structure and quasiperiodic elements are the definitive structural features of nanostructured polycrystals with quasiperiodic grain boundaries, we shall call such polycrystals as quasinanocrystalline materials.

Synthesis of quasinanocrystalline materials is an open technological problem. In any event, however, nanostructured polycrystals synthesized by presently available methods can be effectively treated as composites consisting of the quasinanocrystalline phase and the conventional nanocrystalline phase. Therefore, the notion of quasinanocrystalline materials is of importance in analysis of both the structure and the macroscopic properties of such nanostructured polycrystals. In this context, we will study in next subsection a contribution of quasiperiodic boundaries to the plastic properties of nanostructured polycrystals.

4.4. QUASIPERIODIC BOUNDARIES AND PLASTIC FLOW IN NANOSTRUCTURED POLYCRYSTALS

Let us discuss the features of plastic flow in nanostructured polycrystals related to quasiperiodic boundaries. In nanostructured solids, in which the activity of mobile lattice dislocations in nanocrystallites is low (owing to image forces that act on lattice dislocations near surfaces and hence in confined media [22]), grain boundary sliding provides an essential contribution to plastic deformation processes, e.g. [23–25]. The grain boundary sliding in quasiperiodic boundaries occurs via rearrangements of structural units or via motion of grain boundary dislocations having the features which differ from boundary dislocations in periodic boundaries. In particular, the spectrum of admissible Burgers vectors of boundary dislocations (or, in other terms, DSC lattice) in quasiperiodic boundaries usually is essentially richer than that in periodic boundaries [13, 17]. For instance, as shown in [13], dislocations in a quasiperiodic tilt boundary can have any Burgers vector component b' perpendicular to the tilt axis in the boundary plane (in contrast to periodic boundaries in which dislocation Burgers vectors are quantized). For definiteness, hereafter we confine ourself to analysis of boundary dislocations characterized by small Burgers vectors of b'-type in quasiperiodic tilt boundaries, since (as shown below) they are easily generated and easily move in quasiperiodic tilt boundaries, causing specific plastic properties of the boundaries.

As to details, let us consider a dislocation in a quasiperiodic tilt boundary, having the Burgers vector b' which is perpendicular to the tilt axis in the boundary plane and obeys the inequality: $b' << b$, where b is the characteristic Burgers vector of dislocations in periodic boundaries. Dislocations with (small) Burgers vectors b' can be generated in a quasiperiodic tilt boundary by Frank-Read sources at critical shear stresses $T' = Gb'/L$ (here L denotes the Frank-Read source length and G the shear modulus, e.g. [26]) being essentially smaller than the critical shear stress $T = Gb/L$ which activates Frank-Read sources in a periodic boundary. Say, for $b = a/10$ [27], $b' = a/50$ and $L = 10a$ (here a denotes the crystal lattice parameter), we find $T' = G/500 << T = G/100$.

Dislocations with small Burgers vectors b', generated in a quasiperiodic tilt boundary at low values T' of the shear stress, form pile-ups which are capable of effectively moving in the boundary at the same low values of the shear stress (owing to the effect of stress concentration in dislocation pile-up heads, e.g. [26]). As a result, boundary sliding processes intensively occur (via motion of dislocation pile-ups) in quasiperiodic tilt boundaries at low shear stresses T', in which case such boundaries serve as plastic elements of a nanostructured polycrystal.

Let us estimate a contribution of quasiperiodic tilt boundaries to plastic properties of nanocrystalline materials comprising both periodic and quasiperiodic grain boundaries. In doing so, we model such materials as composites consisting of the matrix, being a nanocrystalline solid with periodic boundaries, and the quasiperiodic tilt boundary phase elements. In the first approximation, the yield stress S of a nanocrystalline material modeled as the composite can be estimated with the help of the standard additive mixing formula for mechanical characteristics of composites as follows:

$$S = \psi S' + (1 - \psi)S'' \tag{16}$$

where ψ is the volume fraction of the quasiperiodic tilt boundary phase, S' and S'' are the yield stresses of the quasiperiodic tilt boundary phase and the matrix respectively.

Let us discuss consequently values of S', S'' and ψ appearing on the r.h.s of formula (16). In the light of our previous analysis, S' is determined by the critical shear stress T' which activates Frank-Read sources in quasiperiodic tilt boundaries: $S' = T'/M$, where M is the standard orientation factor.

In our model all the features of the nanoscale structure of the nanocrystalline material are assumed to be related to only the presence of a high-density ensemble of quasiperiodic tilt boundaries in the material. Therefore, we consider the yield stress S'' of the matrix to be the same as for polycrystalline solids with periodic grain boundaries, in which case S'' depends on the grain size d in the standard Hall-Petch form:

$$S'' = S^* + kd^{-1/2} \tag{17}$$

Here S^* denotes the intrinsic stress resisting crystal lattice dislocation motion and k the Hall-Petch constant.

The volume fraction ψ of the quasiperiodic boundary phase can be represented as $\psi = gh$, where g is the volume fraction of quasiperiodic tilt boundaries in the grain boundary phase and h is the volume fraction of the grain boundary phase in the nanostructured polycrystal. In nanostructured polycrystals $h \approx 9a/d$ [28]. Then we find

$$\psi \approx 9ga/d \tag{18}$$

With (17) and (18) substituted to (16), one obtains the following approximate formula:

$$S \approx S^* + kd^{-1/2} + 9ga(T'/M - S^*)d^{-1} - 9gakd^{-3/2} \tag{19}$$

For characteristic values of $a \approx 3 \cdot 10^{-10}m$, $T' \approx G/500$, $M \approx 0.5$, $G \approx 65GPa$, $S^* \approx 66MPa(\approx G/1000)$, $k \approx 26447MPa \cdot nm^{1/2}$ (see data [29] for Fe) and $g = 0.1, 0.5$ and 1, the dependences $S(d)$ defined by formula (19) are shown in Fig.7, where d ranges from 3 to $500nm$. The curves 1, 2 and 3 correspond to $g = 0.1, 0.5$ and 1, respectively. (The case $g = 1$ describes quasinanocrystalline materials with quasiperiodic tilt boundaries.)

For $d \approx 3 - 20nm$, S (curves 1, 2 and 3 in Fig.7) is the yield stress of nanocrystalline materials with quasiperiodic boundaries. For $d > 20nm$, S (curves 1, 2 and 3 in Fig.7) plays the role as the yield stress of polycrystalline solids with quasiperiodic boundaries. The dependencies S(d) with different values of parameter g have similar features which are as follows. For large values of d, the dependences $S(d)$ (Fig.7) are close to the standard Hall-Petch relationship (dashed line in Fig.7). For small values of d, the dependences $S(d)$ deviate from the Hall-Petch relationship. Similar deviations are inherent to dependencies (observed experimentally; for a review, see [24, 25]) for mechanical characteristics (yield stress, microhardness) of real nanostructured polycrystals. This allows us to think that the presence of quasiperiodic boundaries in nanostructured polycrystals is capable of effectively contributing to the experimentally observed deviations of the yield stress dependence on d from the Hall-Petch relationship.

4.5. INTERFACES IN QUASICRYSTALS

Quasicrystals are solids with long–range quasiperiodic translational ordering and non–crystallographic rotational symmetries, e.g. [1–3]. Quasicrystals (synthesized as aggregates of quasicrystalline grains [30] or nanoparticles [31]) usually contain high–density ensembles of interfaces. Quasiperiodicity in grains/nanoparticles imposes a quasiperiodic translational order to exist in such interfaces [32]. Theoretical and experimental examinations of quasiperiodic interfaces are just at the starting point.

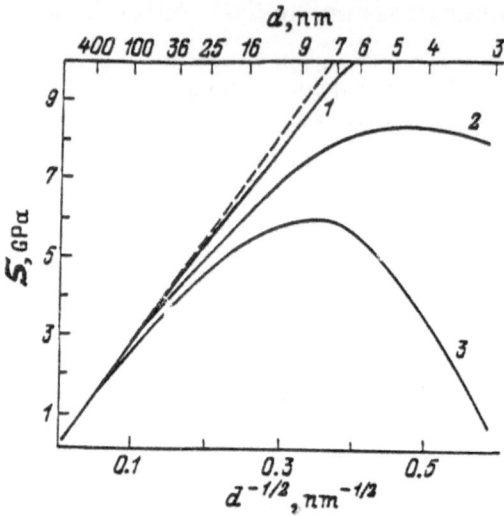

Fig.7. Dependence of the yield stress S on grain size d for $g = 0.1$ (curve 1), $g = 0.5$ (curve 2) and $g = 1$ (curve 3). The standard Hall–Petch dependence $S(d)$ is shown as dashed line.

4.6. CONCLUDING REMARKS

To briefly summarize this section, theoretical analysis indicates about quasiperiodic tilt boundaries as inherent elements of nanostructural polycrystals. Of course, this statement needs to be verified experimentally. However, at that moment, existence of quasiperiodic grain boundaries in nanostructured polycrystals can not and should not be ignored.

5. Disordered Interfaces in Nanostructured Materials

Disordered interfaces are usually formed in nanostructured polycrystals just synthesized in highly nonequilibrium conditions, e.g. [23]. Such interfaces contain high–density ensembles of interfacial defects and serve as sources of internal stresses in just synthesized nanostructured polycrystals [23]. After some relaxation period, disordered interfaces transform into ordered, periodic or quasiperiodic, ones. This process is accompanied by relaxation of internal stresses, leading to decrease of the elastic energy stored in a nanostructured polycrystal. For more details about disordered interfaces in nanostructured polycrystals, see lecture [33] and references therein.

Now let us turn to discussion of nanoamorphous alloys (firstly synthesized in 1989 [34]) which represent a new class of nanostructured materials. Nanoamorphous alloys are synthesized as high–pressure compacted aggregates of amorphous nanoparticles divided by disordered (amorphous) interfaces characterized by higher free–volume density [34, 35]. Just synthesized nanoamorphous alloys specified by nano–scale inhomogeneities of free–volume density can gradually transform into homogeneous amorphous alloys [35, 36].

So, disordered interfaces are unstable elements of the nanostructured materials structure. Disordered interfaces in nanostructured polycrystals transform into periodic or quasiperiodic interfaces. Disordered interfaces in nanoamorphous alloys gradually disappear (diffusionally spread), in which case the nanoamorphous phase transforms into the homogeneous amorphous one.

6. Conclusion

The basic conclusions of this lecture are as follows:

(a) Quasiperiodic film–substrate interfaces and quasiperiodic tilt boundaries are inherent structural elements of nanostructured materials.

(b) Quasiperiodic interfaces represent the special type of ordered interfaces with both the structure and the properties being different from those of periodically ordered interfaces.

(c) Quasiperiodic film–substrate interfaces are capable of contributing to mobility of island films.

(d) Quasiperiodic tilt boundaries serve as plastic deformation elements in nanostructured polycrystals. Their contribution induces deviations of the yield stress dependence on the grain size from the standard Hall–Petch relationship.

(e) Quasinanocrystalline materials (consisting of nanocrystallites and only quasiperiodic interfaces) represent a new class of nanostructured materials.

The following points of development in studies of quasiperiodic interfaces are of primary interest:

(i) Experimental identification of quasiperiodic interfaces in thin films, nanostructured and polycrystalline materials.

(ii) Experimental and theoretical examinations of the properties (in particular, migration, conductivity, magnetic characteristics) of quasiperiodic interfaces as well as contribution of quasiperiodic interfaces to the macroscopic properties of

204

thin films, nanostructured and polycrystalline materials.

(*iii*) Synthesis of quasinanocrystalline materials.

(*iv*) Theoretical analysis of quasiperiodicity in twist and twist–tilt boundaries in nanostructured and polycrystalline materials.

7. References

1. Lubensky, T.C. (1988) Symmetry, elasticity and hydrodynamics in quasi-periodic structures, in M.V.Jaric (ed.), *Introduction to Quasicrystals*, Academic Press, Boston etc., pp.199-280.
2. Janssen, T. (1988) Aperiodic crystals: a contradiction in terms? *Phys.Rep.* **168**, 55-113.
3. Ovid'ko, I.A. (1991), *Defects in Condensed Media: Glasses, Crystals, Quasicrystals, Liquid Crystals, Magnetics, Superfluids*, Znanie, St.Petersburg (in Russian).
4. Jain, S.C., Willis, J.R., and Bullough, R. (1990) A review of theoretical and experimental work on the structure of GeSi structured layers and superlattices, with extensive bibliography, *Adv.Phys.* **39**, 127-90.
5. Ievlev, V.M., Trusov, L.I., and Kholmyanskii, V.A. (1988) *Structural Transformations in Thin Films.* Metallurgiya, Moscow (in Russian).
6. Jesser, W.A., and Kui, J. (1993) Misfit dislocation generation mechanisms in heterostructures, *Mater.Sci.Eng.A* **163**, 101-10.
7. Van der Merwe, J.H. (1991) Strain relaxation in epitaxial interfaces, *J.Electron. Mater.* **20**, 793-803.
8. Kukushkin, S.A., and Osipov, A.V. (1995) Soliton model of island migration in thin films, *Surf.Sci.* **329**, 135-40.
9. Rubets, V.P., and Kukushkin, S.A. (1992) Determination of migration mechanism and the influence of structure of films, *Thin Solid Films* **221**, 267-70.
10. Trusov, L.I., and Kholmyanskii, V.A. (1973) *Island Metallic Films*, Metallurgiya, Moscow (in Russian).
11. Sutton, A.P., and Vitek, V. (1983) On the structure of tilt grain boundaries in cubic metals, *Phil.Trans.Roy.Soc.London A* **309**, 1-68.
12. Nazarov, A.A., and Romanov A.E. (1989) On the average misorientation of general tilt boundaries, *Phil.Mag.Lett.* **60**, 187-93.
13. Sutton, A.P. (1988) Irrational tilt grain boundaries as one–dimensional quasicrystals, *Acta Metall.* **36**, 1291-9.
14. Rivier, N., and Lawrence, A.J.A. (1988) Quasicrystals at grain boundaries,

Physica B **150**, 190-202.

15. Mikaelyan, K.N., Ovid'ko, I.A., and Romanov, A.E. (1997)
Geometric and energetic characteristics of quasiperiodic grain boundaries
in crystals, in V.I.Betekhtin (ed.), *Modern Problems of Physics and
Mechanics of Materials*, St.Petersburg State University, St.Petersburg,
pp.186-93 (in Russian).

16. Mikaelyan, K.N., Ovid'ko, I.A., and Romanov, A.E. Energetic and stress–
field characteristics of quasiperiodic tilt boundaries in polycrystalline
and nanocrystalline materials, submitted to *Mater.Sci.Eng.A*.

17. Gratias, D. and Thalal, A. (1988) Hidden symmetries in general grain
boundaries, *Phil.Mag.Lett.* **57**, 63-8.

18. Ovid'ko, I.A. (1997) Quasinanocrystalline materials, *Nanostruct.Mater.***8**,
149-53.

19. Ovid'ko, I.A. (1997) Quasiperiodic grain boundaries and intergrain sliding
in crystalline and quasinanocrystalline solids, *Phys.Sol.State* **39**, 268-73.

20. Ovid'ko, I.A. (1994) New micromechanism for strengthening in polycrystalline
solids, *Mater.Sci.Eng.A* **188**, 37-41.

21. Ovid'ko, I.A. (1995) Dislocations and quasiperiodic structures, *Phys.Stat.
Sol.(a)* **149**, 389-94.

22. Romanov, A.E. (1995) Continuum theory of defects in nanoscaled materials,
Nanostruct.Mater. **6**, 125-34.

23. Gryaznov, V.G. and Trusov, L.I. (1993) Size effects in micromechanics of
nanocrystals, *Progr.Mater.Sci.* **37**, 239-401.

24. Hahn, H., and Padmanabhan, K.A. (1995) Mechanical response of nanostructured materials, *Nanostruct.Mater* **6**, 191-200.

25. Siegel, R.W., and Fougere, G.E. (1995) Mechanical properties of nanophase
materials, *Nanostruct.Mater.* **6**, 205-16.

26. Vladimirov, V.I. (1975) Einführung in die Physikalishe Theorie der
Plastizität und Festigkeit, VEB Deutscher Verlag für Grundstoffindustrie,
Leipzig.

27. Kaibyshev, O.A., and Valiev, R.Z. (1986) *Grain Boundaries and Properties
of Metals*, Metallurgiya, Moscow (in Russian).

28. Gryaznov, V.G., Gutkin, M.Yu., Romanov, A.E., and Trusov, L.I. (1993)
On the yield stress of nanocrystals, *J.Mater.Sci.* **28**, 4359-65.

29. Armstrong, R.W., Gold, I., Douthwait, R.M., and Petch, N.I. (1962)
The plastic deformation of polycrystalline aggregates, *Phil.Mag.* **7**, 45-58.

30. Schaeffer, R.J., and Bendersky, L.A. (1988) Metallurgy of quasicrystals,
in M.V.Jaric (ed.), *Introduction to Quasicrystals*, Academic Press, Boston
etc., pp.111-42.

31. Inoue, A. (1995) Preparation and novel properties of nanocrystalline and

nanoquasicrystalline alloys, *Nanostruct.Mater.* **6**, 53-64.

32. Ovid'ko, I.A. (1994) Structural geometry of grain boundaries in icosahedral quasicrystals, *Zeit.Phys.B* **95**, 321-6.

33. Romanov, A.E., (1997) Micromechanics of defects in nanostructured materials, paper in this volume.

34. Jing, J., Krämer, A., Birringer, R., Gleiter, H., and Gonser, U. (1989) Modified atomic structure in a $Pd - Fe - Si$ nanoglasses: a Mossbauer study, *J.Non-Cryst.Solids* **113**, 167-70.

35. Gleiter, H. (1991) Nanocrystalline solids, *J.Appl.Cryst.* **24**, 79-90.

36. Wurschum, R., Rollinger, M., Kisker, H., Raichle, A., Damson B., and Schaefer H.-E. (1995) Synthesis of nanoamorphous alloys by particle condensation and compaction, *Nanostruct.Mater.* **6**, 377-80.

MICROMECHANICS OF DEFECTS
IN NANOSTRUCTURED MATERIALS

Alexei E.Romanov

Max-Planck-Institut für Matallforschung, Sestrasse 92,
D-70174, Stuttgart, GERMANY
Permanent address:
A.F.Ioffe Physico-Technical Institute, Russian Academy of Sciences

Polytechnicheskaya 26, 194021 St.Petersburg, RUSSIA

KEYWORDS/ABSTRACT: defects / dislocations / disclinations / nanocrystals / nanoparticles / thin films / interfaces / grain boundaries / mechanical properties

An overview of applications of the theory of defects (dislocations, disclinations, and interfaces) to nanostructured materials is presented. The consideration is restricted by the framework of the continuum approach. First, the defect content of interfaces: grain and phase boundaries is discussed. For grain boundaries a specific nonequilibrium state is analysed. Together with defects generated at grain junctions this state is responsible for strong elastic distortions in nanograin interior. For phase boundaries the transition between coherent and incoherent states of the interface is considered. Individual defects behaviour in nanostructured materials is investigated from a viewpoint of their interaction with different kinds of interfaces. The critical behaviour of dislocations in small particles and nanograins is considered. The role of disclinations in the formation of nanoparticles with pentagonal symmetry is analysed. The properties of dislocations in layered nanostructures are considered. In particular, for heteroepitaxial thin films two problems of technological importance are discussed: (1) prediction of mechanisms of lattice mismatch accommodation and (2) reduction of threading dislocation density.

1. Introduction

It is well established that the notation of defects in solids can be considered as the natural framework for the explanation of so-called structure-sensitive (mainly mechanical) properties of materials [1] . This also means that the theory of defects manifests itself as the language which has to be used for the explanation and modelling of structure-sensitive properties. On the other hand, theory of defects is a bridge connecting materials science with solid state physics (for example, see [2]).

When new phenomena in materials science are discovered or a new class of materials is coming into being, the defect theory should be able to explain the properties depending on the structure. In many cases the existing level of the theory of defects is usually enough for the description of such new phenomena. Sometimes, however, additional theoretical approaches and models need to be proposed which were not used at the previous stage of theoretical research in materials science. The last situation is typical for

G.M. Chow and N.I. Noskova (eds.), Nanostructured Materials, 207–242.
© 1998 *Kluwer Academic Publishers.*

the applications of the defect theory to the nanostructured materials.

Nanostructured materials were broadly invented in materials science during last decade. For a review in this field, comprehensive papers by Gleiter [3], Siegel [4], Gryaznov and Trusov [5] together with the proceedings of two recent international conferences on nanostructured materials [6] and [7] can be recommended. Nanostructured materials include nanoparticles (in particular, fullerenes and carbon nanotubes), nanocrystals, nanophase materials, multilayer nanoscale films, and quantum dots (semiconductor nanoislands on a substrate). These novel materials possess a unique complex of physical and mechanical properties [3-7]. They have a wide range of possible applications with significant technological importance. For example, nanophase films form the basis for modern electronic and optoelectronic devices [8,9]. The stability of devices directly depends on mechanical stability of the films, which is primarily determined by defect behaviour in nanosize solids.

As it was pointed by Gryaznov and Trusov [5], conventional physical models for crystal plasticity and traditional approaches of the mechanics of solids should be revised and modified to include size effects due to the presence of a high density of grain and phase boundaries in nanostructured materials [1]. In the previous review article on the properties of defects in nanostructured materials [13], we have discussed the twofold role of boundaries and interfaces. Firstly, grain and phase boundaries and their junctions may be considered as obstacles to deformation process when the deformation develops in the interior of grains. In this way interfaces contribute to the strengthening of the material. Secondly, plastic deformation in nanophase materials can take place mainly at interfaces which under special conditions can be softer than the bulk crystal. Both of the above effects can be understood by analysing the defect interactions with interfaces and junctions of interfaces together with the investigation of intrinsic properties of interfaces themselves.

The aim of the present paper is to give an introduction to micromechanics of defects (dislocations, disclinations, grain and phase boundaries) in nanostructured materials and to consider specific examples of defect behaviour in such materials.

---

[1]Manifestation of size effects can be found for the conventional polycrystalline materials, too. Well-known example is the Hall-Petch relation for the yield stress of the polycrystals in dependence of the mean grain size [10]. However, for nanosize scale the usual inverse square root dependence does not work [11] and the models have to be modified [12] to explain the experimental data.

2. Continuum approach in the defect theory

The properties of defects in solids can be investigated theoretically in the framework both atomistic and continuum approaches. In atomistic approach the main tools are molecular dynamics simulations based on analytical or empirical potentials for interatomic interactions, and *ab initio* calculations based on electron density functional or methods of quantum chemistry. The centre of attention of the atomistic approach is the short and mid-range order and deviations from this order in the material structure. In the case of defects, it is used to describe the so-called "core regions". At present the domain of atomistic approach applications is usually restricted to thousands of atoms. in case of *ab initio* calculations — to hundreds of atoms. The atomistic approach was successfully used in the description of the structure and properties of isolated atom clusters [14] and individual interfaces (in particular, grain boundaries [15]). However, it was not yet fully applied to the problem of defect behaviour in nanostructured materials.

The continuum approach was historically the first one used in the defect theory. It has its origin in the pioneering works of Vito Volterra on dislocation elasticity in the beginning of our century. The calculations of elastic fields and energies of different defect configurations are still in the centre of attention of the continuum approach, since they allow to determine long-range interactions between defects [16,17]. The other two important parts of the continuum approach are thermodynamics and physical kinetics. In the continuum approach a material (mono-, poly-, nanocrystalline, or amorphous) is considered as continuous medium with given properties (i.e. elastic, diffusional etc. properties). Defects in this case may be treated as quasiparticles which possess their own properties depending however on the properties of the continuum. Core regions of defects, in general, can not be treated in the framework of continuum approach. However, their presence can be included in the models with the help of some additional phenomenological parameters [2].

In the continuum approach nanostructured solids may be considered as composite heterophase materials. The presence of a high density of grain and phase interfaces is then taken into account in two ways: by considering these interfaces as boundaries between material domains with different elastic, plastic or diffusional properties (Fig.1a) or by treating the interfaces as new phases (Fig.1b). In many cases it is also useful to consider the possible

---

[2]An example of a such parameter is the coefficient α which is used in the relation for dislocation energy and accounts for core region contribution to the total energy [16].

nonuniformity of material properties along the interface (Fig. 1c). The last case also includes (as a variant) the presence of different defects (point and linear ones) in the interface structure (Fig.1d).

Therefore, the analysis of interface structure and properties is central to nanomaterials science research. In this analysis, a background of knowledge on grain boundaries and interfaces in coarse grain materials can be used. However, new theoretical models based on the continuum approach or its combination with atomistic calculations appear to be useful for the understanding of interface properties. Two such models, the first one for disclination content of grain boundaries, and the second one for coherent to incoherent transition at phase boundaries are considered in the Section 3.

Interfaces and their imperfections are strong sources of internal elastic stresses. Due to this fact, they interact with defects which are present in grain or phase interior. Interfaces may also serve as sinks or sources for defects. The above interactions can be classified as first-order effects. It is also possible to consider the other type of interactions between defects and interfaces. Interfaces modify, in general, the elastic field of defects, due to the fact that at interfaces specific boundary conditions for the elastic fields need to be fulfilled. As a result the energy of a defect depends on its position with respect to interfaces giving rise to corresponding configurational forces. These interactions can be classified as the second-order screening effects.

A simple scheme of defect interaction will include one defect (dislocation, disclination etc.) and interfaces of different geometry: plane interface (Fig.2a), spherical or cylindrical interface (Fig.2b), junction of interfaces (Fig.2c) and so on. For all these possibilities both first and second order effects and associated configurational forces $\vec{F}$ should be analysed. We discuss as examples the stability of dislocations and disclinations in small particles and nanograins, and the properties of misfit and threading dislocations in growing nanoscale films in the Section 4.

3. Peculiarities of structure and properties of interfaces

3.1. Disclination model for equilibrium and nonequilibrium grain boundaries

The disclination is a linear defect which bounds the surface of a cut in a continuous body, with the faces of the cut undergoing the displacement produced by mutual rotation around a fixed axis [17,18]. Due to their rotational nature disclinations are the most suitable defects for the modelling of grain boundaries [19,20] and grain boundaries junctions [21,22] in polycrystals. Recently the disclination concept was directly applied to understand the

influence of grain boundary junctions on the properties of nanocrystalline
materials [23,24].

In the disclination model a tilt grain boundary with misorientation ϕ is
modelled as a sequence of well defined structure subsets alternating along
the interface and having misorientations $\phi_1 < \phi$ and $\phi_2 > \phi$ (Fig.3a). The
length of the boundary sections l_1 and l_2 can be variable but not smaller
than the size of so-called structural units. Structural units are defined on the
atomistic level as the smallest repeating building blocks of grain boundary
structure [15]. It is assumed that the so-called favoured boundaries (ϕ_1 and
ϕ_2 in our case) with low specific surface energies γ_1 and γ_2 are built only by
structural units of one type. The average misorientation $\bar{\phi}$ (which must be
equal to ϕ) is determined by the average values $\bar{l}_1$ and $\bar{l}_2$ for the lengths of
boundary sections:

$$\sin \frac{\bar{\phi}}{2} = \frac{\bar{l}_1 \sin \dfrac{\phi_1}{2} + \bar{l}_2 \sin \dfrac{\phi_2}{2}}{\bar{l}_1 + \bar{l}_2}. \tag{1}$$

By definition, disclinations of strength $\pm\omega = \pm(\phi_2 - \phi_1)$ appear at the
junctions of the parts of different misorientation. They form a disclination
dipole chain at the grain boundary (Fig.3b). Dipole chain generate an
elastic field and therefore is associated with stored elastic energy E_{el}. In
the framework of disclination model the specific energy of a grain boundary
E^{GB} may be written as:

$$E^{GB} = E_\gamma + E_c + E_{el}, \tag{2}$$

where $E_\gamma = \dfrac{\bar{l}_1 \gamma_1 + \bar{l}_2 \gamma_2}{\bar{l}_1 + \bar{l}_2}$ is average surface energy and E_c is the contribution
of disclination cores.

In the continuum approach, the values of γ_1, γ_2 and E_c can not be found.
They should be considered as parameters of the model and can be taken
from experiments and/or computer simulations. However, the elastic part
of the energy E_{el} (depending on bulk material elastic moduli) can be easily
calculated for the given distribution of disclinations. The dipole distribution corresponding to the lowest possible energy of the grain boundary with
the given misorientation ϕ can be also found in the framework of the model
[20,25]. This distribution corresponds to practically uniform spacing between disclination dipoles (Fig.3b). Using Eq.(2), the dependence $E^{GB}(\phi)$
was calculated for symmetrical equilibrium tilt boundaries in f.c.c. materials [20, 26], this calculation is in good agreement with experimental results
for grain boundaries in Al.

Recently, a new multiscale modelling approach was proposed for calcu-lating energies of tilt boundaries in covalent materials from first principles over an entire misorientation range for a given tilt axes [27,28]. The method uses, as input into disclination model, the energies from density-functional calculations for a few key structures (i.e. γ_1 and γ_2). By simulating one ad ditional grain boundary with fixed misorientation ϕ_3, the method allows to determine the contribution of disclination cores (i.e. E_c). The advantages of the multiscale modelling approach were demonstrated by calculating en ergies of < 001 > symmetrical tilt boundaries in diamond.

The disclination model permits one to investigate the so-called nonequi-librium state of grain boundaries [25]. This state is typical for nanocrys-talline materials produced by severe plastic deformation. It has been shown [25], that the variance D_l of disclination dipole spacing is a good quantita-tive measure of the degree of grain boundary structure nonequilibrium. In a nonequilibrium state, disclination dipoles are distributed non-uniformly along the grain boundary (Fig.3c), however the average misorientation ϕ remains the same. The excess energy of nonequilibrium boundaries is pro-portional to D_l and can reach 50% of the energy of a ground state.

The transition to a nonequilibrium state may be caused, for example, by the absorption of lattice dislocations in the grain boundaries in the process of plastic deformation. The result of the absorption is the formation of the disordered array of disclination dipoles and an array of gliding grain bound-ary dislocations. A disclination dipole distribution can be also considered as a disordered dislocation wall. Grain boundaries in nonequilibrium state become powerful sources of internal stresses. The decay law for the root mean square (r.m.s.) stresses depends on the character of disorder in a dis-location wall [29]: for quasi-equidistant walls, the r.m.s. stresses diminish as $x^{-3/2}$ and for completely random walls as $x^{-1/2}$. Nonequilibrium grain boundaries can exist in nanocrystalline materials due to a high density of triple junctions - the places which are the obstacles for relaxation by sliding along interfaces.

The continuum model for the consideration of distortions in nanocrys-talline materials with nonequilibrium grain boundaries includes disclina-tions in grain boundary junctions [23], disordered walls sessile dislocations (which are equivalent to disordered distributions of disclination dipoles) [30] and arrays of gliding dislocations [31] (Fig.4). The contributions of these three components to the r.m.s. strain, stored energy and volume expansion of nanostructured materials is discussed in detail by Nazarov [31]. It has been shown that all three components of defect structure of nonequilibrium grain boundaries play a comparable role in the formation of properties of nanostructured materials. The estimated values of strain, stored energy and dilatation are in good agreement with experimentally measured characteristics for metalicl materials (Cu, Ni) with a grain size of the order of 100 nm.

3.2. Coherent to incoherent transition in mismatched interfaces

Interfaces between different crystalline phases can be coherent, incoherent, or semi-coherent [15,32]. The most important point is that the contacting phases have mismatching lattice parameters a_1 and a_2. In the case of a coherent interface, the mismatch is accommodated completely by straining the lattices of both phases. In the idealised case, both semi-infinite parts of a body turn out to be uniformly elastically deformed. This deformed state can be modelled with help of virtual coherency dislocations (CDs) uniformly distributed along the interface. In the case of semi-coherent interface in addition to CDs, localized misfit dislocations (MDs) appear at the interface to compensate for far-field elastic strains (Fig.5a). The separation l between MDs is related to the magnitude of their Burgers vector b and the absolute value of misfit parameter $f = 2\dfrac{a_1 - a_2}{a_1 + a_2}$ by $l = \dfrac{b}{|f|}$.

A completely incoherent interface can be obtained as a result of rigid contact of two crystal lattices. The transition from a semi-coherent state of the interface to an incoherent state can be described as delocalization of misfit dislocation cores. A quantitative mesoscopical model for coherent to incoherent transition at an interface with a mismatch is proposed in [33]. The model considers the intermediate partly-incoherent state of the interface where the core of each MD is delocalized over a distance m along the interface. Using a dislocation description, an incoherent region m is modelled by a distribution of anti-coherency dislocations (ADs) (Fig.5b). This approach is similar to those used in the Peierls-Nabarro dislocation model, see for example [16]. The ADs are assumed to be distributed uniformly inside each incoherent region. The last assumption permits us to find elastic strains and stresses associated with partly-incoherent interface. The energy of interface E^{PI} per unit area consists of two contributions: one connected with elastic stresses and the other one associated with short-range distortions which appear due to faults in chemical binding at the interface (this consideration is similar to those used for disclination model of grain boundary):

$$E^{PI} = E_{el} + E_c. \tag{3}$$

It has been shown in [33] that E_{el} in (3) may be found in closed analytical form. In its turn E_c can be associated with constant surface energy $\Delta\gamma$: $E_c = \frac{m}{l}\Delta\gamma$. Here $\Delta\gamma$ represents the difference in short-range (chemical distortions) energies of the coherent and incoherent state of the interface; it is treated as phenomenological parameter of the model. An important property of the dependence $E^{PI}(\frac{m}{l})$ (which can be found for different values of $\Delta\gamma$) is the presence of a minimum corresponding to an optimal (equilibrium) width $\left(\frac{m}{l}\right)^*$ of incoherent regions. Fig.5c gives the dependence of this equilibrium width on the parameters of the problem combined in

$$C = \frac{\Delta\gamma}{\dfrac{G|f|b}{4\pi(1-\nu)}}, \tag{4}$$

where G is an effective shear modulus and ν is Poisson ratio.

The proposed model predicts the transition to the incoherent state of the interfaces when the parameter C is smaller than 10, i.e. in this region $\frac{m}{l} > 0.1$. Taking $\frac{G}{(1-\nu)} = 10^{11}$Pa, $b = 3.10^{-10}$m, $\Delta\gamma = 0.5$Jm^{-2}. and varying the absolute value of misfit $|f|$ from 1% to 15%, one gets a range of values for C from 1 to 15. This estimate confirms the validity of the model for the analysis of experimentally investigated interfaces. There is a possibility for changing the state of interface by varying the material properties $G, b, \Delta\gamma$, and f. It is clear that it is not possible to increase or decrease drastically the values of elastic constants or crystal lattice parameters. Therefore, the state of interface can be most effectively influenced by the value of the misfit f and also by the interface energy change $\Delta\gamma$. The last effect can be achieved, for example, by impurity segregation at the interface. For a quantitative application of the proposed model to real interfaces (for example, to metal-ceramics interfaces like Nb/α-Al$_2$O$_3$ semi-coherent interface [34] or Cu/TiO$_2$ incoherent interface [35]) some modifications should be made. The model should deal with the interfaces between anisotropic phases with different elastic constants (in metal-ceramics interfaces this difference can be 500%). It should operate not only with tangential misfit, but also with shear and rotational misfit. The first problem can be tackled by applying results of the theory of dislocations obtained for defects in non-uniform and anisotropic media. The second problem will include an analysis of screw dislocations (shear misfit) and wedge disclinations (misorientation at interfaces). Additionally, the distribution of anti-coherency dislocations in incoherent regions can be non-uniform. In this case, $\Delta\gamma$ will depend also on this non-uniformities. In general, however, $\Delta\gamma$ can not be found in the framework of mesoscopic considerations. The progress in this direction is possible by applying computer simulations.

4. Defects in nanosize particles and films

4.1. Stability of dislocations and disclinations in small material volumes

The problem of stability of dislocations in small particles and nanocrystals has been addressed in [36,37]. In these studies, the existence of critical size Λ_i for dislocation instability in nanoparticles or nanograins has been predicted. Below this size which depends on such material parameters as elastic (shear) modulus (G) and lattice resistance (σ_f) to the dislocation motion, gliding dislocations are unstable in the nanovolume interior. The reason for such an instability are the configurational forces $\vec{F}$ of an elastic nature that strongly depend on nanocrystallite size.

The key to the understanding of dislocation stability in small volumes can give the solution of a set of elasticity boundary value problems for dislocations. Starting point here is the type of boundary conditions which should be fulfilled for dislocation elastic field. The following boundary conditions correspond to important physical situations:

a) traction free surface S_F, with stresses σ_{ik} satisfying the following condition:

$$\sigma_{ik}n_k|_{S_F} = 0, \tag{5}$$

where n_k are components of the unit vector normal to the surface S_F;

b) interface S_P between domains PI and PII with different elastic constants at which the continuity of displacements $\vec{u}$ and forces takes place:

$$u_l^I|_{S_P} = u_l^{II}|_{S_P},$$

$$\sigma_{ik}^I n_k|_{S_P} = \sigma_{ik}^{II} n_k|_{S_P}; \tag{6}$$

c) Slipping interface S_S at which all tangentional component of forces are equal to zero, but continuity of normal components of displacement and forces is still present:

$$u_l^I n_l|_{S_S} = u_l^{II} n_l|_{S_S},$$

$$n_i \sigma_{ik}^I n_k|_{S_S} = n_i \sigma_{ik}^{II} n_k|_{S_S},$$

$$\left[\sigma_{ik}^{\{I,II\}} n_k - n_m \sigma_{mk}^{\{I,II\}} n_k n_i \right]|_{S_S} = 0. \tag{7}$$

The next step is the choice of the geometry for both dislocation line and boundary surface. The choice is usually dictated by symmetry considerations (Fig.6). For example, this can be a straight dislocation in spherical particle with free surface (Fig.6a) and the exact solution for a screw dislocation in such a geometry was found in [38]. It can be a prismatic dislocation loop in a spherical particle embedded in the matrix with different elastic properties (Fig.6b), or it also can be a straight dislocation parallel to the axis of cylindrical domain with a slipping interface (Fig.6c). To investigate the behaviour of dislocations in nanograins, the influence of the surrounding material is taken into account by introducing average effective elastic moduli $G^{<m>}$ for matrix and/or slipping boundary conditions at the nanograin-matrix interface [5,36,37].

As the result of the solution of boundary value problem, the elastic energy of dislocation configuration E_{el} can be found as a function of set of parameters Ψ defining the geometry of the problem: $E_{el} = E_{el}(\Psi)$. For example, this set Ψ includes the position and shape of a dislocation line. Generalised derivatives of E_{el} with respect to Ψ define so-called configurational forces acting on a dislocation $\vec{F} = -\partial_\Psi E_{el}$. Thus, one can investigate the configurational force acting on a small part of a dislocation line depending upon its position inside particle or grain. The force in this case is called as an image force $\vec{F}^{image}$ [16]. A comparison of the image force with the friction force to a dislocation motion in a crystal lattice gives the condition of dislocation instability in small volumes

$$F^{image}(L) = b\sigma_f, \qquad (8)$$

where b is the value of dislocation Burgers vector, and the frictional stress σ_f in simplest cases can be taken to be equal to Peierls stress σ_P. In Eq. (8) the dependence of the configurational image force on nanocrystallite size L is specifically indicated.

The numerical analysis of the Eq. (8) for different types of dislocation configurations, different types of boundary conditions and various geometry of nanograins gives the critical length Λ_i for dislocation stability in nanoparticles and nanograins [5,36,37]. Typical values for Λ_i are presented in Table 1. The main results concerning the dislocation stability in nanocrystals indicate that in the nanometer length scale, dislocation arrangements and densities drastically change. Configurational forces become big enough to overcome the resistance of crystal lattice to dislocation motion. Mobile dislocations can easily annihilate in the grain interior or at grain or phase boundaries. In the last case they also will contribute to the transition of interface in nonequilibrium state.

When the size of grains is smaller the volume fraction of triple junctions is larger, their influence on material properties becomes more pronounced [39]. As it was mentioned, interface junctions can be considered as a source of elastic stresses in nanostructured materials due to the disclinations generated at these junctions (see, Fig.2c). This causes the first-order effect of dislocation interaction with triple junctions. Triple junctions also modify the distribution of dislocation elastic fields due to the modulus effect (second-order screening effect). To investigate the stability of dislocations near interface junctions the boundary value problem for dislocations in three-phase continuum has been solved [40-42].

There is a lot of experimental data on the presence in small atom clusters and nanoparticles of fivefold symmetry axes (for a review see [5] and [17]). Mainly these are decahedral and icosahedral [3] particles or prismatic whiskers with pentagonal cross-section of the materials with f.c.c. crystal structure (Fig.7a-c). The formation of such particles and whiskers is a result of multiple twinning. Therefore, the intrinsic defects in these nanoobjects are twin boundaries and positive wedge disclinations of strength

$$\omega_p = 2\pi - 10\sin^{-1}\frac{1}{\sqrt{3}} \approx 7°21' \qquad (9)$$

at junction of five twin boundaries (Fig.7d) [17,44]. In a decahedral particle or pentagonal whisker there is one disclination ω_p laying along the fivefold symmetry axis and in an icosahedral particle there are six such disclinations connecting the opposite vertexes of the icosahedron.

To take advantage of the cintinuum description, decahedral and icosahedral particles are modelled as spherical particles containing disclinations (Fig.2b) and pentagonal whiskers as long cylinders with disclinations inside. Due to the presence of disclinations, nonuniform elastic distortion are generated in pentagonal nanoparticles. The exact solution of elasticity boundary value problem for wedge disclination in spherical particle with boundary conditions of the type given by Eq.(5) has been found in [45]. With the growth of a particle size (radius of sphere A or cylinder R) the stored elastic energy E_{el} associated with disclinations rapidly increases

$$E_{el} = \zeta_A G\omega_p^2 A^3, \qquad E_{el} = \zeta_R G\omega_p^2 R^2, \qquad (10)$$

[3]One recent example is the paper [43] dealing with the properties of Pd icosahedral nanoparticles supported by MgO substrates.

where coefficients ζ_A and ζ_R account for a disclination position with respect the centre of a sphere or a cylinder and also for dependence on Poisson ratio.

In a small particle, the expense in elastic energy is balanced by a presence of low energy close-packed crystallographic planes forming particle shape. For decahedral and icosahedral particles these are {111} planes, and for pentagonal whiskers these are {111} planes in the caps and {100} planes in the sides (fig.7). At larger particle sizes, it becomes energetically favourable to have particles with monocrystalline undeformed structure and average surface energy $\bar{\gamma}$ that is larger than surface energy in close-packed faces, i.e. γ_{111} or γ_{100}. This is due to the fact that the total gain in the surface energy $\Delta\Gamma$ increases slower than the excess of the elastic energy:

$$\Delta\Gamma_A = \xi_A(\bar{\gamma} - \gamma_{111})A^2, \qquad \Delta\Gamma_A = \xi_R(\bar{\gamma} - \gamma_{100})R, \qquad (11)$$

where ξ_A and ξ_R are coefficients accounting for the real geometry of a particle or whisker and also for the presence of internal twin boundaries with the surface energy γ_t.

Therefore there exists a critical size Λ_p below which decahedral, icosahedral particles and pentagonal whiskers remains stable. This size can be determined from the balance between E_{el} and $\Delta\Gamma$. For example for pentagonal whiskers and icosahedral particles one finds [5,17]

$$\Lambda_p^W \approx \frac{16\pi(1-\nu)}{G\omega_p^2}(2\bar{\gamma} - 10\gamma_{100}\sin\frac{\pi}{5} - 5\gamma_t),$$

$$\Lambda_p^P \approx \frac{3}{32G\omega_p^2}\frac{1-\nu}{1+\nu}(\bar{\gamma} - \gamma_{111}). \qquad (12)$$

Estimated with help of Eq.(12)$_2$, the values for Λ_p^P are between 5 and 50 nm, depending on the material, which agrees well with experimental data. Numerical estimates of Λ_p^W give values of the order of hundreds of nanometers which also agrees with experiment.

Disclination elastic stresses may relax in nanoparticles or nanowhiskers in other ways. The main relaxation mechanisms has been discussed in [46] and they are schematically shown in Fig.8. First, this can be the generation of dislocation (Fig.8a) having Burgers vector perpendicular to the cylinder radius. Disclination stresses do negative work during such dislocation formation that can significantly lower the stored elastic energy. The same influence has the displacement of the pentagonal axis (motion of the disclination) (Fig.8b), splitting of the pentagonal axis (formation of new partial disclinations) (Fig.8c), and generation of a pore in the centre of particle

(Fig.8d). The last case has been investigated in detail [47]. It is important to point here that the above described relaxation mechanism are associated with well defined critical sizes Λ_r of pentagonal nanoparticles. They can operate when a particle (or a whisker) has its characteristic size L larger than Λ_r. The other comment concerns the experimental observation of the pentagonal nanoparticles with relaxed structure. All the described mechanisms were experimentally observed, see references in [46] and [47].

The results of the present section are summarised in the Table 2, where it is shown that several scales of the instabilities for the geometry and structure of a particle can be assigned to nanosize material volumes: instability of the particle shape (in relation to the effect of electrostatic forces (characteristic scale Λ_e) [36]); image instability of glissile dislocations in free particles and nanograins (characteristic scale Λ_i); the pentagonal instability in relation to the transition twinned particle — single crystal (characteristic scale Λ_p); relaxation instability of pentagonal particles in relation to formation of additional structural defects in nanoparticle interior (characteristic scale Λ_r); and misfit instability of nanosize films in relation to misfit dislocation formation at the film/substrate interface, this case is considered in the next section, where the critical film thickness h_c is defined. All the above mentioned instabilities have their characteristic scales Λ in the range from 1 to 100 mn. This fact makes the defect micromechanics in nanostructured materials different on those known for conventional crystalline materials.

4.2. Misfit and threading dislocations in growing films

The mismatch, f, in lattice parameters at a coherent film/substrate interface leads to generation of elastic strains and stresses in contacting layers (Fig. 9a,b). In the Section 3.2 we have already discussed the mechanism of stress relaxation associated with misfit dislocation (MD) formation and their core delocalization at interface between two semi-infinite parts of different phases. When the thickness of contacting layers are finite there exists a possibility for elastic bending (Fig.9c) which also gives a diminishing of stored elastic energy. In the most relevant technological case of thin films on thick substrates, the internal stress in the substrate is negligibly small and all relaxation processes take place in the film. In an alternative to MD formation (Fig.9d) the following deformation relaxation modes can operate in growing nanoscale films: roughening of the film surfaces (Fig.9e) [48] and formation of elastic domain patterns (Fig.9f) [49,50]. If the material of the film is brittle or cohesion at film/substrate interface is low two types of fracture modes [51]: normal to interface crack opening (Fig.9g) and loss of coherency along the interface (Fig.9h) manifest themselves depending on the sign of misfit stress (tension or compression). In this paper we will, however, concentrate on dislocation mechanism of stress relaxation in growing films.

The energetics of MDs has been investigated in the framework of continuum approach in [52,53] by taking into account the interaction of MDs with misfit stress, free surface of the film and interaction between misfit dislocations [4]. It has been shown in [52] that starting at critical film thickness $h_c(f)$ the minimum of the film-substrate energy will correspond to the presence of MDs at the interface. The critical thickness then can be found from the following relation [52,54]

$$h_c = \beta \frac{b}{f} \ln \frac{\alpha h_c}{b}, \qquad (13)$$

where b is magnitude of dislocation Burgers vector, a factor β accounts for Burgers vector $\vec{b}$ orientation with respect to dislocation line and film/substrate interface (for pure edge dislocations with Burgers vector lying in film/substrate interface $\beta = \dfrac{1}{8\pi(1 + \nu)}$ [54]), and the coefficient α describes both the core radius approximation invoked in dislocation energy calculations and the boundary conditions at the film free surface. Estimates for critical thickness obtained from Eq.(13) give, for example for $f = 0.01$, $h_c \approx 5$nm.

For a fully relax (thick) films, the linear density of MDs is $\rho_{MD}^{\infty} = \dfrac{1}{l} = \dfrac{f}{b}$. For films of finite thickness that are grown on semi-infinite substrates, the equilibrium linear MD density may be readily shown to scale with the film thickness h in the following manner [55]

$$\rho_{MD} = \rho_{MD}^{\infty}(1 - \frac{h_c}{h}). \qquad (14)$$

The conventional versions of continuum consideration of misfit strain relaxation deal with uniform distribution of dislocations with their positions directly in the interface. However, recently, new effects in misfit dislocation arrangements have been predicted theoretically and found experimentally: the displacement of dislocations in an equilibrium stand-off position [56] (Fig.10a) and the formation of nonuniform dislocation distribution with characteristic length λ along the interface [57] (Fig.10b).

---

[4] Reviews on misfit dislocations [54,8,9] together with classical works of Matthews [32] present a state of art in this field.

The shift of MDs is of an elastic origin [56,58]. The image force repels MDs into elastically softer phase (with lower value of elastic constants) due to the second order modulus effect. On the other hand, the MDs departure leads to a formation of elastically strained interlayer between a new position of the MDs and phase boundary (the first order interaction). The corresponding attractive misfit force is in equilibrium with the image force. The magnitude of the shift c depends on the ratio of elastic shear moduli of contacting phases $g = \dfrac{G_2}{G_1}$ and misfit f:

$$c \approx |\frac{A + B}{2\pi(4 - 3A - B)}|\frac{b}{f} \tag{15}$$

where $A = \dfrac{1 - g}{1 + \kappa g}$, $B = \kappa\dfrac{1 - g}{\kappa + g}$, $\kappa = 3 - 4\nu$ and the Poisson ratio for both phases, for simplicity, is considered to be the same: $\nu = \nu_1 = \nu_2$. The estimated with the help of Eq.(15) value of the shift c reaches several lattice parameters. In the case of two-phase films, the interaction of dislocations with free surfaces also contributes to the equilibrium stand-off position [59].

The formation of misfit dislocation groups can be explained in continuum approach by the consideration of back stresses produced by finite dislocation arrays at the surface of a film [57]. These back stresses can lock the dislocation sources operating at the film surface. The energetics of finite dislocation groups have been considered in [60,61]. Finally, the problem of MD group formation has been treated in the framework of dislocation kinetics [62,63].

The continuum approach demonstrates its effectiveness in the solution of more sophisticated problems concerning MD behaviour in thin films. For example, the behaviour of MDs in multilayer structures (superlattices) is considered in [64-66]. For a two-phase film, the general solution of elasticity problem for an edge dislocation has been found which satisfy boundary conditions at both free surfaces (condition of the type given by Eq.(5)) and interface (conditions of the type given by Eqs.(6)) [67,68]. This solution has been used for the explanation of MD behaviour in the systems having comparable thicknesses of film and substrate [58]. The analysis of critical thickness for MD formation in a compliant substrate/film system has been recently given in [69].

Relaxation of crystal lattice mismatch by the formation of MDs is usually accompanied by the appearance of so-called threading dislocations (TDs). TD lines go through the film interior and end at the free surface of the film.

For example for layer-by-layer film growth, the assumption is that dislocation half-loops nucleate at steps or impurities on the growing free surface
[70], and the loops expand to form two threading segments with antiparallel
Burgers vectors, and misfit segment at the film/substrate interface. Other
film growth modes also eventually lead to TDs formation. Commonly the
TD densities up to $\rho_{TD} = 10^{10} \text{cm}^{-2}$ are reported (for overview of some
experimental results, see [71]). As opposite to MDs, threading dislocations
remain in active regions of electronic and optoelectronic devices. Generally, TDs degrade the physical properties of devices. Thus it is of great
importance to find effective means for the elimination or reduction of TD
density.

TDs are nonequilibrium defects that raise the free energy of the film (on
the contrary to MDs which are defects with the equilibrium density given
by Eq.(14)). Therefore, there is a thermodynamic driving force to diminish
the TD density. The high densities of TDs facilitate the development of a
kinetic approach for TD reduction [72]. The kinetic approach considers the
reactions between TDs in relation to their densities and relative motion;
this is effectively is a "reaction kinetics" treatment. The TD motion r is
defined as the lateral movement of the intersection point of the TD with the
free surface. It depends on the parameters of the problem: film thickness h
and misfit f, but also additional internal and external factors (for example
the concentration of point defects). Changes in film thickness h give rise
to lateral TD motion (Fig.11a). A vacancy or interstitial supersaturation
may lead to TD climb (Fig.11b). Misfit strain may lead to MD generation
and concurrent TD motion (Fig.11c). This effect has special features in
superlattices where TDs can have a set of successive displacements (Fig.11c).

Together with relatively slow TD motion, the other more rapid process
can be defined in a TD ensemble, namely, dislocation reactions. One characterizes the reactions by the radius r_I. The reaction radius represents the
distance at which the interaction between TDs is sufficient to overcome the
Peierls barrier σ_P for TD glide or climb. Once initiated, this movement results in reaction product which depends on the reacting dislocation Burgers
vectors. The possible reactions between TDs include annihilation, fusion,
and scattering. In the annihilation reaction, TDs with opposite Burgers
vectors that fall in annihilation radius r_A react and stop the propagation
of both TDs to the film surface. In a fusion reaction with a characteristic
reaction radius r_F, two TDs react to produce a new TD with Burgers vector that is the sum of Burgers vectors of the reacting TDs. The possible
reactions between TDs in the case of (001) epitaxial film growth have been

analysed in details in [71,73].

The above outlined approach was applied for the first time [70] for the analysis of the TD reduction in homogeneous buffer layers to explain experimentally observed dependence of TD density on the buffer layer thickness h. As an example, consider the possibility of growing a relaxed (i.e. free from misfit strain) layer with finite lateral dimension λ as it might be realized in selective are growth over a relaxed film with a high TD density. For this case, the change in TD density with film thickness may be written as [71,74]

$$\frac{d\rho_{TD}}{dh} = \frac{\rho_{TD}}{\Lambda} - K\rho_{TD}^2, \tag{16}$$

where $\Lambda = \dfrac{\lambda}{J}$, $K = 2Jr_I$, and $J = \dfrac{dr}{dh}$ is a geometric factor that describes TD motion during film growth and commonly $J \approx 1$ for inclined TDs in (001) cubic semiconductors films [72]. Eq.(16) gives the following solution if ρ_o is the initial density of TDs at $h = h_o$:

$$\rho_{TD} = \frac{\rho_o}{(1 + K\Lambda\rho_o)\exp[\dfrac{h - h_o}{\Lambda}] - K\Lambda\rho_o}. \tag{17}$$

This solution has two asymptotes that are relevant for TD reduction. In the first limit, the mesa size is large in comparison with film thickness, $\Lambda >> h - h_o$, and solution predicts the inverse proportional dependence of TD density with film thickness [5]. This is an agreement with experimental observations of TD density in homogeneous buffers. In the case $\Lambda \to 0$, which is a limiting case of small mesas, there is exponential decay of TD density with film thickness [71,74].

It has been demonstrated that the continuum approach is helpful in treating the problem of TD behaviour in stressed layers [71], superlattices [77] and also in the case of diffusion assisted motion of dislocations [78]. Under these conditions, the effect of TD bending is very important for cleaning dislocations from the working zones of semiconductor devices. Considering change of the superlattice system energy during misfit dislocation formation [65] and the probability of TD annihilation [77], one can find the conditions for superlattice cleaning that connect number and thickness of layers, misfit between layers, and initial density of TDs.

---

[5]The full crystallographic details of this problem have been included into consideration in [71,73,75]. Computer simulation approach has also been used [74,76] in addition to the continuum description of TD reduction. It has been found a very good agreement between results of continuum and computer simulation approaches.

5. Concluding remarks

We have considered a set of the problems for defects in nanostructured materials. It has been demonstrated that the continuum models serve as a powerful tool for the explanation of defect behaviour and mechanical properties in nanosize solids. Different critical parameters with the dimension of length have been defined that are of nanometer length scales: critical particle or nanograin size of dislocation instability, critical thickness of thin films for the generation of misfit dislocations, critical size of the stability of pentagonal small particles with disclinations and so on (see Table 2). The reduction law for threading dislocation density has been also connected with geometrical properties of the growing film i.e. the thickness and the size of mesas.

Based upon the presented results, we will expect the appearance of size effects in mechanical and other structure sensitive properties of solids in region of nanometer scale. For example, dislocation instability can explain low deform-ability of small particles. Anomalies in flow stress and micro-hardness of nanocrystals may be connected with the peculiarities of defect structure and the character of internal stress distributions. The diffusional phenomena are also strongly modified for nanoscale materials. Finally, grain boundaries in nonequilibrium state and grain boundary junctions contribute to the stored energy and volume change of the material with a strong dependence on the crystallite size.

Acknowledgements

The present work was supported by the grant # 97-3006 from the Russian Research Council "Physics of Solid Nanostructures", Russian Foundation of Basic Researches grant # 96-2-16807-A,   INTAS   Program grant # 94-4380, and Max-Planck-Gesellschaft (Germany).

Figure Captions

Figure 1. Continuum models for interfaces.
a) sharp phase boundary between domains PI and PII with different elastic, plastic or diffusional properties (G_i, $\sigma^i_{0.2}$ and D_i are shear modulus, yield stress and diffusion coefficient for the phase i);
b) interphase layer PIII (a new phase) with different properties;
c) phase nonuniformities along the boundary between grains GI and GII;
d) crystal lattice defects (dislocations) at the boundary between grains GI and GII.

Figure 2. Scheme for defect interactions with interfaces in nanostructured materials.
a) a dislocation in a film on a substrate (interaction with phase boundary and a free surface);
b) a disclination in a spherical small particle;
c) a dislocation interacting with the triple junction of grain boundaries where a disclination is placed.
$\vec{F}$ indicates the configurational force acting on a defect.

Figure 3. Disclination model of grain boundaries.
a) Decomposition of the grain boundary of misorientation ϕ into sections with misorientations ϕ_1 and ϕ_2;
b) Practically uniform distribution of disclination dipoles for equilibrium grain boundary;
c) Strongly nonuniform distribution of disclination dipoles in nonequilibrium grain boundary.

Figure 4. Components of grain boundary defect structure in nanocrystals.
(1) Junction disclinations, (2) disordered walls of sessile dislocations (they are equivalent to disordered arrays of disclination dipoles), and (3) arrays of gliding dislocations.

Figure 5. Dislocation content of the mismatched interface.
a) semi-coherent state of interface, CD denotes coherency dislocations, MD denotes localized misfit dislocations;
b) partly-incoherent state of interface, AD denotes delocalized misfit (anti-coherency) dislocations;

c) equilibrium width of incoherent regions $\left(\dfrac{m}{l}\right)^*$ as function of the parameter C defined by Eq.(4).

Figure 6. Possible dislocation configurations in nanocrystallites.
a) straight dislocation with Burgers vector $\vec{b}$ in a spherical particle with a
free surface S_F;
b) prismatic dislocation loop in a spherical grain embedded in matrix with
different elastic properties (phase boundary S_P);
c) straight dislocation in a cylindrical domain surrounded
by slipping interface S_S.

Figure 7. Disclination content of pentagonal nanoparticles
and whiskers.
a) a decahedral particle with single wedge disclination ω_p.
b) an icosahedral particle with six disclinations ω_p;
c) pentagonal needle-like prismatic crystal;
d) internal structure of multiple-twinned particles and whiskers in a cross
section perpendicular to five-fold axis (in insert the configuration about
the central atom before introduction of disclination is shown).

Figure 8. Main mechanisms of internal stress relaxation
in pentagonal nanoparticles.
a) edge dislocation generation;
b) displacement of the disclination from the centre of a particle;
d) splitting of a disclination core $\omega_1 + \omega_2 = \omega_p$;
c) pore formation in a disclination core.

Figure 9. Relaxation of atomic misfit in film/substrate systems.
a) film and substrate in unconstrained state with their lattice parameters
$a_f \neq a_s$;
b) constrained state of film/substrate system resulting in internal stress field
σ.

Different modes of internal stress relaxation:
c) elastic bending;
d) misfit dislocation (MD) generation;
e) roughening of the free surface of a strained film;
f) domain pattern formation;
g) cracking due to tensile stresses;
h) loss of coherency at the interface due to compressive stresses.

Figure 10. Some peculiarities of misfit dislocations in nanoscale
heterostructures.
a) stand-off position c of dislocations near the interface;
b) nonuniform MD distribution along interface with characteristic length λ.

Figure 11. Reasons for lateral threading dislocation motion in film
nanostructures.
a) differential lateral motion dr of an inclined threading dislocation (TD)
due to changing film thickness dh;
b) dislocation climb motion due to condensation of nonequilibrium point
defects;
c) dislocation glide motion due to relaxation of misfit strain and by misfit
dislocation (MD) formation;
d) annihilation of TDs in strained superlattice.

Table 1. Critical length Λ_i of dislocation stability for metal
nanopaticles and nanocrystals [5].

		Cu	**Al**	**Ni**	α-**Fe**
Free particle Λ_i^F (nm)	sphere	250	60	140	23
Nanograin with slipping interface Λ_i^S (nm)	sphere	38	18	16	3
	cylinder	24	11	10	2

Table 2. Hierarchy of instability scales for defects in
nanostructured materials.

Type of instability	Characteristic scale	Type of boundary conditions	Size of a particle A, R or film thickness h
Electrostatic	Λ_e	Particle on substrate	$A < \Lambda_e = 1 \div 2\text{nm}$
Image	Λ_i	Free particle	$A < \Lambda_i^F = 20 \div 200\text{nm}$
		Nanograin	$A < \Lambda_i^S = 3 \div 40\text{nm}$ $R < \Lambda_i^S = 2 \div 52\text{nm}$
Pentagonal	Λ_p	Free particle	$A < \Lambda_p^P = 5 \div 50\text{nm}$
		Whisker	$R < \Lambda_p^W = 50 \div 500\text{nm}$
Relaxation	Λ_r	Free particle or whisker	$A, R > \Lambda_r = 1 \div 100\text{nm}$
Relaxation	h_c	Epitaxial film on substrate	$h > h_c = 1 \div 100\text{nm}$

Figure 1.

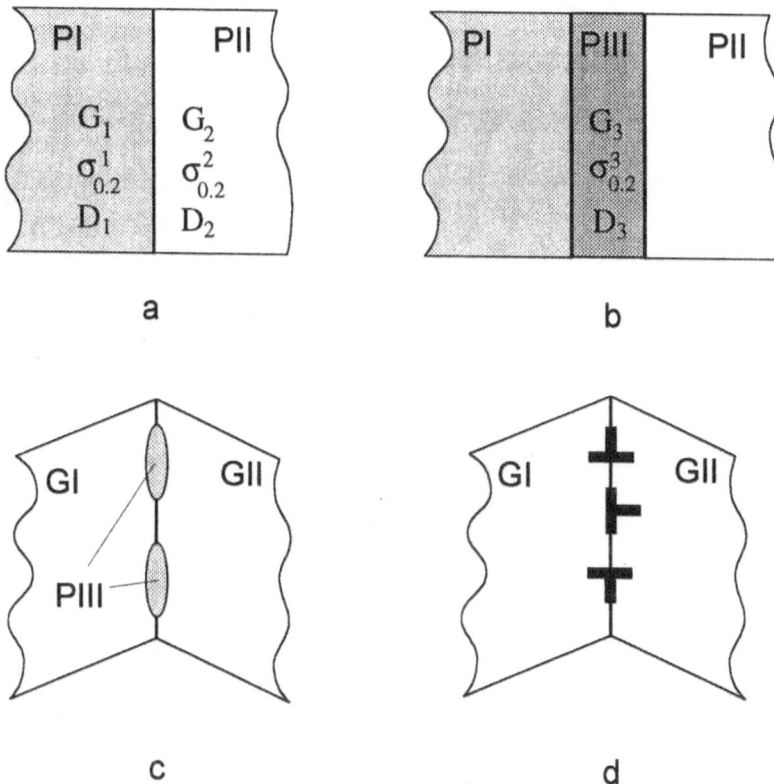

a

b

c

d

Figure 2.

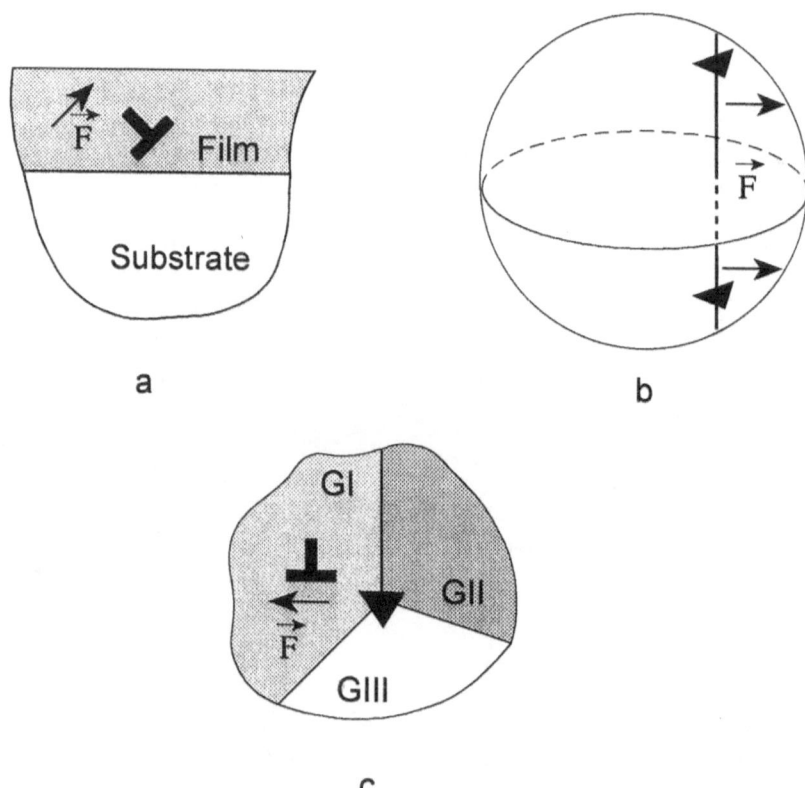

a

b

c

Figure 3.

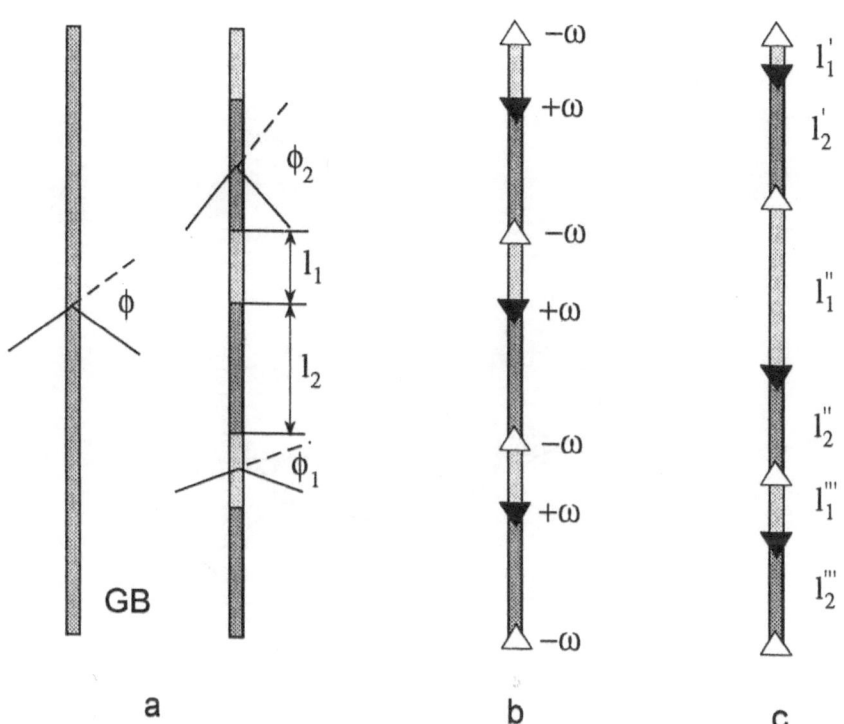

a   b   c

Figure 4.

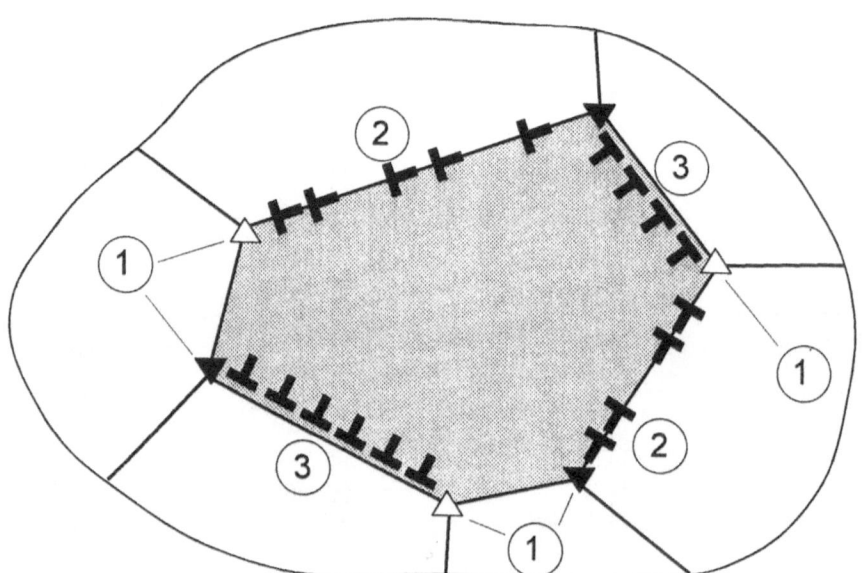

Figure 5.

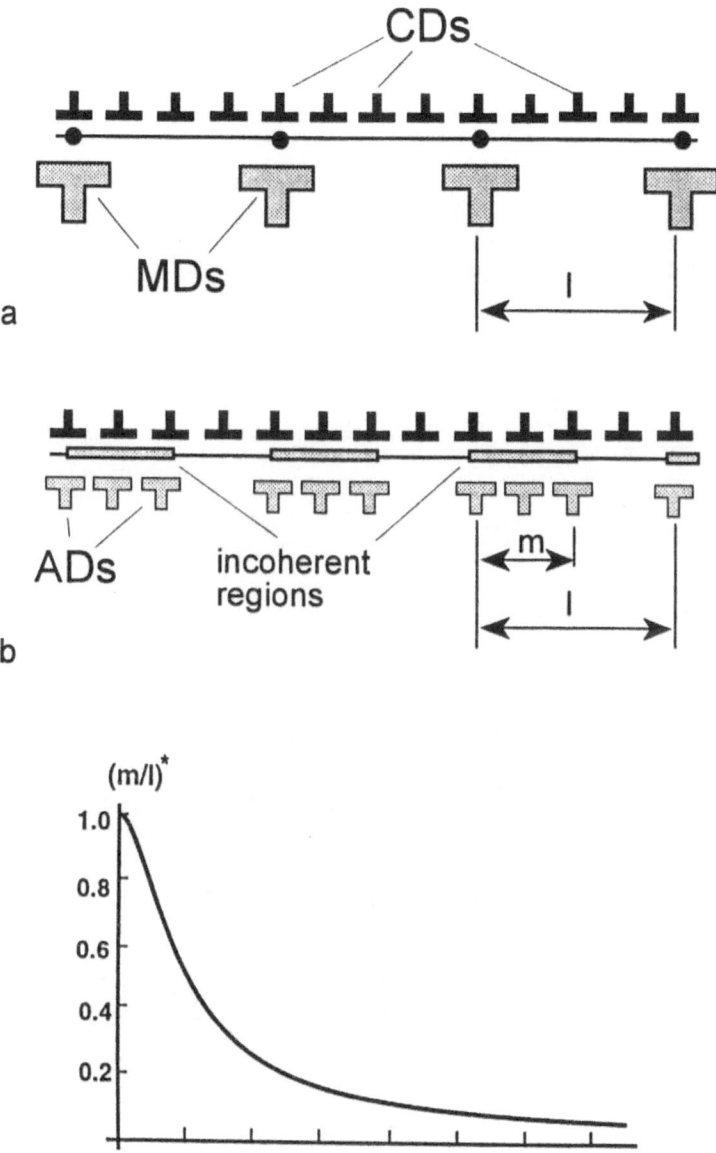

Figure 6.

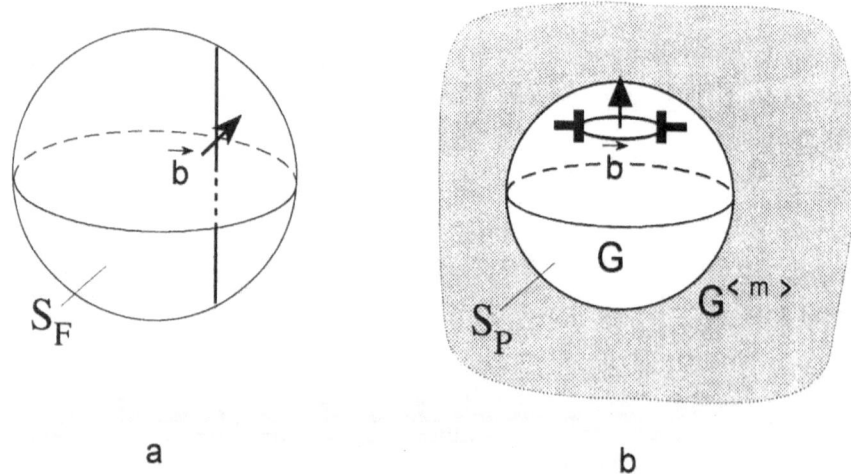

a

b

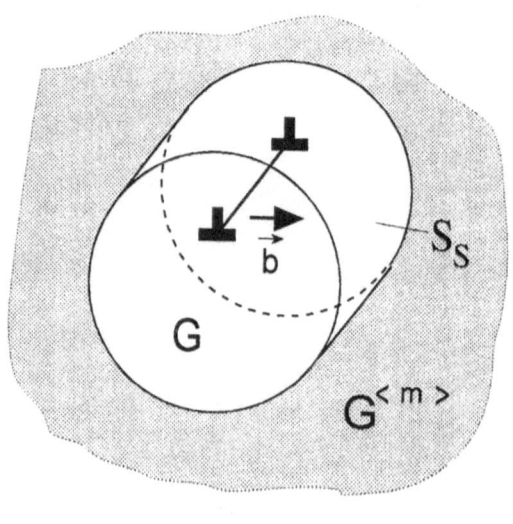

c

Figure 7.

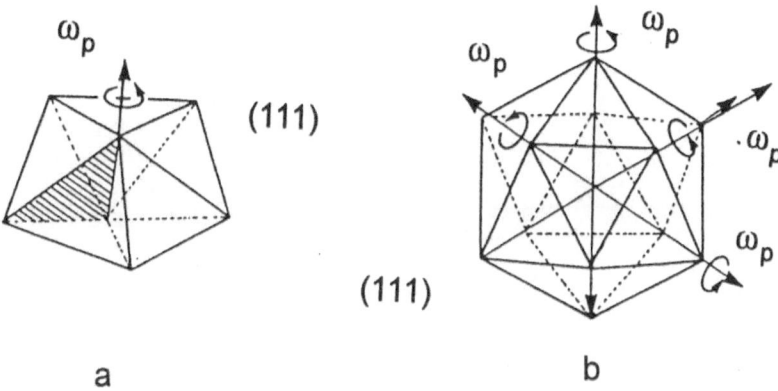

ω_p

(111)

ω_p

ω_p

ω_p

ω_p

(111)

a

b

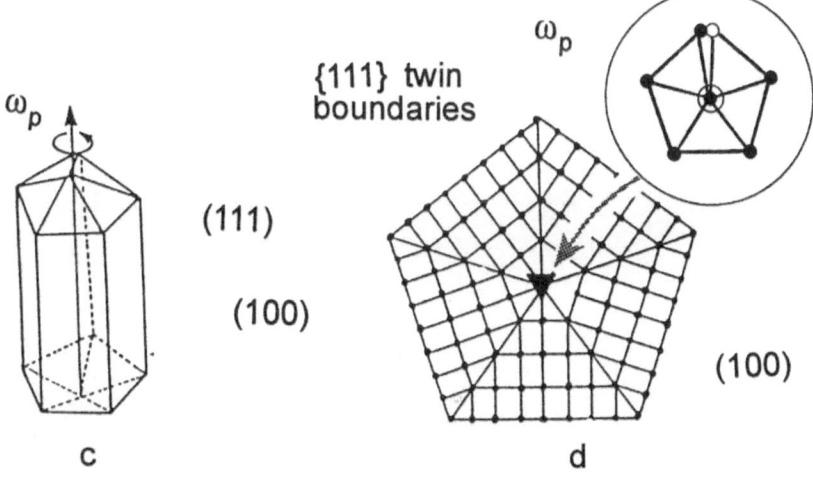

ω_p

(111)

(100)

c

ω_p

{111} twin
boundaries

(100)

d

Figure 8.

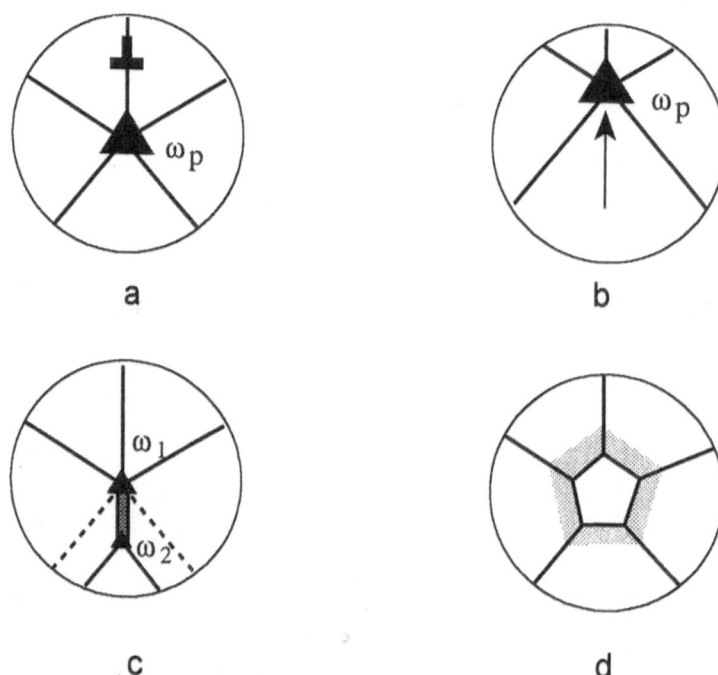

a

b

c

d

237

Figure 9.

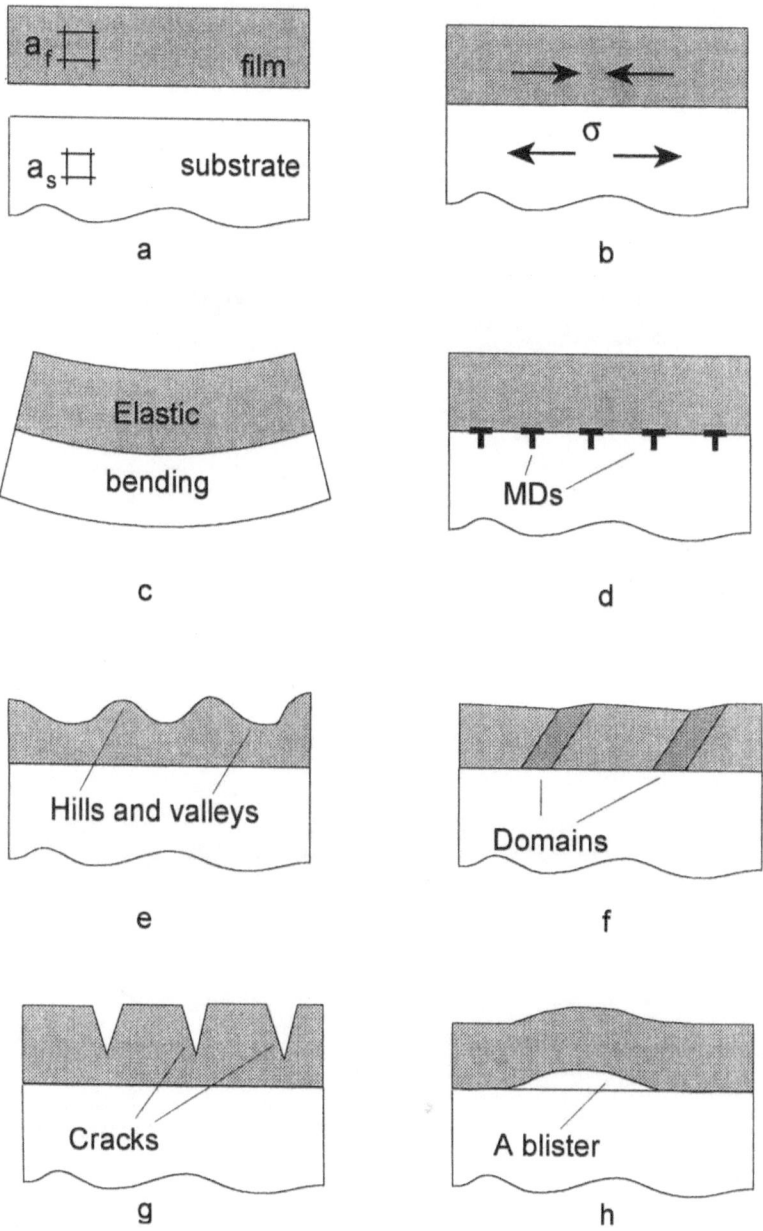

Figure 10.

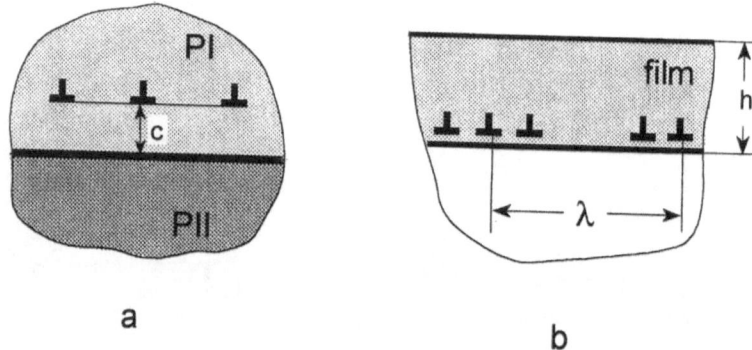

a

b

Figure 11.

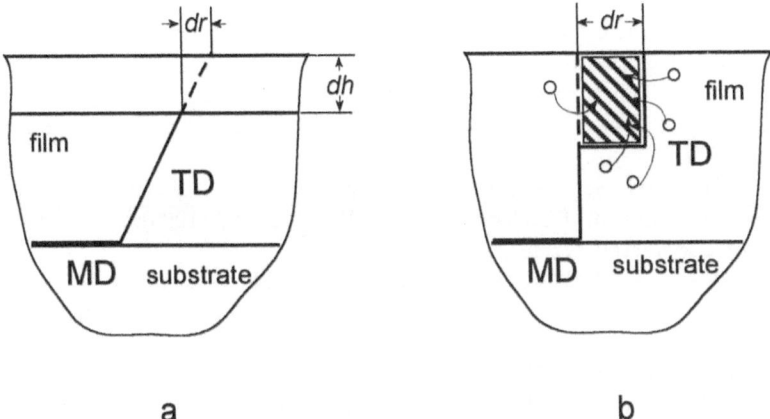

a   b

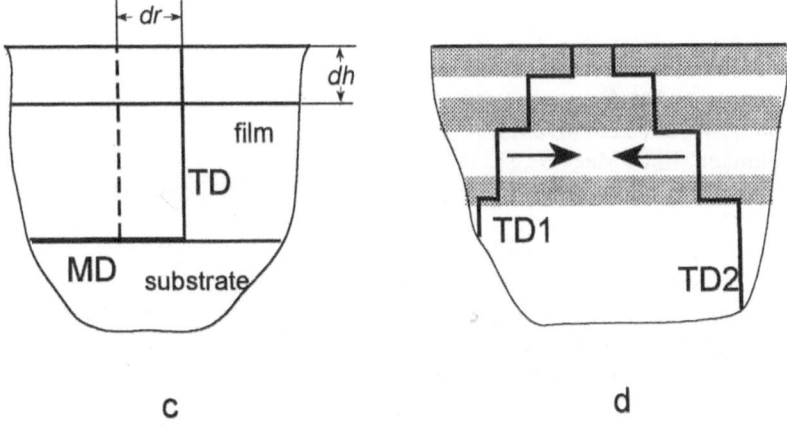

c   d

References

1. Dislocations in Solids, F.R.N.Nabaro Ed., vols. 1-10, North-Holland (1979-1996).
2. Physics of Defects, R.Balian, M.Kleman and J.-P.Poirier Eds. North- Holland (1981)
3. H.Gleiter, Progr. Mater. Sci., **33**, 223 (1989).
4. R.W.Siegel, Mater. Sci. Eng., **A 166**, 189 (1993).
5. V.G.Gryaznov and L.I.Trusov, Progr. Mater. Sci., **37**, 289 (1993).
6. Proceedings of the Second International Conference on Nanostructured Materials, H.-E.Schaefer, R.Wuerschum, H.Gleiter, and T.Tsakalakos Eds., NanoStruct. Materials, **6**, N.1-4 (1995).
7. Proceedings of the Third International Conference on Nanostructured Materials, M.L.Trudeau, V.Provenzano, R.D.Shull, aND J.Y.Ying Eds., NanoStruct. Materials **9**, N.1-8 (1997).
8. E.A.Fitzgerald, Mater. Sci. Reports, **7**, 87 (1991).
9. R.Beanland, D.J.Dunstan, P.J.Goodhew, Adv. Phys., **45**, 87 (1996).
10. R.W.Armstrong, Metall. Trans., **1**, 1169 (1970).
11. K.Lu, W.D.Wei and J.T.Wang, Scripta Met. Mater., **24**, 2319 (1990).
12. V.G.Gryaznov, M.Yu.Gutkin, A.E.Romanov and L.I.Trusov, J. Mater. Sci., **28**, 4359 (1993).
13. A.E.Romanov, NanoStruct. Mater., **6**, 125 (1995).
14. Y.Sasajima, T.Arakawa, M.Ichimura and M.Imabayashi, Jap. J. Appl.Phys., **28**, 1669 (1989).
15. A.P.Sutton and R.Balluffi, Interfaces in Crystalline Materials, Clarendon Press, Oxford (1995).
16. J.P.Hirth J.Lothe, Theory of Dislocations, Wiley, New York (1982)
17. A.E.Romanov and V.I.Vladimirov, in Dislocations in Solids (ed. F.R.N.Nabarro), vol.9, 191, North-Holland (1992).
18. A.E.Romanov, Mater. Sci. Eng. A, **164**, 58 (1993).
19. J.C.M.Li, Surf. Sci., **31**, 12 (1972).
20. V.Y.Gertsman, A.A.Nazarov, A.E.Romanov, R.Z.Valiev, V.I.Vladimirov, Phil. Mag. A, **59**, 1113 (1989).
21. W. Bollmann, Phil. Mag. A, **57**, 181,637 (1988).
22. V.V.Rybin, A.A.Zisman, N.Yu.Zolotarevsky, Acta Metal. Mater., **41**, 2211 (1993).
23. A.A.Nazarov, A.E.Romanov, R.Z.Valiev, Scripta Mater., **34** 729 (1996).
24. P.Müllner and W.-M.Kuschke, Scripta Mater., **36**, 1451 (1997).
25. A.A.Nazarov, A.E.Romanov and R.Z.Valiev, Acta Met. Mater., **41**, 1033 (1993).
26. R.Z.Valiev, V.I.Vladimirov, V.Yu.Gertsman, A.A.Nazarov, and A.E.Romanov, Phys. Met. Metall., **69**, No3, 30 (1990).
27. O.A.Shenderova, D.W.Brenner, A.A.Nazarov, A.E.Romanov and L.H.Yang, Phys. Rev. B, (1998), in press.
28. D.W.Brenner, O.A.Shenderova, L.H.Yang, A.A.Nazarov and A.E.Romanov, in Computer Aided Design of High Temperature Materials, A.Pechenik, R.Kalia, P.Vashishta Eds., Oxford University press, (1997), in press.
29. A.A.Nazarov, A.E.Romanov and B.Baudelet, Phil. Mag. Lett., **68**, 303 (1993).
30. A.A.Nazarov, A.E.Romanov and R.Z.Valiev, NanoStruct. Mater., **4**, 94 (1994).
31. A.A.Nazarov, Scripta Mater., **37**, 1155 (1997).

32. J.W.Matthews, in Dislocations in Solids, Ed. F.R.N.Nabarro, Norht- Holland, **2**, 459 (1979).
33. A.E.Romanov, T.Wagner and M.Rühle, Scripta Mater. (1998), in press.
34. G.Gutekunst, J.Mayer, V.Vitek, and M.Rühle, Phil. Mag. A, **75**, 1357 (1997).
35. M.Wagner, T.Wagner, D.L.Caroll, J.Marien, D.A.Bonell, and M.Rühle, MRS Bulletin, **22**, 42 (1997).
36. V.G. Gryaznov, A.M.Kaprelov and A.E.Romanov, Scripta Met., **23**, 1443 (1989).
37. V.G.Gryaznov, I.A.Polonsky, A.E.Romanov, L.I.Trusov, Phys. Rev. B, **44**, 42 (1991).
38. I.A.Polonsky, A.E.Romanov, V.G.Gryaznov and A.M.Kaprelov, Czech. J. Phys. **41**, 1248 (1991).
39. N.Wang, G.Palumbo, Z.Wang, U.Erb and K.T.Aust, Scripta Metall. Mater., **28**, 253 (1993).
40. M.Hecker and A.E.Romanov, Phys. Stat. Sol. (a), **130**, 91 (1992).
41. M.Hecker and A.E.Romanov, Mater. Sci. Eng., **A164**, 433 (1993).
42. M.Yu.Gutkin, A.E.Romanov and E.C.Aifantis, Phys. Stat. Sol. (a), **153**, 66 (1996).
43. B.Bartuschat and J.Urban, Phil.Mag.A, **76**, 783 (1997).
44. R. de Wit, J. Phys. C, **5**, 529 (1972).
45. I.A.Polonski, A.E.Romanov and V.G.Gryaznov, Phil.Mag.A, **64**, 281 (1991).
46. V.G. Gryaznov, A.M.Kaprelov, A.E.Romanov and I.A.Polonsky, Phys. Status Solidi (b), **167**, 441 (1991).
47. A.E.Romanov, I.A.Polonsky, V.G.Gryaznov, S.A.Nepijko, T.Junghaus and N.I.Vitrykhovski, J. Cryst. Growth **129**, 691 (1993).
48. H.Gao, J.Mech. Phys. Solids, **39**, 443 (1991).
49. J.S.Speck, A.C.Daykin, A.Seifert, A.E.Romanov and W.Pompe, J. Appl.Phys., **78**, 1696 (1995).
50. A.E.Romanov, W.Pompe and J.S.Speck, J. Appl. Phys., **79**, 4037 (1996).
51. M.Ohring, Material Science of Thin Films, Academic Press, New York (1992).
52. V.I.Vladimirov, M.Yu.Gutkin and A.E.Romanov, Poverkhnost' (Surface), **6**, 46 (1988).
53. M.Yu. Gutkin, A.L.Kolesnikova, A.E.Romanov, Mater. Sci. Eng. A, **164**, 433 (1993).
54. L.B.Freund, MRS Bulletin, **17**, 52 (1992).
55. J.Y.Tsao, Materials Fundamentals of Molecular Beam Epitaxy, Academic Press, New York (1993).
56. M.Yu.Gutkin, M.Militzer, A.E.Romanov and V.I.Vladimirov, Phys. Stat. Sol.(a), **113**, 337 (1989).
57. V.I.Vladimirov, M.Yu.Gutkin and A.E.Romanov, Sov. Phys. Sol. State, **29**, 1581 (1987).
58. M.Yu.Gutkin and A.E.Romanov, Phys. Stat. Sol. (a), **144**, 39 (1994).
59. M.Yu.Gutkin and A.E.Romanov, Phys. Stat. Sol. (a), **129**, 117 (1992).
60. M.Yu.Gutkin and A.E.Romanov, Sov. Phys. Sol. State, **32**, 751 (1990).
61. A.Atkinson, S.C.Jain, J. Appl. Phys., **72**, 2342 (1992).
62. A.E.Romanov, E.C.Aifantis, Scripta Met. Mater., **30**, 1581 (1994).
63. M.Yu.Gutkin, A.E.Romanov and E.C.Aifantis, Phys. Stat. Sol. (a), **151**, 281 (1995).
64. S.V.Kamat and J.P.Hirth, J. Appl. Phys., **67**, 6844 (1990).

242

65. M.Yu.Gutkin and A.E.Romanov, Sov. Phys. Sol. State, **33**, 874 (1991).

66. X.Feng and J.P.Hirth, J. Appl. Phys., **72**, 1386 (1992).

67. M.Yu.Gutkin and A.E.Romanov, Phys. Status Solidi (a), **125**, 107 (1991).

68. M.Yu.Gutkin and A.E.Romanov, Phys. Status Solidi (a), **129**, 363 (1992).

69. L.B.Freund and W.D.Nix, Appl. Phys. Lett., **69**, 173 (1996).

70. G.Beltz and L.B.Freund, Phys. Stat. Sol. (b), **180**, 303 (1993).

71. J.S.Speck, M.A.Brewer, G.Beltz, A.E.Romanov and W.Pompe, J. Appl. Phys., **80**, 3808 (1996).

72. A.E.Romanov, W.Pompe, G.E.Beltz and J.S.Speck, Appl.Phys. Lett., **69**, 3342 (1996).

73. A.E.Romanov, W.Pompe, G.E.Beltz and J.S.Speck, Phys. Stat. Sol. (b), **198**, 599 (1996).

74. G.E.Beltz, M.Chang, M.A.Eardley, W.Pompe, A.E.Romanov and J.S.Speck, Mater. Sci. Eng. A, **234-236**, 794 (1997).

75. A.E.Romanov, W.Pompe, G.E.Beltz, and J.S.Speck, Phys. Stat. Sol. (b), **199**, 33 (1997).

76. G.E.Beltz, M.Chang, J.S.Speck, W.Pompe and A.E.Romanov, Phil. Mag. A, **76** 807 (1997).

77. M.Yu.Martisov and A.E.Romanov, Sov. Phys. Sol. State, **32**, 1101 (1990).

78. M.Yu.Martisov and A.E.Romanov, Sov. Phys. Sol. State, **33**, 1173 (1991).

THE PROPERTIES OF Fe-Ni FCC ALLOYS HAVING A NANOSTRUCTURE PRODUCED BY DEFORMATION, IRRADIATION AND CYCLIC PHASE TRANSFORMATION

V.V. SAGARADZE

Institute of Metal Physics, Ural Branch of RAS, 18 S.Kovalevskaya St., 620219 Yekaterinburg, Russia

1. Introduction

The best known methods used for production of nanocrystalline states [1, 2] include fabrication and compaction of powders, refinement of the grain structure by a strong plastic deformation, deposition of the elements from a gaseous or liquid phase, and crystallisation of amorphous alloys. This paper analyses the changes in some physical and mechanical characteristics of submicrocrystalline alloys produced mainly by other nontraditional methods, such as irradiation and phase transformations. Specifically, a high-dose irradiation of alloys with high-energy particles at 773-973 K [3] causes formation of dislocation loops and their rearrangement to dislocation subboundaries and boundaries of nanocrystals. The structure may also be strongly refined as a result of the "multiplication" of the γ-orientations in metastable austenitic alloys during direct and reverse martensitic $\gamma \rightarrow \alpha \rightarrow \gamma$ transformations, which lead to appearance of $24^2 = 576$ orientations of the γ-phase in each initial austenitic grain [4, 5]. The deformation method was also used for production of the nanocrystalline structure. However, this paper focuses mainly on the anomalous phase transformations, which take place during severe cold deformation and which alter the alloy properties.

2. Materials and Methods

The structure refinement caused by the $\gamma \rightarrow \alpha \rightarrow \gamma$ transformations was studied using metastable austenitic alloys 32 Ni (Fe + 32.3 mass % Ni) and 26Ni-Cr-Ti (Fe + 26 mass % Ni, 1.3% Cr and 1.5% Ti) having the martensite start temperature M_s at 183 K and 143 K respectively. A nanostructure was formed mainly in the austenitic steels types 16Cr-15Ni-3Mo-Ti and 16Cr15Ni-3Mo, which were irradiated up to 200 displacements per atom (dpa) with 1.5-MeV Kr^+ ions at 773-973 K [3] in an installation combining an electron microscope and an ion accelerator [6]. The formation of a nanocrystalline structure and the anomalous phase transformations (dissolution of intermetallics and carbides and redistribution of the alloying elements) during cold deformation at 77-293 K were studied mainly on FCC iron-nickel alloys

G.M. Chow and N.I. Noskova (eds.), Nanostructured Materials, 243–262.

244

types 35Ni-3Ti, 35Ni-3Zr, 34Ni-5Al, 34Ni, 0.5C-30Ni-2V and 30Ni-12Cr0. Cold deformation was performed by rolling and a high-pressure (up to 8 GPa) shear in Bridgman anvils [7]. The electron and X-ray diffraction was used for the phase and crystallographic analysis of the alloys. The linear expansion coefficient was measured by a Shevenar dilatometer. The Mössbauer method was used to determine the concentration variations of the iron-nickel alloys during phase transformations, deformation and irradiation.

3. Results and Discussion

3.1. THE STRUCTURE AND PROPERTIES OF THE Fe-Ni ALLOYS AFTER THE $\gamma \rightarrow \alpha \rightarrow \gamma$ TRANSFORMATIONS

3.1.1. Formation of a Nanostructure during the $\gamma \rightarrow \alpha \rightarrow \gamma$ Transformation
A nanocrystalline structure can be produced in metal samples or parts having actually any dimensions subject to a thermal or thermomechanical treatment only if these samples or parts are made of a metastable alloy allowing direct and reverse martensitic transformations (some Fe-Ni invars, metastable stainless steels, metastable FCC Fe-Mn alloys, and alloys with the shape memory effect, etc.). This paper deals with metastable Fe-Ni austenitic alloys, which have the martensite start temperature M_s below 273 K. The mechanism, the structural features and crystallography of the direct $\gamma \rightarrow \alpha$ and

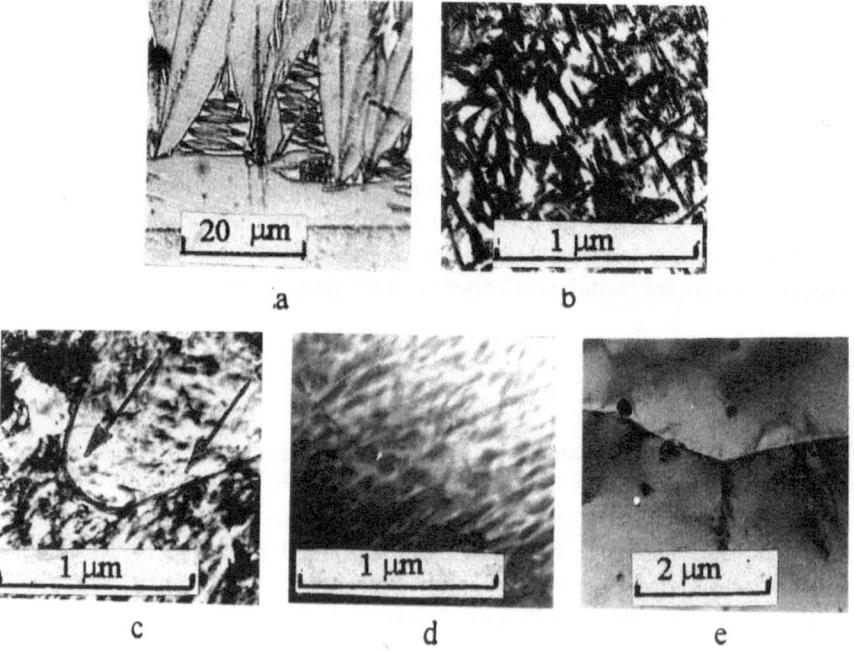

.a   b

c   d   e

Figure 1. Structure of the 32Ni steel in the initial martensitic state (a) and after the $\alpha \rightarrow \gamma$ transformation during a slow heating at a rate of 0.2 deg./min. to 753 K (b), 778 K (c) and 823 K (d) and heating to 823 K (0.2 deg./min) + 1023 K (40 min.) (e)

reverse $\alpha \rightarrow \gamma$ martensitic transformations have been described in detail elsewhere [4, 5].

The 32Ni and 26Ni-Cr-Ti alloys [4, 5] quenched from 1323 K possess a polyhedral austenitic structure with the grains 30-50 μm in size. Cooling in liquid nitrogen causes the direct martensitic $\gamma \rightarrow \alpha$ transformation. About 90% and 65% crystals of the lenticular martensite having 24 orientations and the habit $\{3.10.15\}_\gamma$ (Fig. 1a) appear in the alloys respectively. If the reverse martensitic $\alpha \rightarrow \gamma$ transformation is realised during heating up to 973 K at a rate exceeding 3 deg./min., the dimensions and the shape of the initial austenitic grain are preferably restored and a small number of thin γ twins is formed [4, 5]. These generally accepted conditions of treatment do not lead to a marked refinement of the initial austenitic structure, a fact which is due to the nucleation of the γ-phase at the boundary with the retained austenite and to the reproduction of its orientation. To ensure that 24 orientations of the γ-phase nucleate inside each martensitic α-crystal during heating and the γ orientations multiply during the $\gamma \rightarrow \alpha \rightarrow \gamma$ cycle, one should provide a low-nickel "buffer" layer at the boundaries of the martensitic α-crystals [8]. This layer inhibits the nucleation and growth of the γ-phase on the retained austenite as on a substrate. The low-nickel buffer layer may be formed through a redistribution of nickel between the martensite and the austenite in accordance with the equilibrium diagram of the Fe-Ni system under a very slow heating, which provides conditions for a considerable diffusion of nickel from the surface layers of the martensitic α-crystals to the retained austenite. Calculations show [4, 8] that at 670 K (holding time of 10 hours) a layer containing ≤25 mass % nickel and about 1 nm deep is formed at the boundaries of martensitic crystals. In this nickel-depleted region the temperature of the reverse martensitic $\alpha \rightarrow \gamma$ transformation is much higher than in the central regions of the α-crystals containing 32.3% nickel. For this reason, during the $\alpha \rightarrow \gamma$ transformation the austenite does not nucleate at the boundary with the retained austenite nor restores its initial orientation: it appears inside the martensitic lens and forms all the 24 γ-orientations that are possible.

After the $\gamma \rightarrow \alpha \rightarrow \gamma$ transformation cycle $24^2 = 576$ orientations of the γ-phase may appear. If coinciding orientations are taken into account, 501 different orientations of the γ-phase [4] may be formed in each initial γ grain after a single cycle of the $\gamma \rightarrow \alpha \rightarrow \gamma$ transformations with the Kurdyumov-Sachs orientation relationships. Figure 1b shows superthin crystals of the γ-phase, which were formed as a result of the $\gamma \rightarrow \alpha \rightarrow \gamma$ transformation in the 32Ni alloy. The reverse $\alpha \rightarrow \gamma$ transformation was realised under a slow heating at a rate of 0.3 K/min. Depending on particular conditions of the $\alpha \rightarrow \gamma$ transformation, the γ-crystals may be 10 to 20 nm thick, a value which is several orders of magnitude lower than the dimensions of the initial austenitic grains (30-50 μm or larger). In the 26Ni-Cr-Ti alloy superfine austenitic crystals may be formed also during the isothermal $\alpha \rightarrow \gamma$ transformation at 813 K (Fig. 2a). Typical diffraction patterns for all the 24 orientations of the γ-phase in the initial α-crystal appear (Fig. 2b) [4].

At the beginning of the martensitic $\alpha \rightarrow \gamma$ transformation fine γ-crystals have a coherent BCC/FCC interphase boundary. Then the coherence is disturbed and the γ-crystals of 24 orientations can grow through diffusion until they touch one another as is observed in the Ni26CrTi1 steel during the isothermal $\alpha \rightarrow \gamma$ transformation (Fig. 2a). Since the γ-crystals are connected with the initial α-crystal by relationships similar to

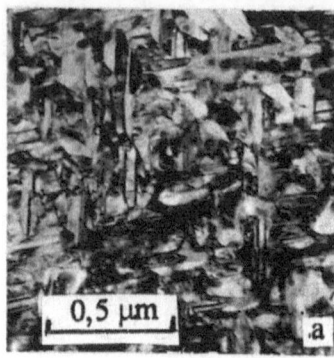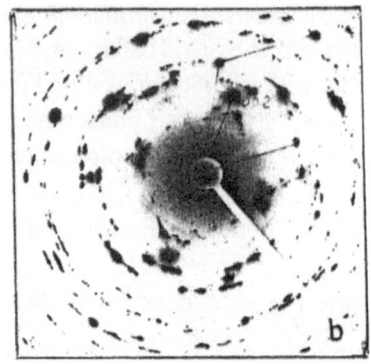

Figure 2. Nanocrystalline γ-phase (a) and the corresponding electron diffraction pattern (b) in the 26Ni-C₁-Ti steel. Treatment: isothermal α→γ transformation at 813 K for 2 h

the Kurdyumov-Sachs relationships, $\{111\}_\gamma \| \{110\}_\alpha$ and $<110>_\gamma \| <111>_\alpha$, the misorientation angles between neighbouring nanocrystals of the γ-phase are not random, but they may be of assorted values in accordance with the number of random pairs comprising a group of 24 γ-crystals. The γ-crystals have predominantly large-angle boundaries, but some of them may have low-angle boundaries corresponding, in particular, to the rotation angle between neighbouring γ-crystals of −4°18' about an axis of the $<111>_\gamma$ type [4]. Under certain treatment [4, 5] the γ-phase nanocrystals growing in the α-matrix can be absorbed by much larger austenitic grains (Fig. 1c-e).

3.1.2. Variation of the Mechanical Properties of the Alloys in the Submicrocrystalline State

The formation of a nanocrystalline structure suggest an improvement of the mechanical properties. It was shown [4, 5] that the control of the α→γ transformation mechanism in the γ→α→γ cycle permits the strength characteristics of the metastable Fe-Ni alloys to be significantly enhanced.

The formation of 25 to 50% dispersed γ-phase inside martensitic α-crystals during the α→γ transformation, which inhibits movement of dislocations, allows increasing the yield stress $\sigma_{0.2}$ of the 30Ni and 32Ni alloys up to 1150-1050 MPa [4], a value which is higher than $\sigma_{0.2}$ of these alloys in the initial austenitic ($\sigma_{0.2}$ = 250 MPa) and martensitic ($\sigma_{0.2}$ = 700 MPa) states. The development of the α→γ transformation in the α-matrix deformed by 80% leads to strengthening of the martensite to 1400 MPa ($\sigma_{0.2}$), which cannot be achieved by plastic deformation of the 30Ni alloy. Plasticity (specific elongation) decreases to 3-7%. Further heating, which causes appearance of a coarser globular austenite (Fig. 1c,d), decreases the strength characteristics almost to the initial level ($\sigma_{0.2}$ = 350 MPa) [4].

Thus, the widely adopted strengthening treatment of structural steels involving the formation of the α-martensite in the austenitic matrix by a sharp quenching cooling is replaced in our case by a reverse process, that is, strengthening treatment of the α-martensite involving the formation of a nanocrystalline austenite in this martensite during heating.

3.1.3. Redistribution of Ni during the α→γ Transformation in FCC Fe-Ni Alloys

The formation of a nanocrystalline γ-structure during the α→γ transformation in slowly heated Fe-Ni alloys is accompanied by a very substantial diffusion redistribution of nickel and appearance of a strong concentration inhomogeneity. This situation cannot be realised by other thermal treatment methods and may be used to control thermal and magnetic properties of alloys.

A sluggish "low-temperature" redistribution of nickel between the α- and γ-phases during a slow heating from 473 K to 773 K at a rate of 0.2-0.3 K/min. is observed in iron-nickel invars only within a few nm from the α/γ boundary. This nickel redistribution has little if any effect on the properties of a coarse-crystal austenitic-martensitic alloy having α-crystals 0.1 to 10 μm in size. However, if γ-crystals ~10 nm thick are formed inside the α-martensite, enrichment of the γ plates with nickel just to a depth of 5 nm on both sides will change completely the composition of all the γ-crystals. As a result, the physical properties will be altered.

The variation in the nickel concentration, which takes place during the α→γ transformation in invar-type alloys, is reliably registered by the Mössbauer method [4]. Figure 3 depicts some Mössbauer spectra of the 32Ni alloy in the austenitic (a) and martensitic (b) states obtained at room temperature. The same figure shows spectra of the samples heated during the α→γ transformation to a temperature from 673 K to 1073 K (c-l). A weakly resolved sextet of the ferromagnetic γ_0 austenite corresponds to the quenched alloy. The spectrum obtained for the alloy, which was treated in liquid nitrogen, is a superposition of two sextets: the ferromagnetic α_0 martensite (the field at the [57]Fe core is ~335 kOe) and a small amount of the retained γ_0 austenite, which has a much smaller field. During a slow heating to 673 K the dispersed austenite is enriched with nickel inside the martensitic crystals at

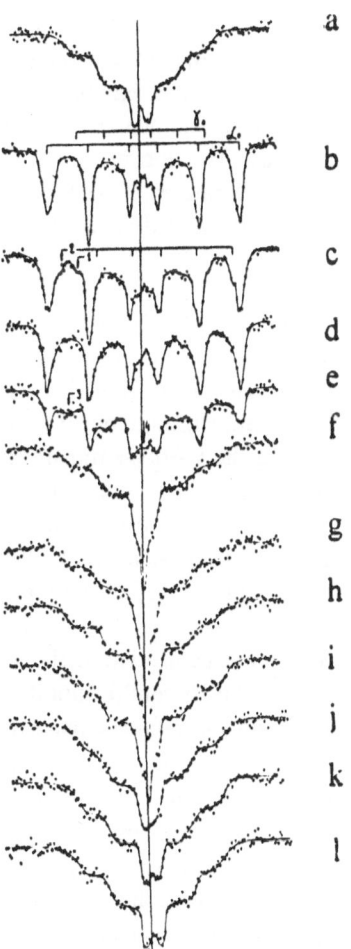

Figure 3. Mössbauer spectra of the 32Ni alloy at 298 K: a - initial austenite (quenching from 1323 K); b - cooling-induced martensite; c-g - α→γ transformation during a slow heating (0.2 deg./min.) to 673 K (c), 723 K (d), 763 K (e), 793 K (f), 823 K (g); h to l - heating to 823 K (0.2 deg./min.) + annealing (40 min.) at 873 (h), 923 (i), 973 (j), 1023 (k) and 1073 (l)

the stage of the nanocrystalline γ-phase formation (Fig. 1b). This is followed by the appearance of the component with the field ~305 kOe in the spectrum (Fig. 3c). This field corresponds to the nickel concentration of the austenite equal to 40-50 mass % A certain increase of the field at the ^{57}Fe core in the α-state from 335 to 342 kOe suggests a simultaneous depletion of the retained austenite in nickel. Heating to 723 K increases the amount of the nickel-enriched superfine austenite. When the temperature is raised to 763-793 K, the spectrum changes qualitatively: a single line appears at the centre of the spectrum. This line indicates that a paramagnetic low-nickel austenite is formed from the nickel-depleted martensite (Fig. 3e,f). The appearance of the low-nickel γ-component corresponds to the replacement of the $\alpha \rightarrow \gamma$ transformation mechanism under a slow heating: a thin-plate γ-phase is substituted for a dispersed globular austenite, which absorbs the two-phase $\alpha+\gamma$ mixture and inherits the concentration inhomogeneity of this mixture. The contrast oscillations of the austenitic structure appear in the electron microscopic photographs. These oscillations follow the location of nickel-enriched γ-plates absorbed by the globular or polyhedral austenite (Fig. 1c,d).

Thus, upon completion of the $\alpha \rightarrow \gamma$ transformation (793 K), the 32Ni alloy comprises a concentrationally inhomogeneous austenite with the nickel-enriched γ_1-component (instead of the thin-plate austenite), the nickel-depleted γ_2-component (instead of the depleted retained martensite), and the γ_0 component having the initial nickel concentration (instead of the retained austenite). The obtained concentrational inhomogeneity is rather stable at elevated temperatures. The nickel concentration becomes homogeneous during a 1-hour holding time at temperatures above 973 K (Fig. 3j). Under these heating conditions the grains grow and the banded contrast vanishes almost completely in the structure (Fig. 1e).

3.1.4. Control of the Thermal Expansion Coefficient (TEC) in FCC Fe-Ni Alloys

The creation of the nanocrystalline state as a result of the cyclic $\gamma \rightarrow \alpha \rightarrow \gamma$ transformation is not the end as itself. We have to gain advantage from this method, which is applicable only to alloys allowing martensitic transformations. It is hardly surprising that a sharp refinement of the structure during the $\gamma \rightarrow \alpha \rightarrow \gamma$ transformation leads to a considerable improvement in the strength of the alloys (see subsection 3.1 2). However, this study showed that the formation of a nanocrystalline structure during the $\gamma \rightarrow \alpha \rightarrow \gamma$ transformation opens up the possibility of producing austenitic Fe-Ni alloys possessing a new property: the TEC can be controlled within broad limits thanks to a redistribution of nickel in the γ-matrix.

The degree of the nickel redistribution in the alloy can be controlled. If the heating temperature is changed in the two-phase $\alpha+\gamma$ region of the Fe-Ni equilibrium diagram, the degree to which the dispersed γ-phase is enriched and the retained martensite is depleted in nickel is altered. Moreover, the relative amount of the α- and γ-phases changes too. If a slow heating in the $\alpha+\gamma$ region is followed by a fast heating up to 873 K to the single-phase γ region accompanied by the formation of a "massive" globular austenite, the concentration redistribution of the nickel to the low- and high-nickel γ-regions is fixed in the single-state austenitic state. This nickel redistribution, which depends on the final slow-heating temperature in the $\alpha+\gamma$ region, causes the austenite

TEC to change over rather broad limits (from 2 to 10.5×10^{-6} deg.$^{-1}$) at 223-323 K. Table 1 shows changes in the TEC of the 32Ni depending on the $\alpha \rightarrow \gamma$ transformation conditions under a heating to 703,743 and 763 K at a rate of 0.2 K/min.

TABLE 1. TEC $\times 10^6$ (deg.$^{-1}$) of the 32Ni alloy depending on the heat treatment conditions

No.	Treatment	Interval, K	
		223-323	523-623
1.	Cooling from 1323 K in water	2.5	16.5
2.	The $\gamma \rightarrow \alpha$ transformation under cooling to 77 K, the $\alpha \rightarrow \gamma$ transformation under slow heating up to 703 K at a rate of 0.2 K/min. + fast heating up to 873 K	7.9	15.6
3.	Slow heating, 743 K + fast heating, 873 K	10.5	15.6
4.	Slow heating, 763 K + fast heating, 873 K	6.0	15.6

Thus, by controlling the formation conditions of the nanocrystalline γ-phase during the $\alpha \rightarrow \gamma$ transformation, one may largely alter the physical properties, specifically the TEC, of one and the same Fe-Ni alloy in the austenitic state. Earlier such unlike values of TEC were obtained for different alloys containing different amounts of alloying elements. In our case the nanocrystalline state of the γ-phase serves as an intermediate condition, which is used subsequently to produce a concentration-inhomogeneous polyhedral Fe-Ni austenite having the grain size of a few micrometers.

3.1.5. Production of an Alloy Possessing the Properties of a Thermal Bimetal

The possibility to change the TEC of the 32Ni invars due to the diffusion redistribution of nickel during the formation of the nanocrystalline γ-phase allows, in particular, the fabrication of solid materials possessing the properties of a thermal bimetal. They are double-layer composites whose layers have the same chemical and phase compositions and differ in the TEC value only. In a 32Ni plate the nickel-homogeneous austenite should play the role of the active layer with a high TEC, while the chemically homogeneous invar austenite having a low TEC should act as the passive layer.

One of the alternative ways of producing the thermal bimetal consists in transformation of the 32Ni invar strip into the noninvar components under the regime described in subsection 3.1.4 and homogenisation of one of the surface layers of the strip using, e.g., a high-speed surface heating to high temperatures. The mean TEC of the homogenised layer of the 32Ni sample is ~2.5×10^{-6} K^{-1} (over the temperature interval from 173 to 373 K) and that of the inhomogeneous layer is 10.5×10^{-6} K^{-1}.

The specific bending of the thermal bimetal was $6 \cdot 10^{-6}$ deg.$^{-1}$. This method of the thermal bimetal production does not require mechanical joining of different alloys but is realised on a single alloy. The material withstands high bending forces without any disturbance of the continuity between the layers.

3.1.6. The Use of the Nanocrystalline γ-Phase to Enhance the Magnetic Hysteresis Properties of a High-Strength Maraging Steel

High-speed hysteresis motors where the rotor runs at ~10^5 r.p.m. require development of a high-strength steel whose hysteresis loop would be nearly rectangular in shape.

250

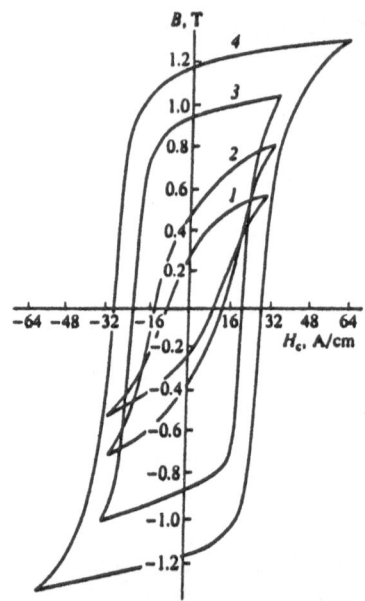

Figure 4. Minor hysteresis loops measured for the 18Ni-9Co-5Mo-Ti steel in the field H_m = 32 A/cm (1-3) and H_m = 65 A/cm (4). Treatment: 1 - quenching from 1273 K in water; 2 - quenching + cold rolling to ε = 40%; 3, 4 - quenching + rolling to ε = 40% + ageing at 823 K for 2 h + 843 K for 1 h

Then it would be possible to effectively magnetise the steel in extremely low fields of 20 to 60 A/cm. It turned out [9] that the appearance of the nanocrystalline austenite during the cyclic $\gamma \rightarrow \alpha \rightarrow \gamma$ transformation favours the formation of a magnetic texture and improves the magnetic hysteresis properties of the high-strength marageing steel type 18Ni-9Co-5Mo-Ti.

High mechanical and magnetic hysteresis properties of the textured cold-rolled steel type 18Ni-9Co-5Mo-Ti are achieved when the intermetallic ageing is accompanied by a partial reverse martensitic $\alpha \rightarrow \gamma$ transformation during heating to 823-843 K [9]. Against the background of the deformed lath martensite having a high density of dislocations the structure exhibits thin crystals of the HCP Ni_3Ti phase and dispersed austenitic plates 20-40 nm thick and 100-400 nm long, which break the martensitic laths to smaller sections. The steel, which has this structure, possesses high mechanical properties: the yield stress $\sigma_{0.2}$ = 1700 MPa; the ultimate strength σ_B = 1840 MPa; specific elongation δ = 10%; reduction of area ψ = 40%.

Figure 4 illustrates minor hysteresis loops of the 18Ni-9Co-5Mo-Ti steel [9], which were measured in small fields. Quenching and 40-% cold rolling of the sheets do not increase the rectangularity of the hysteresis loop (curves 1 and 2, Fig. 4), which is determined by the ratio between the residual induction B_r and the maximum induction B_m. The ageing alone accompanied by precipitation of intermetallics increases little the ratio B_r/B_m too. It is only a simultaneous formation of intermetallics and a superdispersed austenite during annealing of the cold-rolled steel at 823-843 K that leads to appearance of an optimal magnetic structure and improves the magnetic hysteresis properties in small fields (curves 3 and 4 in Fig. 4): rectangularity of the hysteresis loop B_r/B_m = 0.80; residual induction B_r = 0.96 T; coercive force H_c = 24 A/cm in the field of 32 A/cm. It was found [9] that a sharp increase in B_r and B_r/B_m is due to a magnetic anisotropy, which appears during ageing and a partial $\alpha \rightarrow \gamma$ transformation. Probably, the magnetostatic energy of the grains is decreased because low-magnetic plates of the γ-phase break the textured ferromagnetic martensite to partially isolated ferromagnetic fragments and the magnetoelastic energy lowers owing to the stretching stresses arising in the direction of the martensitic laths under the shear $\alpha \rightarrow \gamma$ transformation.

Thus, if a nanocrystalline austenite is formed in the structure of the dislocation martensite during the $\alpha \rightarrow \gamma$ transformation, the high-strength maraging steel type 18Ni-18Ni-9Co-5Mo-Ti turns into a semi-hard magnetic material possessing high magnetic hysteresis properties.

3.2. THE FORMATION OF A NANOCRYSTALLINE STRUCTURE UNDER A HIGH-DOSE IRRADIATION

3.2.1. Structural Changes under Irradiation

It is known [10] that irradiation of austenitic stainless steels with high-energy particles, which causes formation of vacancies and interstitials, leads to appearance of the Frank dislocation loops in the $\{111\}_\gamma$ planes. If the steels are irradiated up to 10-15 dpa, the Frank loops turn into ordinary dislocations and dislocation networks (a/3<111> + a/6<112> = a/2<110>) [11]. The density of the irradiation-induced dislocations reaches $(3...9) \times 10^{10}$ cm^{-2} [11], a value which is 2 orders of magnitude higher than in the annealed material. If the steels are irradiated further, the dislocations may form boundaries of subgrains and superfine grains.

In different steels a ultrafine structure is formed at different temperatures and fluences. The rearrangement of dislocations into subgrains and subsequent formation of superfine grains, which proceeds by the recrystallisation mechanism, depend on the mobility and pinning of dislocations by dispersed particles [3]. Specifically, in the 16Cr-15Ni-3Mo-Ti steel at 823-873 K the dislocations are pinned by the γ'-particles of Ni$_3$Ti during irradiation and fine subgrains and grains are formed at higher fluences than in the nonageing steel type 16Cr-15Ni-3Mo, which has a similar composition.

When the 16Cr-15Ni-3Mo-Ti steel is irradiated up to 30 dpa with 1.5-MeV Kr$^+$ ions (823-873 K), many Frank loops are formed and preserved. A small number of subgrains 30-200 nm in size appear at the damaging dose of 130-200 dpa. A different situation is observed in the nonageing austenitic steel type 16Cr-15Ni-3Mo. Dispersed

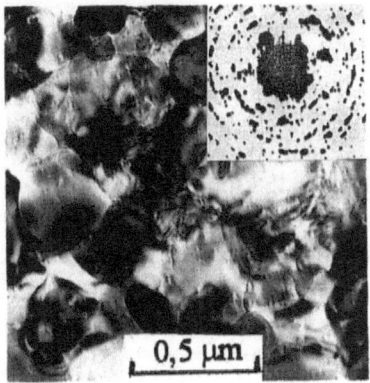

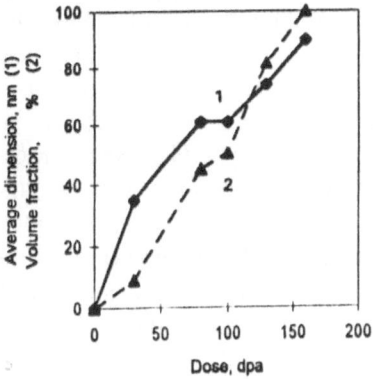

Figure 5. Submicrocrystal structure and the corresponding electron diffraction pattern of the 16Cr-15Ni-3Mo steel after irradiation with 1.5-MeV Kr$^+$ ions at 873 K up to a dose of 200 dpa

Figure 6. Average dimension of the subgrains and grains (1) of the 16Cr-15Ni-3Mo steel and their volume fraction (2) as a function of the irradiation dose with Kr$^+$ ions at 873 K [3]

subgrains ~20 nm in size are seen even after irradiation up to 10-30 dpa (873 K). The azimuthal smearing of the reflexes in the electron diffraction patterns suggests the rotation of the subgrains through 1-2° at a dose of 10-30 dpa [3]. If the ion fluence is increased, the subgrains grow and their misorientation attains 3-6°. Irradiation up to 160-200 dpa at 873 K gives rise to recrystallisation processes and new grains up to 50-200 nm in size are formed (Fig. 5). The electron diffraction pattern becomes a ring one as is the case with a polycrystalline alloy. Figure 6 shows mean dimensions of subgrains (grains) and their volume fraction in the 16Cr-15Ni-3Mo steel depending on the damaging dose under irradiation with krypton ions at 873 K. As may be seen from Fig. 6, the structure of the titanium-free 16Cr-15Ni-3Mo steel, where the dislocations are not pinned by the γ'-phase and are highly mobile, is strongly fragmented into crystallites having the mean dimension of 35-90 nm when the steel is subject to a high-dose irradiation. The fragmentation is comparable with refinement of grains under a very strong plastic deformation by pressure shear [7].

3.2.2. Redistribution of Alloying Elements under Irradiation

It may be thought that the formation of a superdispersed crystalline structure in Fe-Ni-based FCC alloys under irradiation with high-energy particles in the two-phase α+γ region of the equilibrium diagram will be accompanied by a redistribution of alloying elements. As a result, the physical and mechanical properties will be changed (for example, the TEC as was observed in the Fe-Ni invar-type alloys during the α→γ transformation) and it will be possible to realise the radiation modification of the alloys.

Figure 7 depicts one of the latest equilibrium diagrams of the Fe-Ni system [12]. This diagram has two temperature regions of the diffusion redistribution of nickel: The known low-temperature α+γ region of the ordered high-nickel austenite and a new high-temperature region with domes of the thermal and radiation-induced nickel redistribution of the austenite. One of the significant features of the radiation effect is the possibility to realise fast transformations or transformations induced by radiation point defects (ageing, redistribution of the solid solution, ordering, etc.) at low temperatures (373-473 K), where little, if any, phase transformations take place in iron alloys under ordinary thermal treatment conditions. In particular, when iron alloys containing 30 to 36% nickel were irradiated with high-energy electrons at 393-473 K, the austenite redistributed to low-nickel and high-nickel ($L1_0$-type ordering) components [13-14]. Some Mössbauer spectra and the corresponding density distribution functions of the magnetic fields, P(H), for the initial Ni35 alloy and the same alloy irradiated with 5.5-MeV electrons at 393 K (fluence 5×10^{18} e/cm^2) have been given elsewhere [14]. The irradiation causes appearance of a low-nickel γ-phase and a high-nickel γ-component with large field H_m = 290 kOe, which corresponds to the nickel concentration of over 40 mass %. This radiation-induced nickel redistribution should cause the TEC of the 35Ni alloy to change from 1×10^{-6} to 10×10^{-6} K^{-1}, as was the case with the 32Ni alloy after the thermally induced redistribution of the austenite (see subsection 3.1.4).

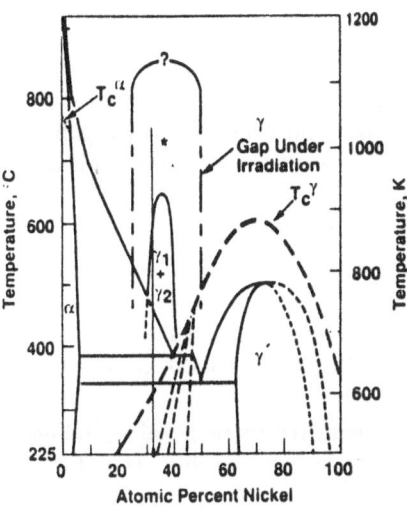

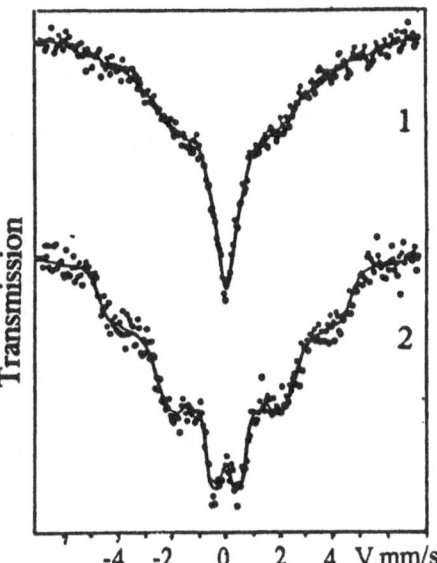

Figure 7. Fe-Ni phase diagram [12]. The 32Ni alloy is marked with the vertical line

Figure 8. Mössbauer spectra of the 32Ni alloy (298 K). Treatment: 1 - nickel redistribution during the α→γ transformation under heating to 823 K at a rate of 0.2 K/min. + cold rolling with 75-% reduction; 2 - homogenisation at 793 K (15 days) after the treatment as per p. 1

A low-temperature irradiation, which leads to formation of vacancies, accelerates the equilibrium diffusion processes. Specifically, when the quenched 35Ni-2Ti alloy is irradiated with 5.5-MeV electrons at 393-473 K (fluence 5×10^{18} e/cm^2), the Ni$_3$Ti intermetallic precipitates. A simultaneous depletion of the austenitic matrix in nickel is easily registered by the Mössbauer method: the mean field at the [57]Fe core decreases by ~40 kOe, a fact which suggests that about 1 mass % nickel goes from the 35Ni-2Ti alloy matrix to the intermetallic [14].

Of great interest is the radiation-induced high-temperature nickel redistribution at 823-973 K (Fig. 7). Here it is possible to form a nanocrystalline structure as the alloy undergoes the radiation modification. Nickel concentration oscillations were detected [12] in 35Ni and 35Ni-Cr iron alloys subject to neutron irradiation at elevated temperatures up to 1000 K. The observed spinodal-like redistribution of the austenite may result not from the radiation-enhanced equilibrium processes but may be due to the radiation-induced segregation phenomena taking place at the dislocation loops or grain and subgrain boundaries, which are sinks of point defects [10, 12]. However, the Fe-Ni diagram (Fig. 7) has a high-temperature region (773-973 K) of the equilibrium thermal redistribution of the austenite. In accordance with [12], this region may be realised during annealing without irradiation. Considering that a long holding time of single-phase iron-nickel FCC alloys in this region [13] does not provide an unambiguous answer as to the development of the thermal nickel redistribution, the following verifying experiment was performed. A nickel microconcentration inhomogeneity (see subsection 3.1, Fig. 1c,d), which gives a banded oscillation contrast with a step of ~50-100 nm in electron diffraction patterns, was produced in two iron alloys containing 31.5 at.% and 33 at.% nickel by the $\alpha \rightarrow \gamma$ transformation under a slow heating to 500 K at a rate of 0.2 K/min. A complicated Mössbauer spectrum (Fig. 3e,f) corresponds to the concentration-inhomogeneous austenite in the 32Ni alloy. This inhomogeneous austenite was annealed at 793 K. To accelerate the diffusion processes, the alloys were deformed by 75% prior to annealing. A 15-day annealing at 793 K decreased the nickel redistribution in the undeformed inhomogeneous austenite but did not lead to a full homogenisation of the alloys. In the deformed alloys the composition was fully nickel-homogeneous even during a shorter holding time. The low-nickel and high-nickel components vanished from the NGR spectrum (Fig. 8). Recrystallised grains without the banded contrast are seen in the structure (Fig. 1e).

Thus, the presence of the high-temperature dome of the thermal redistribution in the Fe-Ni equilibrium diagram (Fig. 7) is thought to be questionable. Therefore the properties of the alloys can be altered through the thermal nickel redistribution only in the low-temperature region during the $\alpha \rightarrow \gamma$ transformation (see subsection 3.1). The experimental data [12] suggest that the radiation-induced modification of Fe-Ni-based alloys may be performed at elevated temperatures (800-1000 K) too.

3.3. ANOMALOUS PHASE TRANSFORMATIONS TAKING PLACE WHEN THE NANOSTRUCTURE IS PRODUCED BY DEFORMATION AND CHANGES IN THE PROPERTIES OF IRON-BASED FCC ALLOYS

3.3.1. The Formation of a Nanocrystalline State in Alloys Subject to a Cold Plastic Deformation

It is known [15-20] that a nanocrystalline structure of iron alloys may be produced by special methods of a strong cold deformation. Specifically [18], the shear under a pressure of 6-16 GPa, which is realised in Bridgman anvils (the true deformation $\varepsilon = 8.2$ at 36 revolutions of the anvil, the temperature ~300 K), provides formation of nanocrystals having the mean diameter of ~10 nm in samples of the austenitic steel type 110Mn13. In this case the microhardness of the nanocrystalline austenitic steel increases to 9-10.5 GPa.

In accordance with the concepts described elsewhere [21-23], the formation of nanocrystals less than 100 nm in size under cold-rolling deformation may be visualised as follows. The plastic deformation begins with an ordinary slipping of dislocations in favourable planes {111} and appearance of a cellular structure in alloys possessing a high energy of stacking faults. As the degree of deformation is increased, microbands and microtwins begin developing in parallel with the operating slip planes. The microbands and the microtwins turn themselves along the rolling plane and the main deformation takes place at the boundaries of the microbands. When the deformation is relatively large (60% for copper), noncrystallographic shear bands, which go across the sample, emerge at an angle of ~35° to the rolling direction and the sheet plane. This is accompanied by appearance of rotation deformation modes (disclinations) and generation of a large number of dislocations and point defects. Under these conditions a nanostructure in the form of strongly misoriented crystals 10-100 nm in size appears locally in the noncrystallographic shear bands. If the deformation is increased further, the number of the noncrystallographic shear bands with the nanostructure rises. A strong pressure-shear deformation further makes for the appearance of misoriented nanocrystals.

The "deformation" vacancies and dislocations have a different effect on the phase transformations, which take place in Fe-Ni-based FCC alloys subject to cold deformation at 77-493 K [20, 24-26]. The vacancies favour equilibrium processes of the solid solution precipitation, for example, intermetallic ageing [24]. On the contrary, the dislocations generated under cold deformation cause the strain-induced dissolution of Ni_3Me intermetallics in the austenite at the same temperatures [17, 20, 24, 26]. Let us discuss in detail the anomalous phase transformations, which take place when Fe-Ni FCC alloys undergo a low-temperature deformation.

3.3.2. Dissolution of Ni_3Me Intermetallics in Austenite under Cold Deformation

Using the electron microscopy method, the authors [27] have been the first to show that the Guinier-Preston zones and the metastable θ' phase may dissolve in the Al-Cu alloy during cold deformation. This observation was not quite surprising since normal diffusion processes are possible in aluminium alloys at room temperature. However, a low-temperature (293 K) strain-induced dissolution of intermetallics upon their interaction with dislocations was found later [17, 20, 24-28] in materials having a higher melting temperature (iron, nickel), where the ordinary diffusion of substitutional elements is not the case. The dissolution was explained qualitatively by the drift of atoms in the stress field of dislocations, where the atom binding energy in a particle is less than the binding energy of these atoms and a dislocation.

Model Fe-Ni FCC alloys of the invar type (35Ni-3Ti, 35Ni-3Zr, 34Ni-5Al) were taken to ascertain accurately the strain-induced dissolution of intermetallics. In these alloys even a small change in the nickel concentration of the Fe-Ni matrix caused by dissolution or precipitation of the Ni_3Ti, Ni_3Zr or Ni_3Al intermetallic is easily registered using the Mössbauer method by variations in the Curie temperature T_C and the mean effective magnetic field at the ^{57}Fe core, H. Figure 9 shows the mean magnetic field vs. the nickel concentration of the Fe-Ni austenitic alloys [29]. From this dependence it is seen that if the nickel concentration of the iron alloys containing 30-35% nickel is changed even by 0.1%, H is altered considerably (by 3 kOe). The Mössbauer measurements were supplemented with electron microscopic studies.

Both rather coarse plates of the Ni_3Ti η-phase having a HCP lattice (ageing temperature 1073 K, 72-hour holding time) and dispersed coherent γ'-particles of Ni_3Ti about 6-8 nm in size (773-823 K) may be formed in the aged 35Ni-3Ti alloy [17]. The structural changes, which occur during cold deformation at 293 K, are more readily observed in the Ni35Ti3 alloy containing coarse plates of the η-phase up to 50 nm thick. Dissolution of the intermetallics begins as soon as the dislocations cut the η-plates during cold deformation with the reduction over 28% [17]. An intensive rolling deformation to 97% leads to a strong fragmentation of the alloy and breakage of the η-plates. The shear under a pressure of 8 GPa (the true deformation $\varepsilon = 5.9$) results in the formation of nanocrystals 20-100 nm in size and causes an almost complete dissolution

256

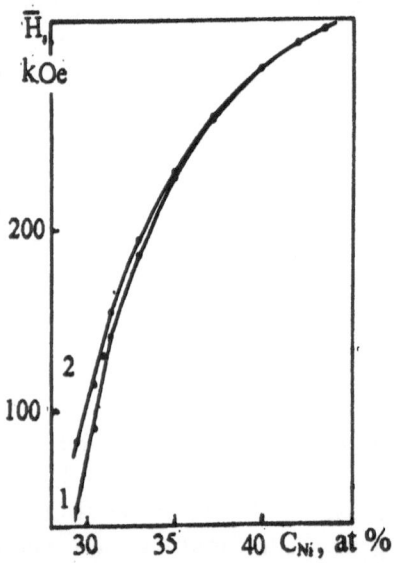

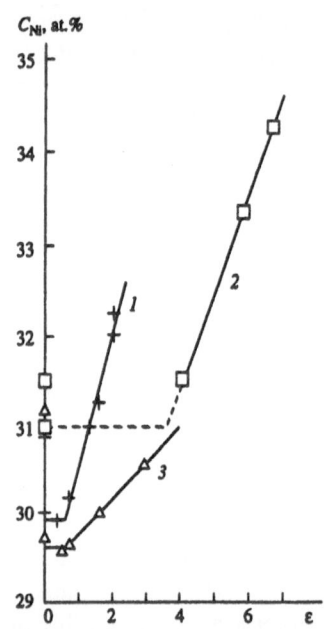

Figure 9. Mean internal field $\overline{H}$ vs. the nickel concentration C_{Ni} in Fe-Ni alloys: 1 - without paramagnetic component; 2 - with paramagnetic component

Figure 10. Nickel content C_{Ni} of the austenitic matrix vs. the true deformation ε in steels aged beforehand at 823 K for 3 h: 1 - 35Ni-3Ti; 2 - 35Ni-3Zr; 3 - 34Ni-5Al

of the intermetallic plates. This is confirmed by diffraction and Mössbauer methods [17, 29]. Dissolution of fine γ'-particles may be observed mainly by the Mössbauer method. Some NGR spectra and the corresponding magnetic field distribution functions P(H) for the 35Ni-3Ti alloy subject to ageing, rolling at 293 K (ε = 4.2), and pressure-shear deformation with ε = 5.9 can be found in Ref. [29]. An examination of the spectrum shows that when the aged 35Ni-3Ti alloy is deformed, the central line of the low-nickel austenite ($\sim$30% Ni) turns into a well resolved sextet, which corresponds to about 35 mass % nickel in the γ-matrix. This fact attests to a complete dissolution of the γ'-particles of Ni$_3$Ni in the austenite under shear.

The dissolution kinetics of the Ni$_3$Ti, Ni$_3$Zr and Ni$_3$Al intermetallics in 35Ni-3Ti, 35Ni-3Zr and 35Ni-5Al alloys respectively is illustrated in Fig. 10 in the "austenite nickel concentration C_{Ni} – true deformation ε" coordinates [17, 24]. The γ'-particles dissolve upon reaching a certain true deformation ε_0, when, according to [24, 26], edge dislocations actively cut the particles and Ni, Ti and other atoms drift in the dislocation stress field from the particles to the matrix for one interatomic distance after every dislocation cutting the particles. It has been shown elsewhere [29] that the intensity of the nickel concentration variation ΔC_{Ni} in the Fe-Ni solid solution, which is caused by dissolution of the Ni$_3$Ti particles interacting with the dislocations, depends on the number of the dislocations passing per unit surface area q, the mean dislocation length L and the Burgers vector b. This, in turn, determines the true deformation ε:

$$\Delta C_{Ni} = K_1 qLb = K\varepsilon.$$

These inferences are confirmed experimentally (Fig. 9) by a rectilinear behaviour of the dependence $\Delta C_{Ni} = f(\varepsilon)$. The variation of the nickel concentration in the γ-matrix during the strain-induced dissolution of the nickel-containing particles is given in a more rigorous notation as $\Delta C_{Ni} = K(\varepsilon - \varepsilon_0)$, where ε_0 is the critical dissolution strain of the particles, which depends on the nature, dimensions and shape of the particles of the dissolved phase. In addition, the dissolution intensity of the particles is characterised by the proportionality coefficient K. The dissolution intensity of coarse and sparse particles is lower than that of closely spaced dispersed phases [29], because in the former case many dislocations bypass the particles or interact with some of the particles only.

The observations, which adduce evidence to the enhancement of the intermetallic dissolution in the austenite when the deformation temperature is decreased from 473 to 293 K and even to 77 K, are quite remarkable [24]. This effect may be explained as follows. Two competing phenomena take place during deformation: 1) an equilibrium precipitation of the intermetallics, which is activated by deformation vacancies; 2) a nonequilibrium strain-induced dissolution of the γ'-particles when they interact with dislocations as a result of the drift of atoms in the stress field of the dislocations. The low-temperature (293 K) intermetallic ageing of the 35Ni-3Ti alloy was shown [14] to exist when a large number of radiation-induced vacancies ($\sim$0.01%) is introduced by irradiation of the alloy with 5.5-MeV electrons. Obviously, the diffusion precipitation of the γ'-phase, which is activated by the deformation vacancies, is hardly seen against the background of an intensive strain-induced dissolution of the γ'-particles. The former process shows itself up only in a lower intensity of the strain-induced dissolution of the Ni_3Ti phase when the deformation temperature is increased.

3.3.3. The Kinetics of Mechanical Alloying of the Fe-Ni Austenite when a Nanocrystalline State is Formed under the Pressure Shear

A high-pressure shear deformation not only allows producing nanodimensional crystals but also provides mechanical alloying, which is followed by formation of solid solutions from mechanical mixtures of alloys and pure metals. As distinct from treatment in ball mills, this deformation treatment rules out uncontrolled heating, gives a precise deformation determined by the turning angle of samples in the Bridgman anvils, and eliminates dissolution of some elements from the protective liquid material (toluene, alcohol, etc.), which is used during mechanical alloying in ball mills.

Powders containing 90%

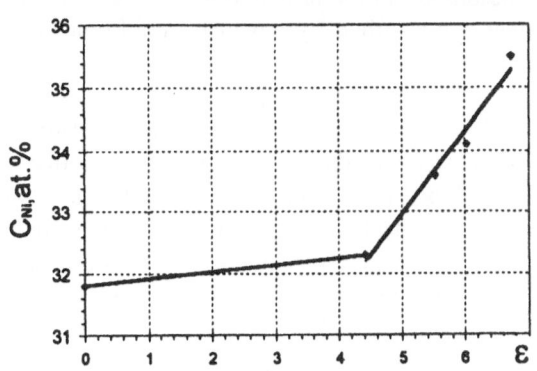

Figure 11. Mechanical alloying kinetics of the Fe-Ni alloy depending on the true deformation ε under shear at a pressure of 8 GPa. The initial components: 33Ni alloy (90 weight %) and Ni (10 weight %).

austenite (33Ni alloy) and 10% nickel, which was doped with 1% Mössbauer isotope ^{57}Fe [30], were taken to examine the strain-induced formation of iron-based FCC solid solutions. The powder mixture was compressed in the Bridgman anvils under a pressure of 8 GPa and underwent a strong torsional shear deformation at 1 revolution to 10 revolutions in the anvils. The initial mechanical mixture of the iron-nickel Fe-33Ni alloy and nickel whose NGR spectrum contained the corresponding components turned into a FCC solid solution. The spectrum, which corresponds to 10 revolutions in the anvils ($\varepsilon = 6.7$), suggests mechanical alloying, which is accompanied by the formation of a Fe-Ni solid solution where the nickel content exceeds 35 at. %. Figure 11 shows the incremental kinetics of the nickel concentration in the γ solid solution [30]. The concentration growth is determined from the increase in the mean magnetic field at the ^{57}Fe core as the degree of deformation increases with the pressure shear. One may see that the dependence is nearly rectilinear and exhibits a critical deformation, which corresponds to the beginning of Ni dissolution, as with the strain-induced dissolution of the intermetallics in the Fe-Ni austenite (see subsection 3.3.2).

3.3.4. On the Dissolution of VC Carbides in the Fe-Ni Austenite under Cold Deformation

It was shown [31] that the Fe_3C cementite may decompose if it interacts with dislocations when a carbon steel having the initial pearlitic structure is subject to cold deformation. The authors [31] think that carbon from the Fe_3C carbide does not dissolve in the BCC lattice of the ferrite but goes to the dislocations and microcracks. The dissolution of VC carbides in the FCC iron alloy type 0.5C-31Ni-2V under cold deformation at 293 K was studied [32]. The increase in the carbon concentration of the austenite during the strain-induced dissolution of carbides was determined by the Mössbauer method observing the increase in the quadrupole component of the NGR spectrum of the deformed sample.

The analysis of the difference NGR spectra of the deformed samples after preliminary quenching and preliminary ageing showed that rolling with a 97-% reduction ($\varepsilon = 4.1$) causes a partial dissolution in the austenite of the most dispersed VC carbides having the crosswise dimension less than 5 nm, which were formed under the low-temperature ageing regime (873 K, 3 hours). VC particles having the dimensions of 10-15 nm, which precipitated during ageing at 973-1073 K, do not dissolve at all under cold deformation. The shear ($\varepsilon = 5.9$) under a high pressure of 8 GPa leads to a partial dissolution of coarser vanadium carbides. The fact that the deformation dissolution of vanadium carbides is hampered compared to the strain-induced dissolution of the intermetallics is probably due not only to a higher binding energy of atoms in VC. The point is that dislocations typically do not cut but bypass VC particles. A large difference in the FCC lattice constants of the particles and the matrix leads to a loss of the coherence of small-size particles. As a result, cutting of the particles by dislocations and dissolution of the particles in the austenite is hampered.

3.3.5. Redistribution of Alloying Elements in the FCC Alloys under Cold Deformation

It is known that the presence of radiation-induced vacancies in austenitic iron alloys may give rise to a low-temperature intermetallic ageing [14] or a segregation Ni- and

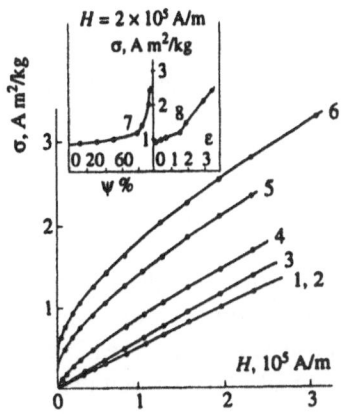

Figure 12. Specific magnetisation of the 30Ni-12Cr steel at 297 K as a function of the magnetic field in samples subject to various strains ψ (%): 1 - 0; 2 - 10; 3 - 50; 4 - 86; 5 - 95.6; 6 - 97.7. The insert shows the specific magnetisation as a function of the strain ψ (%) (curve 7) and the true deformation (curve 8) in the magnetic field H = 2×10^5 A/m

Ti-enrichment of the point defect sinks represented by grain and subgrain boundaries in the neutron-irradiated steel type 16Cr-15Ni-3Mo thanks to Kirkendale's reversed effect [10]. Such effects may be due to strain-induced vacancies. However, under a strong cold deformation any intermetallic phase or microsegregation will dissolve in the austenite when it is cut by dislocations (see subsection 3.3.2). To examine the segregation enrichment of point defect sinks in alloying elements during formation of strain-induced vacancies, one needs to produce strain-resistant sinks in the form of unchangeable boundaries of fragments and micrograins. This is the case with the strain-induced formation of nanocrystals, when refinement of the crystallites stops with increasing strain and further deformation is due mainly to the rotation of the crystallites [23]. In these conditions the boundaries and near-boundary regions of nanocrystals are stable .

Figure 12 [25] shows the effect of cold deformation at 293 K on the specific magnetisation of the quenched single-phase austenitic steel type 30Ni-12Cr, which is resistant to the martensitic transformation. One may see that at small and medium degrees of deformation (up to 50-76%) the field dependence of the magnetisation, $\sigma(H)$, remains to be linear and the paramagnetic state of the steel is preserved. A special situation occurs at a deformation exceeding 86%, when nanocrystals appear. The dependence $\sigma(H)$ becomes nonlinear in small fields (curves 4 to 6, Fig. 11). This is typical of a paramagnetic having small local ferromagnetic regions. After deformation to ~98% magnetisation of the steel in the field H = 2×10^5 A/m increases by 2.5 times. The ferromagnetic regions are not martensite since their Curie temperature T_C = 401 K, which is much lower than T_C of the Fe-Cr-Ni α-martensite (T_C > 800 K).

The observed effect is explained by a redistribution of alloying elements, which is initiated by deformation vacancies, as is the case with the formation of radiation-induced vacancies (see subsection 2.2). The neutron irradiation is known to cause [10, 33] an anomalous nickel enrichment and chromium depletion of the sinks (grain boundaries) in the Fe-Cr-Ni austenite. Point defects move to the sinks, that is, grain boundaries and dislocations. Interstitial atoms having different radii possess a different mobility and therefore the sinks are enriched with more mobile elements. On the other hand, the flows of vacancies meet a counter-flow (from the sinks) of the atoms, which migrate at different speeds. The resulting effect is enrichment or depletion in nickel, chromium or iron of the most stable sinks, for example, boundaries of the nanocrystals in the noncrystallographic shear bands caused by a strong deformation. The appearance of ferromagnetic regions in the 30Ni-12Cr steel having $T_C = 401$ K subject to a cold deformation may be related to enrichment of these near-boundary regions with nickel up to the compositions 40Ni-12Cr-, 37Ni-5Cr, etc. Homogenisation of the deformed steel, which is accompanied by vanishing of the ferromagnetic component, takes place within the temperature interval of the subsequent thermal annealing at 473-873 K [25] where the recovery and recrystallisation processes come into play.

3.3.6. Variation of the Physical and Mechanical Properties of FCC Alloys when a Nanostructure is Produced by Deformation

When FCC metals and alloys operate under strong deformation effects, for example, under sliding friction, a nanocrystalline structure appears. It is known [34] that when copper undergoes sliding friction, the deformation in surface layers is realised through rotation of the formed nanocrystals about the axis normal to the friction direction and parallel to the friction surface. Various phase transformations occur in the surface layers of these alloys. As a result, physical and mechanical properties (specifically, wear resistance) of steels and alloys are altered.

The wear resistance is typically enhanced through strengthening of the surface layers thanks to the formation of the strain-induced BCC α-martensite in metastable carbon steels [35]. The strain-induced carbonless HCP ε-martensite, which is formed in iron alloys containing 20-35% manganese, improves the wear resistance for another reason, namely at the expense of a sharp decrease in the friction coefficient from 0.4-0.6 to 0.1-0.2 [35, 36]. This is characteristic of HCP materials, for example, cobalt and cobalt alloys where the basal slip takes place. Therefore it is possible to transform certain FCC iron alloys to antifriction materials.

Anomalous phase transformations, which occur during friction, such as dissolution of strengthening phases (intermetallics or dispersed carbides), may lead to softening of the surface layers of aged FCC alloys and degradation of their wear resistance. The dissolution of Ni_3Ti intermetallics during sliding friction was reported elsewhere [24]. An effective strengthening of the austenite during the intermetallic ageing does not improve the wear resistance [35]. For the wear resistance of austenitic alloys to be considerably improved at the expense of the precipitation hardening, one needs to precipitate such strengthening phases that would not dissolve if a nanocrystalline structure is formed in the surface layers of parts operating under conditions of dry sliding friction. For example, austenitic steels of the 0.7C-20Mn-V-Mo type have a high wear resistance under friction at a speed of 0.05-4.0 m/s and a specific pressure of 2 MPa [5]. When these steels undergo ageing at 923 K for 10 hours, VC and Mo_6C carbides are formed, which do not dissolve under deformation.

The physical properties of Fe-Ni austenitic alloys may change considerably if Ni_3Ti intermetallics undergo strain-induced dissolution. For example, cold deformation (96%) of the aged 35Ni-5Ti alloy increases the Curie temperature T_C from 223 K to 423 K [28] thanks to dissolution of the γ'-phase. A similar deformation of the binary 34Ni alloy changes T_C little if at all. It is known [5] that quenching of Fe-Ni-Ti and Fe-Ni-Cr-Ti austenitic alloys does not eliminate initial stages of the intermetallic ageing, which is accompanied by changes in the nickel concentration of the matrix about the particles. In this case a homogenised single-phase FCC solid solution can be produced by cold deformation only. The strain-induced homogenisation of high-strength Fe-Ni-Co-Ti invars permits enhancing their invar properties. Cold plastic deformation of the 36Ni -10Co-3Ti invar increases the upper limit of the invar range from 493 K to 583 K [37].

4. Conclusion

Taking austenitic and austenitic-martensitic steels, it was shown that a superdispersed or nanocrystalline structure containing crystals 20 to 85 nm in dimension can be produced as a result of a high-dose (10-200 dpa) irradiation with 1.5-MeV krypton ions at 823-873 K or cyclic martensitic $\gamma \to \alpha \to \gamma$ transformations, which give rise to more than 500 different orientations of the γ-phase in the initial austenitic grain. The nanocrystalline state, which is formed during the $\alpha \to \gamma$ transformation in different thermal treatment conditions and which leads to the concentrational redistribution of the material, allows adjusting the thermal expansion coefficient of the 32Ni alloy over broad limits at temperatures from 223 K to 323 K. The superdispersed γ-phase formed during the $\alpha \to \gamma$ transformation largely improves the strength of the martensite and enhances the magnetic hysteresis properties of marageing steels. The formation of the nanostructure under a strong cold deformation of the austenite is followed by anomalous phase transformations – dissolution of intermetallics and carbides and a redistribution of alloying elements in the solid solution, which largely affect the wear resistance, magnetic and invar properties of FCC alloys.

This study has been performed with a partial support from the Russian Foundation for Fundamental Research (projects No. 96-15-96515 and No. 95-02-03539a).

5. References

1. Gleiter, H. (1995) Nanostructured materials: state of the art and perspectives. *Nanostructured Materials* **6**, No. 1-4, 3-14.
2. Proceedings of the 3rd international conference on nanostructured materials. (1997) M.I.. Trudeau. V. Provenzano, R.D. Shull, J.Y. Ying (ed.), *Nanostructured Materials* **8**, No. 1-8, 771 p.
3. Goshchitskii, B.N., Kirk, M.A., Sagaradze, V.V., and Lapin, S.S. (1997) Formation of a submicrocrystal FCC structure under irradiation with high-energy particles. *Nanostructured Materials* **9**, No. 1-8, 189-192.
4. Malyshev, K.A., Sagaradze V.V., et al. (1982) *The Transformation Hardening of Fe-Ni-Based Austenitic Alloys*, Nauka Publishers, Moscow, 260 p.
5. Sagaradze, V.V., and Uvarov, A.I. (1989) *Strengthening of Austenitic Steels*, Nauka Publishers, Moscow, 270 p.
6. Allen, C.W. (1994) In-situ ion- and electron-irradiation effects studies in transition electron microscopies. *Ultramicroscopy* **56**, 200-210.
7. Sagaradze, V.V., Morozov, S.V., Shabashov, V.A., Romashev, L.N., and Kuznetsov, R.I. (1988) Dissolution of spherical and lamellar intermetallics in Fe-Ni-Ti austenitic alloys subject to cold plastic deformation. *Phys. Metal. Metallography* **66**, No. 2, 328-338.
8. Sagaradze, V.V., Malyshev, K.A., Stchastlivtsev, V.M., Vaseva, Yu.A., and Proleyeva, L.M. (1975) The effect of the heating rate on the reverse martensitic transformation in the iron alloy containing 31.5% nickel. *Phys. Met. Metallography* **39**, No. 6, 1239-1250.
9. Belozerov, E.V., Sagaradze, V.V., Popov, A.G., and Pastukhov, A.M. (1995) The formation of a magnetic texture in a high-strength marageing steel. *Phys. Met. Metallography* **79**, No. 6, 58-67.

262

10. Garner, F.A. (1993) Irradiation Performance of Cladding and Structural Steels in Liquid Metal Reactors, in Br. Frost (ed.) *Mat. Science and Technol.* **10**, 419-543.
11. Neklyudov, I.M., and Voevodin, V.N. (1994) Features of structure-phase transformations and segregation processes under irradiation of austenitic and ferritic-martensitic steels. *J. Nucl. Mat.* **212-215**, 39-44.
12. Garner, F.A., McCarthy, J.M., Russel, K.C., and Hoyt, J.J. (1993) Spinodal-like decomposition of Fe-35N and Fe-Cr-35Ni alloys during irradiation or thermal aging. *J. Nucl. Mater.* **205**, 411-425.
13. Chamberod, A., Laugier, J., and Penisson, J.M. (1974) Electron irradiation effects on iron nickel invar alloys. *J. Magn. Magn. Mater.* **10**, No. 2-3, 139-144.
14. Sagaradze, V.V., Shabashov, V.A., Lapina, T.M., and Arbuzov, V.L. (1994) Phase transformations during a low-temperature electron irradiation of Fe-Ni and Fe-Ni-Ti austenitic alloys. *Phys. Met. Metallography* **78**, No. 4, 414-419.
15. Sagaradze, V.V., and Morozov, S.V. (1992) Mechanical alloying of Fe-Ni-Ti austenite during low-temperature deformation, in P.H. Shingu (ed.) *Mechanical Alloying. Material Science Forum* **88-90**, 147-154.
16. Valiev, R.Z., Korznikov, A.V., and Mulyukov, R.R. (1993) Structure and properties of nanocrystalline materials prepared by severe plastic deformation. *Mater. Sc. Eng.* **A168**, 141-149.
17. Sagaradze, V.V., and Shabashov, V.A. (1997) Deformation-induced anomalous phase transformations in nanocrystalline FCC Fe-Ni based alloys. *Nanostructured Materials* **9**, No. 1-8, 681-684.
18. Teplov, V.A., Korshunov, L.G., Shabashov, V.A., et al. (1988) Structural transformations of high-manganese austenitic steels during deformation by shear under pressure. *Phys. Met. Metallography* **66**, No. 3, 135-143.
19. Valiev, R.Z., Ivanisenko, Yu.V., Rauch, E.F., and Bandelet, B. (1996) Structure evolution in α-Fe subjected to severe plastic deformation. *Acta Metal. Mater.* **44**, 971-979.
20. Sagaradze, V.V., and Shabashov, V.A. (1997) Deformation-induced anomalous phase transformations in nanocrystalline FCC Fe-Ni-based alloys. *Nanostructured Materials* **9**, 681-684.
21. Hatherly, M., and Malin, A.S. (1984) Shear bands in deformed metals. *Scripta Met.* **18**, No. 5, 449-454.
22. Morii, K. (1984) Development of shear band in FCC single crystals. *Acta Met.* **33**, No. 3, 379-386.
23. Rybin, V.V. (1986) *Large Plastic Deformations and Destruction of Metals*, Metallurgiya Publishers, Moscow, 223 p.
24. Sagaradze, V.V., Shabashov, V.A., Lapina, T.M., et al. (1994) Low-temperature strain-induced dissolution of the intermetallic phase Ni_3Al (Ti, Si, Zr) in FCC Fe-Ni alloys. *Phys. Met. Metallography* **78**, No. 6, 619-628.
25. Zavalishin, V.A., Deryagin, A.I., and Sagaradze, V.V. (1993) Redistribution of alloying elements and variation of the magnetic properties induced by cold strain in stable austenitic chromium nickel steels. *Phys. Met. Metallography* **75**, No. 2, 173-179.
26. Gleiter, H., and Hornbogen, E. (1968) Die Forwanderung von Ausscheidungen durch Diffusion im spannungsfeld von Versetzungen. *Acta Met.* **16**, No. 3, 455-464.
27. Rakin, V.G., and Buinov, N.N. (1961) Effect of plastic deformation on the stability of particles precipitated in aluminum-copper alloy. *Phys. Met. Metallography* **11**, No. 1, 59-73.
28. Zemtsova, N.D., Sagaradze, V.V., Romashev, L.N., et al. (1979) The increase in the Curie temperature of aging alloys during plastic deformation. *Phys. Met. Metallography* **47**, No. 5, 937-942.
29. Shabashov, V.A., Sagaradze, V.V., Morozov, S.V., and Volkov, G.A. (1990) Mössbauer study of the kinetics of strain-induced dissolution of intermetallic compounds in Fe-Ni-Ti austenite. *Metallofizika* **12**, No. 4, 107-114.
30. Mukoseyev, A.G., Shabashov, V.A., Pilyugin, V.P., and Sagaradze, V.V. (1998) Strain-indused solid solution in the Fe-Ni system..*Phys.Met.Metallography* **85**, No. 5, In press.
31. Gavrilyuk, V.G. (1987) *Carbon Distribution in Steels*, Naukova Dumka Publishers, Kiev, 208 p.
32. Shabashov, V.A., Sagaradze, V.V., Morozov, S.V., et al. (1991) The effect of cold plastic deformation on the behavior of the carbide phase in the aged austenitic steel type 50Ni31V2. *Phys. Met. Metallography* **12**, 119-129.
33. *Phase Transformations during Irradiation*, (1983) F.V. Nolfi (ed.), Applied Science, London, 311 p.
34. Heilman, P., Clark, W.A., and Rigney, D.A. (1983) Orientation determination of subsurface cells generated by sliding. *Acta Met.* **31**, No. 8, 1293-1305.
35. Korshunov, L.G. (1992) Structural transformations during friction and wear resistance of austenitic steel. *Phys. Met. Metallography* **74**, No. 8, 150-162.
36. Korshunov, L.G., Sagaradze, V.V., Tereschenko, N.A., and Chernenko, N.L. (1983) The effect of ε-martensite on friction and wear of high-manganese alloys. *Phys. Met. Metallography* **55**, No. 2, 341-348.
37. Uvarov, A.I., Sagaradze, V.V., Menshikov, A.Z., Kazantsev, V.A., and Gasnikova, G.P. (1995) The effect of strengthening treatments on the thermal and mechanical properties of iron-nickel invars doped with cobalt and titanium. *Phys. Met. Metallography* **80**, No. 5, 155-165.

THE-STATE-OF-THE-ART OF NANOSTRUCTURED HIGH MELTING POINT COMPOUND-BASED MATERIALS

R.A.ANDRIEVSKI

Institute for New Chemical Problems, Russian Academy of Sciences
Chernogolovka, Moscow Region 142432, RUSSIA

1. Introduction

High-melting compounds (HMC) are carbides, nitrides, borides, oxides and other compounds with the melting point (T_m) above 2000°C (or even 2500°C). These limits are very conditional because there are no physical reasons for this selection but only considerations of convenience. As cited in [1], two-component HMC systems number at least 130, with $T_m > 2500$°C, and about 240, with $T_m > 2000$°C. The number of well-studied and practically used HMCs is much less. This overview concerns HMCs that were most extensively studied such as TiN, TiC, TiB$_2$, WC, AlN, Al$_2$O$_3$, Si$_3$N$_4$, SiC, BN, B$_4$C, ZrO$_2$, MgO, CeO$_2$, Y$_2$O$_3$ and some others. These compounds may be described as advanced ceramics and their promising properties and wide application are well known.

It should be noted that the above-mentioned HMCs can be divided into three groups according to their interatomic bonding nature: (1) metallic interstitial phases based on transition metals and metalloids; (2) ionic compounds (oxides), and (3) covalent compounds (nitrides and carbides of B, Si, and Al). To be sure the bonding character in these groups is not in the pure state but rather is in the very mixed one. This is to the most extent manifested in the case of interstitial phases. On one hand, the latter are hard and brittle like covalent and ionic solids, and, on the other, their electrical, magnetic, and optic properties are similar to those of metals. Brittleness and high values of the brittle-to-ductile transition temperature are characteristic of all HMCs. However, the information on HMCs properties in the nanocrystalline (nc) state is very limited [2]. It seems interesting to reveal the specific features of the size effect in all these cases using the data on the influence of chemical bonding at the transition from the conventional crystalline to nc state.

Besides, nanostructured materials (NM), which are normally characterised by the grain size (L) below 100 nm should be considered. They belong to so-called low-dimensional subjects such as clusters, dusts, ultrafine powders, nanotubes, colloidal subjects, catalysts, films, and so on. Some of them have a long history, for example, colloidal materials known in ancient Egypt. However, the results of Gleiter *et al.* [3,4] were undoubtedly an important impact in the proper NM problem (see also review [5]). The great attention enjoyed by NM everywhere for the past 10-15 years is connected at

G.M. Chow and N.I. Noskova (eds.), Nanostructured Materials, 263–282.
© 1998 *Kluwer Academic Publishers.*

least with two reasons. Firstly, this is caused with a hope to realise the unique mechanical, physical, and chemical properties (and therefore performance ones) in the nc state. Secondly, this topic revealed many gaps, not only in our understanding of the features of this state, but in the technology of its realisation too. The development of NM emerged as an important step in creating a new generation of materials. So foregoing circumstances gave rise to many investigations, Conferences, and publications. Since 1992, the Journal "Nanostructured Materials" is being published in the USA. Three specialised Conferences were held (the fourth one will be held in Sweden in 1998) and total their annual quantity is about 20-25 for a period of 5-7 years. This NATO ASI "Nanostructured Materials: Science and Technology" is one of these events. Two others NATO ASI have been taken [6,7].

It should also be stopped on the NM classification. Leaving aside the structural aspect (see classifications of Gleiter and Siegel [8, 9]), it seems reasonable to divide all NMs into three main types according to the methods of their preparation: particulate materials, materials obtained by controlled crystallisation from amorphous state (CCAS) as well as films and coatings. Table 1 shows this classification [10, 11].

TABLE 1. Methods of NM preparation

Types of NM	Methods and versions

Particulate materials	Conventional powder technology
	  - compaction and sintering, rate-controlled sintering
	  - hot pressing, HIP
	High-energy technique
	  - hot forging, extrusion
	  - electro-discharge compaction
	  - high static pressures and high temperatures
	  - high dynamic pressures (shock compaction, etc.)

Materials obtained by CCAS	Crystallisation under ordinary conditions Crystallisation under high pressures
Films and coatings	CVD PVD Electroplating Sol-gel technique

This classification includes only principal preparation methods and is also conventional because, for example, both particulate materials and films can be prepared in amorphous state and CCAS is an integral part of the process cycle. It should be possible to point severe plastic deformation for the NM preparation but mainly in connection with metals and intermetallics. It is generally supposed that the boundary between different preparation methods and their versions is often diffuse and poorly defined.

With only few exceptions, all methods listed in Table 1 may be used for preparation of NM based on HMCs. From Table 1 it can be seen that to accomplish these ends there are many routes which enlarges the technical possibilities but also creates the competition. The powder technology methods both in the common and advanced high-energy versions seem to be more universal. Although the presence of residual porosity in particulate NM often stands in the way. Films, coatings, and materials obtained by CCAS can be prepared in the dense state, but their compositions and sizes are not so comprehensive as in the case of particulate NM. So it is obvious that in some cases different types of NM and their preparation methods do not only compete but are mutually complementary. Let us consider the features of preparation and properties of these three types of NM.

2. Particulate Materials

2.1. ULTRAFINE POWDERS

The problems concerning ultrafine powders (UFPs) have been known for many years in powder metallurgy, ceramics, catalysis, and other fields. According to HMCs this question has been covered elsewhere (e.g. [2, 12]) therefore our description will be brief. Table 2 shows different methods of UFP preparation which may be classified into physical and chemical groups, with some variants. This classification is very conventional, because in many cases the preparation method is based on more than one basic principle (e.g. gas-condensation technique in reactive atmospheres and mechanical alloying are inextricably connected with chemical reactions). Once again notice there are many routes for UFP preparation. The competition and interchangeability presence is also evident. Preparation of UFP is one of the long-explored areas in the NM problem. Production of UFP is realised not only in the laboratory scale, but also in to larger units for many HMCs. Gas-condensation technique, mechanical alloying (mechanosynthesis), plasma and thermal syntheses, and thermal decomposition of condense precursors are the most popular methods. Many International Companies such as Nanophase Technologies Corp., USA (Y_2O_3, Al_2O_3, CeO_2, etc.), Nanodyne Inc., USA (WC-Co), H.C.Starck, Germany (TiN, Si_3N_4), some Japan and Chinese Firms (Si_3N_4, SiC, ZrO_2, Al_2O_3, etc.) are specialising on the production of carbide, nitride and oxide UFPs. To our knowledge, only boride UFPs are available in the limited laboratory scale.

We notice some interesting new tendencies in the UFP preparation, such as chemical vapour condensation (CVC) process and combustion flame-CVC one [13-15], synthesis via metathesis reaction between solid-state precursors [16], and electrical explosion of wire [17]. As compared to the traditional gas-condensation technique, CVC processes permit of considerable increase of the output. Oxide UFPs quantities in excess of 20 g/hr were readily produced in a small laboratory reactor [14]. Besides the increased quantities obtained, it was demonstrated that carbide and nitride UFPs could be also synthesised in these continuous CVC processes. It is important that the oxide and other UFPs are non-agglomerated powders, with controlled particle size in the range 2-20 nm.

TABLE 2. Methods of UFP preparation [2]

Group	Method	Variation	Compound
Physical	Gas-condensation	In inert gas or vacuum	MgO, Al_2O_3, Y_2O_3, ZrO_2, SiC
		In reactive gas	TiN, ZrN, NbN, VN, AlN, ZrO_2, Y_2O_3, Al_2O_3
	High-energy destruction	Milling	ZrC, Si_3N_4, AlN, SiC
		Mechanical alloying	TiC, WC, SiC, TiB_2, TiN
		Detonative treatment	BN, TiC, MgO, SiC, Al_2O_3
Chemical	Synthesis	Plasma	TiN, TiC, Ti(C,N), TiB_2, AlN, Si_3N_4, SiC, Si_3N_4/SiC, Al_2O_3
		Laser	TiB_2, SiC, Si_3N_4
		Thermal	BN, SiC, WC-Co, AlN, ZrO_2
		Electrolysis	WC, CeO_2, ZrO_2, Al_2O_3
		In solutions	SiC, TiB_2, W_2C, Mo_2C
	Thermal decomposition	Condense precursors	Si_3N_4, SiC, BN, AlN, ZrO_2, Y_2O_3, Si_3N_4/SiC,
		Gaseous precursors	TiB_2, ZrB_2, BN

Chemical exchange (metathesis) reactions driven in a high energy mill rise to the formation composite UFP. So the interesting TiB_2/TiN nanocomposites may be obtained not only by pyrolysis of polymeric precursors [18], but they also may be prepared by mechanosynthesis using, for example, following reactions resulting in an intimate mixture of the metal nitride and boride:

$$2TiCl_3 + 3/2MgB_2 + 3NaN_3 \rightarrow TiB_2 + TiN + 3/2MgCl_2 + 3NaCl + B + 4N_2, \quad (1)$$

$$3Ti + 2BN \rightarrow TiB_2 + 2TiN. \quad (2)$$

Reaction (1) is a self-propagating one as the result of exothermic interaction between $TiCl_3$ and NaN_3 [16]. Some other examples of using iron-containing compounds for the nanocomposites synthesis (e.g. AlN/Fe, Si_3N_4/Fe, etc.) were also reported [19]. Mechanosynthesis was shown to be capable of producing a wide variety of HMCs in the nc state including many nanocomposites. This method is promising at production scale, but the problem of significant contamination from the milling media or atmosphere remains to be solved [2, 20].

It might be well to point out that preparation of HMCs in the nc state based on using both traditional wet technique and organometallic precursors is rapidly progressing [2, 21]. As indicated earlier [2], the true value of any powder and any preparation method is defined by the results of very careful processing, its economy and ecology, and the properties of not only powders bur also materials based on them. To date, in most cases, such plausible information is not available and the necessity of the different methods comparison is evident.

2.2. CONSOLIDATION AND PROPERTIES

The purpose of all preparation methods of particulate NM listed in Table 1 is not only full densification but also retention of the nc structure as well. This problem is considered to be essential in NM preparation because of intensive recrystallization processes assisting UFP sintering. In many cases, recrystallization provides a conventional structure with the L values of about 1 μm and larger [2]. In this connection high-energy consolidation such as hot forging, high static and dynamic pressures, etc., seem to be the most useful for NM preparation. This problem is also discussed in some reports of the NATO ASI (e.g. [22-25]) and the present author restricts his consideration to some principal comments:

1. Cold compaction of oxide, nitride, and boride UFPs even at the pressures up to 10GPa does not lead to obtaining dense compacts. Residual porosity was about 10-20% and more (e.g. [2, 10, 11, 26]. The poor compressibility of HMC UFPs is consequence of both poor ductility of HMC and significant interparticle friction.
2. Regular sintering and hot pressing (HIP) do not conserve the nc structure [2, 27].
3. Sintering under special rate-controlled conditions (rate-controlled sintering) is characterised by no noticeable pores coalescence and this assists in retaining the nc structure [24, 28]. However, in the majority of cases, the L values of the obtained specimens are only on the upper level (about 100 nm).
4. Electro-discharge compaction (plasma-activated sintering) of UFPs tends also to the near similar results [29, 30].
5. The features of liquid- and solid-phase sintering, as applied to WC-Co UFP, were discussed in [31-33]. It was demonstrated that the nc structure could be obtained by using:(i) very small times at sintering;(ii) grain growth inhibitor such as small VC additions; (iii) hot mechanical processing in solid state (T=1200-1300°C). It is interesting that in the case of UFP the main part of shrinkage was in the solid state [33].
6. Hot forging and technique of high pressures and high temperatures revealed suitable results in obtaining NM [5, 10, 11, 23, 25, 34-36]. However, even in these cases obtaining dense specimens with the nc structure is not so simple and it is necessary to strike a definite compromise between densification and recrystallization in each specific example of consolidation.

The fracture surfaces of TiN UFP with the initial mean particle diameters (d_o) of 70-80 nm and 16-18 nm after consolidation at 1200°C (P=4 GPa) are compared in Figure 1 [37]. The specimens with the different L values are also characterised by the different hardness. In terms of porosity θ, the effect of relative density on microhardness H_V can be written as

$$H_V \cong H_{VO} (1 - 4.7\theta) \quad \text{for } Ti(N_{0.96}C_{0.01}O_{0.03})_{0.88-0.92} \ (d_o=70\text{-}80 \text{ nm}), \quad (3)$$
$$H_V \cong H_{VO} (1 - 2.1\theta) \quad \text{for } Ti(N_{0.92}C_{0.04}O_{0.04})_{0.84-0.89} \ (d_o=16\text{-}18 \text{ nm}), \quad (4)$$

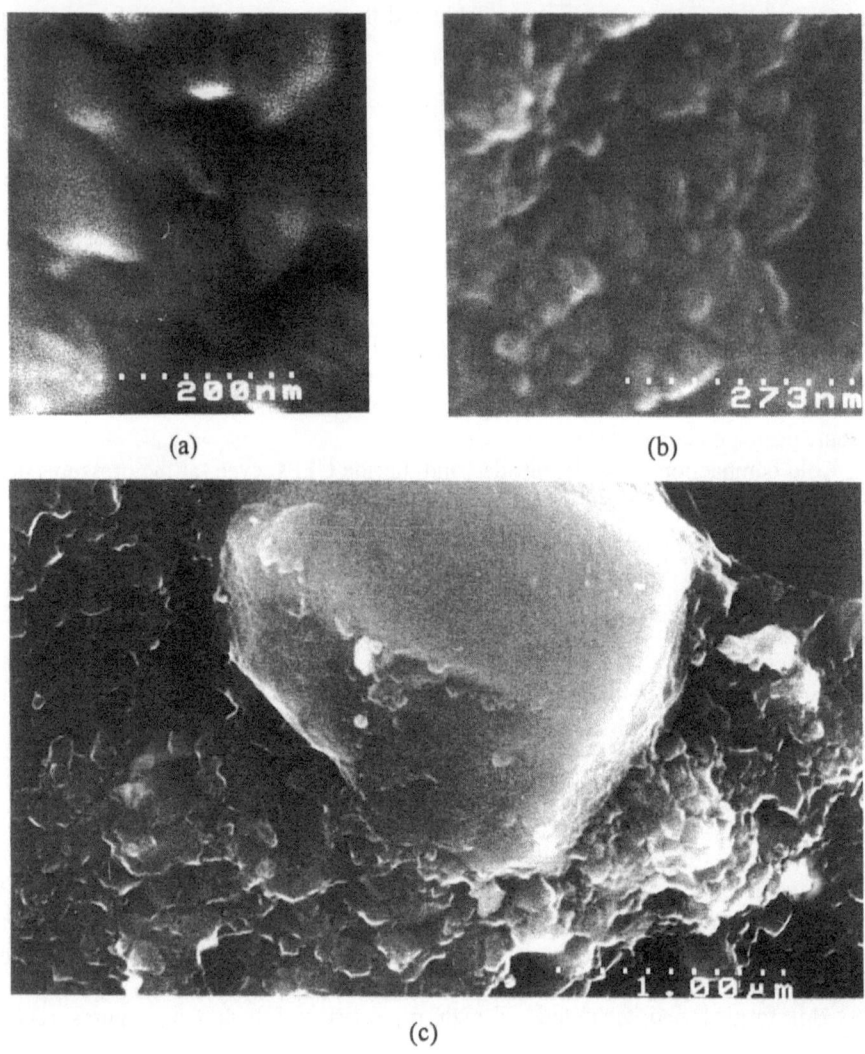

(a) (b)

(c)

Figure 1. FE-SEM micrographs of fracture surfaces of TiN UFP consolidated at 1200°C (P=4 GPa): (a) d_0~80 nm (L~70-150 nm); (b) d_0~18 nm (L~30-50 nm); (c) d_0~80 nm (with some large particles).

where H_{VO} is microhardness of the pore-free material, which is correspondingly equal to 26.2 GPa for TiN (d_0=70-80 nm) and 28.8 GPa for TiN (d_0=16-18 nm). The linear character of relationships (3,4) is usual for many particulate porous materials in the case of small interval of θ [1]. As demonstrated by detail analysis and comparison with the published data [10, 37], our results on H_{VO} are two-three times higher as compared with the values for the conditional hot-pressed and/or sintered TiN specimens and one-and-half times higher again as compared with the data on pore-free TiN. So undoubtedly these results reflect the nc structure effect. It was also revealed that elastic properties of nc TiN did not depend on the grain size.

Noteworthy also is the intercrystalline fracture that is evident from fractography study (Figure 1). At the same time, the transition from intercrystalline to transcrystalline fracture is found to occur during the increase the L values up to a micrometer size (Figure 1, c).

Table 3 shows the hardness values of some NM based on HMCs. The size of NM specimens are normally not so big and so the hardness measurements are most popular for NM characterisation because of the ease of realisation and the usefulness of the information obtained.

TABLE 3. Vickers microhardness of some HMC-based NMs

NM	Preparation method	Relative density	Grain size (nm)	Hardness H_V (GPa) at load (N)
Fe/63vol%TiN [38]	Dynamic compaction (P= 39 GPa; T=20°C)	0.92	~12	~13.5 (3)
Ni/64vol%TiN [38]	Dynamic compaction (P=39 GPa; T= 580°C)	0.975	~10	~13 (3)
Ag/76%MgO [39]	Compaction at 1 GPa		2-50	~2.5 (1-2)
ZrO$_2$-3%Y$_2$O$_3$ [40]	Sintering (T=1100°C)	~0.99	~160	~13 (0.5-1)
SiC [27]	HIP (P=350 MPa; T=1700°C)	~0.9	200-400	22-26 (100)
WC-6%Co [32]	Hot mechanical working (T=1200°-1300°C)	~1.0	100-300	21-22
WC-6%Co- -0.8%VC [32]	Liquid phase sintering	~1.0	100-300	21.3
TiN [11, 37]	High-pressure sintering (P=4 GPa; T=1400°C)	~1.0 ~1.0	50-80 30-40	26.2 (0.3) 28.8 (0.3)
BN [41]	Sintering in shock waves	~0.96	25	43-80 (5)

As it was noted [2], in the case of nc metals (copper, palladium, silver, nickel, titanium, etc.), the hardness increased on average, by four to six times, as a result of the transition to the nc state. However, in the case of HMCs the increase is not so large and equal to only about two times. The nature of such difference is to be further clarified and it is desirable to study the availability of Hall-Petch-like behaviour for nc HMC in the wide range of the grain size. The effect of porosity, deviations from stoichiometry, and admixtures on properties must be also considered in this investigation.

Other physical-mechanical and physical-chemical properties of HMCs in the nc state (creep rate, strength, etc.) were discussed elsewhere [2] and not so many new results are available. The fracture toughness (K_{IC}) and hardness of nc and submicron 3Y-ZrO$_2$ was found to be independent of the grain size [40]. On the other hand, high values of K_{IC} (about 16-23 MN m$^{-3/2}$) were pointed for the nc ZrO$_2$-Al$_2$O$_3$ system [42]. In general, two-phase nanocomposites attract a great attention in connection with both the superplasticity and superhardness problems. It is a common knowledge that the recrystallization rate in two-phase systems is lower than that in one-phase subjects. Some interesting results on mechanical properties of submicron ceramic composites (Al$_2$O$_3$/SiC, MgO/SiC, 8YSZ/SiC, Al$_2$O$_3$/nano-Al$_2$O$_3$, etc.) were obtained by Niihara *et al.* [43-45], Bellosi *et al.* [46], and Zhang *et al.* [47]. Table 4 shows the change in the H_V and K_{IC} values in the TiN/TiB$_2$ composites obtained by high-pressure sintering of TiN UFP and TiB$_2$ submicron powders [48, 49].

TABLE 4. Hardness (H_V; load of 5 N) and fracture toughness (K_{IC}) of TiN/TiB$_2$ composites obtained by high-pressure sintering (T=1200-1400°C; P= 4 GPa)

Composition (wt%)	Porosity (%)	H_V (GPa)		K_{IC} (MPa m$^{1/2}$)
		T_{TEST}=20°C	T_{TEST}=900°C	
TiN (d_0~80 nm)	~2	14.5±0.5	7.1±0.2	3.8±1.8
75TiN/25TiB$_2$	~1	17.0±1.2	7.2±0.7	4.7±0.4
50TiN/50TiB$_2$	~2	18.0±1.8	9.0±1.2	4.5±0.5
25TiN/75TiB$_2$	~3	18.8±1.85	9.6±0.5	5.2±0.5
TiB$_2$ (d_0~400 nm)	~4	18.2± 1.9	9.1±0.85	4.5±1.1

As is evident, some small unmonotonous change in H_V and K_{IC} is observed but the effect of the composition is not so high. The further investigations with the smallest particles and their different mixes as well as the increase in the test temperature seem to be interesting for revealing superplasticity and superhardness. Some new results on ceramics superplasticity (Si$_3$N$_4$, ZrO$_2$, etc.) were published (e.g. [23, 50-52]) but they covered mainly submicron subjects; real nanocomposite materials with the L values lower than 100 nm remain essentially uninvestigated.

As a summary of this section it may be pointed that a molecular-dynamics study of nc silicon nitride was performed by Nakano *et al.* [53]. Million-atom molecular-dynamics simulations were realised to investigate the structure, mechanical behaviour, and dynamic fracture in cluster-assembled Si$_3$N$_4$ which contained highly disordered interfacial regions with 50% undercoordinated Si atoms.

3. Materials Obtained by Controlled Crystallisation from Amorphous State

The data on this NM type are not so comprehensive as compared to those on particulate materials and are exhausted by some examples regarding Si$_3$N$_4$, ZrO$_2$, ZrB$_2$, Al$_2$O$_3$, etc. in powder form (e.g. [2, 54-64]). It is well known that plasma and laser syntheses of

Si_3N_4 UFP with particles size below ~20 nm result in the nc state. As far as we know, the preparation of bulk amorphous silicon nitride was not described and all information is concentrated on UFP or porous compacts. Mo *et al.* [55] revealed that crystallisation of the amorphous Si_3N_4 UFP (d_0~15 nm) obtained by Laser Induced Chemical Vapour Deposition started between 1300 and 1400°C. After crystallisation, the particles grew very quickly. The nucleation process was closely connected with the role of the internal interfaces. On the other hand, Hu *et al.* [56] pointed that the nano-amorphous Si-N-C powders began to crystallise at 1500°C to form α-Si_3N_4 and β-SiC. Properties of these amorphous UFPs, studied by positron lifetime measurements, TEM, XPS, NMR, etc., were described by several authors (e.g. [54, 57-61]; see also numerous refs. in [62]). Based on the results of high- pressure sintering of Si_3N_4 UFP [63] and the data of [55, 56], preparation of amorphous bulk silicon nitride could be performed at conditions of high-pressure sintering (T=1300-1400°C; P=8-10 GPa).

Preparation and characterisation of nc zirconia and alumina by crystallisation of amorphous precursors was described by Chaim [64]. The precursors could be obtained in a powder form by an electrochemical method. However, the regimes of sintering and hot compaction of these UFP do not ensure the high hardness values and need further specifying.

There are some results concerning the problems of amorphous films and coatings based on SiC_xN_y, BN, TiB_2/TiC, $MoSi_2N_x$, etc. (e.g. [62, 65-69]). Structural features of amorphous SiC_xN_y chemical vapour deposited films were studied in detail by XPS, FT-IR, EELS, MAS-NMR, and Raman spectrometry [62]. From these numerous results it is obvious that many circumstances of the SiC_xN_y microstructure are still lacking precision although it is clear that the interatomic bonding and local ordering are much more complicated than those of crystallised Si_3N_4/SiC mixtures. A very interesting example was demonstrated by Holleck and Lahres [65] in the case of the TiB_2/TiC films deposited by magnetron nonreactive sputtering and studied in the as-deposited (amorphous) and annealed (crystalline) states. It is notable that while hardness of amorphous TiB_2 and TiC was lower than that of crystalline ones, the opposite occurred for the mixed annealed and crystalline TiB_2/TiC compositions deposited from a 50/50 target. It is possible that in the latter case the superplasticity effect was revealed.

As is evident from the foregoing, the situation with the materials obtained by CCAS as applied to HMC is in the initial stage of the data accumulation. In the case of other NM (for example, different nc alloys and intermetallics) the progress seems to be more appreciable (e.g. [70-73]), to say nothing of commercialised FINEMET. It is also interesting that the CCAS method was successful by used for single nc phase bulk materials as applied to Se and $NiZr_2$ [74,75].

4. Films and Coatings

Films and coatings based on HMCs are well known and have wide application in the tool industry and electronics. Some books and reviews are devoted to this subject (e.g. [68, 69, 76-78]). Since the films and coatings are usually deposited under different conditions, the variety of their structures is very large and includes amorphous, nc, polycrystalline, and single crystal states. Many versions of main preparation methods (see Table 1) are known: thermal evaporation by electron beam and laser, arc deposition, magnetron sputtering, ion-beam-assisted deposition, implantation, plasma activated or assisted CVD processes, etc. As applied to the nc films, the most popular methods are arc deposition, magnetron sputtering, and ion-beam technique because of possibility of low-temperature deposition far from the equilibrium conditions. The formation of the nc structure is fixed by TEM, SEM, XRD, STM, and AFM examination. Figure 2 shows the dark field TEM micrograph of the Ti-B-N film [79]. It can be seen that the L value ranges up to about 5-10 nm.

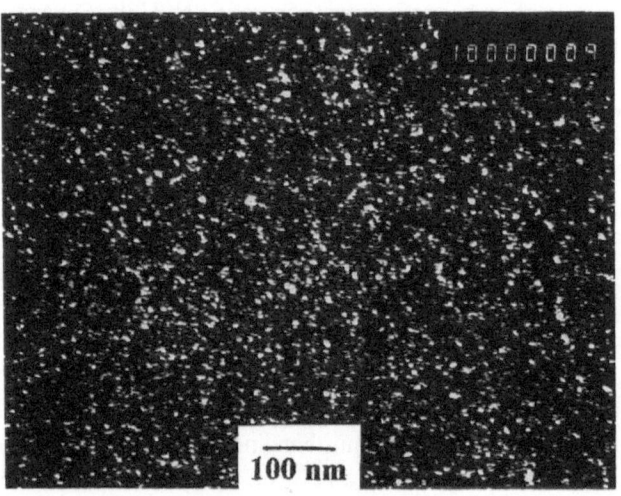

Figure 2. Dark field TEM micrograph of the $Ti(B_{0.73}N_{0.2}O_{0.05}C_{0.02})_{1.56}$ film obtained by d.c. magnetron non-reactive sputtering ($T_{substr} = 150°C$; $N = 1.5$ kWt; bias voltage of -30 V).

Figure 3 shows the AFM images of the TiN film deposited onto the Si substrate. The surface topography exhibits rugged fine morphology. The distribution of the rugged mountains seems to be near uniform. However, the detail peculiarities of the nc films surface topography and agreement between different methods of structure examination are far from being settled [68, 69].

Nevertheless, it is possible to formulate some principal and general comments based on a detailed analysis of structure and different physical-mechanical properties of nitride, boride and carbide films [68, 69, 76]:

1. Nitride films and partly boride ones, especially those of TiN and TiB_2, are now better understood. There are numerous data on their hardness, electrical, elastic and other properties although good agreement between the results of different authors is not observed in many cases.

2. The regimes and methods of deposition such as substrate temperature, bias voltage, ion energy, annealing temperature, etc. exert a pronounced effect on the films structure and properties.

(a)

(b)

Figure 3. Three-dimensional (a) - and two-dimensional (b) - AFM images of the TiN film obtained by r.f. magnetron sputtering.

3. In many cases the solubility of alloying elements and the quantity of nonthermal vacancies in the nc films differ essentially from those in the equilibrium phases.

4. Evaluation of the size effect in films requires close control over admixtures, stoichiometry deviation, and the strain level.

5. Nevertheless, some authors pointed the validity of the Hall-Petch-like relation (e.g.[80, 81]) and the high hardness values (about 35-40 GPa for TiN) were obtained. Especially high ones were observed in the case of multilayer films (superlattices) [68, 69, 76, 82].

Some interesting results on the films fracture features were obtained by the cross-sectional SEM observation of the indentation impressions (e.g. [83-87]). Figure 4 shows a typical fully dense columnar structure of a TiN film after load indentation [83]. The example of plastic deformation of brittle TiN is very impressive.

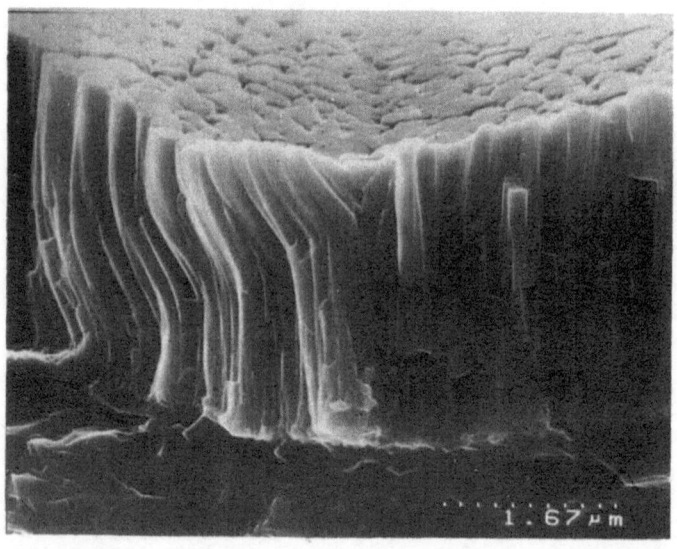

Figure 4. High resolution FE-SEM image of a fractured cross-section of a TiN film through an indentation imprint showing occurrence of microbuckling behaviour at high indentation load (courtesy of Ma and Bloyce [83]).

The availability of two types of deformation mechanism is evident from Figure 5. In the case of the Ti-B-N films with the hexagonal structure (the AlB_2 type), the formation of shear bands is recognised and localisation of deformation is given in an explicit form (Figure 5, a, b). In the case of the Ti-N-B films with the NaCl structure, the formation of shear bands is not observed and the deformation seems homogeneous although the cracks in films occur especially at high loads (Figures 4 and 5, c, d). Analogous homogeneous deformation was also fixed by Shiwa *et al.* for TiN films [85]. At inhomogeneous deformation, the band width and the step height equal respectively to about 70 nm and 25 nm. It is very interesting that such types of deformation are also an inherent feature of metallic glasses (e.g. [88-91]). As it was pointed earlier (Figure 2) [79], the L value in our Ti-B-N films is very low and equalled about 5-10 nm in the case

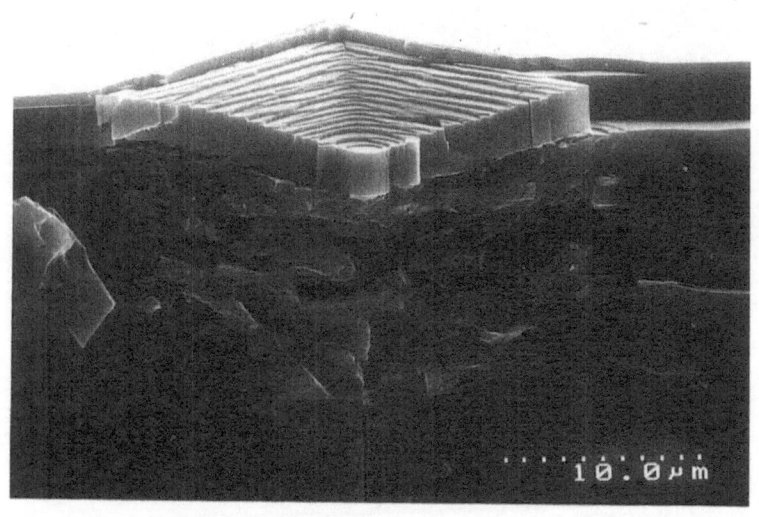

(a)

(b)

276

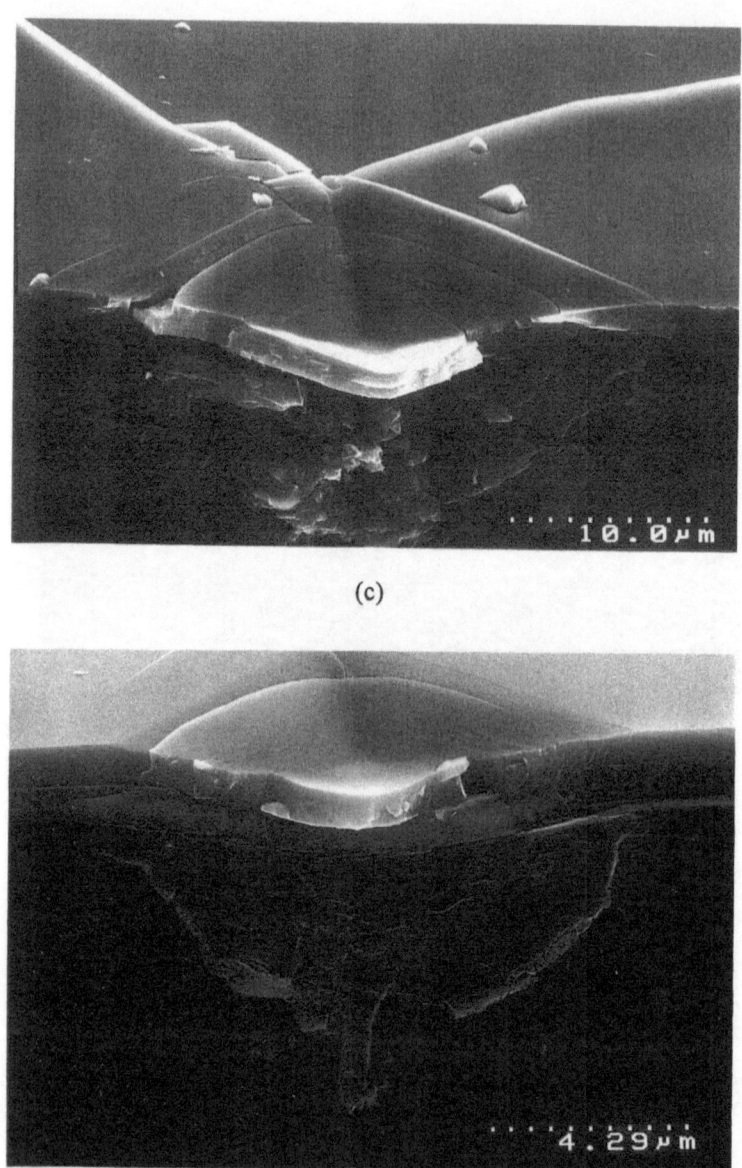

(c)

(d)

Figure 5. Fracture high resolution FE-SEM images through indentations on the Ti-B-N films deposited onto the Si substrate: a) and b) hexagonal d.c. sputtered Ti $(B_{0.73}N_{0.2}O_{0.05}C_{0.02})_{1.56}$ film, applied load of 5 N; c) and d) cubic r.f. sputtered Ti$(N_{0.49}B_{0.34}O_{0.12}C_{0.05})_{1.49}$ film, applied loads of 5 N (c) and 1 N (d).

of hexagonal films and about 3-6 nm for cubic ones. From simple considerations, for these L values the volume fraction of atoms in the grain boundaries (this interfacial environments is similar to an amorphous state) can be as much as 40-60 % (see, for example, [2, 12]). In this connection the analogy with the deformation behaviour of metallic glasses seems to be very likely. The different behaviour of hexagonal and cubic films may be connected with known difference in anisotropy and slip systems for the hexagonal TiB_2 and cubic TiN [1]. Primary slip systems for TiB_2 are only $\{10\bar{1}0\}<11\bar{2}0>$ and $\{10\bar{1}0\}<0001>$. Face-centered cubic TiN has $\{111\}<1\bar{1}0>$ polyslip. However, the nature of sliplike inhomogeneous deformation of amorphous solids remains obscure [88-91]. Many questions such as the size and composition effect on the deformation feature, the influence of temperature and stress, the nature of the fracture mechanism, etc. are yet to be studied.

5. Conclusions

As is evident from the foregoing, the information on HMCs in the nc state is not very comprehensive and/or exhaustive. The largest body of the results is related to UFP and films, whereas very little was published on the materials obtained by CCAS. Understanding of many fundamental problems such as the nature of the grain boundary , the size effect, the mechanism of deformation, etc. is far from being satisfactory. Particular emphasis must be taken to the effect of oxygen and other impurities both in solving and inclusions forms. Taking into consideration a wide range of promising applications of NMs such as tool materials, microelectronics, and decorative techniques it is clear that further efforts seem to be necessary and warrant.

6. Acknowledgements

The author would like to express his deep thanks to Drs. A.Bloyce, G.V.Kalinnikov, K.J.Ma, A.V.Ragulya, and V.S.Urbanovich as well as to Prof. Zhang Lide for their active help. The supports from the NSF (grant No MTF000), RBRF (grant No 95-02-03518) and INTAS (grant No 94-1291) are much appreciated.

7. References

1. Andrievski, R.A. and Spivak, I.I. (1989) *Strength of High-Melting Compounds and Materials on Their Base* , Metallurgiya, Chelyabinsk (in Russian).
2. Andrievski, R.A. (1994) Review - Nanocrystalline high melting compound-based materials, *J. Mater. Sci.* **29**, 614-631.
3. Gleiter, H. (1981) Nanocrystalline materials, in N.Hansen, T.Leffers, and H.Lilholt (eds.), *Deformation of Polycrystals: Mechanisms and Microstructures*, Riso National Laboratory, Roskilde, pp.15-21.
4. Birringer, R., Herr, U, and Gleiter, H. (1986) Processing and properties of nanocrystalline materials, *Trans. Jpn. Inst. Met. Suppl.* **27**, 43-52.

5. Andrievski, R.A.(in press) The-state-of-the-art and perspectives in the field of particulate nanostructured materials, *Powder Metallurgy(Minsk)* (in Russian).

6. Nastasi, M., Parkin, D.M., and Gleiter, H.(eds.) (1993), *Mechanical Properties and Deformation Behaviour of Materials Having Ultra-Fine Microstructures*, Kluwer Academic Publishers, Dordrecht.

7. Hadjipanayis, G.C. and Siegel, R.W. (eds.) (1994), *Nanophase Materials. Synthesis - Properties - Applications*, Kluwer Academic Publishers, Dordrecht.

8. Gleiter, H. (1995) Nanostructured materials: state of the art and perspectives, *Nanostruct. Mater.* **6**, 3-14.

9. Siegel, R. (1996) Recent progress in nanophase materials, in C.Suryanarayana, J.Singh, and F.H.Froes (eds.), *Processing and Properties of Nanocrystalline Materials*, The Minerals, Metals & Materials Society, Warrendale, pp.3-10.

10. Andrievski, R.A. (1996) Processing and properties evolution of nanocrystalline particulate and films materials based on nitrides and borides, in C.Suryanarayana, J.Singh, and F.H.Froes (eds.), *Processing and Properties of Nanocrystalline Materials*, The Minerals, Metals & Materials Society, Warrendale, pp.135-142.

11. Andrievski, R.A. (1996) Possibility of powder technology in processing advanced nanocrystalline particulate materials, in A.Clayton and L.Youngberg (comps.), *Advances in Powder Metallurgy and Particulate Materials -1996*, vol.1, Metal Powder Industry Federation, Princeton, pp.(2-79)-(2-88).

12. Suryanarayana, C. (1995) Nanocrystalline materials, *Int. Mater. Rev.* **40**, 41-64.

13. Chang, W., Skandan, G., Danforth, S.C., Kear, B.H., and Hahn, H. (1994) Chemical vapour processing and applications for nanostructured ceramic powders and whiskers, *Nanostruct. Mater.* **4**, 507-520.

14. Hahn, H. (1997) Gas-phase synthesis of nanocrystalline materials, *Nanostruct. Mater.* **9**, 3-12.

15. Chen, Y., Glumac, N., Kear, B.H., and Skandan, G. (1997) High rate synthesis of nanophase materials, *Nanostruct. Mater.* **9**, 101-104.

16. Gillan, E.C. and Kaner R.B. (1996) Synthesis of refractory ceramics via rapid metathesis reactions between solid-state precursors, *Chem. Mater.* **8**, 333-343.

17. Kotov, Yu.A., Azarkevich, E.I., Beketov, I.V., Demina, T.M., Murzakaev, A.M., and Samatov, O.M. (1997) Producing Al and Al_2O_3 nanopowders by electrical explosion of wire, *Key Eng. Mater.* **132-136**, 173-176.

18. Nowakowski, M., Su, K., Sneddon, L., and Bonnel, D. (1993) Synthesis, processing and phase evolution of TiN/TiB_2 composites from polymeric precursors, in S.Komarneni, J.C.Parker, and G.J.Thomas (eds.), *Nanophase and Nanocomposite Materials*, Materials Research Society, Pittsburgh, pp.425-430.

19. Matteazzi, P., Le Gaer, G., and Mocellin, A. (1997) Synthesis of nanostructured materials by mechanical alloying, *Ceram. Int.* **23**, 39-44.

20. Koch, C.C. (1997) Synthesis of nanostructured materials by mechanical milling: problems and opportunities, *Nanostruct. Mater.* **9**, 13-22.

21. Bill, J. and Aldinger, F.(1995) Precursor-derived covalent ceramics, *Adv. Mater.* **7**, 775-787.

22. Lavernia, E.J. (1998) Thermal spraying of nanocrystalline materials, in G.M.Chow (ed.), *Nanostructured Materials: Science and Technology*, Kluwer Academic Publishers, Dordrecht.

23. Mayo, M. (1998) Nanocrystalline ceramics for structural applications: proces-sing and properties, in G.M.Chow (ed.), *Nanostructured Materials: Science and Technology*, Kluwer Academic Publishers, Dordrecht.

24. Skorokhod, V.V. (1998) Features of nanocrystalline structure formation on sintering of ultrafine powders, in G.M.Chow (ed.), *Nanostructured Materials: Science and Technology*, Kluwer Academic Publishers, Dordrecht.

25. Urbanovich, V.S. (1998) Consolidation of nanocrystalline materials at high pressures, in G.M.Chow (ed.), *Nanostructured Materials: Science and Technology*, Kluwer Academic Publishers, Dordrecht.

26. Ivanov, V., Paranin, S., and Nozdrin, A. (1997) Principles of pulsed compaction of ceramic nano-sized powders, *Key Eng. Mater.* **132-136**, 400-403.

27. Vassen, R., Kaiser, A., Forster, J., Buchkremer, H.P., and Stover, D. (1996) Densification of ultrafine SiC powders, *J. Mater. Sci.* **31**, 3623-3637.

28. Ragulya, A.V., Skorokhod, V.V., and Andrievski, R.A. (1997, in press) Rate-controlled sintering of nanocrystalline TiN powder, *Nanostruct. Mater.* **8**.

29. Risbud, S.H., Shan, C.-H., Mukherjee, A.K., Kim, M.J., Bow, J.S., and Holl, R.A.(1995) Retention of nanostructure in aluminium oxide by very rapid sintering, *J. Mater. Res.* **10**, 237-239.

30. Schneider, J.A., Risbud, S.H., and Mukherjee (1996) Rapid consolidation processing of silicon nitride powders, *J. Mater. Res.* **11**, 358-362.

31. Kear, B.H. and McCandish, L.E. (1993) Nanostructured hard alloys, *Nanostruct. Mater.* **3**, 19-25.

32. Mohan, K. and Strutt, P.R. (1996) Observation of Co nanoparticle dispersions in WC nanograins in WC-Co cermets consolidated from chemically synthesised powders, *Nanostruct. Mater.* **7**, 547-555.

33. Porat, R., Berger,S. and Rosen, A.(1996) Dilatometric study of the sintering me-chanism of nanocrystalline cemented carbides, *Nanoctruct. Mater.* **7**, 429-436.

34. Hague, D.C. and Mayo, M.J. (1995) Modelling densification during sinter-forging of yttria-partally-stabilized zirconia, *Mater. Sci. Eng.* **A204**, 83-89.

35. Boutz, M.M.R., Winnubst, L., and Burggraaf, A.J.(1995) Low-temperature sinter-forging of nanostructured Y-TZP and YCe-TZP, *J. Am. Ceram. Soc.* **78**,121-128.

36. Hirai, H. and Kondo, K. (1994) Shock-compaction Si_3N_4 nanocrystalline cera-mics: mechanisms of consolidation and of transition from α- to β- form, *J. Am. Ceram. Soc.* **77**, 487-492.

37. Andrievski, R.A. (1997) Physical-mechanical properties of nanostructured TiN, *Nanostruct. Mater.* **9**, 607-610.

38. Ogino, Y. (1996) Mechanical nitriding and its application to production of nano-crystalline metal-nitride dual phase alloys, in C.Suryanarayana, J.Singh, and F.H Froes (eds.), *Processing and Properties of Nanocrystalline Materials*, The Minerals, Metals & Materials Society, Warrendale, pp.81-92.

39 Kizuka, T., Ichinose, H., and Ishida, Y.(1994) Structure and mechanical proper-ties of nanocrystalline Ag/MgO composites, *J. Mater. Sci.* **29**, 3107-3112.

40. Cottom, B.A. and Mayo, M.J. (1996) Fracture toughness of nanocrystalline ZrO_2-$3\%Y_2O_3$ determined by Vickers indentation, *Scripta Mater.* **34**, 809-814.

41. Kovtun, V.I., Kurdiumov, A.V., Zeliavskiy, V.B., Ostrovskaja, N.F., and Trefilov, V.I. (1992) Sintering of BN in shock waves, No12, 38-44 (in Russian).

280

42. Inamura,S., Miyamoto, M., Imaida,Y., Takagawa, M., Hirota, K., and Yamaguchi, O.(1993) High fracture toughness of ZrO_2 solid solution ceramics with nanometre grain size in the system ZrO_2-Al_2O_3, *J. Mater. Sci. Lett.* **12**, 1368-1370.

43. Jeong, Y.K. and Niihara, K. (1997) Microstructure and mechanical properties of pressureless sintered Al_2O_3/SiC nanocomposites, *Nanostruct. Mater.* **9**, 193-196.

44. Bamba, N., Choa, Y.H., and Niihara, K. (1997) Fabrication and mechanical properties of nanosized SiC particulate reinforced yttria stabilized zirconia composites, *Nanostruct. Mater.* **9**, 497-500.

45. Choa, Y.H., Kawaoka, H., Sekino, T., and Niihara, K. (1997) Microstructure and mechanical properties of oxide based nanocomposites fabricated by spark plas-ma sintering, *Key Eng. Mater.* **132-136**, 2009-2012.

46. Scitti, D., Fabbriche D.D., and Bellosi, A. (1997) Fabrication and characteristics of Al_2O_3/SiC nanocomposites, *Key Eng. Mater.* **132-136**, 2001-2004.

47. Li, G., Jiang, A., and Zhang L. (1996) Mechanical and fracture properties of nano-Al_2O_3 alumina, *J. Mater. Sci. Lett.* **15**, 1713-1715.

48. Andrievski, R.A., Kalinnikov, G.V., and Urbanovich, V.S. (1997) Consolidation and evolution of physical-mechanical properties of nanocomposite materials based on high-melting compounds, in S. Komarneni, J.C. Parker, and H.J. Wollenberger (eds.), *Nanophase and Nanocomposite Materials II*, vol. 457, Materials Research Society, Pittsburgh.

49. Andrievski, R.A., Urbanovich, V.S., and Shipilo, V.B. (in press) Fracture toughness of nitride/boride nanocomposites obtained by high-pressure sintering, *Powder Metallurgy (Kiev)* (in Russian).

50. Andrievski, R.A., Ivannikov, V.T., and Urbanovich, V.S. (1994) Creep studies in Si_3N_4-TiB_2 materials, *Key Eng. Mater.* **89-91**, 445-448.

51. Wakai, F., Kondo, N., Ogawa, H., Nagano, T., and Tsurekawa (1996) Ceramics superplasticity: deformation mechanisms and microstructures, *Mater. Charac-ter.* **37**, 331-341.

52. Burger, P., Duclos, R., and Crampon, J. (1997) Superplastic behaviour of low-doped silicon nitride, *Mater. Sci. Eng.* **A222**, 175-181.

53. Kalia, R.K., Nakano, A., Tsurita, K., Vashishta, P. (in press) Morphology of po-res and interfaces and mechanical behaviour of nanocluster-assembled silicon nitride ceramic, *Phys. Rev. Lett.*

54. Andrievski, R.A. (1995) Silicon nitride: synthesis and properties, *Russ. Chem. Rev.* **64**, 291-308.

55. Wang, T., Zhang, L., and Mo, C. (1994) A study on growth and crystallisation behaviour of nanostructured amorphous Si_3N_4, *Nanostruct. Mater.* **4**, 207-213.

56. Li, Y.-L., Liang Y., and Hu, Z.-Q. (1994) Crystallisation and phase development of nanometric amorphous Si-N-C powders, *Nanostruct. Mater.* **4**, 857-864.

57. Zhang, L., Mo, C., Wang, T., Cai, S., and Xie, C. (1993) Structure and bond properties of compacted and heat-treated silicon nitride particles, *Phys. Stat. Sol.* **136**, 291-300.

58. Wang, T., Zhang, L., Mo., C., Hu, J., and Xie, C. (1993) A study of defects in nanostructured amorphous silicon nitride, *Phys. Stat. Sol.* **139**, 303-307.

59. Wang, T., Zhang, L., and Mou, J. (1993) Anomalous dielectric behaviour in nanometer-sized amorphous silicon nitride, *Chin. Phys. Lett.* **10**, 676-679.

60. Wang, T., Zhang, L., Hu, J., and Mo, J. (1993) Study of dangling bonds in nano-meter-sized granulate silicon nitride by electron-spin resonance, *J. Appl. Phys.* **74**, 6313-6316.

61. Leone,E.A., Curran,S., Kotun, M.E., Carrasquillo,G., Weeren,R., and Danforth,C. (1996) Solid-State ^{29}Si NMR analysis of amorphous silicon nitride powder, *J. Am. Ceram. Soc.* **79**, 513-517.

62. Bendeddouche, A., Berjoan, R., Bache, E., Merle-Mejean, T., Schamm, S., Tail-lades, G., Pradel, A., and Hillel, R.(1997) Structural characterisation of amorph-ous SiC_xN_y chemical vapour deposited coatings, *J. Appl. Phys.* **81**, 6147-6154.

63. Andrievski, R.A., Konyaev, Yu.S., Leontiev, M.A., and Pivovarov, G.I. (1989) The influence of high pressures on structure and properties of silicon nitride, *High Pressure Research* **1**, 329-331.

64. Chaim, R. (1992) Fabrication and characterisation of nanocrystalline oxides by crystallisation of amorphous precursors, *Nanostruct. Mater.* **1**, 479-489.

65. Holleck, H. and Lahres, M. (1991) Two-phase TiC/TiB_2 hard coatings, *Mater. Sci. Eng.* **A140**, 609-615.

66. Kester, D.J., Ailey, K.S., Davis, R.F., and More, K.L. (1993) Phase evolution in boron nitride thin films, *J. Mater. Res.* **8**, 1213-1216.

67. Kung, H., Jervis, T.R., Hirvonen, J-P., Mitchel, T.E., and Nastasi, M. (1996) Syn-thesis, structure and mechanical properties of nanostructured $MoSi_2N_x$, *Nano-structur. Mater.* **7**, 81-88.

68. Andrievski, R.A. (1997) The synthesis and properties of interstitial phase films, *Russ. Chem. Rev.* **66**, 53-72.

69. Andrievski, R.A. (1997, in press) Films of interstitial phases: synthesis and pro-perties, *J. Mater. Sci.* **32**.

70. Kim, L.S., Chang, H., and Averback, R.S. (1993) Nanophase processing of amor-phous alloys, *J. Alloys Comp.* **194**, 245-249.

71. Trudeau, M.L. (1995) Engineering nanocrystalline materials from amorphous precursors, *Mater. Sci. Eng.* **A204**, 233-239.

72. Lu, K. (1996) Nanocrystalline materials crystallised from amorphous solids: nanocrystallisation, structure, and properties, *Mater. Sci. Eng. R* **16**, 161-221.

73. Greer, A.L. (1998) Changes in structure and properties associated with the tran-sition from the amorphous to the crystalline state, in G.M.Chow (ed.), *Nano-structured Materials: Science and Technology*, Kluwer Academic Publishers, Dordrecht.

74. Zhang, H.Y., Lu, K., and Hu, Z.Q. (1995) Transformation from the amorphous to the nanocrystalline state in a pure selenium, *Nanostruct. Mater.* **5**, 41-50.

75. Lu, K., Liu, X.D., and Yuan, F.H. (1996) Synthesis of the $NiZr_2$ initermetallic compound nanophase materials, *Physica* **B 217**, 153-159.

76. Sundgren, J. and Hultman, L.(1995) Growth, structure and properties of hard nit-ride based coatings and multilayers, in Y.Pauleau (ed.), *Materials and Proce-sses and Interface Engineering*, Kluwer Academic Publishers, Dordrecht, pp.453-474.

77. Hocking, M.G., Vasantasree, V.S., and Sidky, P.S. (1989) *Metallic and Ceramic Coatings: Production, High-Temperature Properties and Applications*, Long-man, Harlow.

282

78. Konuma, M. (1992) *Film Deposition by Plasma Techniques*, Springer Verlag Berlin.
79. Andrievski, R.A., Kalinnikov, G.V., Kobelev, N.P., Soifer, Ya.M., and Shtansky, D.V. (in press) Structure and physical-mechanical properties of nanostructured boride/nitride films, *Phys. Solid State* (in Russian).
80. Sue, J.A. (1993) Development of arc evaporation of non-stoichiometric titanium nitride coatings, *Surf. Coat. Technol.* **61**, 115-120.
81. Bendavid, A., Martin, P.J., Netterfield, R.P., and Kinder, T.J. (1994) The properties of TiN films deposited by filtered arc evaporation, *Surf. Coat. Technol.* **70**, 97-104.
82. Barnett, S.A. (1993) Deposition and mechanical properties of superlattice thin films, in M.H.Francombe and J.L.Vossen (eds.), *Physics of Thin Films. Mecha-nic and Dielectric Properties,* vol.17, Academic Press, Boston, pp.1-77.
83. Ma, K.J. and Bloyce, A. (1995) Observations of deformation and failure mechanisms in TiN coatings after hardness indentation and scratch testing. *Surf. Eng.* **11**, 71-74.
84. Ma, K.J., Bloyce, A., and Bell, T. (1995) Examination of mechanical properties and failure mechanisms of TiN and Ti-TiN multilayer coatings, *Surf. Coat. Technol.* **76-77**, 297-302.
85. Shiwa, M., Weppelmann, E., Munz, D., Swain, M.V., and Kishi, T. (1996) Acoustic emission and precision force-displacement observations of pointed and spherical indentation of silicon, *J. Mater. Sci.* **31**, 5985-5991.
86. Ma, K.J., Bloyce, A., Andrievski, R.A., and Kalinnikov, G.V. (in press) Microstructural response of mono- and multilayer hard coatings during indentation microhardness testing, *Surf. Coat. Technol.*
87. Andrievski, R.A., Bloyce, A., Kalinnikov, G.V., and Ma, K.J. (in press) Observations of deformation features in nanostructured T-B-N films after indentation testing, *J. Mater. Sci. Lett.*
88. Zielinski, P.G. and Ast, D.G. (1983) Slip bands in metallic glasses, *Phil. Mag.* A **48**, 811-824.
89. Donovan, P.E. (1989) Plastic flow and fracture of $Pd_{40}Ni_{40}P_{20}$ metallic glass un-der an indentor, *J. Mater. Sci.* **24**, 523-535.
90. Glezer, A.M. and Molotilov, B.V. (1992) *Structure and Mechanical Properties of Amorphous Alloys*, Metallurgiya, Moscow (in Russian).
91. Bobrov, O.P. and Khonik, V.A. (1995) Inhomogeneous flow via dislocations in metallic glasses: a survey of experimental evidence, *J. Non-Cryst. Sol.* **192/193**, 603-607.

THERMAL SPRAY PROCESSING OF NANOCRYSTALLINE MATERIALS

E.J. LAVERNIA, M.L. LAU, AND H.G. JIANG
Department of Chemical and Biochemical Engineering and Materials Science
University of California, Irvine, Irvine, CA 92697-2575, USA

1. Introduction

Technological advancements in many sectors of modern society depend strongly on the materials science and engineering community's ability to conceive of novel materials with attractive combinations of physical and mechanical properties. For instance, in the aerospace industry, the ever increasing demand to manufacture lighter aircraft that can travel at higher speeds and can withstand a higher payload capacity has fueled the development of high strength/low density materials with improved damage tolerance and enhanced temperature capabilities. Driven in part by this critical need, research in materials science and engineering has shifted towards the study and application of non-equilibrium processes. The significant departure from thermodynamic equilibrium associated with these types of processes allows material scientists and engineers to develop materials with unusual combinations of microstructure and physical attributes.
  Some notable examples of non-equilibrium processes include mechanical alloying, rapid solidification (e.g., atomization, melt spinning, etc.), chemical and vapor deposition, and spray processes (e.g. plasma spraying, thermal spraying, and spray atomization and deposition) [1-3]. The objective of this paper is to provide an overview of thermal spray processes, and present recent preliminary results on the application of this technology to the manufacture of nanostructured coatings.

2. Thermal Spraying

Thermal spraying is a technique in which molten or semi-molten particles are deposited onto a substrate to form a two-dimensional coating or three-dimensional self-standing structure, the microstructure of which depends on the thermal and momentum characteristics of the impinging particulates [1]. Thermal spraying combines particle melting, quenching and consolidation in a single operation. Original interest in thermal spraying stemmed from the ability of this technology to generate coatings that are chemically and metallurgically homogeneous. Consequently, early thermal spray technology found an immediate application in corrosion resistant zinc coatings and

G.M. Chow and N.I. Noskova (eds.), Nanostructured Materials, 283–302.
© 1998 *Kluwer Academic Publishers.*

coatings for other refractory metals [1, 4]. Today, thermal spray technology may be utilized in many coating applications which include:

- arc plasma spray (APS) coating of Cr_2O_3 on hardened steel drilling components in petroleum mining to improve the service lifetime [1];
- high velocity oxy-fuel (HVOF) coating of stainless steel 316L onto the chemical refinery vessels to provide good protection against sulfur and ammonia corrosion [1, 5];
- WC-cermet coating produced by HVOF onto the rolling surface of steel rolls utilized in steel industry rolling mills to increase the abrasion and friction resistance of the rolls [6];
- Al_2O_3 coatings by APS on aluminum mid-plate for diode assembly in automotive alternators to provide resistance against salt corrosion and moisture absorption [1, 7];
- thermal barrier coatings (TBCs) of ZrO_2-Y_2O_3 (top ceramic layer) /CoCrAlY (bond coating) by plasma spraying to reduce heat transfer and thus increase engine efficiency for piston crown and cylindrical head in adiabatic diesel engines [1, 8];
- wear resistant coatings of WC-M (M= Ni, Co, or Co-Cr) by HVOF or APS on compressor fan and disc mid-span stiffeners in aero-engines [9, 10], and
- tungsten coatings by plasma vacuum spray for plasma-facing components of nuclear fusion devices for longer service life due to corrosion processes on tungsten components being longer than that of the components composed of low-atomic weight elements [11].

Other applications of various thermal sprayed coatings are widely used in the electronic industries [12], powder generation plants [13], marine gas-turbine engines in marine industries [14], ceramic industries [1], and printing industries.

2.1. THERMAL SPRAYING TECHNIQUES

In principle, powders, rods, and wires can be used as spraying materials. Metals and alloys in the form of rods or wires are commonly used in arc spraying (AS) and flame spraying (FS). Powders of metals, alloys, ceramic oxides, cermets, and carbides are often used in thermal spraying to produce a homogeneous microstructure in the resulting coatings. In most cases, the sprayed surface should be degreased, masked and roughened prior to spraying to maximize the bonding strength between the coating and the substrate material. Various techniques for pre-spraying treatment can be described in Ref. [15].

A variety of thermal spray techniques have been developed since the 1900's. Today, flame spraying (FS), atmospheric plasma spraying (APS), arc spraying (AS), detonation gun (D-gun) spraying, high velocity oxy-fuel spraying (HVOF), vacuum plasma spraying (VPS), and controlled atmosphere plasma spraying (CAPS) are widely used to produce various coatings for different industrial applications. The process parameters of existing thermal spray techniques mentioned above are briefly listed in Table 1 [1].

TABLE 1. Process parameters of various thermal spray techniques

Thermal spraying techniques	Working flame	Flame temperature (K)	Flame velocity (m/s)	Powder particle sizes (µm)	Powder injection feed rate (g/min)	Spraying distance (mm)
FS	fuel + O_2 (g)	3000-3350	80-100	5-100	50-100	120-250
APS	Ar, or mixture of Ar+H_2, Ar+He and Ar+N_2 (g)	up to 14,000	800	5-100	50-100	60-130
AS	various electrically conductive wires (e.g. Zn, Al)	arc temperature of 6100 K by an arc current of 280 A [16]	velocity of molten particles formed can reach up to 150 m/s	wires diameter range from 2-5mm	50-300	50-170
DGS	detonation wave from a mixture of acetylene + O_2	up to 4500 K with 45% acetylene	2930 [17]	5-60	16-40 [18]	100 [19]
HVOF	fuel gases (acetylene, kerosene, propane, propylene, or H_2) with O_2	up to 3440 K at ratio of O_2:acetylene (1.5:1 by volume) [20]	2000	5-45	20-80	150-300
VPS	Ar mixed with H_2, He or N_2	temperature expressed in electron temperature of 10,000 to 15,000 K	Velocity of plasma between 1500-3000	5-20	50-100 (spraying in vacuum; during spraying pressure ~ 655 - 133 Pa)	300-400
CAPS	same as APS	same as APS	same as APS	same as APS	same as APS	100-130 in shrouded plasma spray [21]

Flame spray (FS), sometimes referred as combustion flame spraying, involves the combustion of fuel gas in oxygen (1:1 to 1.1:1 in volume ratio) to heat the feedstocks (in the form of powders, wires, or rods) [22]. The flame gases are introduced axially, and the particles travel in the direction perpendicular to the flame gases. The particles are melted in the flame and accelerated toward the target substrate. The typical coating thickness is between 100-2500 µm, and porosity ranges from 10 to 20 %. The bond strength for FS ceramic coatings are approximately 15 MPa and 30 MPa for most materials while NiAl coatings can reach 60 MPa [1].

In atmospheric plasma spray (APS), the flame gas (Ar or mixture of Ar+H_2, Ar+He or Ar+N_2) is heated by a plasma generator (60 kW or more) which produces an electric arc. The advantages of plasma processing include a clean reaction atmosphere, which is needed to produce a high-purity material; a high enthalpy to enhance the reaction kinetics by several orders of magnitude; and high temperature gradients that provide the possibility of rapid quenching and generation of fine-size [12]. Due to the high temperature of the flame gas (up to 14,000 K), APS is commonly used to produce ceramic thermal barrier coatings (TBC) of Y_2O_3-stabilized ZrO_2, Al_2O_3-ZrO_2, and cermet coatings [8, 13]. The bond strength of typical ceramic coatings produced by APS is in between 15-25 MPa and 70 MPa for some bonding alloys (NiAl or NiCrAl) or metals (Mo). The porosity of APS coatings is generally lower (1-7 %) than those produced by FS, and the thickness of the coating ranges from 50-500 μm [1].

Arc spraying (AS) involves two electrically conductive wires (Zn or Al) which are arc melted, and the molten particles are propelled by a compressed gas. The high velocity gas (flow rate of 1-80 m^3/hr) acts to atomize the melted wires and to accelerate the fine particles to the substrate. Alloy coatings can be readily produced if the wires are composed of different materials [1]. The thickness of the coating produced by AS is between 100-1500 μm, and the bond strength is in the range of 10-30 MPa for Zn and Al coatings [14].

Detonation-gun (D-gun) spraying is commonly used in producing WC-Co and Al_2O_3 coatings due to the low porosity in the resulting coatings (~0.5 % for WC-Co) and high bond strength (83 MPa for WC-Co) [1, 15]. In D-gun spraying, a mixture of flame gas (oxygen and acetylene) is fed to a long barrier with a charge of powder. Upon ignition, a detonation wave is produced (1-15 detonation/s) which delivers the powder particles at a velocity up to 750 m/s to the substrate [1].

High velocity oxy-fuel (HVOF) spraying is the most significant development in thermal-spray industry since the development of plasma spray [1]. HVOF is characterized by high particle velocities and low thermal energy when compared to plasma spraying. The applications of HVOF have expanded from the initial use of tungsten carbide coatings to include different coatings that provide for wear or erosion/corrosion resistance [23]. HVOF uses an internal combustion jet fuel (propylene, acetylene, propane and hydrogen gases) to generate a hypersonic gas velocities of 1830 m/s, more than five times the speed of sound. When burned in conjunction with pure oxygen, these fuels can produce a nominal gas temperature greater than 3029 K. The powder particles are injected axially into the jet gas, heated, and propelled toward the substrate. With the relatively low temperatures of the flame gas associated with the HVOF systems, superheating or vaporization of individual particles are often prevented [24]. Furthermore, the lower particle temperatures present lead to carbide coatings which exhibit less carbide loss than that of the plasma sprayed coatings. In essence, the advantages of HVOF process over conventional plasma spraying are higher coating bond strength, lower oxide content, and improved wear resistance due to a homogeneous distribution of carbides [4, 25].

Vacuum plasma spraying (VPS), sometimes referred to as low-pressure plasma spraying (LPPS), consists of a plasma jet stream produced by heating an inert gas by an electric arc generator (more than the power required in APS) [1]. The powders are introduced into the plasma jet in vacuum, undergo melting, and are accelerated towards the substrate material [1, 4]. The position of the injection port in the nozzle plays an important role in VPS system since the pressure of the powder injector must be greater than the pressure in the nozzle in order to propel the powders properly [26]. Vacuum plasma spray has the ability to co-deposit incompatible materials with wide compositional ranges within the deposit to produce layered structures. A wide range of functionally gradient materials have been produced by VPS which include high temperature containment components, graded structures to separate materials for the combinations of thermal and oxidation resistance; and graded structures used in solid oxide fuel cell technology [27]. The advantages of utilizing VPS over conventional APS are lower porosity in the resulting coating caused by incomplete melting, wetting, or fusing together of deposited particles, lower amount of oxide phases in the resulting coating deposit, and higher particle velocities leading to denser deposits than the APS coatings [4, 25].

Any thermal plasma spraying techniques enclosed in a controlled atmosphere other than air or vacuum can be classified as controlled atmosphere plasma spraying (CAPS). Inert plasma spraying (IPS) involves the plasma spraying into an inert gas (He, N_2) chamber. Shrouded plasma spraying was developed for TBCs in which the plasma jet is protected from the atmosphere. The shielding nozzle is connected to the anode of the plasma torch, and the nozzle is in close distance (100-130 mm) with the substrate [1].

3. Spray Parameters

The control of spraying parameters is essential in order to obtain optimal combinations of tensile strength, superficial hardness, microhardness and microstructure in a coating. Crawner et al., for example, conducted experiments to investigate the fundamental parameters determining the physical properties of WC-17at.% Co, CrC-25at.% NiCr, CrC-20at.% NiCr, and Tribaloy 800 coatings sprayed by the Miller Thermal HVOF system [28]. Results from this study indicated that using a 12mm diameter nozzle is the dominant factor for increasing the tensile strength of a CrC-25at.% NiCr coating while the spray distance is the most critical factor for the CrC-20at.% NiCr and Tribaloy 800 coatings [28].

The critical spray parameters for HVOF processes include [1, 28]:
- fuel and oxygen flow
- fuel and oxygen ratios
- carrier gas flow
- combustion nozzle type
- powder feed rate
- particle diameter
- torch linear speed
- gun water temperature
- gun-to-substrate spray distance

3.1. FUEL AND OXYGEN FLOW

To produce consistent quality of coatings, a constant flow of oxygen and fuel is an important factor in the operation of HVOF spraying processes. The total flow of fuel and oxygen affects the microstructure, oxide content, microhardness, and abrasive wear resistance of a coating [29]. The temperature of the combustion flame depends strongly on the stoichiometry of the oxygen-fuel mixture. For instance, the ideal stoichiometric ratio for the combustion of propylene to produce a neutral flame is 4.5:1 of O_2 to C_3H_6, as expressed in the following equation:

$$2C_3H_6 + 9O_2 \rightarrow 6H_2O + 6CO_2 \tag{1}$$

A lean combustion will yield an excess of O_2 which will enhance the formation of oxides in the coating. However, excessive fuel introduced during combustion will lower the flame temperature due to the depletion of O_2. Consequently, high concentration of unmelted particles may result which will lead to an increase in porosity [24].

Creffield and White studied the effects of total gas flow rate and different oxygen to propane ratios on the properties of WC-Co coating sprayed using a Miller Thermal HV2000 system [29]. Liquid propane with two different compositions were selected as fuel in this study to determine the effects of coating properties of WC-Co coating. A greater amount of oxygen is required for complete combustion when the liquid propane composed of higher hydrocarbons is selected. The incomplete combustion leads to the deposition of carbon in the interior of the gun which results in build-up of powders in the gun barrel [29].

3.1.1. Coating Oxidation

Oxidation is a serious problem for corrosion-resistant coatings and thermal barrier coatings. For example, oxidation was reported to be the dominant factor affecting the wear performance and microhardness of WC-C coatings [29]. Low oxide coatings have been reported to improve the performance of the coatings in selective corrosion-resistant applications, although these improvements are not as significant as those offered by carbides and silicides [30]. In thermal barrier coating applications, thermal coefficients mismatch between the oxides and the metal often leads to degradation in the coatings [30]. Therefore, controlling the oxide content becomes a critical factor in obtaining a desirable coating.

In an effort to elucidate the mechanism of coating oxidation which results from the gas dynamics of the HVOF process, the phenomenon of gas shrouding has been studied extensively [30, 31]. Gas shrouding refers to the interaction between the jet gas with the spray particles and the surrounding atmosphere. Various laser scattering visualization devices have been used by Hackett and Settles to study the flow field of the entrainment and mixing between the jet gas and the atmosphere [30]. In this study, the micron-sized particles introduced in the HVOF process are visible by laser illumination, and the particle path lines yield the flow field of the jet gas in the recorded image. In particular, measurements from the planar laser scattering image of the HVOF jet mixing show that the atmospheric entrainment rapidly increases the mass flux of the HVOF jet at spray standoff distances of 30 to 40 cm by a factor of ten above the initial mass flux at the exit of the nozzle [30].

3.2. SPRAY DISTANCE

The distance from the gun to the substrate is critical in determining the thermal stress behavior of the coating. The shorter spray distances increase the temperatures of the coating surface which leads to undesirable thermal stress fracture. Thermal stress fracture can be reduced by increasing the total flow of cooling air onto the coating surface [28]. Varacalle et al. assessed different processing parameters (oxygen and fuel flow, air flow, powder feed rate, standoff distance) on the structure and mechanical properties of Inconel 718 coatings [24]. Results from this study showed shorter spray distances increase the deposition efficiency and superficial hardness in the Inconel 718 coatings [24].

3.3. FEEDSTOCK POWDERS

The materials used for conventional HVOF spraying are usually in the form of powders with spherical morphology. The particle size ranges from 5 to 45 μm, depending strongly on the type of powder feeder used [1]. More recently, investigations have successfully used nanocrystalline particles and clusters of particles, in HVOF thermal spraying [32-34]. Preliminary results are encouraging, and presently several research groups, funded by the Office of Naval Research, and supported by various industries, are actively pursuing this technology. Anticipated benefits of nanocrystalline coatings include [35]:

- extended part performance
- improved energy conservation
- increased useful life of parts
- reduced manufacturing costs

3.3.1. Nanocrystalline Materials

Nanocrystalline materials are characterized by a microstructural length scale in the 1-100 nm regime [36]. More than 50 volume percent of atoms could be associated with grain boundaries or interfacial boundaries when grain size is small enough [37]. Thus, a significant amount of interfacial component between neighboring atoms associated with grain boundaries contributes to the physical properties of nanocrystalline materials [37]. A number of techniques that are capable of producing nanocrystalline materials include gas condensation, mechanical alloying/milling, crystallization of amorphous alloys, chemical precipitation, spray conversion processing, vapor deposition, sputtering, electrodeposition, and sol-gel processing technique [3].

Mechanical alloying/milling techniques have been used to produce large quantities of nanocrystalline materials for possible commercial use [3].

Mechanical Alloying/milling. Mechanical alloying is a high energy ball milling process, in which elemental or pre-alloyed powders are welded and fractured to produce metastable materials with controlled microstructures. Today, mechanical alloying has been widely used to synthesize amorphous alloys, intermetallic compounds and nanocrystalline materials [38-40]. During mechanical milling, particle welding and fracturing result in severe plastic deformation. The continuous process produces flake-shape agglomerates. Several factors can influence the process of mechanical alloying which include milling time, balls to powders charge ratio, milling environment, and the internal mechanics specific to each mill. Aikin et al. [41] investigated how milling time affected the coalescence and fracturing events during mechanical alloying, and concluded that longer milling times decreased the probability of cold welding of the powder particles.

Cryogenic Milling. Cryogenic milling, often referred to as "cryomilling", is a technique in which a liquid nitrogen medium (77K) is introduced continuously to the milling process creating a slurry. The technique was first employed by Luton et al. [42] in the mechanical alloying of aluminum and dilute aluminum alloys; the alloyed powders synthesized by this technique were strengthened by aluminum oxy-nitride particles (2-10 nm in diameter with a mean spacing of 50-100 nm). The dispersoids are formed in-situ during cryomilling by the co-adsorption of nitrogen and oxygen onto clean aluminum surfaces. The advantages of cryogenic milling include the following [43]:

- reduced oxygen contamination from atmosphere
- minimized heat generated during the milling process which favors fracturing over welding of ductile materials during the milling process

For example, the reported increase in thermal stability of cryomilled Fe-Al powders has been attributed in part to the formation of oxy-nitride particles during cryomilling [44-46]. Evidence from selected area diffraction on cryomilled Fe-10wt.% Al for 25 hours suggested the presence of nanometer scale dispersoids composed of γ-Al_2O_3 and AlN [44]. Results from the atomic probe field ion microscopy analysis of bulk nanocrystalline Fe-5.2 at.% Al consolidated by hot isostatic pressing show that oxygen and nitrogen introduced during the cryomilling process result in the accumulation of various oxides and nitrides at grain boundaries. The oxides and nitrides thereby provide thermal stability [47]. Other studies have also shown that cryomilling may provide grain size stabilization in other alloy compositions such as Ni-Al [43].

4. Particle Behavior

The quality of the coating depends on the optimization of the spraying parameters during the spraying process. The powder particles propelled into the flames undergo acceleration and significant amount of heating before contacting the substrate. Therefore, the microstructural evolution of the sprayed coating and the resulting properties of the coatings are influenced by the momentum and thermal transport between the flame gas and the powder particles during flight [1]. Ramm et al. [48] investigated the correlation between the spraying conditions and the resulting microstructure for Al_2O_3 coatings in which porosity is related to the impact velocities of the impinging particles on to the substrate. Direct measurement of particles velocity profile has been made by laser velocimeters such as laser Doppler velocimetry (LDV) and laser-two-focus (L2F). Low impact velocity causes the formation of coarse pores (3-10 µm) due to incomplete melting of the particles and subsequent unfilled interstices between prior layer of deposited particles. On the other hand, fine pores (<0.1µm) are caused by incomplete contact between lamellae [48]. Therefore, the study of the momentum and thermal history of the impinging particles enhances the optimization of the spraying parameters.

Although HVOF is considered to be simpler than PS, the spraying parameters are still very complex due to the characteristic processes associated with the HVOF spraying gun which combine the thermodynamic laws of compressible fluid flow, heat transfer principles, and coating formation [49]. In developing a mathematical model for the velocity and temperature profile of the impinging particles, the following parameters are:
- composition and stoichiometry of the flame gas;
- temperatures of the flame gas and the particles;
- velocities of the flame gas and particles; and
- flight distance

Detailed numerical analysis describing particles behavior can be found in Refs. [1, 50-52]

5. Deposition

The formation of a thermal spray coating results from the build-up of individual molten or semi-solid particles that strike on the surface of a substrate. Particle impingement at the substrate surface is a dynamic process which combines particle deformation and solidification simultaneously. The temperature of the particle at the moment of impact influences the grain size and phase composition of the coating [1]. Thereby, the phenomenon of melted or semi-melted particles impacting on the substrate is crucial in determining the coating characteristics such as porosity, inclusions, and chemical segregation [53].

5.1. PARTICLE DEFORMATION AT IMPACT

The flattening process of individual particles have been studied analytically, numerically, and experimentally [52-59]. Results from these studies show that the kinetics of particle flattening depend on the particle size and impact velocity [52].

Inertia dominates at the initial moment after impact and the effect of viscous flow proceeds which influences particle spreading. The kinetic energy is dissipated to overcome the viscous forces of the flowing particles [60-61].

The presence of porosity in HVOF coatings is likely to emerge as a critical issue in the application of nanocrystalline technology, as confirmed by preliminary results presented in a subsequent section. Porosity in thermal sprayed coatings is closely related to the thermal, fluid flow and solidification conditions that are present. Consequently, the deformation and solidification of one or multiple droplets impinging on a substrate should be addressed in detail.

Work in this area is presently underway at the University of California, Irvine (UCI) using numerical simulation. The numerical simulations have been accomplished on the basis of the full Navier-Stokes equations and the Volume Of Fluid (VOF) function by using a 2-domain method for the thermal field and solidification problem and a new two-phase flow continuum model for the flow problem with a growing solid layer. A new phenomenon during the impingement of droplets is found to be a predominant mechanism governing the formation of micro-pores in the solidified layer.

On the basis of this mechanism, some fundamental trends and effects of important processing parameters on micro-porosity may be reasonably explained, and optimal processing conditions for dense coatings may be determined. The simulations reveal that the spreading liquid separates from the solidified splat [62]. This mechanism is believed responsible for the formation of micro-pores in the solidified layer [63, 64]. It has also been shown that a fully-liquid droplet impinging on a solid substrate may generate good contact between the splat and the substrate whereas it will produce liquid ejection if it strikes onto another splat. Simulations of deforming molten-metal droplets as they interact with other droplets and/or a non-flat substrate provided further insight concerning the fundamental mechanisms governing pore formation in spray-processed materials [65]. If the roughness width is larger than the initial droplet diameter the droplet undergoes a succession of accelerations and decelerations and, eventually, breakup. If the droplet is larger than the roughness width, the spreading process is hindered. In the cases mentioned above, droplet impact velocities are high and the effect of the surrounding gas is neglected. The viscous and surface tension effects at the interface (free surface when gas is neglected) are relatively less important than the inertia force due to high Reynolds and Weber numbers. The results obtained with W, Ni and Ti demonstrate that a droplet spreads uniformly in the radial direction during impinging onto a flat substrate and eventually forms a thin splat. The flattening rate is fast initially and decreases asymptotically.

The spreading process of droplets under thermal spray conditions is essentially governed by the inertia and viscous effects, and the surface tension effect is not significant to the deformation dynamics. Therefore, increasing impact velocity, droplet diameter and material density, or decreasing material viscosity, leads to an increase in the final splat diameter, while a large droplet diameter, a high material density, or a small material viscosity corresponds to a longer spreading time. The inherent dependence of the final splat diameter and the spreading time on the inertia and viscous effects may be approximated by the correlation derived from the regression analyses of the calculated results. A fully liquid droplet impinging onto a flat solid substrate may lead to good contact and adhesion between the splat and the substrate; whereas a fully liquid droplet striking onto the flattening, fully liquid splat produces ejection, rebound and breakup of the liquid. These phenomena may reduce the deposit rate and deteriorate the bonding and deposit integrity. Except for the complete axisymmetric case, a fully liquid droplet colliding with the flattening, fully liquid splat also causes formation of voids within the liquid, which may become inner pores when solidification and subsequent contraction occur. The validity of the droplet deformation simulations has been verified by a test calculation and by examining the viscous and surface tension effects in our earlier work [64-66].

5.2. PARTICLE SOLIDIFICATION

The adhesion of molten particles on the substrate depends, to a large extent, on the solidification history of the particles. Solidification starts at the interfaces between the substrate and the particle in which the interface forms a heat sink for the liquid [1]. In most cases, heterogeneous nucleation occurs due to the presence of carbides or oxides inclusions in the impinging particle [52]. In related studies, Guilemany et al. investigated the bonding behavior of WC-Co coating on Ti-6% Al-4% V substrate and found that a metallic bond of WC-αTi bond forms at the interface as a result of high impact pressure from the initial splat, characteristic of HVOF spraying [67]. Delplanque et al. performed detailed numeric analysis on a multidirectional solidification model describing the micropore formation in solidifying molten droplets impinging on a substrate [68].

5.3. RESIDUAL STRESSES

Residual stresses in coatings originate from the misfit strains that arise during spray deposition and from differential thermal contraction stresses during cooling. Residual stresses ultimately affect the electrical, mechanical, and optical properties of the coatings [69]. Extensive studies were conducted by Clyne [69] on the relationship between stress distributions and release rate of interfacial strain energy in various types of coatings. Direct measurements such as x-ray diffraction and neutron diffraction can be used to study residual stress distributions. However, the accuracy of these techniques is limited to a thin surface layer of the coatings.

Stresses induced from differential thermal contraction result from the differences of thermal expansivities between coating coated material. The misfit in strain can be calculated by the following equation [69]:

$$\Delta \varepsilon = \int_{T_2}^{T_1} (\alpha_s(T) - \alpha_s(T)) dT \tag{2}$$

$\Delta \varepsilon$=misfit strain (unitless)

T_1, T_2=cooling temperatures (K)

α_s, α_d=thermal expansivities of substrate and coating deposit respectively (K^{-1})

Since thermal expansivity increases with rising temperature, large errors associated with the value of misfit strain can occur if thermal expansivities are assumed constant [69].

6. Experimental Results

In the sections that follow, preliminary experimental results are described in details. In these experiments, nanocrystalline powders of various compositions were prepared by milling (both cryomilling and methanol milling), followed by thermal spraying using HVOF. Whereas preliminary results are encouraging, it is also evident that the merger of nanocrystalline materials and HVOF thermal spraying gives rise to a series of technical and scientific challenges that must be addressed for this novel technology to reach its anticipated potential. Some of these issues are discussed in the present work.

6.1. POWDERS PROCESSING

Inert gas atomized Ni, Inconel 718, and 316-stainless steel powders (Sulzer Metco (US) Inc.) with a nominal particle size of 45 ± 11 μm were selected for the study. Mechanical milling in a methanol environment was conducted in a modified Union Process 01-ST attritor mill with a grinding tank capacity of 0.0057 m^3. The drive shaft operated at a speed of 180 rpm with the shaft clearance of 0.635 cm from the bottom of the grinding tank to minimize powder accumulation. Stainless steel balls (0.635 cm in diameter) were used as the grinding media, and the powder to ball mass ratio was 1:20. Cryogenic milling of the atomized powders was performed with the same parameters as those used in the methanol milling experiments. Liquid nitrogen was continuously introduced into the mill to ensure complete immersion of the powders. The temperature was maintained in the range of 83 K to 98 K throughout the experiment. A globe valve was used to control the amount of liquid nitrogen introduced into the mill manually.

After completion, the powders were carefully removed from the grinding tank of the attritor frame and placed in the glove box filled with argon. After all of the liquid nitrogen evaporated, the mixture containing the balls and powders were removed from the grinding tank followed by the separation of the powders from the balls by sieving. The particle size distributions of the milled powders was determined by the Microtrac Standard Range Particle Analyzer.

6.2 HVOF SPRAY OF NANOCRYSTALLINE POWDERS

The mechanically milled powders were thermal sprayed using a Sulzer Metco DJ 2600 HVOF spray system on a 1020-stainless steel substrate. H$_2$ gas was used to generate a hypersonic gas velocity of approximately 2000 m/s and a pressure of 0.28 MPa. When burned with O$_2$, the fuels produce a nominal flame temperature of 2755 K. Through a fluidized bed, the powders are injected axially into the jet gas, in which the powders are heated and propelled towards the substrate to produce a coating with an average thickness of 0.0254 cm. Each material was sprayed with two different carrier gases. Table 2 lists the parameters that characterize the spraying conditions:

TABLE 2. Spraying parameters for the Ni, Inconel 718, and 316-stainless steel powders

| | Pressure (MPa) | | | | Gas flow (m³/hr) | | | |
	O_2	H_2	Air	N_2	O_2	H_2	Air	N_2
Standard	1.17	0.96	0.69	0	13.8	41.0	22.3	0
Low oxide content	1.17	0.96	0	0.76	13	46.4	0	31.1

6.3. METHODS OF ANALYSIS

6.3.1. Scanning Electron Microscopy Analysis

The milled powders and the coatings to be analyzed by scanning electron microscopy (SEM) were mounted in a conductive mold and mechanically ground. The agglomerate size of the milled powders was determined by SEM analysis performed in a Philips XL30 FEG scanning electron microscope equipped with EDAX analysis.

6.3.2. Transmission Electron Microscopy Analysis

The milled powder particles to be analyzed by transmission electron microscopy (TEM) were dispersed in methanol, deposited on carbon film substrates and allowed to dry in air. The coatings were sectioned into 3 mm x 3 mm TEM samples, mechanically ground, and jet polished. Table 3 lists the jet polishing solutions for various coating materials:

Table 3. Jet polishing solutions for the preparations of TEM coatings

Coating material	Jet polishing solution
Ni	20% perchloric acid 80% ethanol
Inconel 718	10% perchloric acid 20% ethanol 70% butanol
316-stainless steel	20% sulfuric acid 80% methanol

TEM analysis was performed in a Philips 200 transmission electron microscope operated at 200 keV. Grain sizes of the milled powders and the coatings were determined from measurements obtained from dark field imagings.

6.3.2. Microhardness

Microhardness measurements were performed on the coatings which have been mechanically polished to provide a smooth surface. A Buehler microhardness tester with a diamond indentor and a 300 g load was used. At least ten measurements were taken and averaged for each sample.

7. Results

7.1. POWDER CHARACTERIZATION

The milled powders conducted in methanol or liquid nitrogen produced flake-shape agglomerates. The aspect ratio, the ratio of length/width of the agglomerate, increases with increasing milling time as a result of continuous welding and fracturing during mechanical milling. Figures 1a-b show the morphological changes of Inconel 718 powders before milling and after 10 hours of milling in methanol. Table 4 lists the agglomerate size, average grain size determined by TEM dark field imaging and aspect ratio of Ni, Inconel 718, and 316-stainless steel powders under different milling conditions. Results obtained from TEM dark field imaging indicate that the average grain size decreases with increasing milling time for Ni, Inconel 718, and 316-stainless steel powders.

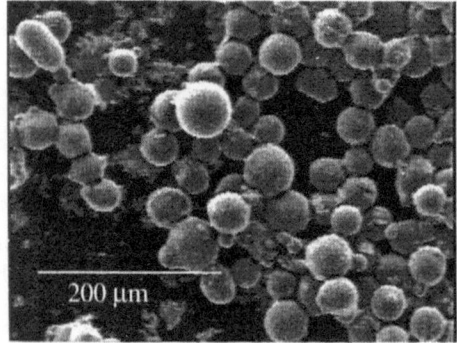 

Figure 1a. As received Inconel 718 powders.   *Figure 1b.* Methanol milled Inconel 718 powders for 10 hrs.

TABLE 4. Characteristic powders properties of mechanical milled Ni, Inconel 718, and 316-stainless steel in different milling conditions

Material	Milling condition (hrs)	Milling environment	Agglomerate size D_{50} (μm)	Grain size (nm)	Aspect ratio
Ni	5	methanol	80	90	1.40
Ni	10	methanol	35	82	1.42
Ni	5	liquid nitrogen	25	26	1.55
Ni	10	liquid nitrogen	43	28	1.50
Inconel 718	10	liquid nitrogen	65	16	1.34
316-stainless steel	5	methanol	67	26	1.43
316-stainless steel	10	methanol	54	24	1.68
316-stainless steel	5	liquid nitrogen	50	38	1.38
316-stainless steel	10	liquid nitrogen	31	21	1.68

7.2. THERMAL SPRAYED NANOCRYSTALLINE COATINGS

Backscattered electron images from SEM analysis performed on the as-sprayed nanocrystalline coatings show higher porosity present than those of conventional coatings with identical spraying parameters. Figures 2a-d compare the coating structures of conventional Ni coatings and nanocrystalline Ni coatings by standard spraying parameters and low oxide content spraying parameters respectively. Chemical composition analysis performed on the Ni coatings by EDAX, as shown in Table 5, indicates that the oxygen content of the Ni coating sprayed using air as the carrier gas is higher than that of the coating sprayed using N_2 as the carrier gas. Table 6 lists the coating characteristics of Ni, Inconel 718, and stainless steel 316 which include porosity and microhardness.

Transmission electron microscopy (TEM) analysis performed on the as-sprayed HVOF Ni coating of methanol milled Ni powders for 10 hrs indicates that the presence of nanocrystalline grains with an average grain size of 15 nm. Regions of nanocrystalline grains were also observed in the TEM analysis on the cross sectional methanol milled Ni (10 hrs.) coating, as shown in Figures 3a-b. Areas with elongated grains with an aspect ratio of 2-3, and a grain size range of 100-200 nm were also observed. Elongated grains resulting from the process of mechanical milling suggest that fractions of the nanocrystalline agglomerates did not melt during the HVOF spraying.

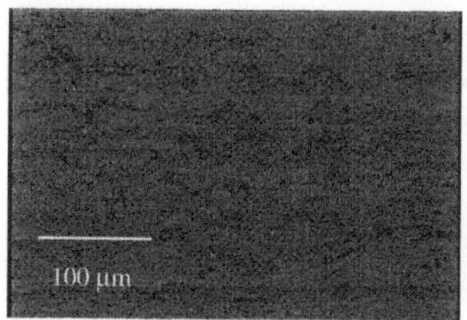

Figure 2a. Conventional Ni coating : standard spray.

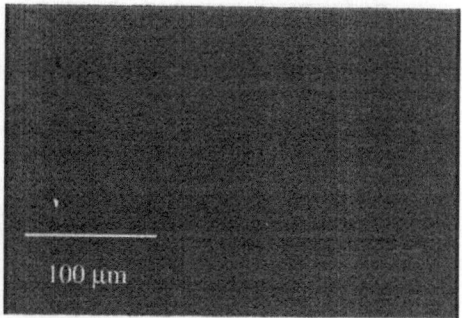

Figure 2b: Conventional Ni coating: low oxide content spray

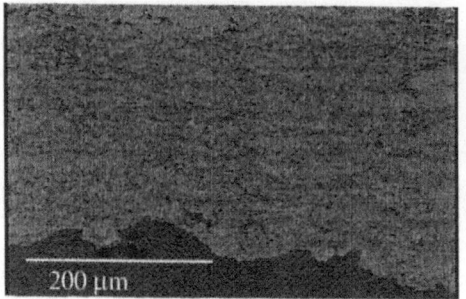

Figure 2c. Methanol milled Ni (5 hrs.) coating: standard spray.

Figure 2d. Methanol milled Ni (5 hrs.) coating: low oxide content spray.

TABLE 5. Chemical composition analysis of methanol milled Ni coatings sprayed with
different spraying parameters

Element	Methanol Milled Ni (5 hrs.) Coatings	
	Standard	Low oxide content
O (wt. %)	3.1	1.9
Ni (wt. %)	96.9	98.1

TABLE 6. Coating characteristics and properties of Ni, Inconel 718, and 316-stainless steel

Material	Milling time/media	Spraying parameter	Porosity (%)	Microhardness 300g load (DPH)
Conventional Ni	not applicable	standard	1.09	447
Conventional Ni	not applicable	low oxide content	<0.5	390
Nanocrystalline Ni	10/methanol	standard	1.42	535
Nanocrystalline Ni	10/methanol	low oxide content	1.42	484
Conventional Inconel 718	not applicable	standard	<0.5	440
Nanocrystalline Inconel 718	30/liquid nitrogen	standard	1.41	712
Conventional stainless steel 316	not applicable	standard	<0.5	451
Conventional stainless steel 316	not applicable	low oxide content	<0.5	411
Nanocrystalline stainless steel 316	10/methanol	standard	1.5	613
Nanocrystalline stainless steel 316	10/methanol	low oxide content	2.4	503
Nanocrystalline stainless steel 316	10/liquid nitrogen	standard	2.1	460
Nanocrystalline stainless steel 316	10/liquid nitrogen	low oxide content	1.9	486

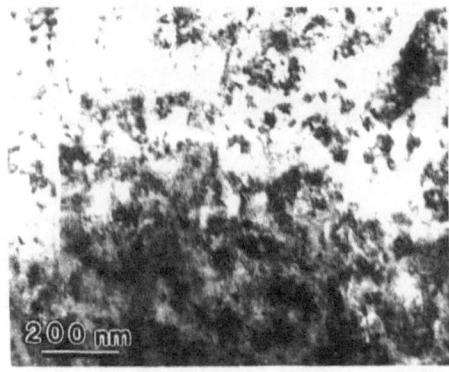
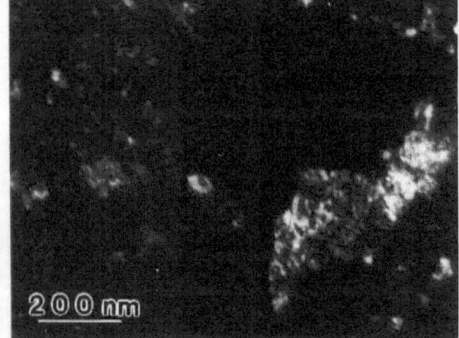

Figure 3a. Bright field image of cross section methanol milled Ni (10 hrs.) coating.

Figure 3b. Dark field image of cross section methanol milled Ni (10 hrs.) coating.

8. Discussion

Thermal spraying of nanocrystalline materials represents a revolutionary approach to exploit the unusual physical attributes of nanocrystalline materials. The spraying of nanocrystalline Ni, Inconel 718, and 316-stainless steel powders prepared by methanol/liquid nitrogen milling, has been shown to yield coating with microhardness values that are approximately 20%, 60%, and 36% higher than those of conventionally sprayed coatings.

High velocity oxy-fuel spraying involves complex processes of combustion and spraying gas dynamics, particle behavior during flight, particle flattening and solidification on the substrate surface to form a coating. The dynamic processes are further complicated for the thermal spraying of nanocrystalline materials as the particle morphology deviates from the conventional spherical powders used for spraying. Results from scanning electron microscopy indicate the porosity of the nanocrystalline coatings is higher than those of the conventional coatings with identical spraying parameters. Furthermore, TEM analysis on the cross sectional methanol milled Ni (10 hrs.) coating has shown that the coating remains nanocrystalline with regions of elongated nanocrystalline grains, suggesting that fractions of the nanocrystalline agglomerates did not melt during the HVOF spraying.

The data presented in Table 6 shows that the microhardness values of the nanocrystalline coatings are higher than those of the coating produced using conventional powders. These results should be analyzed with care, in light of the fact that the presence of pores is likely to influence the microhardness measurements. There are, however, a preliminary observations worth noting. First, it is evident that the spraying parameters exert an influence on the resultant porosity and microhardness of the coating. Second, milling conditions, whether liquid nitrogen or methanol, also influence the resultant porosity and microhardness values. It is likely that both of these observations may be rational on the basis of the chemistry and distribution of phases that are formed during milling as well as during spraying, although other factors, such as the morphology and size of the nanocrystalline agglomerates may also play a role. Evidently, the microhardness increases with increasing aspect ratio. Further studies are underway to investigate this phenomenon.

9. Summary

Nanocrystalline Ni, Inconel 718, and 316-stainless stainless powders prepared by methanol milling/cryomilling were thermally sprayed by HVOF processing to produce nanocrystalline coatings. Coating characteristics of the nanocrystalline coatings are compared to those of the conventional coatings. Significant increase in hardness is observed in nanocrystalline coatings. Further study is underway to optimize the coating parameters and to examine various physical and mechanical properties of the as-sprayed nanocrystalline coatings, and enhance our understanding of the relevant fundamental phenomena.

300

10. Acknowledgments

The authors would like to acknowledge the financial support by the Office of Naval Research under grant No.N00014-94-1-0017. Professor E.J. Lavernia would like to acknowledge the Alexander von Humboldt Foundation in Germany for support of his sabbatical visit at the Max-Planck-Institut Fur Metallforschung, in Stuttgart, Germany.

11. References

1. Pawlowski, L. (1995) *The Science and Engineering of Thermal Spray Coatings,* John Wiley & Sons, England.
2. German, R.M. (1984) *Powder Metallurgy Science,* Metal Powder Industries Federation, Princeton, New Jersey.
3. Suryanarayana, C. (1995) Nanocrystalline materials, *Int. Mat. Rev.* **40**, 41.
4. Srivatsan, T.S. and Lavernia, E.J. (1992) Review--Use of spray techniques to synthesize particulate-reinforced metal-matrix composites, *J. Mat. Sci.* **27**, 5965.
5. Moskowitz, L.N. (1992) Application of HVOF thermal spraying to solve corrosion problems in the petroleum industry, in C.C. Berndt (ed.), *Thermal Spray: International Advances in Coatings Technology*, ASM International, Materials Park, Ohio, 611.
6. Matsubara, Y. and Tomiguchi, A. (1992) Surface texture and adhesive strength of high velocity oxy-fuel sprayed coatings for rolls of steel mills, in C.C. Berndt (ed.), *Thermal Spray: International Advances in Coatings Technology*, ASM International, Materials Park, Ohio, 637.
7. Byrnes, L. and Kramer, M. (1994) Method and apparatus for the application of thermal spray coatings onto aluminum engine cylinder bores, in C.C. Berndt and S. Sampath (eds.), *Thermal Spray Industrial Applications*, ASM International, Materials Park, Ohio, 39.
8. Chen, H., Liu, Z., Zhuang, Y. and Xu, L. (1992) Quality upgrading of plasma-sprayed thermal barrier ceramic coatings by hot isostatic pressing, *Chinese J. of Mech. Eng.* **5**, 183.
9. Nicoll, A.R., Bachmann, A., Moens, J.R. and Loewe, G. (1992) The application of high velocity combustion spraying, in C.C. Berndt (ed.), *Thermal Spray: International Advances in Coatings Technology*, ASM International, Materials Park, Ohio, 149.
10. Niemi, K., Vuoristo, P. and Mantyla, T. (1992) Abrasion wear resistance of carbide coatings deposited by plasma and high velocity combustion processes, in C.C. Berndt (eds.), *Thermal Spray: International Advances in Coatings Technology*, ASM International, Materials Park, Ohio, 685.
11. Malhouroux-Gaffet, N. and Gaffet, E. (1993) Solid state reaction induced by post-milling annealing in the Fe-Si system, *J. Alloys Comp.* **198**, 143.
12. Smyth, R.T., Dittrich, F.J. and Weir, J.D. (1978) Thermal spraying-a new approach to thick film circuit manufacture, in *International Conference on Advances in Surface Coating Technology*, Welding Institute, London, England, 233.
13. Bennett, A.P. and Quigley, M.B.C. (1990) The spraying of boiler tubing in power stations, *Welding & Metal Fabrication*, 485.
14. Wortman, D.J. (1985) Performance comparison of plasma spray and physical vapor deposition BC23 coatings in the LM2500, *J. of Vac. Sci. and Tech.* **A3**, 2532.
15. Roseberry, T.J. and Boulger, F.W. *A plasma flame spray handbook, U.S. Department of Commerce Report No.MT-043*, National Technical Information Service, Springfield, VA.
16. Tucker, R.C. (1982) in R. F. Bunshah (ed.), *Deposition technologies for films and coatings*, Noyes Publications, New Jersey, 454.
17. Marantz, D.R. (1974) in B.N. Chapman and J.C. Anderson (eds.), *Science and Technology of Surface Coating*, Academic Press, London, 308.
18. Smith, C.W. (1974) in B.N. Chapman and J.C. Anderson (eds.), *Science and Technology of Surface Coating*, Academic Press, London, 262.
19. Borisov, Y.S. (1990) *Detonation spraying: equipment, materials, and applications,* Thermische Spritzkonferenz, Essen, Germany.
20. Schwarz, E. (1980) *in 9th International Thermal Spraying Conference,* Nederlands Instituut voor Lastechniek: The Hague, Netherlands, 91.

21. Niederberger, K. and Shciffer, B. (1990) Eigenshaften Verschiedener Gase und Deren Einfluss Beim Thermischen Spritzen, in T. Spritzkonferenz (eds.), Essen, Germany.
22. Okada, M. and Maruo, H. (1968) New plasma sprays and its application, *British Welding J.* **15**, 371.
23. Parker, D.W. and Kutner, G.L. (1991) *Adv. Mat. Process* **139**, 68.
24. Varacalle, D.J., Ortiz, M.G., Miller, C.S., Steeper, T.J., Rotolico, A.J., Nerz, J. and Riggs (II), W.L. (1992) HVOF combustion spraying of Inconel powder, in C.C. Berndt (ed.), *Thermal Spray: International Advances in Coatings Technology*, ASM International, Materials Park, Ohio, 181.
25. Apelian, D., Wei, D. and Farouk, B. (1989) *Metall. Trans.* **20B**, 251.
26. Vinayo, M.E, (1985) *in 7th International Symposium on Plasma Chemistry*, ed., Eindhoven, Netherlands, 1161.
27. Kim, M.R., Smith, R.W. and Kapoor, D. (1996) Vacuum plasma spray forming of tungsten base functionally gradient composites, in C.C. Berndt (ed.), *Thermal Spray: Practical Solutions for Engineering Problems*, ASM International, Materials Park, Ohio, 7.
28. Crawmer, D.C., Krebsbach, J.D. and Riggs (II), W.L. (1992) Coating Development for HVOF Process Using Design of Experiments, in C.C. Berndt (ed.), *Thermal Spray: International Advances in Coatings Technology*, ASM International, Materials Park, Ohio, 127.
29. Creffield, G.K., Cole, M.A. and White, G.R. (1995) The effect of gas parameters on HVOF coatings, in C.C. Berndt and S. Sampath (eds.), *Advances in Thermal Spray Science & Technology*, ASM International, Materials Park, Ohio, 291.
30. Hackett, C.M. and Settles, G.S. (1995) Research on HVOF gas shrouding for coating oxidation control, in C.C. Berndt and S. Sampath (eds.), *Advances in Thermal Spray Science & Technology*, ASM International, Materials Park, Ohio, 21.
31. Hackett, C.M. and Settles, G.S. (1994) Turbulent mixing of the HVOF thermal spray and coating oxidation, in C.C. Berndt and S. Sampath (eds.), *Thermal Spray Industrial Applications*, ASM International, Materials Park, 307-312.
32. Tellkamp, V., Lau, M., Fabel, A. and Lavernia, E.J. (1996) Thermal spraying of nanocrystalline Inconel 718, *NanoStructured Mat.* **9**, 489.
33. Lau, M.L., Jiang, H.G. and Lavernia, E.J. (1997) "Synthesis and characterization of nanocrystalline Ni, Inconel 718, and stainless steel coatings," presented in *Thermal Spray Processing of Nanoscale Materials*, August 3-8, Davos, Switzerland.
34. Kear, B.H. and Strutt, P.R. (1994) Nanostructures: The next Generation of high performance bulk Materials and Coatings, *Naval Research Reviews* **4**, 4.
35. Lau, M.L. (1997) *notes from Advance coating technology development program kick-off meeting*, Arlington, Virginia.
36. Birringer, R. (1994) Structure of nanostructured materials, in G.C. Hadjipanayis and R.W. Siegel (eds.), *Nanophase Materials: Synthesis-Properties-Applications*, Kluwer Academic Publishers, The Netherlands, 157.
37. Birringer, R. (1989) Nanocrystalline Materials, *Mat. Sci. & Eng.* **A117**, 33.
38. Koch, C.C. (1989) Materials synthesis by mechanical alloying, *Annu. Rev. Mat. Sci.* **19**, 121.
39. Gaffet, E., Malhouroux, N. and Abdellaoui, M. (1993) Far from equilibrium phase transition induced by solid-state reaction in the Fe-Si system, *J. Alloys and Compounds* **194**, 339.
40. Koch, C.C. (1993) The synthesis and structure of nanocrystalline materials produced by mechanical attrition: a review, *Nanostructured Mat.* **2**, 109-129.
41. Aikin, B.J.M., Courtney, T.H. and Maurice, D.R. (1991) Reaction rates during mechanical alloying, *Mat. Sci. & Engr* **A147**, 229.
42. Luton, M.J., Jayanth, C.S., Disko, M.M., Matras, S. and Vallone, J. (1989) Cryomilling of nano-phase dispersion strengthened aluminum, *Mat. Res. Soc. Symp. Proc.* **132**, 79.
43. Huang, B., Vallone, J. and Luton, M.J. (1995) The effect of nitrogen and oxygen on the synthesis of B2 NiAl by cryomilling, *NanoStructured. Mat.* **5**, 631.
44. Perez, R.J., Huang, B. and Lavernia, E.J. (1996) Thermal stability of nanocrystalline Fe-10 wt.% Al produced by cryogenic mechanical alloying, *NanoStructured Mat.* **7**, 565.
45. Rawers, J.C. (1995) Microstructure and tensile properties of compacted, mechanical alloyed, nanocrystalline Fe-Al, *Metall. Trans. A* **27A**, 3126.
46. Farrell, K. and Munroe, P.R. (1996) Grain growth in Fe-30at.% Al, *Scripta Metall.* **35**, 615.
47. Melmed, A.J., Tambakis, N.C., Lau, M. and Lavernia, E.J. (1997) APFIM study of a nanocrystalline Fe-Al Alloy, submitted.

302

48. Ramm, D.A.J., Clyne, T.W., Sturgeon, A.J. and Dunkerton, S. (1994) Correlations between spraying conditions and microstructure for alumina coatings produced by HVOF and VPS, in C.C. Berndt and S. Sampath (eds.), *Thermal Spray Industrial Applications*, ASM International, Materials Park, Ohio, 239.

49. Knotek, O. and Schnaut, U. (1992) Process modeling of HVOF thermal spraying systems, in C.C. Berndt (ed.), *Thermal Spray: International Advances in Coatings Technology*, ASM International, Materials Park, Ohio, 811.

50. Liang, X., E.J. Lavernia, Wolfenstine, J. and Sickinger, A. (1995) Microstructure evolution during reactive plasma spraying of MoSi2 with methane, *J. Therm. Spray Tech.* **4**, 252.

51. Pfender, E. and Lee, Y.C. (1985) Particle dynamics and particle heat and mass transfer in thermal plasmas. part i. the motion of a single particle without thermal effects, *Plasma Chem. and Plasma Proc.* **5**, 211.

52. Sobolev, V.V. and Guilemany, M. (1996) Dynamic processes during high velocity oxy-fuel spraying, *Int. Mat. Rev.* **41**, 13.

53. Sobolev, V.V. and Guilemany, J.M. (1995) Formation of chemical inhomogeneity in the coating structure during high velocity oxy-fuel (HVOF) spraying, *Mat. Lett.* **25**, 285.

54. Sobolev, V.V. and Guilemany, J.M. (1996) Influence of solidification on the flattening of droplets during thermal spraying, *Mat. Lett.* **28**, 71.

55. Sobolev, V.V., Guilemany, J.M. and Martin, A.J. (1996) *J. Therm. Spray Tech.* **5**, 207.

56. Sobolev, V.V., Guilemany, J.M. and Calero, J.A. (1996) Investigation of the development of coating structure during high velocity oxy-fuel (HVOF) spraying of WC-Ni powder particles, *Surf. Coat. Tech.* **82**, 114.

57. Sobolev, V.V., Guilemany, J.M. and Martin, A.J. (1996) Analysis of splat formation during flattening of thermally sprayed droplets, *Mat. Lett.* **29**, 185.

58. Sobolev, V.V., Guilemany, J.M. and Martin, A.J. (1996) Formation of powder particles during thermal interaction of liquid and solidified drops in the process of metal atomization, *J. Mat. Proc. Tech.* **62**, 216.

59. Fukanuma, H. (1994) *J. Thermal Spray Tech.* **3**, 33.

60. Trapaga, G. and Szekely, J. (1991) Mathematical modeling of the isothermal impingement of liquid droplets in spraying processes, *Metall. Trans.* **22B**, 904.

61. Trapaga, G., Matthys, E.F., Valencia, J.J. and Szekely, J. (1992) Fluid flow, heat transfer, and solidification of molten metal droplets impinging on substrates: comparison of numerical and experimental results, *Metall. Trans.* **23B**, 710.

62. Liu, H., Lavernia, E.J. and Rangel, R.H. (1993) Numerical simulation of substrate impact and freezing of droplets in plasma spray processes., *J. of Phys. D: (Appl. Phys.)* **26**, 1900.

63. Delplanque, J.-P., Lavernia, E.J. and Rangel, R.H. (1995) Description of a micro-pore formation mechanism in a deforming and solidifying metal droplet, *accepted for presentation at the 1995 ASME Winter Annual Meeting*, San Francisco, California.

64. Liu, H., Rangel, R.H. and Lavernia, E.J. (1994) Modeling of reactive atomization and deposition processing of Ni_3Al, *Acta Metall. Mat.* **42**, 3277.

65. Liu, H., Lavernia, E.J. and Rangel, R.H. (1995) Modeling of molten droplet impingement on a non-flat surface, *Acta Metall. Mat.* **43**, 2053.

66. Kim, G.H., Kim, H.S. and Kum, D.W. (1996) Determination of titanium solubility in alpha-aluminum during high energy milling, *Sci. Mat.* **34**, 421.

67. Guilemany, J.M., Nutting, J. and Dong, Z. (1997) Coating-substrate bonding after HVOF thermally spraying WC-Co on to a Ti-6%Al-4%V Alloy, *J. Mat. Sci. Lett.* **16**, 1043.

68. Delplanque, J.-P., Lavernia, E.J. and Rangel, R.H. (1996) Multidirectional solidification model for the description of micropore formation in spray deposition processes, *Numerical Heat Transfer, Part A* **30**, 1.

69. Clyne, T.W. (1996) Residual stresses in surface coatings and their effects on interfacial debonding, *Key Eng. Mat.* **116-117**, 307.

THE SURFACE CHARACTERIZATION OF NANOSIZED POWDERS: RELEVANCE OF THE FTIR SURFACE SPECTROMETRY

Marie-Isabelle BARATON
University of Limoges
Faculty of Sciences, LMCTS - ESA 6015 CNRS
123, Av. Albert Thomas, F-87060 Limoges (France)
e-mail: baraton@unilim.fr

1. Introduction

Even though the study of the bulk properties of crystalline solids is facilitated by the periodicity existing in the lattice, the control of the surface properties and of the interface behavior still represents a challenge to scientists. Techniques to investigate the specific structure and composition of the first atomic layers are very often derived from bulk analysis methods. As a consequence, the minimum depth that can be analyzed, although adequate for traditional materials, may be too large for nanosized materials in which crystal sizes can be smaller than the depth resolution of the characterization technique. Fourier transform infrared (FTIR) spectrometry, widely used for bulk analyses, is, however, a powerful tool to characterize the very first atomic layer provided specific setups are attached to the spectrometer. Several examples will be discussed in the following showing the specific nature of the surface and the relevance of the FTIR spectrometry for obtaining detailed information on the chemical species and the atoms constituting the first atomic layer as well as the coordination number of the surface atoms. Moreover, because of the important role played by the nanomaterial surface in many industrial applications, the surface modifications are a key issue to tailor the surface properties. To this end, FTIR surface spectrometry is also a performant technique to follow the modification of the surface chemical species and to study *in situ* the selectivity and the behavior of the modifications under various treatments.

2. Specificity of a Surface

The cleavage of perfect crystals under ultra-vacuum conditions yields clean surfaces which, however, already differ from the bulk structure because of relaxation processes. Indeed, the bond breakage implies atomic rearrangements to reach the minimum energy. In addition, the real surface is non-homogeneous and presents macroscopic defects such

G.M. Chow and N.I. Noskova (eds.), Nanostructured Materials, 303–317.

as steps, cracks, corners, as well as point defects such as vacancies [1]. The structure of the clean surface also depends on the cleavage direction with respect to the crystalline planes.

But these clean surfaces only survive in ultra-vacuum environments. The cleavage of the bonds leaves electrical charges in the uppermost layer of either side, that is coordinatively unsaturated anions and cations [2]. Therefore, when exposed to atmosphere, all solid surfaces become covered with various adsorbed species. These latter species partly compensate the surface unsaturation due to the crystal cleavage. The contamination by any surrounding atmosphere is very fast. The adsorbed molecules can cause drastic changes in the surface properties. According to the nature of the adsorbed molecules, the temperature and pressure conditions, chemical reactions can occur. Moreover, the atoms adsorbed on the surface can diffuse and react with the bulk atoms thus yielding a surperficial layer whose chemical composition and morphology may be quite different from those of the bulk. Basic examples are the corrosion and the formation of an oxide layer on a metallic surface.

The surface discontinuity can affect several atomic layers, but the part of the first atomic layer is of critical importance in all the phenomena occurring at the interfaces. It is obvious that the loss of periodicity in the structure as well as the introduction of impurities or defects result in changes of the electronic properties at the surface [1]. Moreover, the presence of coordinatively unsaturated anions and cations results in a specific surface chemistry. On the other hand, the nature of the impurities adsorbed on the surface (referred to as surface species in the following) can be partly related to the synthesis history of the material [2]. For example, on metal oxides prepared by chemical routes, carbonates, nitrates and hydrocarbon residues are among the most probable species to be found on the surface. But the most abundant component of the surface layer is water whose molecules can dissociate to form hydroxyl groups.

Since many technological processes involve surface phenomena, the nature and the properties of the surface itself are key issues to understand the interactions between various media and solid surfaces. A comprehension of the surface structure at the molecular level is required for a systematic control of technical properties such as wetting, adhesion, friction, wear, barrier properties, bio-compatibility, etc [3]. This turns to be critical when the solid is a nanosized powder since its surface to bulk ratio is very high. Indeed, the fraction of atoms at the surface [4] is of the order of 10^{-7} in single crystals considered as 1 cm diameter spheres. This fraction may increase up to 0.5 in nanocrystals considered as 1.5-3 nm diameter spheres. In this case, the properties of the surface compete with those of the bulk. In addition, the concentration of the surface defects greatly increases when the crystal size decreases. On the other hand, due to the small particle size and the strong constraints, the crystalline structure of the surface layer may be different from that of the bulk. The surface structure is often deduced from that of a crystal plane considered as an exposed face. But, actually, it is very difficult to elaborate a model for the real surface, in particular for high-surface area catalysts [5].

Therefore, the important questions to be answered by the surface analysis are the following:

(a) What is the nature of the atoms which compose the first atomic layer?
(b) What are the chemical states of these surface atoms?
(c) What is the distribution of these atoms?

3. Some Techniques Used for Surface Analysis

Most of the techniques commonly used for surface investigation are based on the analysis of the energy emitted from a surface after it has been bombarded with ions, electrons, or X-rays photons. Among the most popular techniques, SIMS, XPS, EXAFS, AFM and SERS may be mentioned [6-8].

The secondary ion mass spectrometry (SIMS) is based on the erosion of a solid surface with an ion beam (1-10 keV) and the ionized groups are analyzed by mass spectrometry. The technique gives the chemical composition as a function of the depth. The surface sensitivity varies from 0.3 to 100 nm, but it is a destructive technique.

The X-ray photoelectron spectroscopy (XPS) allows to determine the binding energy of the electrons to the atomic levels by irradiation with an incident photon beam. Information on the chemical nature and binding of the atoms from the analysis of core electron photoemission peaks can thus be obtained. The surface sensitivity varies from 0.5 to 10 nm in depth, that is about ten atomic monolayers on an average.

In the extended X-ray absorption fine structure (EXAFS), the core electrons are ejected from the atomic potential well by absorption of X-ray radiation. Interatomic distances and coordination numbers can thus be determined. But, this technique is rather expensive since it requires a high power X-ray source such as synchroton sources. Nevertheless, it is used in the catalysis field for the study of supported-metal or metal catalysts essentially, as the observed effects are weak for metal oxides [9]. As for the surface analysis (SEXAFS), it requires an ultra-high vacuum.

Atomic force microscopy is becoming very popular for the study of the surface structures [10]. This technique derives from scanning tunneling microscopy (STM). The great advantage of AFM over STM is that non-conductive samples can be characterized but the sensitivity to the electronic structure is of course lost. Forces are detected by a probe mounted on a flexible cantilever. Imaging of individual atoms is obtained on suitable surfaces.

Upon irradiation of a sample by a monochromatic light, the inelastic light scattering by the solid sample as a result of the excitation of the molecular vibrations is called the Raman effect. The obtained information is similar to that given by infrared spectrometry but, because the selection rules are not the same, both techniques are often complementary. Recent developments in Raman spectrometry are due to the new generation of Fourier transform Raman spectrometers [11]. However, since the Raman effect is usually weak, this technique when applied to surface analysis requires a very sensitive detector and therefore, has not been widely used for surface investigations, so far. The surface enhanced Raman spectroscopy (SERS) is a particular case of the Raman effect in which the signal is enhanced when molecules are adsorbed on some particular metals, such as copper and silver [5].

Thermal desorption spectroscopy (TDS) is a powerful technique for measuring what comes off a surface after adsorption of various species [8]. After adsorption at low temperature, the sample is heated and the species desorbing from the surface are monitored with a mass spectrometer. Information can be obtained about the desorbing species, the desorption kinetics and eventually the path of surface reactions.

This list of techniques used for surface analyses is far from being exhaustive and does not constitute a review of surface characterization methods. According to the nature of the samples to be analyzed and the kind of information required, one specific method may be the most convenient one in a particular case whereas completely inadequate in other cases. But, we must be fully aware that, when comparing the results obtained by different methods, the different analyzed depths as well as the different levels of information obtained may lead to severe discrepancies.

4. Infrared Spectrometry

In fact, very few techniques do analyze the first atomic layer of a surface. Among them is the infrared (IR) surface spectrometry. This technique that we will focuss on is based on the absorption by a sample of infrared wavelengths which excite interatomic vibrations. The energy of the vibrations corresponding to the absorbed wavelengths depends on the nature and the bindings of the chemical groups. The IR transmission mode is essentially known and widely used for bulk characterization. However, under particular conditions which will be discussed in the following, this technique can give access to the chemical composition of the first atomic layer as well as the chemical states of the surface atoms, including surfactants. This surface analysis technique which is well known in the field of catalysis, is particularly suitable for a detailed study of the surface chemistry of nanoparticles.

4.1. EXPERIMENTAL SETUPS

The design of most of the interferometers used for FTIR spectrometry is based on the interferometer originally designed by Michelson [12]. It is a device which can divide a beam of radiation into two paths and recombine the two beams after a path difference has been introduced. Therefore, interferences can occur between the beams. The intensity variations of the reconstructed beam are measured as a function of the path difference (interferogram). When applying the Fourier transform, the intensity is obtained as a function of wavenumbers and this constitutes the spectrum. The standard interferometers operate from 5000 to 400 cm^{-1} and their resolution can vary from 0.5 to 16 cm^{-1}.

When the infrared beam reaches the sample, the light can be absorbed, transmitted, reflected and diffused. According to the chosen technique, the analyzed light will be the transmitted beam or the diffused beam or the reflected beam and will be compared to the IR incident energy in order to get information on the absorbed energy [5].

The reflection-absorption setup is particularly useful for studying adsorbed molecules on a metal surface and finds applications in corrosion and coating technology. In the internal reflection spectroscopy, also called attenuated total internal reflection (ATR), the absorbing layer to be analyzed is deposited on the surface(s) of a prism in which the infrared beam propagates through several internal reflections. This technique is particularly used in electrochemistry and colloid chemistry [5]. But the two methods which are the most often used for powder analyses are the transmission and the diffuse reflection techniques.

The transmission technique is widely used to characterize the bulk of materials in the standard potassium bromide matrixes. But, it is also a powerful technique to analyze the surface species. To this end, the powdered absorbent is slightly pressed into a thin wafer. Obviously, the higher is the specific surface area, the easier is the analysis. However, compared to the bulk, the surface species are minority by far. It must not be forgotten that, because of the "large" amount of powder needed for the surface analysis (compared to bulk characterization), the absorption due to the vibrational modes of the bulk is very intense. Due to the high weight of the atoms constituting the bulk of the materials (for example Al in alumina, Ti in titania, Si in silicon nitride ...), the absorption bands of the bulk vibrations concern the lowest wavenumber range, whereas the vibrations of the surface species absorb in the highest wavenumber range. As a consequence, quite often, the low wavenumber range of the spectra is obscured. Morever, the bands due to overtones and band combinations are no longer negligible. It must be mentioned, however, that this technique is not appropriate for highly conductive materials. In the cells generally used for the surface analysis (examples of design can be found in Ref. 13-15), the sample is under vacuum or under controlled pressures of gases and can be thermally treated. The essential items are the furnace, the connections to vacuum pumps and to gas cylinders or liquid containers.

In the diffuse reflectance technique (DRIFTS) [12,16], the scattering of the radiation by a sample is the essential effect to be considered. With this technique, it is possible to analyze loose powders and bulk samples as well. The cell can be decomposed into two parts. The first one corresponds to the optical set-up. Mirrors focus the beam onto the sample and the scattered beam is collected by other mirrors toward the detector. This setup by itself allows one to analyze samples under atmosphere at room temperature. When a specific cell (second part) is added to the previous apparatus, the sample can be thermally treated under vacuum or controlled pressures of gases in the same way as in the cell for transmission analyses.

An advantage of the diffuse reflectance technique is that the area of contact between gaseous adsorbate and powdered adsorbent is much greater than if the sample were pressed into a pellet. Moreover, diffusion of reactants into and effusion of products from the sample are much faster. On the other hand, the intensity of the absorption bands from the surface species may be enhanced by multiple internal and external reflections. But, like in absorption spectrometry, the diffuse reflectance is nearly zero when the absorption of a sample is too high [12]. However, in particular cases, it is possible to analyze samples which are not transparent to the infrared radiation. Another disadvantage is the poor reproducibility of the band intensities because of the variation of the scattering coefficient each time the sample is loaded in the cell. Nevertheless, according to most of the specialists in the field of FTIR spectrometry, the diffuse reflectance technique should become the most powerful technique for studying the chemical phenomena taking place on the surface of high surface area powders [12].

4.2. CHARACTERIZATION PROCEDURES

The surface characterization procedure by FTIR spectrometry can be considered in two distinct and complementary ways. The first and obvious one is the use of FTIR spectrometry to determine the chemical nature of the adsorbed species. The other way is to characterize the adsorbent by observing the spectrum of carefully chosen adsorbed probe-molecules [17]. Indeed, the vibration frequencies of an adsorbed molecule depend not only on the nature of the surface atoms to which it is bonded but also on the coordination number of these atoms within the surface. A typical example is the $\nu(OH)$ stretching vibration of surface hydroxyl groups which depends on the coordination number of the surface atoms to which these OH groups are bonded.

Two types of adsorption processes can be distinguished : the physical adsorption (or physisorption) and the chemical adsorption (or chemisorption). In the physisorption process, the adsorbent-adsorbate interaction is of Van der Waals type. The binding energy for physisorbed molecules is typically 0.25 eV or less [7]. In chemisorption, the interaction is stronger and may be dissociative, non-dissociative or reactive. However, it must be noted that no clear border between physisorption and chemisorption is defined.

The changes in the vibrational spectrum of the adsorbed molecule are related to the nature and the strength of the adsorption. If the molecule is weakly physisorbed, the spectrum is slightly perturbed and the vibrational frequency shifts can be less than one per cent. The weakly physisorbed species are easily removed by evacuation at room temperature. On the contrary, in the chemisorption process tremendous changes may be noted in the spectrum of the adsorbed molecules, particularly if the process is of reactive or dissociative nature.

Both physical and chemical adsorptions can take place at the same time on any surface. Moreover different types of chemical adsorption mechanisms can simultaneously occur. Therefore, an evacuation at increasing temperature (thermal desorption) makes it possible to discriminate the different adsorbed species according to their thermal stability.

Several types of sites can exist on a surface :
(a) Lewis acid sites or electron acceptors
(b) Lewis basic sites or electron donors
(c) Brønsted acid sites or proton donors
(d) Brønsted basic sites or proton acceptors.

The strength of the acidic or basic sites may vary according to the coordination of the corresponding atom on the material surface.

Figure 1 shows the process of the surface characterization of a nanosized powder [18] and can be summarized as follows:

The first step is the activation of the surface. The activation consists in heating the sample under dynamic vacuum for one or two hours. This treatment clears the surface of all physisorbed molecules and weakly chemisorbed species, according to the temperature. In the following, a so-treated surface will be referred to as "activated surface". It is worth noting that an activated surface is no longer in equilibrium. Therefore, as soon as any chemical species will be in contact with this activated surface, an interaction, possibly a chemical reaction, will occur in order to get back to the equilibrium state.

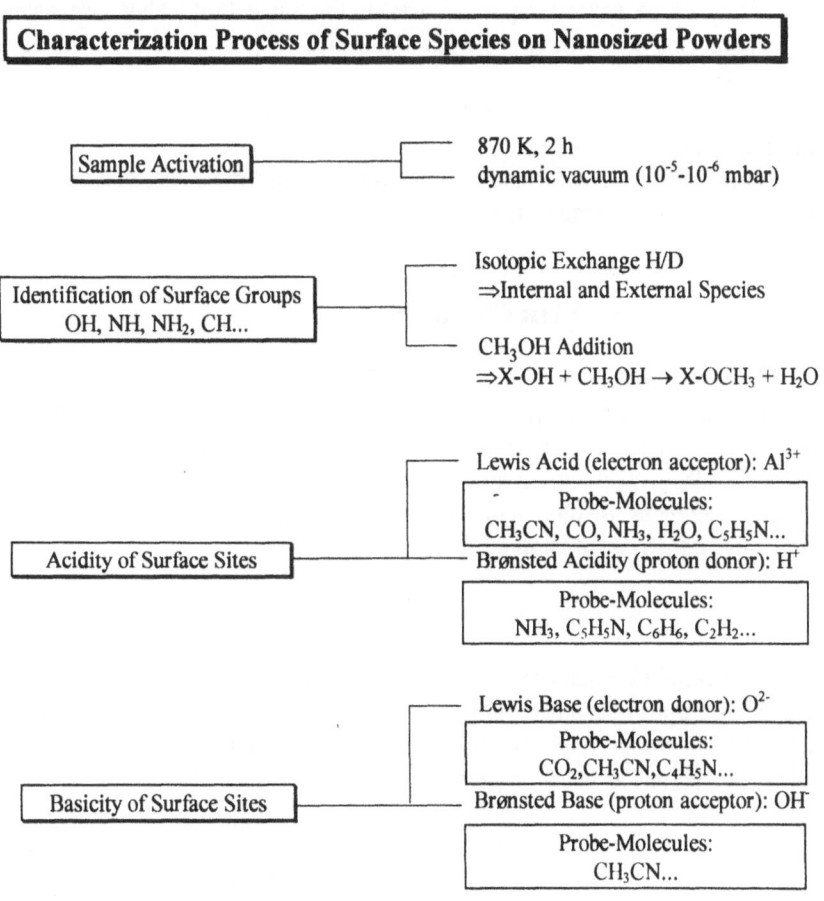

Figure 1. Process of surface characterization

Because hydrolysis is the most probable reaction to get saturated bonds and balanced forces on any surface, many surface bonds involve hydrogen atoms. Consequently, the next step in the surface characterization will be the isotopic exchange H/D by deuterium addition. The exchange can only take place on surface species and therefore, it allows one to discriminate hydrogen-containing groups on the surface from the ones in the bulk. Moreover the vibrational frequencies of the exchanged groups shift toward lower values due to the higher molecular weight of deuterium. Thus, deuterium acts as a marker of the surface hydrogen vibrations.

Methanol is also a very useful probe-molecule. It probes the lability of the OH surface groups. These groups can indeed react with methanol leading to the formation of O-CH$_3$ methoxyl groups.

To probe the acidity and basicity of the surface sites, different other molecules are used [19]. A non-exhaustive list is given in Figure 1. It must be noted that some molecules may probe several types of sites at the same time. Thus, we have to run several experiments with different probe-molecules to get a good knowledge of the surface reactivity.

To summarize, surface FTIR spectrometry can provide information on:
(a) the composition and the structure of the surface species;
(b) the nature of the bonds formed between the adsorbed molecules and the surface;
(c) the various types of active surface centres.

5. Examples of Surface FTIR Characterization

To emphasize the differences between bulk and surface composition, the examples of silicon carbonitride (SiCN) and aluminum nitride (AlN) nanosized powders will be discussed and compared to nanosized silica and alumina respectively. The surface of these latter oxides is well-known and detailed information can be found in Ref. 2, 8, 13, 19-22 and references therein. The experiments presented in the following were all performed using the transmission mode.

5.1. SILICON CARBONITRIDE

The silicon carbonitride nanopowder discussed in this study was obtained by chemical vapor condensation of hexamethyldisilazane [23]. Its specific surface area was found to be 272 m^{2}g^{-1} and the average crystallite size is in the 10 nm range. The spectrum of the raw silicon carbonitride pellet is given in Figure 2a. When the sample is activated at 500°C (Fig. 2b), the broad band around 3370 cm^{-1} decreases concurrently with the increase of a sharp band at 3744 cm^{-1}. The difference between the spectra recorded before and after activation (Fig. 2c) gives a better view of the evolution. The negative bands correspond to decreasing species whereas the positive bands correspond to increasing species. The broad negative band centered at 3355 cm^{-1} along with the 1685 cm^{-1} band correspond to the desorption of water molecules hydrogen bonded to surface

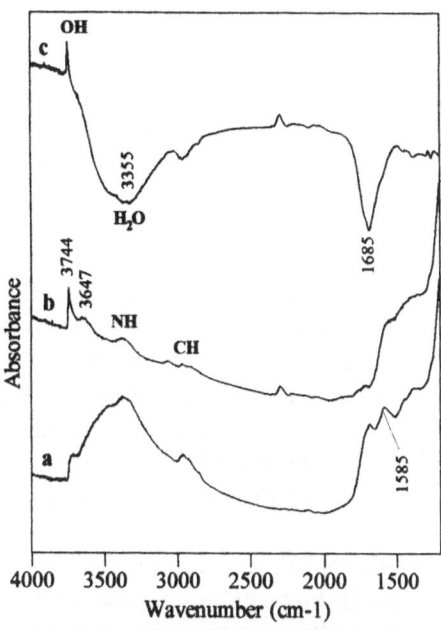

Figure 2. Infrared spectra of a SiCN pellet.
a) room temperature, under vacuum; b) after
activation at 500°C; c) difference spectrum b-a.

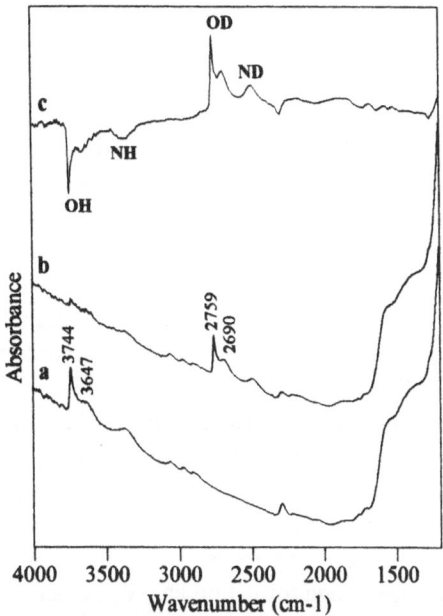

Figure 3. H/D isotopic exchange on SiCN.
a) after activation at 500°C; b) after D₂ addition;
c) difference spectrun b-a.

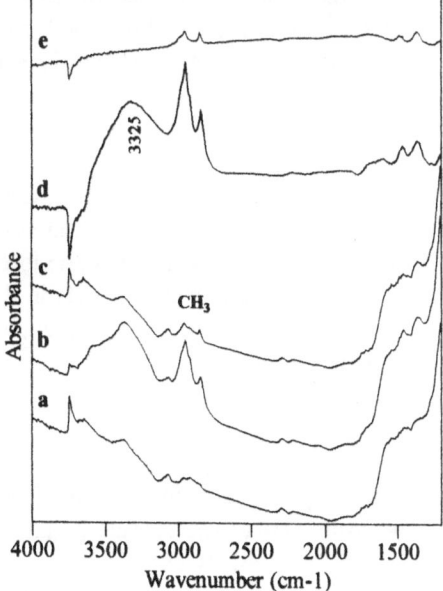

Figure 4. Methanol adsorption on SiCN.
a) after activation at 500°C; b) after methanol
addition at rt; c) after desorption at 300°C; d)
difference b-a; e) difference c-a.

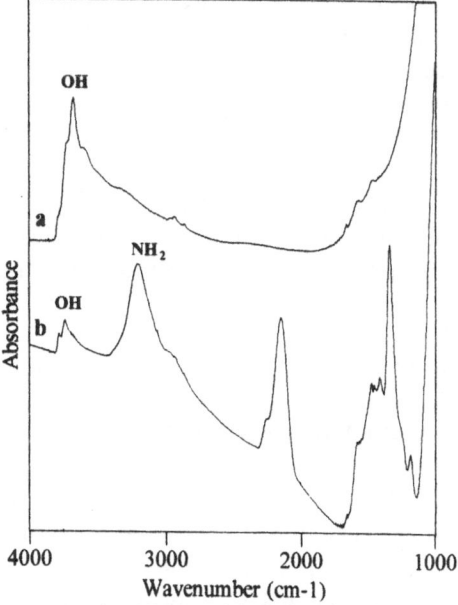

Figure 5. Comparison of the infrared surface
spectra of γ-Al₂O₃ and AlN after activation at
500°C. a) γ-Al₂O₃; b) AlN.

hydroxyl groups as similarly observed in the case of silica. The corresponding freed OH groups absorb at 3744 cm^{-1} as those on the silica surface [21]. The broader band at 3647 cm^{-1} is also assigned to silanol groups. The band at 3370 cm^{-1} revealed by the activation is assigned to the ν(NH) stretching vibration in Si-NH groups and can be compared to similar groups on the silicon nitride surface [24]. Besides, the band at 1585 cm^{-1} which clearly appears in the spectrum of the raw sample (Fig. 2a), disappears by heating (Fig. 2b). It may be assigned to the bending vibration of ammonia molecules coordinated on the surface. The corresponding NH stretching band would be overlapped by the ν(OH) band of adsorbed water. The 2900 cm^{-1} region is characteristic of the ν(CH) stretching vibrations. Thus, this activation process allows one to identify several chemical groups and, in particular in the silicon carbonitride powder, SiOH, SiNH and CH$_x$ groups.

To check the location of these chemical groups, an isotopic exchange H/D was performed by deuterium addition (Fig. 3). All the OH and NH groups are exchanged (Fig. 3b). In Figure 3c, the appearing ν(OD) and ν(ND) stretching bands correspond to the disappearing ν(OH) and ν(NH) bands. Consequently, all SiOH and SiNH groups are on the silicon carbonitride surface.

As previously mentioned (cf. section 4.2), methanol can link on a silica surface by hydrogen bond formation and also react with surface hydroxyl groups to yield methoxyl groups [8]. In this experiment, methanol vapor was added at room temperature to the activated sample (Fig. 4). The broad positive band around 3325 cm^{-1} along with the negative sharp band at 3744 cm^{-1} visible on the difference spectrum (Fig. 4d) are the obvious proof of an hydrogen bond formation on the first type of silanol groups. Concomitantly, the bands appearing in the 2900 cm^{-1} region and below 1500 cm^{-1} correspond to the CH vibrations in methanol. In this difference spectrum (Fig. 4d), it is clear that the 3647 cm^{-1} band which we have previously assigned to the ν(OH) vibration in a different type of silanol groups is not affected by the hydrogen bond. Consequently, these silanol groups are not accessible to methanol. But, since they are exchanged by deuterium we can conclude that they are located inside open pores whose size prevents methanol from approaching. The broadness of this 3647 cm^{-1} band compared to the 3744 cm^{-1} band results from the perturbation of the OH bonds caused by the electrostatic force field inside the pores. After evacuation at 300°C (Fig. 4c,e), the remaining bands in the 2900 cm^{-1} region show the formation of methoxyl groups. The surface is irreversibly modified since these bands are still visible after evacuation at 500°C.

5.2. ALUMINUM NITRIDE

We observed that the surface of silicon carbonitride is hydrolyzed and presents silanol groups similar to those on the silica surface. The hydrolysis of the surface of nitride compounds is a general feature and this makes their surface reactivity close to that of the corresponding oxides. However, the persistent presence of nitrogen in the first atomic layer should bring specificities. Direct consequences should be noted in the deagglomeration properties.

Figure 5 shows the surface spectra of an aluminum nitride nanopowder synthesized by a chemical route [18] (average crystallite size: 60 nm) and of a commercial γ-alumina powder (crystallite size varying from 10 to 50 nm). Both pellets were activated at 500°C under the same conditions. Below 1000 cm^{-1}, the spectra are obscured by the bulk vibrations. The most important difference between these two spectra is the presence of the ν(NH) absorption bands around 3200 cm^{-1} in the spectrum of the aluminum nitride (Fig. 5b). These bands correspond to NH$_2$ groups. It has been proven by deuterium isotopic exchange that these groups are located on the surface [25]. The assignment of the band at 2136 cm^{-1} along with the shoulder at 2247 cm^{-1} in the spectrum of aluminum nitride is still under discussion. The bands in the 1500 cm^{-1} region are assigned to overtones or combination bands superimposed on the δ(NH$_2$) bending vibration should absorb at 1543 cm^{-1}.

In Figure 6, the ν(OD) absorption region is reported for both samples. Because of a better signal-to-noise ratio in this wavelength region, it is indeed easier to study the OD groups obtained by deuterium isotopic exchange of the OH groups on the activated surfaces. According to the Knözinger's model [2,19,21], five possible OH configurations exist at the surface of a spinel alumina. They correspond to:

(a) OH groups bonded to a single aluminum atom in octahedral coordination (type Ib);
(b) OH groups bonded to a single aluminum atom in tetrahedral coordination (type Ia);
(c) OH groups bridged on 2 aluminum atoms in octahedral and tetrahedral coordinations (type IIa);
(d) OH groups bridged on 2 aluminum atoms both in octahedral coordination (type IIb);
(e) OH groups bridged on three aluminum atoms (type III).

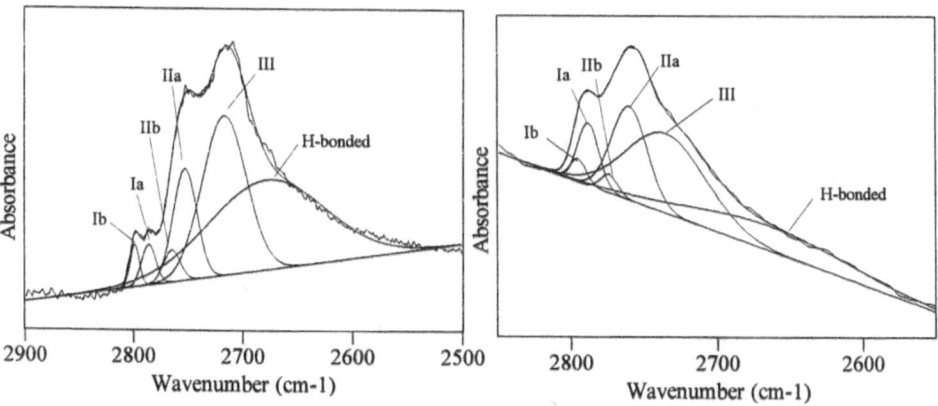

Figure 6. Decomposition of the ν(OD) absorption range for γ-Al$_2$O$_3$ (left) and AlN (right).

Although the Knözinger's model is still under discussion as it does not take the cation vacancies into account, it is the most frequently used one. Based on this model and the corresponding frequency assignments, a decomposition of the ν(OD) absorption range was performed for both aluminum nitride and alumina [26]. We showed that six subbands with a Gaussian profile were necessary to obtain a satisfactory fit of the ν(OH) experimental band. The highest five frequencies correspond to the OH groups defined in the Knözinger's model. As for the sixth subband at the lowest frequency, it is due to residual hydrogen-bonded hydroxyl groups.

The frequencies of the six subbands as well as the band widths are very close for both samples (Table 1). Since the band areas are related to the concentration of adsorbers (Beer's law), it is possible to compare the relative amount of OH groups bonded to a single aluminum atom in octahedral (Al^{VI}) or in tetrahedral (Al^{IV}) sites. The ratioed band areas (Table 2) show that in the case of aluminum nitride, the amount of OH groups bonded to one tetrahedral aluminum is much larger than in the case of gamma-alumina. This is quite expected because of the crystalline structure of aluminum nitride. However, the presence of OH groups bonded to octahedral aluminum reveals that the hydrolysis causes not only the introduction of oxygen in the superficial layer but also changes in the unit cell volume. According to the literature, the presence of aluminum atom in octahedral coordination implies the decrease of aluminum vacancies and an expansion of the unit cell. These crystalline changes result in a decrease of the thermal conductivity of aluminum nitride.

TABLE 1: ν(OD) decomposition for AlN and γ-Al_2O_3 and frequency assignments according to the Knözinger's model. Ib: OH linked to one Al^{VI}; Ia: OH linked to one Al^{IV}; IIb: OH bridged to two Al^{VI}; IIa: OH bridged to one Al^{VI} and one Al^{IV}; III: OH bridged on three Al^{VI}.

		Ib	Ia	IIb	IIa	III	H-bonded
Al_2O_3	ν(OD) (cm^{-1})	2799	2786	2765	2753	2717	2678
	ν(OH) (cm^{-1})	3795	3778	3749	3733	3684	3631
	Band width (cm^{-1})	10	15	18	25	47	123
AlN	ν(OD) (cm^{-1})	2795	2788	2774	2760	2735	2665
	ν(OH) (cm^{-1})	3790	3780	3762	3743	3709	3614
	Band width (cm^{-1})	13	17	13	28	64	130

TABLE 2: Ratioed band areas giving the relative amount of OH linked to a single aluminum atom (types Ia and Ib).

	Ib / (Ia+Ib)	Ia / (Ia+Ib)	Ia / Ib
Al_2O_3	0.42	0.58	1.40
AlN	0.22	0.77	3.53

6. Surface Modifications Followed *in situ* by FTIR Spectrometry

For a long time, scientists have modified surface properties of metals, semiconductors or insulators by adsorbing organic molecules onto their surfaces. Indeed, surface modification is the way to obtain controlled surface properties as well as a controlled surface charge behavior.

Among the applications of the surface modifications either on powders or on bulk materials, we can summarize them as follows:
  (a) improvement of catalytic properties;
  (b) improvement of powder dispersion with a good colloidal stability;
  (c) production of an interface for the growth of well-ordered or mechanically stable self-assembled multilayers;
  (d) production of new electrochemically based sensors;
  (e) resistance improvement to corrosive attack by water, carbon dioxide or other gases;
  (f) production of new microelectrochemical devices;
  (g) sintering improvement.

While there is no doubt on the industrial importance of these applications, the fundamental questions on how the surface is modified have still to be answered [27]. The formation of new chemical groups may be selective or non selective. Moreover, if the modified samples have to be in contact with another chemical compound, this latter competes with the newly-formed surface species and an equilibrium may be reached to the detriment of the original surface modifications. FTIR surface spectrometry is a powerful investigation tool, particularly relevant to this fundamental analysis.

Results obtained by FTIR spectrometry on the chemical grafting of hexamethyldisilazane (HMDS) on titania surface can be found in Ref 28. HMDS was chosen as it is widely used for surface modifications due to its reasonably high vapor pressure at room temperature and as it is easy to handle and environmentally friendly. It usually reacts with the surface hydroxyl groups leading to ammonia formation. It has been demonstrated that the grafting of HMDS on the surface hydroxyl groups of titania is irreversible and selective. The non-modified OH groups were identified as adjacent OH groups bonded to each other by hydrogen bridges. On the other hand, when the grafted titania surface was desorbed at 500°C and then oxidized under dry oxygen at the same temperature, new groups were formed and assigned to Si-OH silanol groups grafted on titania surface. Therefore, from this grafted surface, it appears that the generation of Si-OH groups could be easily monitored. It is worth noting that the Si-OH groups should not have the same acido-basicity as the original hydroxyl groups on titania.

As a summary of its use in the field of surface modification, FTIR surface spectrometry allows one to follow the grafting *in situ*, and to check its thermal stability. Moreover, it is possible to discriminate the grafted and the non-grafted sites and to test the stability when other organic molecules compete with the grafted species. Note that it cannot *a priori* be considered that the surface is only modified by the addition of new chemical groups as smartly chosen as they may be. Independently from any steric hindrance, these new groups may perturb the surface acido-basicity by electronic rearrangement. Moreover, since the acido-basicity of a nanosized powder surface is very dependent on the synthesis and pretreatment conditions, a fundamental analysis must be initially performed in order to determine the optimized and reproducible conditions for subsequent surface modifications and property tailoring.

316

7. Conclusion

The FTIR spectrometry which has been known for a long time for its performance in bulk characterization, proves to be a powerful tool in surface studies of nanosized materials. It appears that the chemical composition of the very first atomic layer can be determined by using specific attachments to the FTIR spectrometer. This technique which is non-destructive and cost-effective, allows the control, modification and tailoring of surfaces of nanosized powders. The experiments can be performed *in situ* so that it is possible to precisely follow the evolution of the surface chemical species under various treatments. The experiments described in this paper were all performed in transmission mode, but the diffuse reflection mode is becoming the preferred method for the study of strong absorbers [29] or transparent coatings on absorbing substrates (metal oxide or metal nanoparticles, for example) when the sensitivity is enhanced by the development of advanced detectors.

Surface FTIR spectrometry should not be a tool reserved for chemists dealing with catalysis. Scientists should also take advantage of this technique in the interdisciplinary field of nanostructured materials and nanotechnology whether they are physicists, metallurgists, material science engineers, electrical or biomedical engineers. It is indeed believed that the development of nanostructured materials for high-added value applications critically depends on the ability of the scientists to fully control and master surface and interface properties on the molecular level. In this regard, any strategy and technique on tool development will help to achieve this goal and contribute to the future of the nanostructured materials for truly marketable products. Thus, they will also sustain the interest of public and private institutions for further research funding.

8. References

1. Alvarez, J. and Asensio, M.C. (1990) Electronic structure and composition of surfaces, in *Spectroscopic characterization of heterogeneous catalysis* (Part A), J.L.G. Fierro Ed, Elsevier, Amsterdam, pp. A79-A160.
2. Morterra, C. and Magnacca, G. (1996) A case study: surface chemistry and surface structure of catalytic aluminas, as studied by vibrational spectroscopy of adsorbed species, *Catalysis Today* 27, 497-532.
3. Rheis, K. (1995) Nanostructures in industrial materials, *Thin Solid Films* 264, 135-140.
4. Veprek, S. (1997) Electronic and mechanical properties of nanocrystalline composites when approaching molecular size, *Thin Solid Films* 297, 145-153.
5. Busca, G. (1996) The use of vibrational spectroscopies in studies of heterogeneous catalysis by metal oxides: an introduction, *Catalysis Today* 27, 323-352.
6. Fierro, J.L.G. (1990) Surface spectroscopic techniques, in *Spectroscopic characterization of heterogeneous catalysis* (Part A), J.L.G. Fierro Ed, Elsevier, Amsterdam, pp. A1-A78.
7. Prutton, M. (1994) *Introduction to surface physics*, Oxford University Press, Oxford (UK).
8. Boehm, H.-P. and Knözinger, H. (1983) Nature and estimation of functional groups on solid surfaces, in *Catalysis* V.4, J.R.A. Anderson and M. Boudart Eds, Springer Verlag, Berlin, pp. 39 -207.

9. Conesa, J.C., Esteban, P., Dexpert, H. and Bazin, D. (1990) Characterization of catalyst structures by extended X-ray absorption spectroscopy, in *Spectroscopic characterization of heterogeneous catalysis* (Part A), J.L.G. Fierro Ed, Elsevier, Amsterdam, pp. A225-A297.

10. Yu, E.T. (1996) Nanoscale characterization of semiconductor materials and devices using scanning probe techniques, *Materials Science and Engineering* **R17**, 147-206.

11. Knözinger, H. (1996) In situ Raman spectroscopy. A powerful tool for studies in selective catalytic oxidation, *Catalysis Today* **32**, 71-80.

12. Griffiths, P.R. and de Haseth, J.A. (1986) *Fourier transform infrared spectrometry*, John Wiley & Sons, New-York.

13. Hair, M.L. (1967) *Infrared spectroscopy in surface chemistry*, M. Dekker, New-York.

14. Basu, P., Ballinger, T.H. and Yates, J.T. Jr (1988) Wide temperature range IR spectroscopy cell for studies of adsorption and desorption on high area solids, *Rev. Sci. Instruments* **59**, 1321-1327.

15. Baraton, M.-I. (1994) Infrared and Raman characterization of nanophase ceramic materials, *High Temperature and Chemical Processes* **3**, 545-554.

16. Brimmer, P.J., Griffiths, P.R. and Harrick, N.J. (1986) Angular dependence of diffuse reflectance infrared spectra. Part I: FTIR spectrogoniometer, *Applied Spectroscopy* **40**, 258-265.

17. Hollins, P. (1992) The influence of surface defects on the infrared spectra of adsorbed species, *Surface Science Reports* **16**, 51-94.

18. Baraton, M.-I., Chen, X. and Gonsalves, K.E. (1996) Application of Fourier transform infrared spectroscopy to nanostructured materials surface characterization, in *Nanotechnology-Molecularly designed materials*, G.M. Chow and K.E. Gonsalves Eds, ACS Symposium Series 622, Washington DC, pp. 312-333.

19. Knözinger, H. (1976) Specific poisoning and characterization of catalytically active oxide surfaces, *Advances in Catalysis* **25**, 184-261.

20. Lavalley, J.C. (1996) Infrared spectrometric studies of the surface basicity of metal oxides and zeolites using adsorbed probe molecules, *Catalysis Today* **27**, 377-401.

21. Morrow, B.A. (1990) Surface groups on oxides, in *Spectroscopic characterization of heterogeneous catalysis* (Part A), J.L.G. Fierro Ed, Elsevier, Amsterdam, pp. A161-A224.

22. Davydov, A.A. (1984) *Infrared spectroscopy of adsorbed species on the surface of transition metal oxides*, John Wiley & Sons, New-York.

23. Baraton, M.-I., Chang, W. and Kear, B.H. (1996) Surface chemical species investigation by FT-IR spectrometry and surface modification of a nanosized SiCN powder synthesized via chemical vapor condensation, *J. Physical Chemistry* **100**, 16647-16652.

24. Ramis, G., Busca, G., Lorenzelli, V., Baraton, M.-I., Merle-Mejean, T. and Quintard, P. (1989) FT-IR characterization of high surface area silicon nitride and carbide, in *Surfaces and Interfaces of ceramic materials*, L.-C. Dufour et al. Eds, Kluwer Academic Publishers, Dordrecht, 173-184.

25. Baraton, M.-I., Chen, X. and Gonsalves, K.E. (1996) FTIR analysis of the surface of nanostructured aluminum nitride powder prepared via chemical synthesis, *J. Materials Chemistry* **6**, 1407-1412.

26. Baraton, M.-I., Chen, X. and Gonsalves, K.E. (1997) FT-IR characterization of the acidic and basic sites on a nanostructured aluminum nitride surface, Proceedings MRS Fall Meeting 1996 Symposium S, MRS Ed, Pittsburgh, in press.

27. Whitesides, G.M. and Laibinis, P.E. (1990) Wet chemical approaches to the characterization of organic surfaces: self-assembled monolayers, wetting, and the physical-organic chemistry of the solid-liquid interface, *Langmuir* **6**, 87-96.

28. Baraton, M.-I., Chancel, F. and Merhari, L. (1997) In situ determination of the grafting sites on nanosized ceramic powders by FT-IR spectrometry, *Nanostructured Materials* **9**, 319-322.

29. Baraton, M.-I., Carlson, G. and Gonsalves, K.E. (1997) DRIFTS characterization of a nanostructured gallium nitride powder and its interactions with organic molecules, *Materials Science & Engineering B*, in press.

ENHANCED TRANSFORMATION AND SINTERING OF TRANSITIONAL ALUMINA THROUGH MECHANICAL SEEDING

MARTIN L. PANCHULA[†] and JACKIE Y. YING[‡*]
[†]*Department of Materials Science and Engineering*
[‡]*Department of Chemical Engineering*
Massachusetts Institute of Technology
Cambridge, MA 02139

1. Introduction

Alumina is one of the most widely used ceramics, encompassing spark plugs, catalysts, heat sinks on computer chips, high temperature insulation, lighting envelopes and milling media. Because of this wide variety of applications, alumina is one of the most thoroughly studied ceramic materials. In particular, a significant amount of research has been performed since the 1950's within the catalysis and ceramic fields with the goal of understanding and controlling such properties as surface acidity and basicity, surface area, crystallite size, agglomeration, thermal stability, phase transformation kinetics, and sinterability. This paper will focus on the phase transformations of alumina and show how the transformation kinetics can be increased to improve the sinterability of compacts prepared from the transitional alumina phases.

2. Background

Aluminum oxide has at least eight readily attainable different crystal structures $(\gamma, \delta, \theta, \eta, \chi, \kappa, \rho, \alpha)$, of which seven are metastable. Generally $\alpha\text{-}Al_2O_3$ is considered the thermodynamically stable phase though recent studies have suggested that at a very fine particle size, $\gamma\text{-}Al_2O_3$ may be the lower energy crystal structure due to surface energy contributions [1,2]. As shown in Figure 1, the hydroxide phases from which the aluminas are derived can have a large effect on the crystal forms observed during thermal treatments. The only hydrated phase which does not produce the transitional alumina phases during heating is diaspore $(Al_2O_3 \cdot H_2O)$ which transforms directly to $\alpha\text{-}Al_2O_3$ after dehydration at temperatures greater than 500°C. Although diaspore would be very useful for making fine-grained, highly reactive $\alpha\text{-}Al_2O_3$, it requires high temperatures and pressures to synthesize and is not found naturally in sufficient quantities and purity to be commercially attractive for advanced applications [3].

---

* To whom correspondence should be addressed.

G.M. Chow and N.I. Noskova (eds.), Nanostructured Materials, 319–333.

320

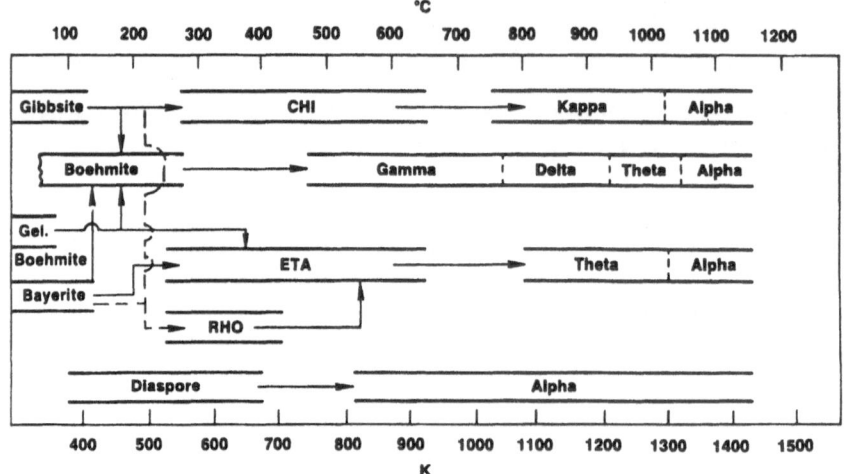

Figure 1. Schematic of alumina phase transformations, from Wefers and Misra [4].

In general, the transitional phases are easily produced as very fine grained materials with crystallite sizes in the nanometer range. Nanocrystalline alumina synthesis methods include precipitation and calcination, gas phase synthesis, arc discharge, laser ablation, and thermal evaporation [5-8]. According to Herring's scaling law [9], these ultrafine crystals should result in a large decrease in the temperatures necessary for sintering. Unfortunately, simply making a compact of the transitional aluminas and pressureless sintering does not ensure that improved densification will occur. As the transitional phases are heated, relatively few alpha alumina nuclei are produced (approximately 10^{10}-10^{11} nuclei/cc [10,11]). These nucleated grains undergo rapid grain growth through the porous transitional alumina matrix phase, resulting in a porous vermicular structure (Figure 2). A contributing factor to this phenomena is the large change in molar volume which occurs when the transitional aluminas transform to the α-Al_2O_3 phase. For instance, the transformation to α-Al_2O_3 from θ-Al_2O_3 requires a 8% molar volume shrinkage. The combination of low nucleation density, high growth rate through the fine-grained transitional alumina matrix, and the change in molar volume produces large porous α-Al_2O_3 grains. During the final stages of sintering the pores are typically trapped within the grains making their removal extremely difficult.

Researchers have investigated several different methods of overcoming the low nucleation density. The use of additives to promote the transformation kinetics are one obvious possibility. For instance, the effect of ten different chemical additives on the transformation temperature of alumina was investigated by Xue and Chen [12]. They reported that B_2O_3 and SiO_2 additions increased the transformation temperature, but ZrO_2 additives retarded the phase transformation most significantly. Additives that promoted a liquid phase, such as ZnF_2, CuO, V_2O_5, and Li_2O, increased the transformation kinetics, perhaps by a solution-precipitation mechanism thereby

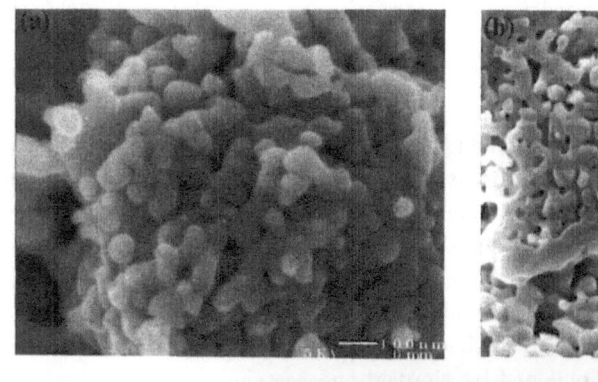
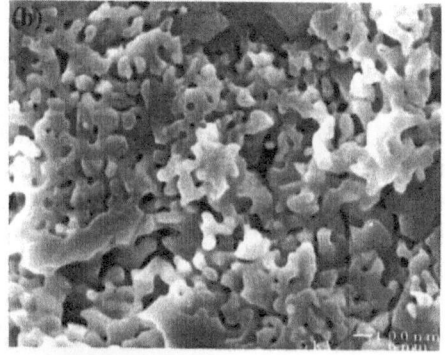
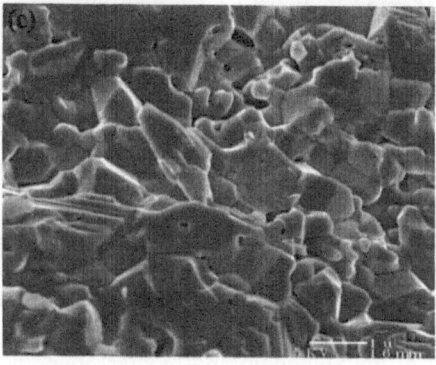

Figure 2. Fracture surface micrographs of unseeded boehmite compacts hot pressed under 50 MPa at (a) 1100°C, (b) 1200°C, and (c) 1400°C for 30 minutes.

speeding up the growth portion of the transformation process. Although they affected the transformation of the alumina, these compounds are typically not acceptable in the final microstructure since they can greatly affect the properties of the material. For example, the additives that form a liquid phase can significantly reduce the high-temperature strength and creep resistance of the alumina and several of the other additives can substantially reduce the transparency and make the material useless for lighting envelope applications.

Heterogeneous nucleation on seeds that are isostructural with α-Al$_2$O$_3$, such as Fe$_2$O$_3$ or Cr$_2$O$_3$, have also been investigated [13,14]. Boehmite gels that had been seeded with α-Fe$_2$O$_3$ showed an increase in the γ-to-α-Al$_2$O$_3$ transformation kinetics corresponding to a decrease in the activation energy of the transformation from 578 kJ/mol to 476 kJ/mol. The increased nucleation density resulted in much finer α-Al$_2$O$_3$ grain sizes, significantly different microstructures, and increased densification. Instead of forming the porous vermicular structure, small dense grains of α-Al$_2$O$_3$ were produced during the transformation which led to enhanced densification. For example, it was reported that fully dense material could be produced by pressureless sintering the

α-Fe$_2$O$_3$-seeded materials (10^{14} seeds/cc) at 1300°C for 2 hours, while only 80% of theoretical density was achieved in samples which had not been seeded [11].

The most effective heterogeneous seeding, however, has come from the addition of α-Al$_2$O$_3$ seeds to the transitional aluminas. Various researchers have shown that significant improvements in the transformation kinetics, microstructure, and densification can be achieved through seeding with submicron α-Al$_2$O$_3$ [10,15-23]. The process generally used is to add submicron α-Al$_2$O$_3$ particles, previously harvested from a coarse mixture of α-Al$_2$O$_3$ by sedimentation, to a boehmite sol. Following gelation and drying, the gel is broken up by light grinding and densified by hot-pressing, sinter-forging or pressureless sintering. The smallest grain sizes and highest densities at the lowest temperatures have been obtained by Nordahl and Messing [23] who obtained dense α-Al$_2$O$_3$ with a grain size of 230 nm after sinter forging under 280 MPa at 1060°C for 30 minutes. This process could also produce transparent alumina with only slight changes in the processing [24].

The various approaches described above rely on the external addition of seed particles or the production of a liquid phase to enhance the transformation kinetics. It is also possible to increase the density of nuclei through mechanical operations on the powder or compact. There are two arguments which have been invoked to describe the effects of pressure and mechanical energy on the nucleation process. The first argument is based upon the fact that the nucleation is most likely to occur in locations where the local packing of the aluminum and oxygen atoms is similar to α-Al$_2$O$_3$, or where defects reduce the energy requirement for nucleation. This atomic packing and/or defect presence will most likely occur at the interface between two particles. This argument is invoked by Pach *et. al.* [25] to describe the influence of particle packing (which was changed in boehmite gels via cold isostatic pressing) on the transformation kinetics. They state that "there is a higher number of potential nucleation sites...at a higher average coordination number of alumina particles." At higher pressures, a higher packing is obtained that increases the number and area of particle-particle contacts, thereby increasing the kinetics of nucleation and growth. A related discussion is used to explain differences observed in α-Al$_2$O$_3$ nucleation for boehmite gels which have been prepared differently [26], but which have not been subjected to significant exterior pressures. In this case, both the microstructure and the atomic arrangement of the gels are different due to the different methods of preparation, so it is difficult to determine which dominates the nucleation kinetics.

The second argument used to explain the effect of pressure on the kinetics of the nucleation process is that the applied pressure directly creates α-Al$_2$O$_3$ nuclei. This argument is based upon the observed effect of pressure on the transformation of transitional alumina to α-alumina [27]. In this study, Ishitobi *et al.* showed that η-Al$_2$O$_3$ could be transformed directly into α-Al$_2$O$_3$ at temperatures as low as 600°C under 1.5 GPa in less than five minutes. While this combination of pressures and temperatures is not easily achieved with conventional compaction techniques, certain high-energy processes such as dynamic shock compaction, explosive compaction [28], and high-energy milling [29-35] have been shown to produce sufficient increases in localized temperature and pressure to generate α-Al_2O_3 directly from the transitional phases.

Depending on the amount and type of energy introduced into the system and the speed at which it is applied, one or more of the above mechanisms may be affecting the nucleation density. Obviously, during these high-energy processes a significant amount of particle rearrangement can occur and even superplastic deformation of the nanometer-sized grains has been reported. Also, some of the alumina can be amorphized during these high energy processes which may enhance not only the nucleation rate but also the subsequent growth rate.

Theoretically, the relative effects of pressure on the phase transformation and the rearrangement of particles to reduce the nucleation barrier could be experimentally determined, however, the complexity of the system makes this a daunting task. In addition to the effects discussed here, it has been reported that the presence of water, impurities [10-12], the degree of crystallinity [36], the crystallite size, previous thermal history, surface area [1,2] and even the presence of large electric fields [37] can influence the phase transformation kinetics of alumina. Given that there are numerous synthesis routes and chemical precursors for alumina, it is not surprising that researchers have come to different conclusions concerning the relative importance of these parameters on the transformations of alumina.

3. Experimental Procedures

3.1. SYNTHESIS
The goals of this study were to quantitatively investigate the effect of the high-energy milling on the transformation rates of γ-Al_2O_3 to α-Al_2O_3, and to examine the sintering behavior of the mechanically seeded powder. The first step in this process, therefore was to create a transitional alumina source which was free of potential seed material. A pseudo-boehmite was synthesized via precipitation using aluminum nitrate for the cation source and ammonium hydroxide as the precipitating agent. Approximately 250 ml of a 0.4 M aluminum nitrate in ethanol solution was added dropwise to a rapidly stirred solution of 200 ml water, 200 ml ethanol, and 100 ml of 30% ammonium hydroxide. The resulting powder was aged overnight and washed with ethanol through centrifugation. The powder was then converted to nanocrystalline γ-Al_2O_3 by calcining at 550°C for four hours in air. This technique also allowed the addition of MgO to the alumina by mixing magnesium nitrate with aluminum nitrate and coprecipitating the cations following the above procedure. Undoped γ-Al_2O_3 produced in this way was used for the phase transformation studies, and both pure and MgO-doped γ-Al_2O_3 were used for the sintering studies. A commercial source of boehmite (Catapal B, Vista Chemical Company) was used as the precursor for the undoped γ-Al_2O_3 sintering studies, after it

was determined that it had the same transformation kinetics as the γ-Al$_2$O$_3$ prepared through precipitation.

3.2. MILLING

Milling of the powder was performed in a SPEX 8000 mixer mill with a WC/Co ball and vial set. This study used WC/Co milling media in order to separate the effects of external seeding and mechanical seeding. If this process were to be used commercially, however, another milling material such as stainless steel, zirconia, or even alumina could be used. However, in order to study the changes in the transformation kinetics due to mechanical seeding, it was necessary to avoid the introduction of external seeds which could come from stainless steel (Fe$_2$O$_3$) or alumina (α-Al$_2$O$_3$) milling media. Zirconia could be used as the milling material, but it would be nearly impossible to remove the zirconia impurities after milling without also dissolving the alumina. Therefore, WC/Co was chosen for this study. Three WC/Co balls, each weighing approximately 10 g, were used and the ball-to-powder ratio was 5:1 by weight. One measure of the effect of short milling times on the seeding of the system is the change in the phase transformation temperature. This was accomplished in a Perkin Elmer System 7 differential thermal analyzer (DTA) with platinum pans using a 10°C/min heating rate under flowing nitrogen. Surface area measurements of the various powder samples were performed on a Micromeritics ASAP 2000 using nitrogen adsorption.

3.3. ISOTHERMAL TRANSFORMATION STUDIES

Thermal treatments of the powders to study the phase transformation kinetics were conducted by placing a small amount of the milled powder, approximately 0.3 g, into small quartz crucibles and placing the crucible directly into the preheated furnace. The time for temperature equilibration was approximately 90 seconds. After the required isothermal treatment time had elapsed, the quartz crucible was removed from the furnace and air quenched so that the cooling rate, especially the initial rate, was very high. The thermally treated powder was then mixed with the internal standard in a 4:1 sample-to-standard weight ratio, and the fraction of α-Al$_2$O$_3$ was obtained from X-ray diffraction (XRD) studies by comparing the ratio of integrated peak intensities to those obtained from a calibration curve. The internal standards used for the quantitative determination of the α-Al$_2$O$_3$ content were elemental silicon or CaF$_2$. XRD studies of the powders and pellets were performed using a Siemens D5000 θ-θ diffractometer (Cu Kα) with a rotating sample stage.

3.4. SINTERING STUDIES

The densification of the mechanically seeded and unseeded samples were compared through pressureless sintering and hot pressing. All of the sintering studies were performed on powders which went through an aqua regia wash, which was used to reduce the tungsten content of the powders from an initial 2.5 wt.% to approximately 0.025 wt.%. The WC removal was, of course, only necessary for the milled samples, however the unmilled samples underwent the same washing procedures in order to keep the comparison accurate. After the powder was stirred in aqua regia for ten hours it was washed via centrifugation with deionized water twice, then ethanol twice, and finally dried at 150°C overnight. The powder was then lightly jar milled in polypropylene with 10 mm zirconia media in ethanol to break up some of the larger agglomerates. Following this step, it was dried and calcined at 500°C for one hour to decompose any hydroxides that had formed during the acid wash.

Specimens to be pressureless sintered were prepared by pressing pellets 8 mm in diameter at 70 MPa followed by cold isostatic pressing at 410 MPa. The pressureless sintering was performed in an alumina tube furnace under flowing oxygen. The heating schedule was 5°C/min to 500°C and 10°C/min to the sintering temperature, followed by a 2-hour dwell. After the soak, the furnace was switched off and allowed to cool. Hot pressing was performed in a Materials Research Furnace with graphite heating elements. Samples were vacuum sintered in a 20 mm graphite die under 50 MPa pressure with a ramp rate of 25°C/min and soaked for 30 minutes at the sintering temperature. Density measurements were performed by Archimedes technique with water.

4. Results and Discussions

4.1. MILLING RESULTS

During the milling process the impact energy transforms small regions of the γ-Al_2O_3 powder into α-Al_2O_3. If the material is milled for a long enough period of time (~12 hours), the powder is fully converted into nanocrystalline α-Al_2O_3 with a grain size of 22 nm and a residual strain of 0.48%, as determined by Kochendorfer's formula (Figure 3). However, long milling times result in significant amount of impurities (as seen by the intensity of the WC peak) due to the abrasive nature of the alumina and the high energy of the milling process. The contamination can be minimized and many of the benefits of the small grained α-Al_2O_3 can be obtained by stopping the milling process after a short period when sufficient nuclei have been created to effectively seed the system. As shown in Figure 4, the milling process has a large effect on the transformation temperature, which is plotted here as the peak temperature of the transformation exotherm as a function of milling time. These results are noteworthy since the lowest exothermic peak temperature reported for externally seeded systems is approximately 1050°C. The fact that the transformation temperature observed after mechanical seeding is approximately 100°C lower suggests that the transformation is significantly enhanced through this process. The presence of an exotherm does not prove a phase transformation is occurring, since the observed exotherm could be due to the release of strain energy in the crystals and/or the crystallization of an amorphous fraction of the sample created during the milling process. However, results obtained from the isothermal treatment studies show that the materials which have been milled for 2 hours undergo a phase transformation from approximately 5% α-Al_2O_3 to

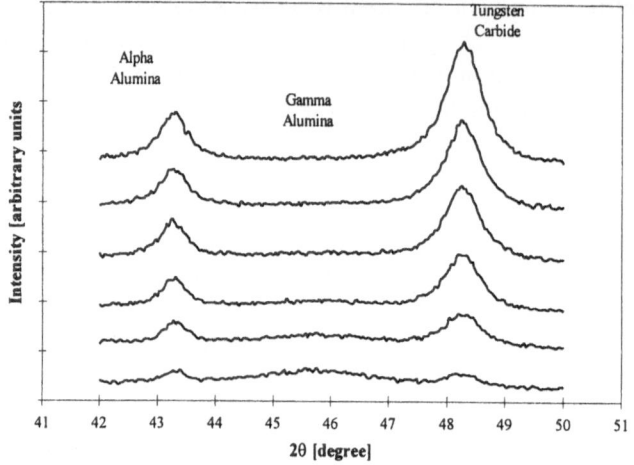

Figure 3. X-ray diffraction of powders milled in WC/Co for different time durations.

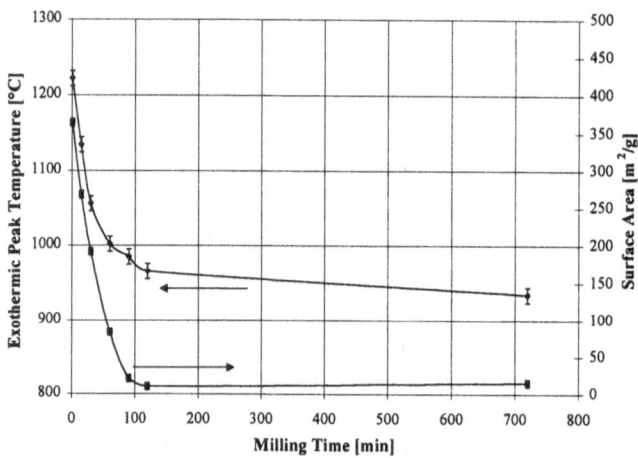

Figure 4. Effect of milling time on surface area and phase transformation temperature.

70% α-Al_2O_3 in 8 minutes at 950°C, confirming that a change in phase composition can occur at such low temperatures. Based on the appearance of α-Al_2O_3 in the X-ray diffraction pattern and the low temperature of the exothermic peak, it is believed that γ-Al_2O_3 which has been high-energy milled for even short periods of time is sufficiently seeded so that nucleation is no longer the rate-limiting step to the transformation and the source of the observed exotherm is the growth of the seed particles to complete the transformation of the sample. That the transformation temperature observed in this system is significantly lower than that noted in externally seeded systems may be due to the fact that these seeds were created *in situ* via milling so that surface nucleation on the seed particles, as described by heterogeneous nucleation theory, is not required before the phase transformation can occur. In other words, the intimate contact between the seeds and the matrix does not require a nucleation barrier, however small, be overcome before growth can begin.

4.2. ISOTHERMAL TRANSFORMATION STUDIES

As expected from the DTA results described above, the rate of the transformation under isothermal conditions increased as milling time increased. The percentage of α-Al_2O_3 as a function of heat treatment was determined from quantitative X-ray diffraction using calcium fluoride or silicon as an internal standard. An interesting observation made in this study was that during the transformation of the seeded systems, the γ-Al_2O_3 appeared to transform directly into α-Al_2O_3 without going through the δ or θ phases. In order to quantify the transformation data, the Avrami model for nucleation and growth was applied. The general Avrami equation (1) can be linearized (2) to allow least squares refinement of the data.

$$X = 1 - e^{-(kt)^n} \tag{1}$$

$$Log(-\ln(1-X)) = nLog(k) + nLog(t) \tag{2}$$

Where X=fraction of α-Al_2O_3 and t=time (sec). From a plot of $Log(-\ln(1-X))$ vs. Log (t), the time exponent (n) and the rate constant (k) of the reaction can be obtained. Figure 5 is an example of the phase transformation data which has been linearized to determine the rate constants and the time exponents. Although it is perhaps the most common transformation equation used in materials science, the Avrami equation has some significant shortcomings in this application. It was derived from a continuum model and does not take into account the effects of particles or particle boundaries on the growth of the nucleated phase. It also does not differentiate between the nucleation kinetics and the growth kinetics, so that the resulting rate constant, and therefore the activation energy, are some combination of the two. However, most models which

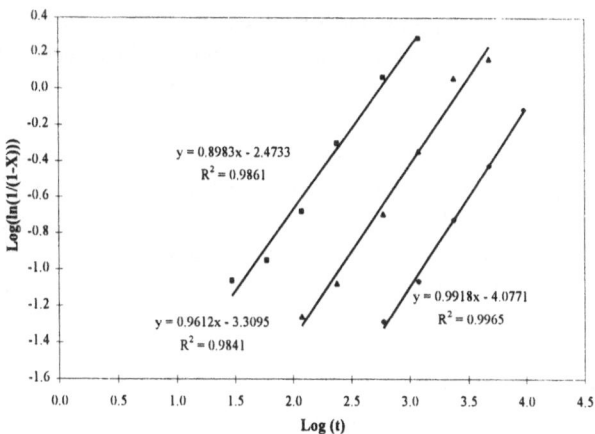

Figure 5. Linearized data from 950°C isothermal transformation for powders milled for (◆)30 minutes, (◆) 1 hour, and (■) 2 hours.

328

attempt to overcome these difficulties and accurately describe the process of nucleation and growth involved with phase transformations are extremely cumbersome to apply and do not offer any additional insight into the mechanics of the transformation.

Assuming an Arrhenius relationship for the reaction rate, a plot of ln(k) vs. 1/T can give the activation energy for the phase transformation (Figure 6). The slopes of the lines are effectively equivalent within the error of these data, and correspond to an activation energy of 330 kJ/mol. In comparison, Shelleman *et. al.* [17] reported an activation energy of 352 kJ/mol for γ-Al$_2$O$_3$ which had been externally seeded with 2.6 x 10^{13} seeds/cc. As expected from the low transformation temperatures observed with the DTA, the activation energy for these mechanically seeded samples is lower than that observed for the externally seeded systems.

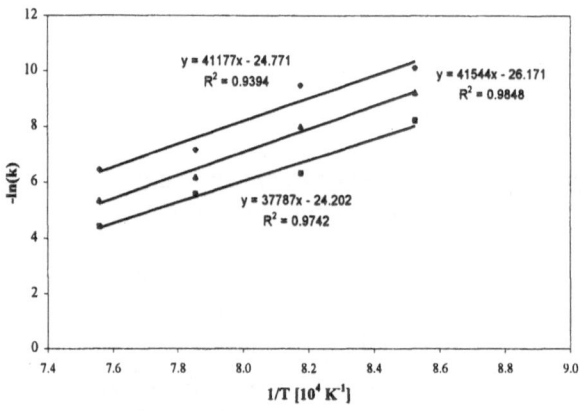

Figure 6. Arrhenius plot of rate constants as a function of temperature for different milling times:($\blacklozenge$)30 minutes, ($\blacklozenge$) 1 hour, and ($\blacksquare$) 2 hours.

4.3. SINTERING RESULTS

The densification curves of the pressureless sintered and hot pressed samples are shown in Figures 7 and 8. The pressureless sintered samples show very little difference between the unseeded and mechanically seeded samples. The approximately 8% difference in density can be traced back to differences in the green density of the compacts from the milled and unmilled powders. These samples display classical boehmite sintering behavior with high densities (>90%) not attained until temperatures of 1600°C are used and even then only a gradual increase in densification is observed. On the other hand, the hot pressed samples show significant differences between the mechanically seeded and unmilled samples. Densities greater than 98% were attained for the milled samples at 1300°C while the unseeded samples were only 60% densified. The cause of this large difference can be understood by examining the microstructure of the hot pressed specimens (Figure 9). In this series of micrographs, one can observe the creation of the large vermicular porous α-Al$_2$O$_3$ grains in the unseeded system that resist further densification. The mechanically seeded material, on the other hand,

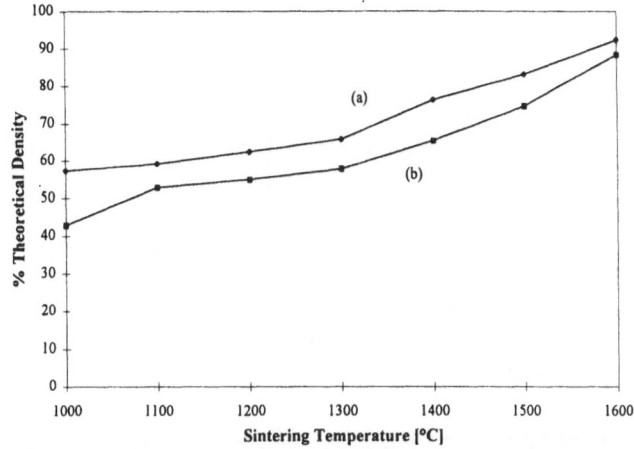

Figure 7. Pressureless sintering of (a) mechanically seeded and (b) unseeded samples.

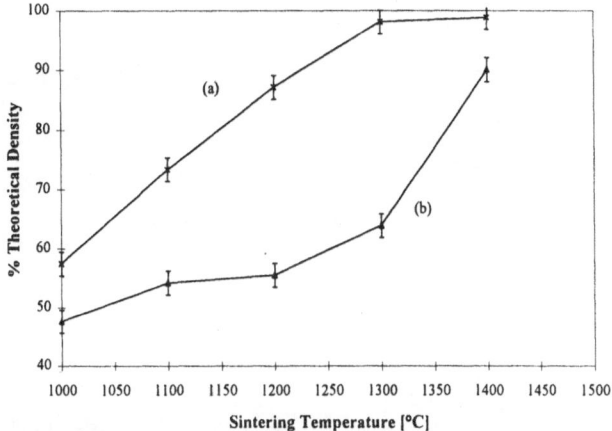

Figure 8. Pressure-assisted densification of (a) mechanically seeded and
(b) unseeded γ-Al$_2$O$_3$ with a 50 MPa load and 25°C/min ramp rate.

transforms into small dense α-Al$_2$O$_3$ particles. Presumably, these smooth dense spherical grains can rearrange more easily under the applied pressure and densify more readily than the interlocking porous particles of the unseeded system. By 1300°C, the mechanically seeded samples are nearly fully dense with the residual porosity present at the triple junctions and grain boundaries. The unseeded system, however, has pores trapped within the α-Al$_2$O$_3$ grains, which will be extremely difficult to remove. The sintering behavior of the 0.1 wt.% MgO-doped alumina was not significantly different from the undoped material. Several authors have reported that for unimodal, high-purity alumina, MgO doping is not necessary to prevent abnormal grain growth or to achieve high densities. Further low-temperature densification studies remain to be

330

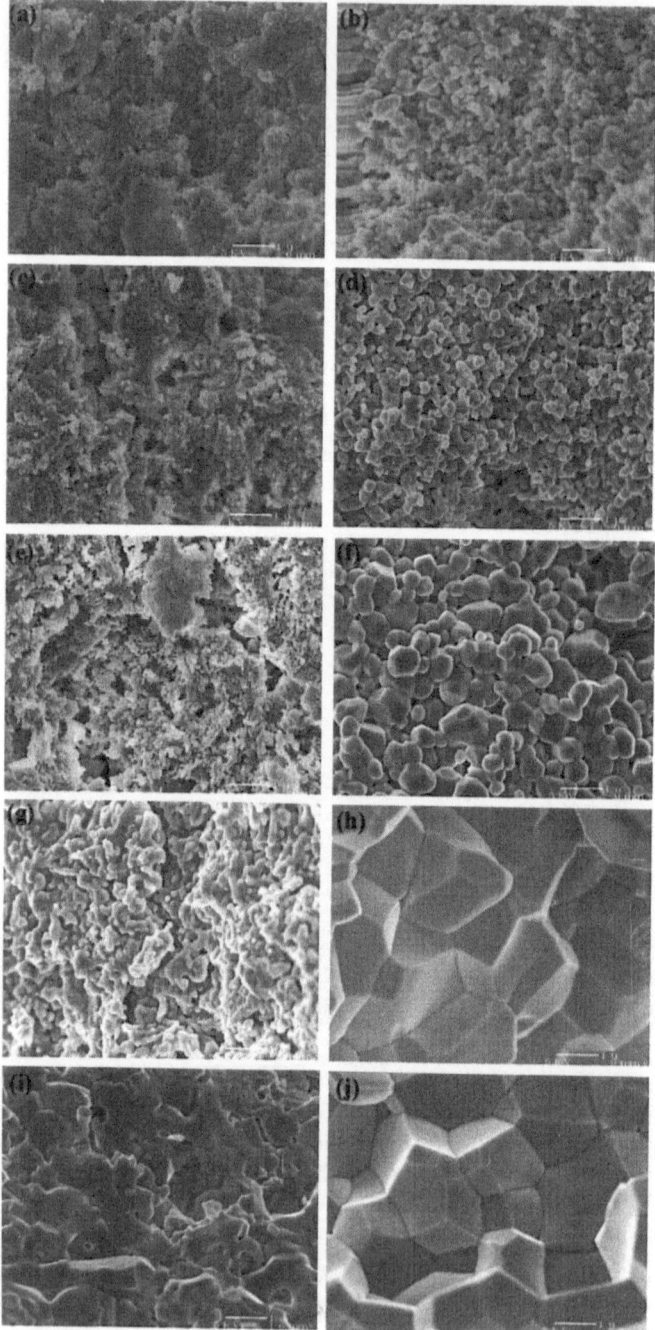

Figure 9. Fracture surfaces of pellets hot pressed under 50 MPa at (a,b) 1000°C, (c,d) 1100°C, (e,f) 1200°C, (g,h) 1300°C, and (i,j) 1400°C. Pellets in (a), (c), (e), (g), and (i) are unseeded samples, while pellets in (b), (d), (f), (h), and (j) have been seeded by high-energy ball milling for two hours. Scale bar equals 1 μm.

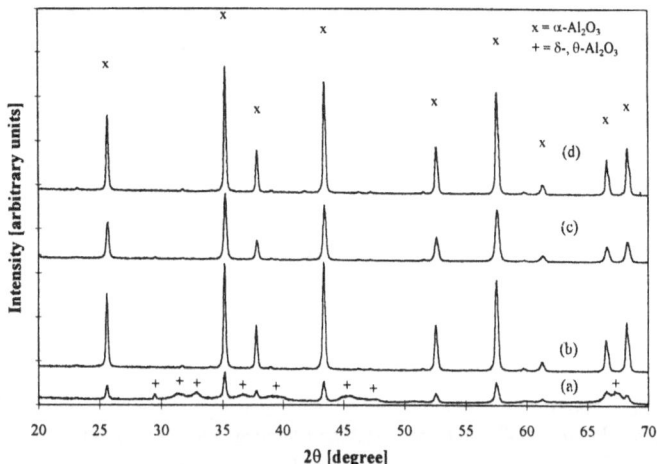

Figure 10. X-ray diffraction patterns of pellets that have been hot pressed under 50 MPa at 1000°C (a,b) and 1100°C (c,d). Samples (a) and (c) were unseeded and samples (b) and (d) were mechanically seeded.

performed to elucidate the effect of MgO on the densification and grain growth of our system. Figure 10 also illustrates the effectiveness of the mechanical seeding; the X-ray diffraction patterns are from the hot pressed pellets which were sintered at 1000°C and 1100°C. The mechanically seeded system has fully transformed into α-Al$_2$O$_3$ by 1000°C while the unseeded system still contains some θ- and δ-Al$_2$O$_3$. Note that the application of pressure did reduce the transformation temperature of the unseeded system from the nominal 1200°C to less than 1000°C, but the kinetics were significantly slower than the mechanically seeded system, which has completely transformed by 1000°C. After heating to 1100°C for 30 minutes, the unseeded system finally became fully transformed, but the resulting microstructure, as described earlier (Figure 9), is quite different from that of the mechanically seeded system.

5. Summary

The effectiveness of mechanical seeding on enhancing the phase transformation and improving the sinterability of transitional aluminas has been established in this study. This process produced powders that gave lower transformation temperatures and faster transformation kinetics than those obtained with the other external seeding processes. Large differences in densification were observed between the mechanically seeded and unseeded specimens, especially in pressure-assisted sintering. It is hypothesized that the sintering difficulties in the pressureless sintered specimens can be attributed to agglomerate formation during powder processing, leading to poor particle packing in the green state. Although the particle packing is improved somewhat during sintering *via* hot pressing, both the pressureless and pressure-assisted sintering rates could be significantly enhanced by more carefully preventing particle agglomeration before and during powder compaction. Optimizing the powder processing and handling conditions will help us achieve the ultimate goal of creating an inexpensive and simple method of transforming the transitional aluminas into low-temperature sinterable materials.

6. Acknowledgments

The authors would like to thank Robert Bosch GmbH and the MIT Sloan Fund for supporting this research. MLP also acknowledges the support of a National Science Foundation Graduate Fellowship.

7. References

[1] McHale, J.M., Auroux, A., Perrotta, A.J., and Navrotsky, A. (1997) Surface Energies and Thermodynamic Phase Stability in Nanocrystalline Aluminas, *Science*, 277 788-791.

[2] McHale, J.M., Navrotsky, A., and Perrotta, A.J. (1996) Effects of Increased Surface Area and Chemisorbed H_2O on the Relative Stability of Nanocrystalline γ-Al_2O_3 and α-Al_2O_3, *J. Phys. Chem.*, in press.

[3] Tsuchida, T. and Kodaira, K. (1990) Hydrothermal Synthesis and Characterization of Diaspore, β-$Al_2O_3 \cdot H_2O$, *J. Mater. Sci.*, 25 [10] 4423-4426.

[4] Wefers, K. and Misra, C. (1987) Oxides and Hydroxides of Aluminum, *Alcoa Technical Paper 19, Rev.* Alcoa Technical Center, p. 47.

[5] Hirayama, T. (1987) High-Temperature Characteristics of Transition Al_2O_3 Powder with Ultrafine Spherical Particles, *J. Am. Ceram. Soc.*, 70 [6] C-122-124.

[6] Warble, C.E. (1985) Surface Structure of Spherical Gamma-Alumina, *J. Mater. Sci.*, 20 2512-2516.

[7] Borsela, E., Botti, S., Giorgi, R., Martelli, S., Turtu, S., and Zappa, G. (1993) Laser-driven Synthesis of Nanocrystalline Alumina Powders from Gas-Phase Precursors, *Appl. Phys. Lett.*, 63 [10] 1345-1347.

[8] Johnston, G.P., Muenchausen, R., Smith, D.M., Fahrenholtz, W., and Foltyn, S. (1992) Reactive Laser Ablation Synthesis of Nanosize Alumina Powder, *J. Am. Ceram. Soc.*, 75 [12] 3293-3298.

[9] Herring, C. (1950) Effect of Change of Scale on Sintering Phenomena, *J. Appl. Phys.*, 21 301-303.

[10] Yarbrough, W.A. and Roy, R. (1987) Microstructural Evolution in Sintering of AlOOH Gels, *J. Mater. Res.*, 2 [4] 494-515.

[11] Messing, G.L. and Huling, J.C. (1993) Transformation, Microstructure Development and Sintering in Nucleated Alumina Gels, in P. Duran and J.F. Fernandez (eds.), Proc. 3rd Europ. Ceram. Soc. Conf. Vol. 1 *Processing of Ceramics* Faenze Editrice Iberica S.L. pp. 669-679.

[12] Xue, L.A. and Chen, I.-W. (1992) Influence of Chemical Additives on the γ-to-α Transformation of Alumina, *J. Mater. Sci. Lett.* 11 [8] 443-445.

[13] McArdle, J., Messing, G., Tietz, L., and Carter, C. (1989) Solid-Phase Epitaxy of Boehmite-Derived α-Alumina on Hematite Seed Crystals, *J. Am. Ceram. Soc.*, 72 [5] 864-867.

[14] McArdle, J. and Messing, G. (1993) Transformation, Microstructure Development, and Densification in α-Fe_2O_3-Seeded Boehmite-Derived Alumina, *J. Am. Ceram. Soc.*, 76 [1] 214-222.

[15] Kumagai, M. and Messing, G.L. (1984) Enhanced Densification of Boehmite Sol-Gels by α-Alumina Seeding, *J. Am. Ceram. Soc.*, C-230-231.

[16] Kumagai, M. and Messing, G.L. (1985) Controlled Transformation and Sintering of a Boehmite Sol-Gel by α-Alumina Seeding, *J. Am. Ceram. Soc.*, 68 [9] 500-505.

[17] Shelleman, R.A., Messing, G.L., and Kumagai, M. (1986) Alpha Alumina Transformation in Seeded Boehmite Gels, *J. Non-Cryst. Solids*, 82 277-285.

[18] Shelleman, R.A. and Messing, G.L. (1988) Liquid-Phase-Assisted Transformation of Seeded γ-Alumina, *J. Am. Ceram. Soc.*, 71 [5] 317-322.

[19] Pach, L., Roy, R., and Komarneni, S. (1990) Nucleation of Alpha Alumina in Boehmite Gel, *J. Mater. Res.*, 5 [2] 278-285.

[20] Prouzet, E., Fargeot, D., and Baumard, J.F. (1990) Sintering of Boehmite-Derived Transition Alumina Seeded with Corundum, *J. Mater. Sci. Lett.*, 9 779-781.

[21] Kilbride, I.P. and Barker, A.J. (1994) Enhanced Densification by Seeding of Extruded Boehmite Gels Derived by Hydrothermal Decomposition of Basic Aluminium Acetate, *Brit. Ceram. Trans.*, **93** [5] 187-191.

[22] Messing, G.L. and Kumagai, M. (1994) Low-Temperature Sintering of α-Alumina-Seeded Boehmite Gels, *Am. Ceram. Soc. Bull.*, **73** [10] 88-91.

[23] Nordahl, C.S. and Messing, G.L. (1996) Transformation and Densification of Nanocrystalline θ-Alumina during Sinter Forging, *J. Am. Ceram. Soc.*, **79** [12] 3149-3154.

[24] Kwon, O., Nordahl, C.S., and Messing, G.L. (1995) Submicrometer Transparent Alumina by Sinter Forging Seeded γ-Al_2O_3 Powders, *J. Am. Ceram. Soc.*, **78** [2] 491-494.

[25] Pach, L., Kovalik, S., Majling, J., and Kozankova, J. (1993) Effect of Pressure on α-Alumina Nucleation in Boehmite Gel, *J. Eur. Ceram. Soc.*, **12** 249-255.

[26] Nishio, T. and Fujiki, Y. (1994) Phase Transformation Kinetics of Precursor Gel to α-Alumina, *J. Mater. Sci.*, **29** [13] 3408-3414.

[27] Ishitobi, Y., Shimada, M., and Koizumi, M. (1979) Sintering of Dense Alumina by Direct Transformation from Eta to Alpha Al_2O_3 Under High Pressure, *Proc. Round Table Meet. Spec.*, 113-133.

[28] Beauchamp, E.K. and Carr, M.J. (1990) Kinetics of Phase Change in Explosively Shock-Treated Alumina, *J. Am. Ceram. Soc.*, **73** [1] 49-53.

[29] Andryushkova, O.V., Ushakov, V.A., Kryukova, G.N., Kirichenko, O.A., and Poluboyarov, V.A. (1996) Solid Phase Transformation of Mechanically Activated Alumina During Thermal Treatment, *Chem. Sust. Dev.*, **4** 15-26.

[30] Kacsalova, L. (1979) Transformation of Bayerite into α-Al_2O_3 Under Mechanical Impact, *Acta Chim. Academ. Sci. Hung.*, **99** [2] 115-120.

[31] Panchula, M.L. and Ying, J.Y. (1996) Mechanical Synthesis of Nanocrystalline α-Al_2O_3 Seeds for Enhanced Transformation Kinetics, *Nanostruct. Mater.*, **9** [1-8] 161-164.

[32] Tonejc, A., Kosanovic, C., Stubicar, M., Tonejc, A.M., Subotic, B., and Smit, I. (1994) Equivalence of Ball Milling and Thermal Treatment for Phase Transitions in the Al_2O_3 System, *J. Alloys Comp.*, **204** L1-3.

[33] Tonejc, A., Stubicar, M., Tonejc, A.M., Kosanovic, K., Subotic, B., and Smit, I. (1994) Transformation of γ-AlOOH (Boehmite) and Al(OH)₃ (Gibbsite) to α-Al_2O_3 (Corundum) Induced by High Energy Ball Milling, *J. Mater. Sci. Lett.*, **13** [7] 519-520.

[34] Zielinski, P.A., Schulz, R., Kaliaguine, S., and Van Neste, A. (1993) Structural Transformations of Alumina by High Energy Ball Milling, *J. Mater. Res.*, **8** [11] 2985-2992.

[35] Zdujic, M.V., Milosevic, O.B., and Karanovic, Lj.C. (1992) Mechanochemical Treatment of ZnO and Al_2O_3 Powders by Ball Milling, *Mater. Lett.*, **13** 125-129.

[36] Mehta, S.K., Kalsotra, A., and Murat, M. (1992) A New Approach to Phase Transformations in Gibbsite: The Role of Crystallinity, *Therm. Acta*, **205** 191-203.

[37] Mackenzie, K.J.D. and Hosseini, G. (1976) Effect of Electric Fields on the Transformation of Gamma to Alpha Alumina, *Trans. J. Brit. Ceram. Soc.*, **77** [6] 172-176.

NANOSTRUCTURED MATERIALS FOR GAS-REACTIVE APPLICATIONS

V. Provenzano

Physical Metallurgy Branch,
Materials Science and Technology Division
Naval Research Laboratory, Washington, DC   20375-5343, USA

Abstract

Nanostructured materials with their small grain size, large number of grain boundaries and surfaces, together with their strong reactivity with gaseous species, are very interesting materials to study and develop for gas-reactive applications. These applications include: gas-sensors and getters and hydrogen storage. In this chapter, the basic characteristics of nanostructured materials, especially those related to gas reactions will be presented. This will be followed by a review of the research activities on nanostructured materials in gas sensors, getters, and hydrogen storage devices. The chapter will conclude by briefly discussing the challenges and the opportunities associated with this very interesting and promising area of research.

1. Introduction

Previous studies on nanostructured metallic-based materials have clearly shown that these materials are very reactive with gaseous species even at low temperatures (1). For example, both cold pressed nanocrystalline copper or nickel produced by the inert gas condensation method absorb large amounts of oxygen when exposed to air at room temperature. On the other hand, the same coarse-grained metals are non-reactive at room temperature. The high reaction rates of these fine-grained materials with gaseous species is a direct consequence of their high specific surface area (surface area per unit mass or volume).   The high specific surface area together with the potential of controlling both the microstructure and chemistry, makes these materials very attractive for gas-reactive applications, such as: gas sensors, getters, and hydrogen storage devices. The microstructural features that may be controlled include: grain size, morphology, distribution, defect structure, concentration of alloying addition, and phases.

This chapter on nanostrucured materials for gas-reactive applications is outlined as follows. First, in a schematic and generic fashion, the principles of gas-reactive processes associated with gas sensing, gettering and hydrogen storage are briefly reviewed. Secondly, some pertinent details on mechanisms of solid state

G.M. Chow and N.I. Noskova (eds.), Nanostructured Materials, 335–359.

gas-sensors (namely, semiconducting oxides) are presented and discussed within the context of a brief background on solid state physics that pertains to the basic reactive principles between gas species and solid surfaces. These basic principles on gas-solid surface interactions will be applied to explain the fundamental operating principles of stannic oxide (tin oxide, SnO_2), a gas sensor pioneered by Taguchi about thirty years ago (7-8). This explanation includes a brief history on the development of this important class of gas-sensors. The section on sensors will conclude by briefly reviewing the recent results reported in the literature on nanocrystalline oxide gas sensors highlighting the results that were obtained on nanocrystalline tin oxide sensors and emphasizing the promising trends and potential for improved sensing properties through the use of the nanostructured materials.

The next two sections of this lecture will be devoted to nanostructured materials for gas getters and hydrogen storage devices. The presentation of these latter two topics will be similar to one employed for gas-sensors. As in the previous section, the intent here is to emphasize the promising property trends observed nanocrystalline devices together with the potential for further property improvement and optimization.

The lecture will be concluded by presenting some the research challenges and technological opportunities offered by nanostructured materials for gas-reactive applications. These challenges and opportunities are given in terms of suggested future research and development efforts directed in realizing the full potential of the nanoscale materials in the area of gas-reactive technology.

2. Background on Gas Reactions and Nanostructured Materials

The three gas-reactive applications that are discussed in this lecture have in common some basic features. For example, the initial step involves the interaction of gaseous species with a solid surface. That is, the gas molecules impinge on the surface, and are then adsorbed by the surface. For the cases of getters and hydrogen storage, gas adsorption is followed by the diffusion of the gas species in the bulk of the solid where they are either absorbed preferentially at the grain boundaries (for the case of gas gettering) or form stable hydrides (for the case of hydrogen storage). On the other hand, for semiconducting gas sensors, the adsorption process, that also includes gas ionization, results in a change in the electronic state of the solid. It is the change (in the resistance or conductance of the solid) that is used as the basis for the gas sensing process. As schematically illustrated in Fig. 1, the various steps and different paths for the three gas-reactive devices are as follows. (1) The gas molecules impinge on the surface of the solid, where (2) are adsorbed and dissociate. (3) The dissociated molecules may diffuse into the bulk of the solid, where the can either remain as interstitials or substitutional or they can (4) chemically react with solid forming second phases. However, in order to become operative, the three gas reactions require some thermally activated processes. In the following sections of this lecture, details of the thermally activated processes together with the mechanisms that are associated with each gas-reactive application will be discussed. However, before discussing these mechanisms, it is useful to consider some generic properties of nanostructured materials that, when applied to the different gas-reaction applications, are expected not only to enhance the properties but also to increase our fundamental understanding of the complex gas-solid reactions that are associated with each gas reactive application device.

As stated in the introduction, nanostructured materials are characterized by small grain size and by large surface/volume ratios. A direct result of the small grain size is that in these materials a large number of grain boundaries and, consequently, a large fraction of the atoms reside at or close to the grain boundaries. The large number of grain boundaries due to the small grain size results in smaller diffusion distances, making the diffusion kinetics in nanocrystalline materials faster than those than those of their coarse-grained counterparts. Of particular interest for gas reactions is the fact that, when these materials are alloyed or doped, the solubility limits can be significantly higher than those of the corresponding coarse-grained materials (2). Nanostructured materials are quite interesting to study and attractive for developing the next generation of gas-reactive devices because of the increased solubility, faster diffusion kinetics, and the ability to closely control the doping level and defect structure.

3. Metal Oxide (Semiconducting) Gas Sensors

The economies of industrialized and developing modern societies, based on intense industrial production for a large variety of consumer goods, have created the need for mass produced simple and reliable gas sensors that can be produced at a low cost. The applications for gas sensors include the monitoring of air quality with respect of pollutant levels and/or the potential release of toxic or flammable gases such as NO_x, HS_2, CH_4, CO, and H_2 into the atmosphere. In this brief review of metal oxide gas sensors, the gas sensors based on tin oxide (SnO_2) will be used as the model to describe, discuss, and illustrate the various aspects of gas sensing. This is done for two main reasons. Tin oxide has played an important role in the development of gas sensors and was the first sensor to be commercialized. Secondly, because the bulk of the research and development of solid state gas sensors has been concentrated on this metal oxide. In the following paragraphs a synopsis of the historical development as well the basic operating principles of a solid state gas sensor are presented using tin oxide as the model material. This will be followed by a review and a discussion of the promising results recently reported in literature concerning improved gas sensitivity and selectivity that were obtained through the use of nanostructured tin oxide-based sensors. Finally, this section will conclude by presenting some of the technical challenges and opportunities offered by the these fine-grained materials to further improve the gas sensing properties.

About 40 years ago it was observed that the adsorption of a gas on the surface of a semiconductor resulted in a sizable change in its electrical resistance or conductance (3-5). Soon after this observation, it was suggested that this effect could be employed for gas detection. Seiyama and his coworkers (6) in 1962 were the first suggested to use metal oxide semiconductors as gas sensors. This sensor concept is shown schematically in Fig. 2. Also in the same year, Taguchi, using a purely empirical approach and working independently, started experimenting with metal oxides to develop a gas sensor as a gas leak detector for bottled gas (7). It took Taguchi six years of total dedication to develop the gas sensor that bears his name (7-8). In fact, in 1968, in Japan, the Taguchi Gas Sensor (TGS) based SnO_2 was commercialized by Figaro Engineering, Inc., as domestic gas leak detector. Later, this class of gas sensors was also applied to detect alcohol content and incomplete combustion in boilers, and as sensors in microwave ovens.

Figure 3 illustrates a typical tin (stannic) oxide TGS-type sensor perfected and marketed by Figaro Engineering based on the original design by Taguchi. As shown in the schematic, the sensor consists of a sintered SnO_2 sensor element on an alumina tube substrate, a heating coil inside the tube, and two electrodes and their respective lead wires for measuring the difference in the electric resistance of the sensor element between the reference atmosphere and the atmosphere containing the gas to be detected. For an reducing (deoxidizing) gas, the sensitivity, S, of the sensor is usually expressed as the ratio of the resistance of the element in the reference atmosphere, R_a, to the resistance, R_g, of the atmosphere containing the toxic or polluting gas. As it will be explained in the next section, the inverse resistance ratio is used to express the sensor sensitivity in the presence of an oxidizing gas. The reader is referred excellent review by Chiba (7) and the text by Ihokura and Watson (8) for additional details about the history of the development of the stannic oxide sensor pioneered by Taguchi and its subsequent commercialization by Figaro Engineering.

The solid state principles that are involved in the TGS-type SnO_2 gas sensor are illustrated by the schematics diagrams presented in Figs. 4-6. SnO_2 is n-type semiconductor, implying that conducts by the conduction band electrons. When a surface of the semiconductor is exposed to an air atmosphere, a bending in the energy level in the conduction band occurs due the ionization of the oxygen molecules. This gives rise to the build up of a space charge layer (refer to Fig. 4). The removal of some of the conduction electrons needed to ionize the oxygen molecules results in an increase the electrical resistance of the semiconductor. On the other hand, the presence of a deoxidizing gas such as hydrogen or carbon monoxide in the atmosphere produces a decrease in the resistance of the semiconductor due to an increase in the number of electrons in the conduction band. The change in the resistance is proportional to the amount of deoxidizing gas in the atmosphere. The opposite effect (an increase in the semiconductor resistance) takes place if an oxidizing gas such as nitrous oxide (NO_x) is present in the atmosphere. This is because the ionization of the oxidizing gas results in an additional removal of electrons from the conduction band. As in the case of the deoxidizing gas, the change (the increase) in the electrical resistance is proportional to the concentration of the oxidizing gas.

As previously stated, solid state gas sensors make it possible to detect the presence of polluting or toxic gases in the atmosphere and to quantify their concentration by monitoring (measuring) the changes in the resistance relative to the resistance value of the sensing element in a "standard" atmosphere (usually, relative to a dry atmosphere) at a given operating temperature. Some of the challenges in sensor research and development concern the accurate quantification of the gas, the selectivity of the sensor to distinguish different gases, its sensitivity for a particular gas, the response time, the effect of water vapor in the atmosphere, and the long term sensor stability.

The schematic diagrams shown in Figs. 5 and 6 illustrate the catalytic reaction taking place at the surface of a sensing element in the presence of a deoxidizing gas (H_2). Also, shown in this figure are the important roles played by grain size, grain boundaries and defect structure (such as oxygen vacancies) in the sensing process. For example, the average size of the grains compared to that of the Debye length (defined as the depth of the space charge layer, refer to Fig. 4), the number of grain boundaries available for the receptor and transduction functions, as well as the level of oxygen vacancies (expressed as the degree of off-stiochiometry) directly affect sensor response. This is because, as it is illustrated in Fig. 4, the number of oxygen vacancies determines the level of the energy bending, relative to that of the energy conduction value at a surface in the absence of oxygen gas (the energy barrier height). As shown in Fig. 5, the grain boundaries determine the number of sensing sites (i. e., each grain boundary can be viewed as a small sensing element), whereas the size of the grain relative to Debye length affects the energy barrier height of the charge double layer (Fig. 6).

As stated in the introduction, even though nanostructured materials offer the potential to significantly improve the properties of solid state (semiconductor metal oxide) gas sensors and provide an opportunity to gain valuable insights into the sensing mechanisms, however, to date the number of studies on this important research area has been somewhat limited. Below a brief summary of the results that have been reported in the literature involving the use of a nanostructured materials in gas sensor research and development is given. Similar to the historical development of thick film (conventional) metal oxide gas sensors, the majority of the studies in nanostructured gas sensors have also been concentrated on tin oxide. As it is discussed below, the focus of these studies has been mostly directed in examining the effects of grain, grain size stabilizers, surface doping by noble metals (Pd and Pd) on the sensor properties.

Crescenti (9), Barbi (10-11), and Rickerby (12) and their respective coworkers have examined the effects of grain size, microstructure, morphology, and catalytic metal (Pd and Pt) surface modification on the sensitivity of thin-film SnO_2 gas sensors when exposed to air atmosphere containing either carbon monoxide or nitrous oxide gas. For the most part, these investigators reported that the grain size affects the sensitivity of the sensors. However, the effect of Pt or Pd-surface modification was not conclusive. Rickerby et al. (12) have hypothesized that, the columnar texture of the sputter deposited SnO_2 films and the influence of shear planes on oxygen vacancies also have an effect on the sensitivity. All these investigators appear to agree that the activation temperature for the optimum sensitivity in the thin film sensors is 100-150°C lower than that of the corresponding thick-film, coarse-grained counterpart (9-12).

The effects of grain size on the properties of SnO_2 gas sensors was also investigated by Xu and Yamazoe and their coworkers (13-14). However, in these *latter studies, the grain size effect on properties was investigated within the broader* context of studying the complex relationships among: grain size, oxide grain stabilizers, receptor function modification by catalytic metal surface modification, and valence control by doping with substitutional elements. Furthermore, these studies related the sensor properties to the ratio of the grain size (D) to that of the Debye length, L, (depth of the space-charge layer). Generally, the sensor sensitivity increases significantly when D was comparable to 2L. In addition, D could be kept small by using oxide stabilizers, whereas the depth of the space charge layer could be controlled by Pd and Ag surface modifiers and by substitutional element doping, such Al^{3+} and Pb^{5+}, all of which alter the concentration of electron donor states in the space-charge layer.

Recently, Schweizer-Berberich and coworkers (15) have further elaborated the influence Pt and Pd surface doping on the sensing properties of nanocrystalline SnO_2 thick and thin-type gas sensors when exposed to carbon monoxide gas. This study confirmed that surface doping significantly increases the sensitivity to the same gas, in both types of sensors, However, it was observed that the largest increase in the sensitivity for the thick film sensors was due to the porous structure of the sensing element, whereas surface doping had a smaller effect on the sensitivity for the case of thin film-type sensors. These results are consistent with the results of an earlier study by Chang (16) on tin oxide thin film sensors modified by a 3.5 nm thick Pd-Au vapor deposited layer (16). Chang reported that the surface modified sensors could detect the presence of C_3H_6 or H_2 in an air environment in concentrations as low as 10ppm.

In the past, the properties of thin films tin oxide sensors produced either by sputtering or by evaporation have been studied by a number investigators (17-21). The majority of these studies were directed at examining the role microstructure, morphology, deposition techniques, deposition parameters, surface doping by catalytic metal layers and post-deposition oxidation treatment on sensor properties. Even though these studies were not explicitly directed in examining the effects of the nanoscale microstructure on properties, they indirectly dealt with some the aspects of the nanoscale structure since the above deposition techniques normally results in nanoscale microstructures. An excellent review article by Gopel and Schierbaum on SnO_2 sensors, published in 1995 (22), includes, among other aspects, a brief survey of the basic scientific issues that are associated with the use of nanocrystalline materials for improving both the properties and the fundamental understanding of the receptor and transduction mechanisms of gas sensors. Finally, recent studies by Jiang (23-24), Sun (25), and Chadwick (26) and their respective coworkers focused on investigating the gas properties of different nanostructured metal oxide gas sensors such as: $Fe_2O_{3x}-SnO_{2(1-x)}$ nanocomposites and nanocrystalline ZnO and Fe_2O_3 that were synthesized by different techniques, such as mechanical alloying and sol-gel techniques. These investigators have confirmed and further elaborated the results and observations that have been reported by the earlier studies on SnO_2-based nanocrystalline gas sensors. That is, the synthesis and processing methods and the resulting microstructure, grain size, defect structure, morphology, and doping, all affect the sensor response to a varying degree.

4. Getters

Another technological area in which nanostructured materials are very promising is the area of gas getters. However, before presenting the limited experimental results that have been recently obtained on nanostructured gas getters, it is instructive to first define and describe gas getters, briefly review their history and their many uses, especially in the area of vacuum technology. According to the comprehensive review article by della Porta (27) on the historical and technical development of getters, the term "getter" refers to materials that chemically absorb (sorb) active gases, such as oxygen, hydrogen, water vapor in a vacuum environment. In the same review article della Porta makes clear the distinction between two classes of getters: evaporable and non-evaporable getters. In the late '40's and early '50's both SAES Getters, Inc., in Italy and Telefunken, in Germany, pioneered the development of barium-based evaporable getters for CRTs and for electron vacuum tubes. In the evaporable-type getters the gas molecules are absorbed by the evaporated barium film when the getter is activated by heating. The use of evaporable getters reached its peak production in the '60's with the widespread use of radio and television receivers, when, for example, each TV set required more than 30 electron tubes. However, due to some limitations with evaporable getters (for example, incompatibility of Ba-based getters for some applications, geometric constraints, higher operating temperature and voltage) combined with the development of the transistor and its use in many electronic applications, there arose the need for a more efficient and versatile getter design.

Starting in the late '50's, this need led to the development of the non-evaporable getter. At the present time non-evaporable getters are used in a variety of applications. These applications include: vacuum systems, CRTs, X-ray tubes, lamps, gas handling and purification, semiconductor manufacturing, and other industrial applications (27). Since its conception and early development, the non-evaporable getter has been constantly modified and adapted for new applications, and this development is still ongoing. In the following paragraphs, the operating principles of the evaporable getter will be reviewed. This will be followed by a discussion concerning the use of nanostructured materials for improving the getter properties. This section will then conclude by presenting some initial and preliminary results obtained on nanostructured getters.

The basic principles of the non-evaporable getter are illustrated in the schematic diagrams presented in Figs. 7 and 8. The basic components of a non-evaporable getter are constituted by a porous zirconium-based alloy getter element (pellet) to which a small resistor heater is attached (Fig. 8). The getter assembly (pellet plus heater) is attached to a vacuum device where it is ready to "get" (absorbed) the active gases that may be present. For activation, the pellet is heated by the heating element to a set temperature that ranges from 250 to 700°C. It is hypothesized that during the activation process, the passivating oxide layer formed at the grain boundaries of the getter alloy diffuses inside the crystallite cores leaving fresh grain boundary surfaces free to absorbed the active gases (27). The redistribution of the grain boundary oxides inside the crystallite cores is schematically shown in Fig. 8B. The important parameters for getter efficiency are: getter capacity for a given volume, activation temperature and time, absorption speed, and reversibility. For improved gettering characteristics, it is highly desirable to have: higher capacity, lower activation temperature and time, higher absorption speed, and faster reversibility. Nanostructured materials hold the promise to be able to greatly improve the properties of getters. What follows is a summary of some initial results obtained at the Naval Research Laboratory (NRL) using a nanostructured approach both to improve the gettering properties and to gain a better understanding of the gettering mechanisms. This research effort at NRL is being performed in close collaboration with SAES Getters, Inc., in Milan and the University of Bologna, Italy.

Two separate synthesis and processing techniques have been used to prepare the nanocrystalline samples for the gettering study: inert gas condensation method (magnetron sputtering or PVD) and high vacuum ball milling. At NRL, nanocrystalline zirconium-nickel powders produced by magnetron sputtering were first deposited onto a nitrogen-cooled rotating stainless steel drum ('the cold finger'), scraped and then cold compacted in-situ without breaking vacuum. The getter response of the compacted samples were measured at SAES Getters. The microstructural features and gettering results obtained from the nanocrystalline samples are shown in Figs. 9 and 10, respectively. The SEM micrographs presented in Fig. 9 show a fine-grained porous structure that is typical of vapor deposited cold pressed nanocrystalline metals. The nanoscale structure is consistent with the accompanying X-ray diffraction spectrum also included in Fig. 9C. The gettering data of Fig. 10 show that for one of the vapor-deposited nanocrystalline samples the gettering response (speed and sorbed gas capacity) to the CO test gas is enhanced relative to the conventional ST707 coarse-grained getter at the higher activation temperature ($300^{\circ}C$) but it is smaller at the lower activation temperature ($250^{\circ}C$), whereas the response of the second sample is smaller than the conventional getter for both activation temperatures. It is believed that significant oxidation of the nanocrystalline zirconium-nickel samples due to handling after the cold compaction and during the transport to Italy, is primarily responsible for the mixed results obtained in the nanocrystalline samples. In fact, as confirmed from X-ray diffraction analysis, although to a varying degree, the two nanocrystalline gettering samples were heavily oxidized.

As previously discussed, nanostructured metals are very promising for significantly improving the properties of getters. However, for reasons that were evidenced by the results presented above, a serious technical challenge must be overcome before the potential of the nanoscale getters can be fully realized. This challenge is the protection of the nanostructured getter before it is used in an actual application. That is to say, a passivation scheme must be developed in order to protect the getter from being contaminated, mostly through oxidation. Before the nanocrystalline zirconium-nickel samples were sent to SAES Getters for evaluation, they were exposed to nitrogen gas. The exposure to nitrogen was used to protect the surface of the zirconium-nickel samples. The concept of using N_2 as a passivating atmosphere was based on previous studies with vapor-deposited nanocrystalline copper (28). Cold compacted nanocrystalline copper pellets that had previously absorbed a large amount of gas (mostly oxygen) after only a short time exposure (less than 30 minutes) to air at room temperature, became completely passivated if, prior to air exposure, they were treated in nitrogen for about an hour (28). It is believed that in nanocrystalline copper, nitrogen forms an oxynitride self-protecting (passivating) surface layer. The gettering data discussed above suggests that the nitrogen treatment is not as effective in protecting zirconium-nickel as it is in protecting copper.

In order to solve the oxidation problem associated with nanocrystalline zirconium-based getter materials, a different processing approach was employed. Pre-alloyed (ST707 alloy) micron size powder was first ball-milled and then consolidated at the University of Bologna under high vacuum. The alloy is mostly zirconium with small amounts of vanadium and iron (27). The resulting grain size of the ball-milled samples was controlled by carefully varying the milling parameters, such as ball size and speed. The SEM micrographs presented in Fig. 11 typify the microstructural evolution of the ST707 powder as a function of milling time. As it may be seen in the micrographs, both the crystallite size and the size of the powder agglomerates decrease with milling time. Figure 12 shows the gettering response of two ST707 ball-milled samples that were milled for five and nine hours, respectively. The pumping speed and getter capacity was quite good for both samples. However, the sample that was ball-milled for nine hours exhibited the better response than the one that had been milled for five hours. Current efforts on this research are being focused in varying the ball-milling parameters in order to optimize the gettering properties.

5. Hydrogen Storage

The last section of this book chapter deals with potential use of nanostructured materials for hydrogen storage applications. The characteristics of nanostructured materials that make them very attractive for hydrogen storage are similar to those already discussed in association with other two gas-reactive applications (small grain size resulting in large surface to volume ratios, short diffusion distances, and the ability to engineer the material properties by controlling the morphology, defect structure and doping). Therefore, with regard to hydrogen storage devices, nanostructured materials offer the potential for improving the following properties: higher storage capacities, lower activation temperatures, and faster storage and retrieval rates. In this review, the experimental results, for this third gas-reactive application, are first presented and then briefly discussed. These results were obtained within the context a an ongoing larger research effort being conducted at the Naval Research Laboratory on hydrogen storage both for naval and dual-use applications.

However, before presenting and discussing the NRL results, it is useful to go over the basic features of a hydrogen storage device, as shown in the schematic diagram presented in Fig. 13. Hydrogen molecules impinge and dissociate at the surface, diffuse through the bulk where they form metal hydrides. The desirable properties of such a device have been listed in the previous paragraph.

The following is a synopsis of the NRL effort to date on nanocrystalline materials for hydrogen storage applications. The SEM micrographs and X-ray diffraction spectra in reported in ref. 30 described the microstructure and phases of sputtered deposited Mg-50Ni (in wt. %) nanocrystalline samples before and after the hydriding treatment . The as-deposited material also exhibited a porous, nano-size grain structure. After the hydriding treatment, the material underwent some grain growth but the structure was still in the nanoscale regime and different hydride phases were present together with a small magnesium oxide peak. In order to examine the effects of processing on the hydrogen storage properties of nanocrystalline magnesium-nickel, some of the samples were also prepared by ball milling using mineral spirit as the milling medium. As in the case of the sputtered deposited samples, the microstructure of the as-milled samples is nanoscale and it is retained after hydriding (However, the microstructure in the milled material is not as fine as the one of sputtered deposited samples. The corresponding X-ray diffraction spectra showed the two elemental peaks, Mg and Ni, and an intermetallic phase in the as-milled condition and two separate hydride phases after hydriding treatment (30-32). However, contrary to what had been observed and in the sputtered deposited samples, no oxide phases were present in the hydrided samples. The stability of the ball-milled material against oxidation it believed to be the result of the protective effect of the mineral spirit and the coarser grain structure in the ball milled samples. The hydrogen capacity of nanocrystalline magnesium as a function of processing and nickel concentration has been reported and discussed in ref. 32. These investigators reported that the capacity of the ball-milled samples with low amounts of nickel are close to theoretical maximum, whereas the capacity of the samples prepared by sputtering are either lower or follow the theoretical curve. The generally higher capacity observed in the ball-milled samples are believed to be partially due to the beneficial effect of the defect structure (primarily dislocations) that results from the milling process (31 and 32).

What has been reviewed above constitutes only a brief highlight of the results that have been obtained at NRL on nanocrystalline Mg-Ni . For additional details the reader is referred to the various published articles by Holtz and his coworkers (29-32) on this interesting research area.

6. Concluding Remarks

In the previous sections of this book chapter, a number of studies were cited and a rationale was given to illustrate the potential offered by nanostructured materials to both enhance the properties of gas-reactive devices (sensors, getters, and hydrogen storage) as well as improving the fundamental understanding on the complex mechanisms governing the different gas-solid reactions. Both in the introduction and in the background section of this chapter, the properties of nanostructured materials that are very attractive for gas-reaction applications were reviewed. Therefore, in order to avoid repetitions and also for conciseness, these properties are not reiterated here. However, nanostructured materials together with their novel and attractive properties present some unique technical and scientific challenges. In a large measure, these challenges must be successfully met before their full potential can be realized. One of these challenges is directly associated with their strong reactivity with gaseous species. The mestability of these materials present also another challenge. The strong reactivity makes these materials prone to contamination, especially by oxide formation. A direct consequence of this strong reactivity is the adsorption of large quantities of gases, such as oxygen and water vapor when these material are exposed to air even at room temperature. Gas adsorption leads to contamination through the formation oxides and other compounds. Further, the metastable nature of nanostructured materials make it so that these materials, especially metallic-based materials, are not very stable against grain growth even at moderate temperatures.

The strong reactivity coupled with the inherent metastability directly affect the response of these fine-grained materials to gas reactions. For example, in getter applications, the adsorption of large quantities of oxygen and its eventual conversion to oxides, neutralizes the gettering capacity of these materials. A way must be found to protect the surfaces of a nanocrystalline getter by some passivation scheme, so that the surface of the getter can be thermally activated to perform its gettering function at the right time. The apparent passivation by exposure to nitrogen gas observed on nanocrystalline copper may be exploited in other getter alloys. Gas adsorption followed by oxidation can also adversely affect the performance of nanostructured materials in hydrogen storage devices. Oxide contamination is thought not to be a serious problem for gas sensors being that these materials are already oxides. However, gas adsorption, especially water vapor, could present a potential problem in the sensor performance. Perhaps, more of a challenge is the microstructural instability of nanocrystalline sensors, since the sensor element must be thermally activated so it can perform its designed function. Microstructural stability in these materials can be improved by the incorporation of grain size stabilizers. It would be very attractive to find some minor element additions that not only would act as grain growth inhibitors but also enhance both the catalytic and transduction properties of the sensor. The noble metal additions discussed earlier in the text might serve this purpose.

In summary, the characteristics of nanostructured materials that make them very appealing for gas-reactive applications are the same ones that present the biggest challenges. Therefore, to fully harness their potential, it is important to carefully control their properties by a number of means. The most important and interrelated parameters are: the control of microstructure and chemistry.

Acknowledgment

First, I like to thank Dr. Lynn Kurihara for reviewing the manuscript and for her help and support in the course of writing this chapter. Further, I like to thank Dr. Gan-Moog Chow for inviting me to deliver this lecture at the NATO ASI on the Science and Technology of Nanostructured Materials, held this past Summer (1997) in Saint Petersburg, Russia, that provided the basis for this chapter. I would like to also thank: Drs. Ron Holtz, of NRL, Giovanni Valdre, of the University of Bologna, and Claudio Boffito of SAES Getters, Inc., Milan, Italy, for allowing me to include in the manuscript unpublished results on nanocrystalline getters; Dr. Michel Trudeau of Hydro-Quebec Research Laboratory, Montreal, Canada, for the solubility data on nanocrystalline metals.

346

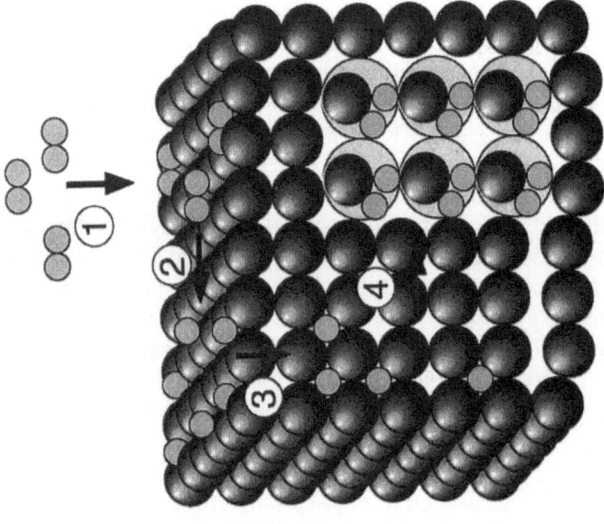

Figure 1. Schematic diagram illustrating the possible gas-solid interactions involved in: gas sensors, getters or hydrogen storage devices (steps 1-4 shown in the figure are explained in the text).

347

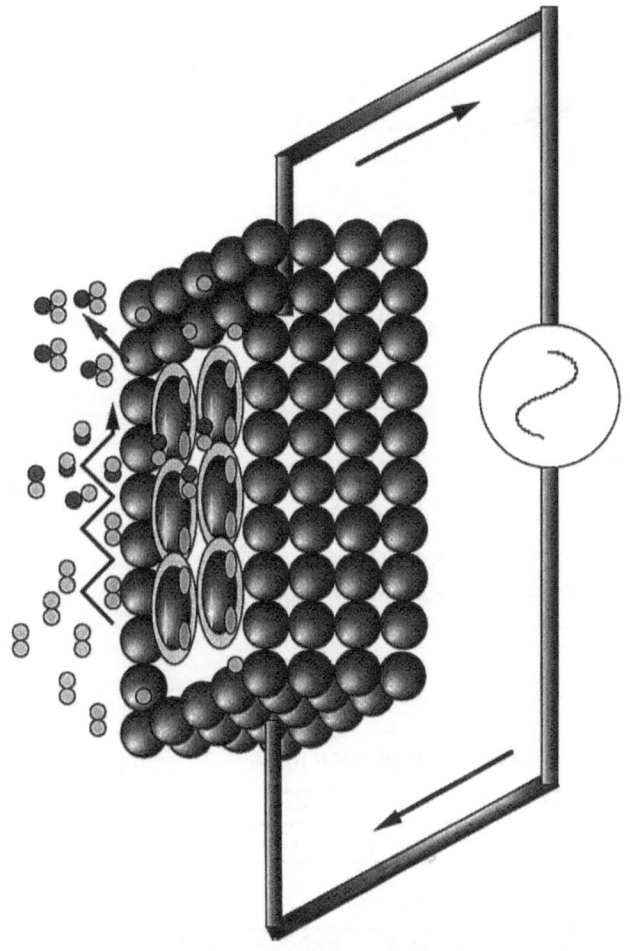

Figure 2. Schematic diagram showing the basic principles of a solid state gas sensor.

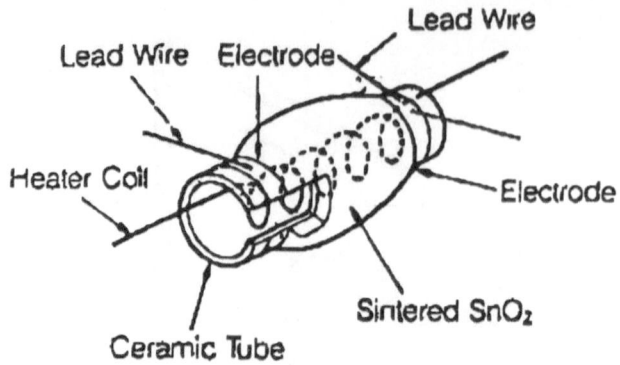

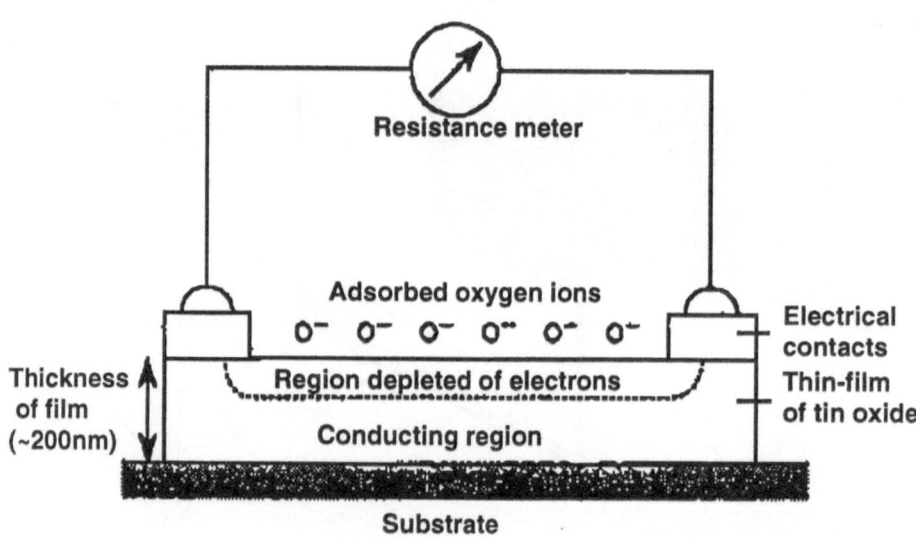

Figure 3. Schematic diagrams of thick-film (top) and thin-film (bottom) of tin oxid
semiconductor gas sensors.

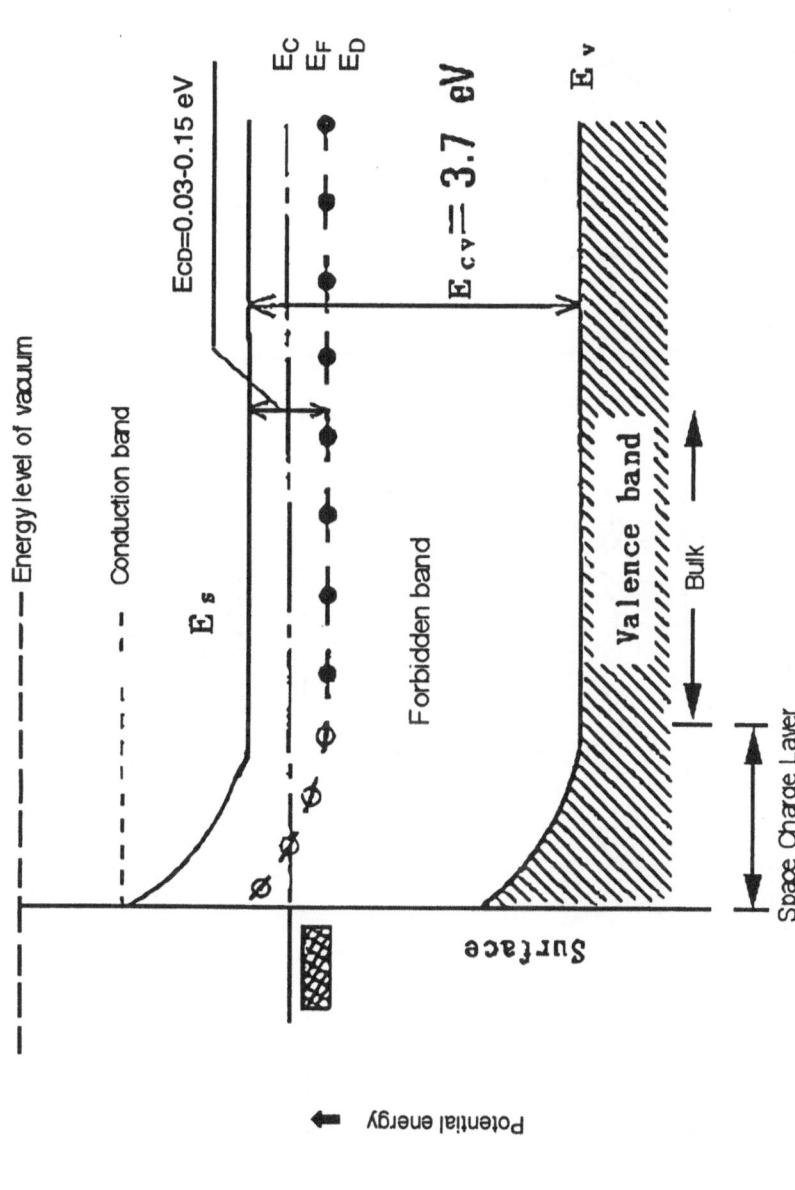

Figure 4. Band diagram for SnO_{2-x} with negatively charged adsorbed oxygen ions. Potential energy barrier, E_S; Fermi level, E_F; donor level, E_D; lowest level of conduction band, E_C; highest level of valence band, E_V; depth of donor level, E_{CD}; and, energy gap between E_C and E_V, E_{CV}.

350

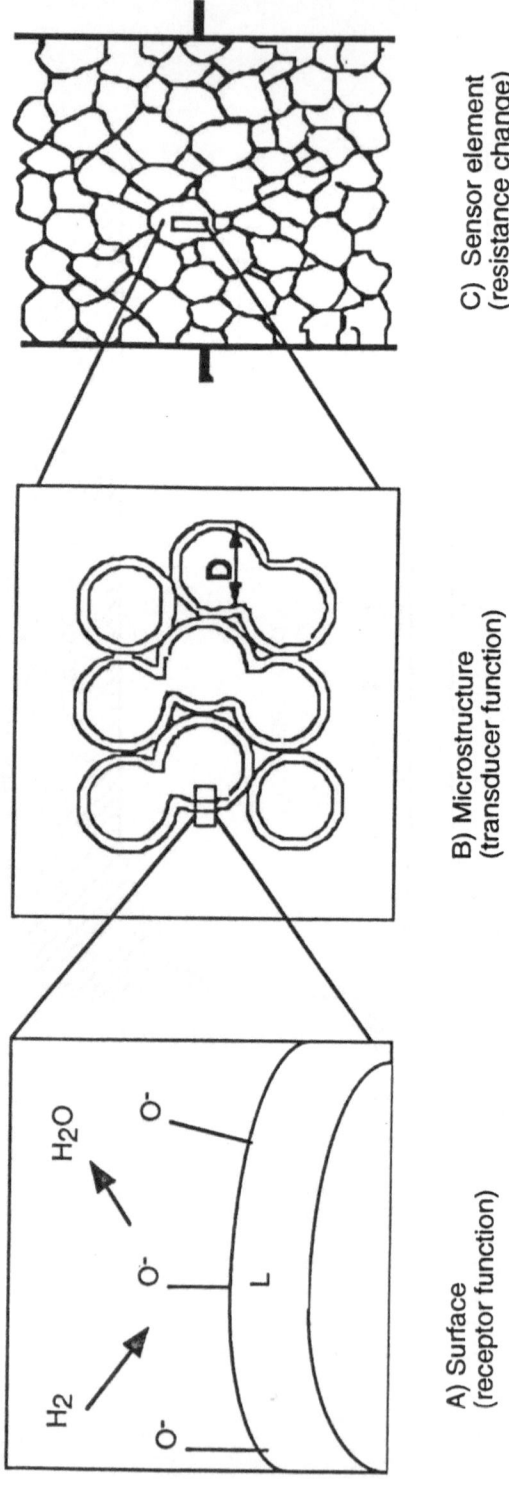

A) Surface
(receptor function)

B) Microstructure
(transducer function)

C) Sensor element
(resistance change)

Figure 5.   Schematic of receptor and transducer functions of a semiconductor gas sensor.
Particle size: D; space charge layer depth: L.

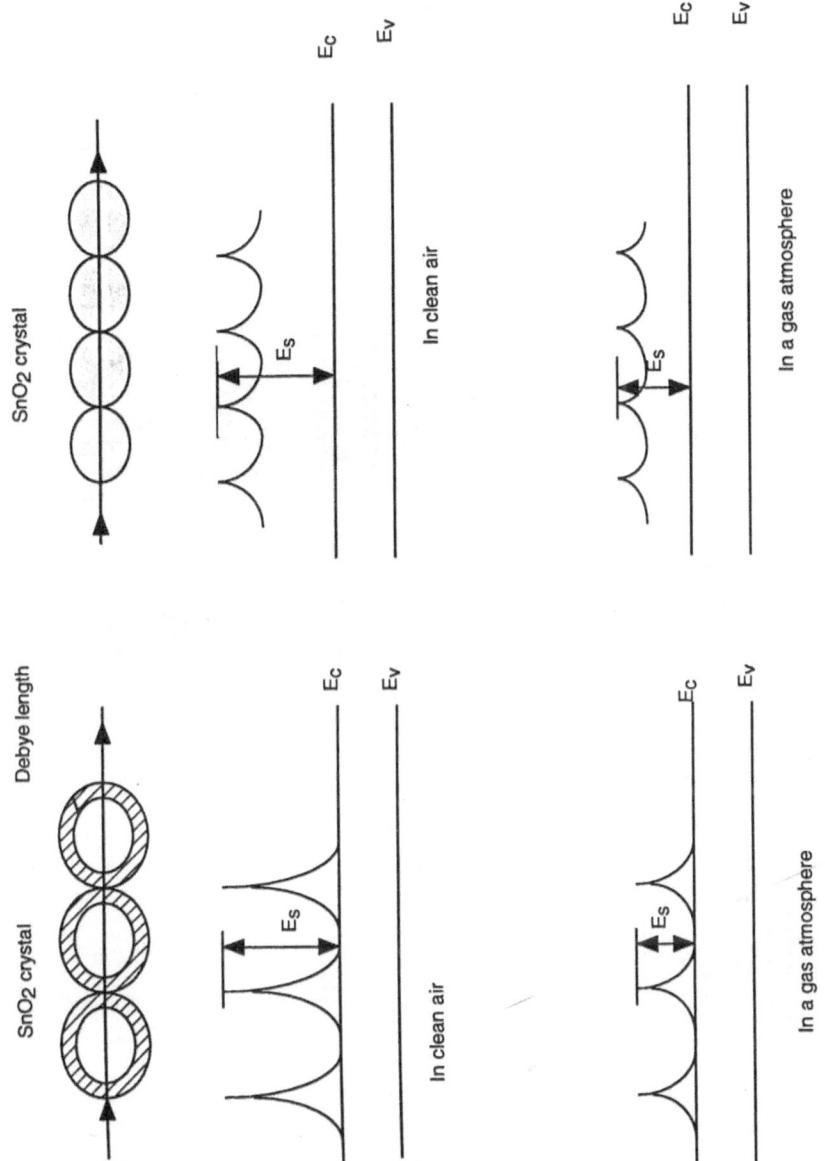

Figure 6. Schematic diagram showing barrier height variation as a function of grain size as compared to space charge layer depth.

352

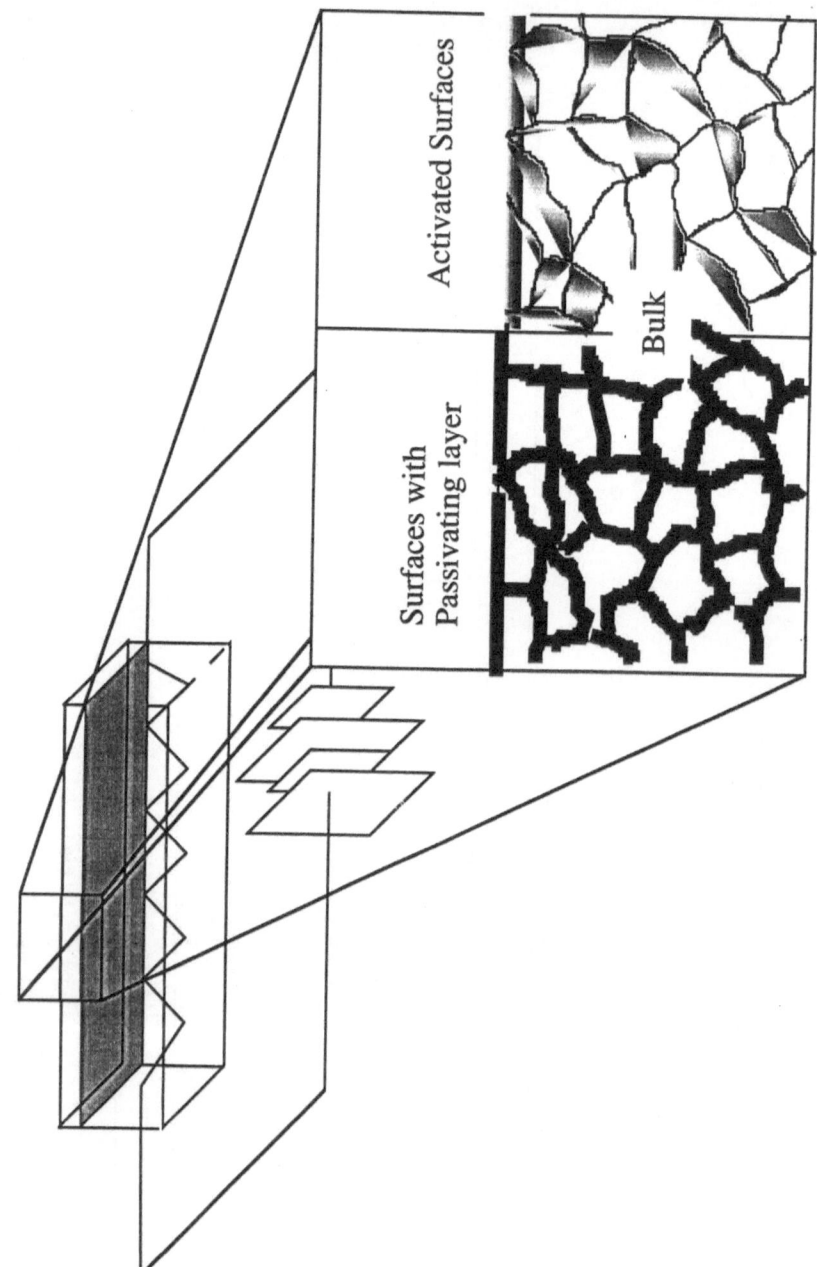

Figure 7. Shematic diagram showing the passivating and activations processes associated with the operation of a getter.

353

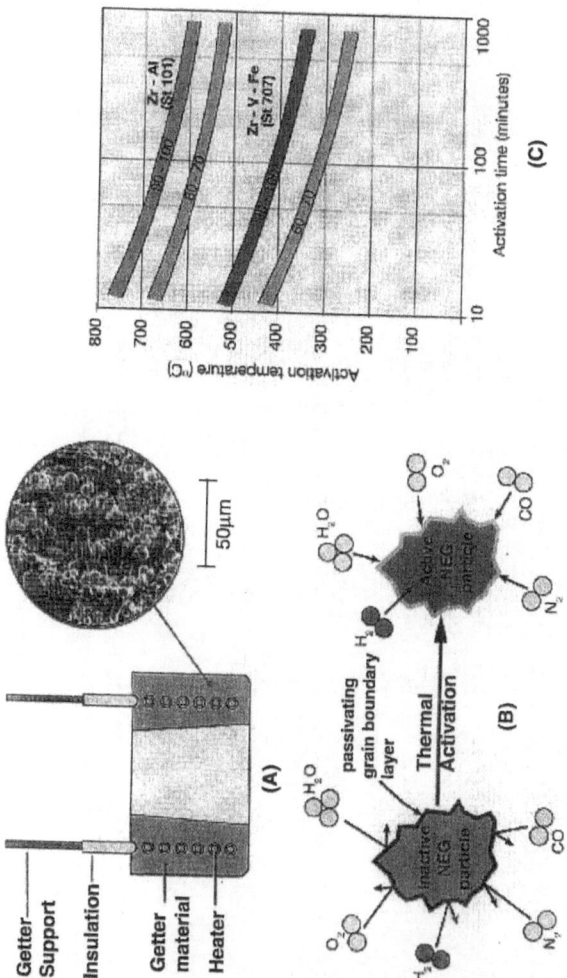

Figure 8. Schematic diagram of non-evaporable getter (A); Schematics of activation and gettering processes (B); activation efficiency (in percent of full activation) for different temperature/time combinations for ST707 and ST101 Zirconium-based alloys (C).

354

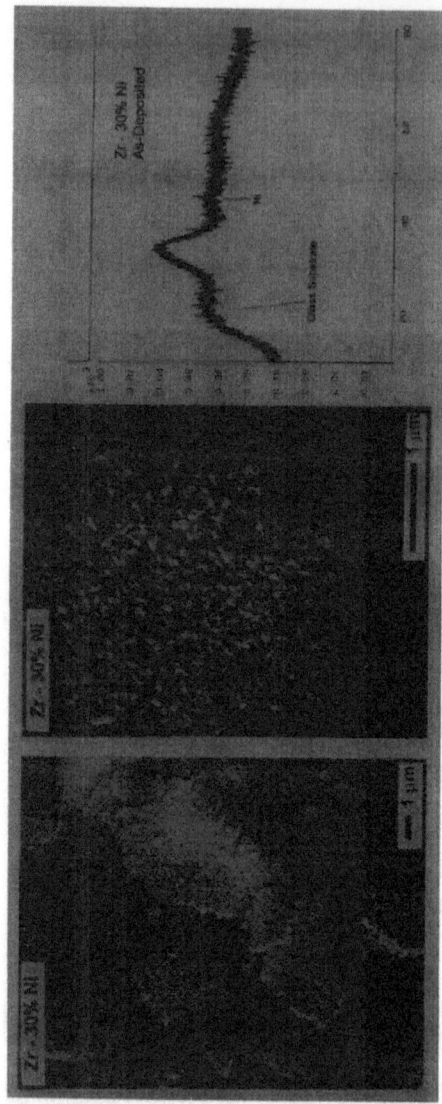

Figure 9. SEM micrographs and corresponding X-ray diffraction spectrum of cold pressed sputtered deposited gettering Zr-Ni alloy showing as- deposited microstructure and phases.

Gas absorption test performed according ASTM standard F 798-82

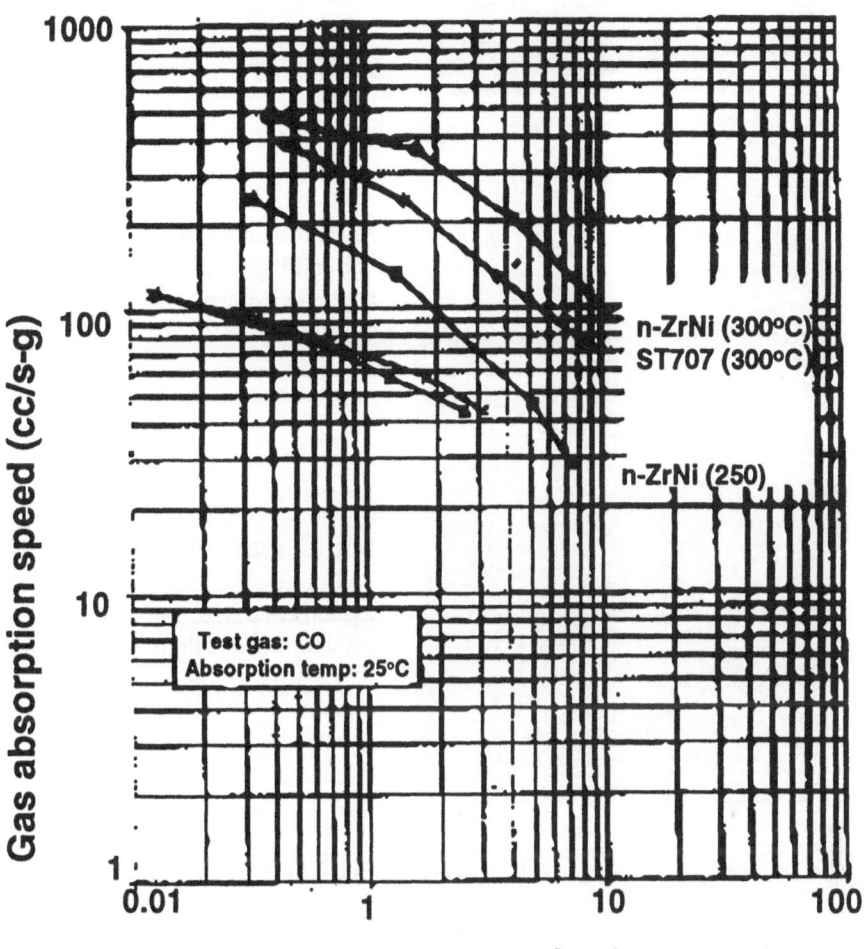

Figure 10. Gettering response of sputtered deposited nanocrystalline
zirconium-nickel compared that of ST707 getter
at two activation temperatures.

356

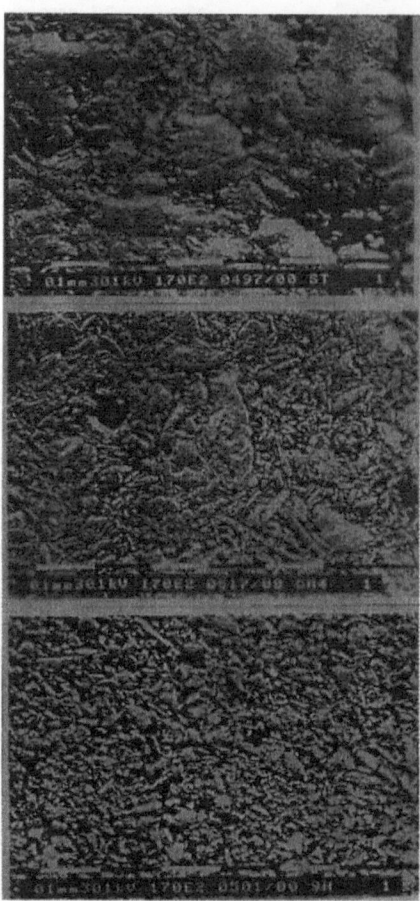

Starting microstructure
agglomerate size: 10-300μm
crystallite size: 400nm

Microstructure after
ball-milling for 5hrs
agglomerate size: 10-150μm
crystallite size: 70nm

Microstructure after
ball-milling for 9hrs
agglomerate size: 5-40μm
crystallite size: 50nm

Figure 11. Micr᠁᠁᠁᠁᠁e evolution of ST707 alloy as a function of milling time

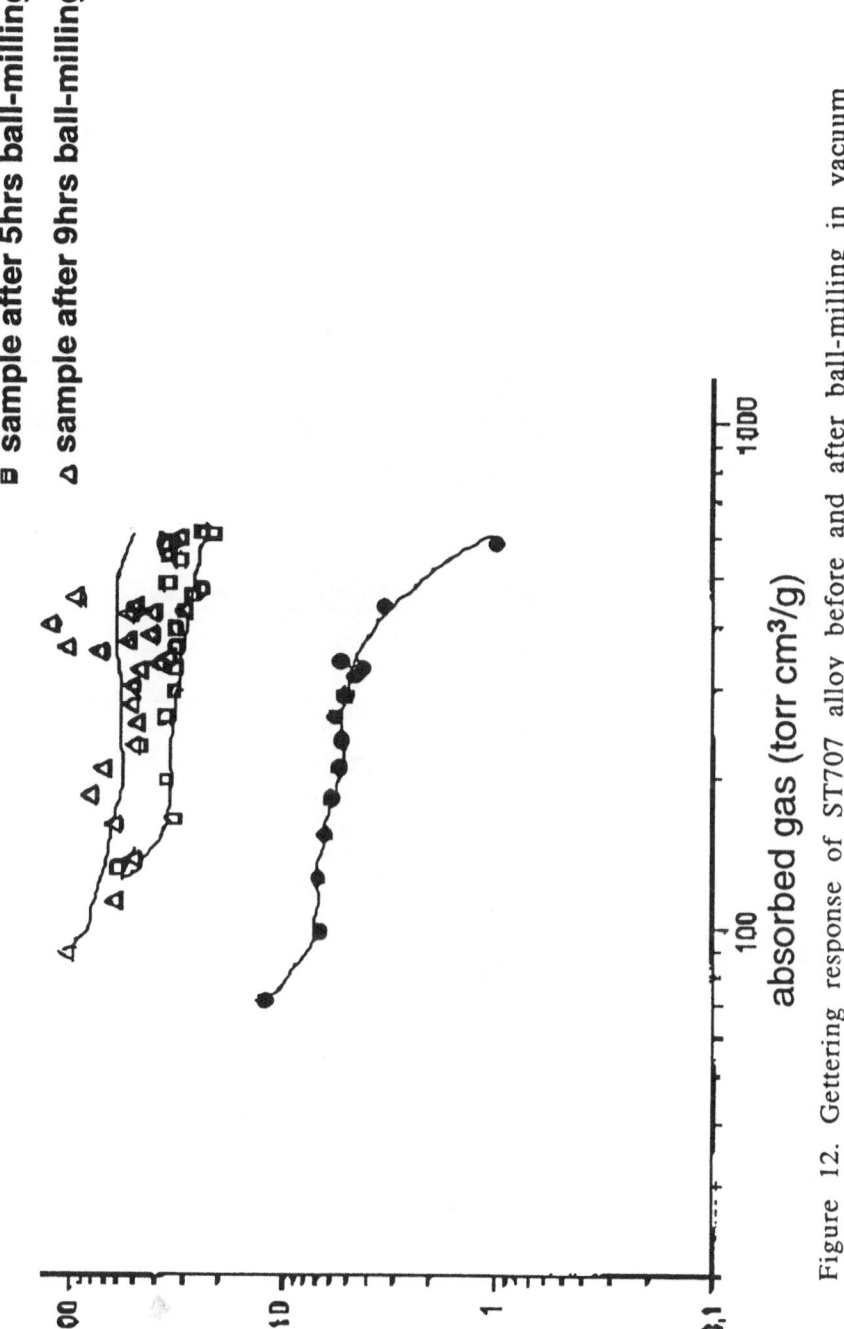

Figure 12. Gettering response of ST707 alloy before and after ball-milling in vacuum

358

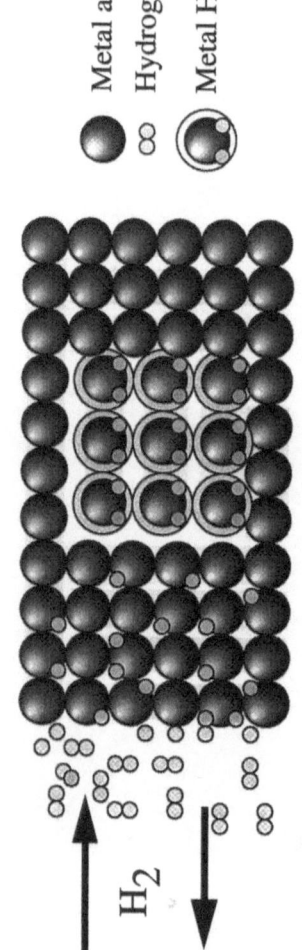

Metal atom

Hydrogen Molecule

Metal Hydride

H_2

Figure 13. Schematic diagram illustrating the basic principles of hydrogen storage.

References

1. V. Provenzano and R. L. Holtz, *Mater. Sci. Engng.* **A204** (1995) 125.
2. Michel L. Trudeau, private communication.
3. W. Brattain and J. Bardeen, *Bell Systems Technical J.* **32**, (1953) 1.
4. S. R. Morrison, J. Phys. Chem. 57 (1953) 860
5. W. S. Morrison, *Adv. Catl.* **VII** (1955) 259
6. T. Seiyama, A. Kato. K. Fujiishi and M. Nagatami, *Anal. Chem.*, **34** (1962) 1502-1503
7. A. Chiba, Chemical Sensor Technology, Vol. 4, Shigeru Yamauchi. Ed. (1992) 1-18
8. K. Ihkura and J. Wilson, *The Stannic Oxide Gas Sensor, Principles and Applications*, text published by CRP Press (1994)
9. A. Criscenti, R. Generosi, M. A. Scarselli, P. Perfetti, P. Siciliano, A. Serra, A. Tepore, C. Coluzza, J. Almeidal and G. Margaritondo, *J. Phys. D: Appl. Phys*, **29** (1996) 2235-2239.
10. G. B. Barbi and J. Santos Blanco, *Sensors and Actuators B*, **15-16** (1993) 372-378.
11. G. B. Barbi, J. P. Santos, P. Serrini, P. N. Gibson. M. C. Horrillo, and L. Manes, *Sensors and Actuators B*, **24-25** (1995) 559-563.
12. D. J. Rickerby, M. C. Horrillo, J. P. Santos and P. Serrini, , Vol. 9, **1-8** (1997) 43-52.
13. C. Hu, J. Tamaki, N. Miura and N. Yamozoe, *Sensors and Actuators B*, **3** (1991) 147-155.
14. N. Yamazoe, *Sensors and Actuators B*, **5** (1995) 7-19.
15. M. Schweizer-Berberich, J. G. Zheng, U. Weimar., W. Göpel, N. Bârsan, E. Pentia, and A. Tomescu, Sensors and Actuators B, 31 (1996) 71-75.
16. Shih-Chia Chang, J. Vac. Sci. Technol. A Vol. 1, 2 (1983) 296.
17. M. Di Giulio, G. Micocci, A. Serra, A. Tepore, R. Rella and P. Siciliano, *Sensors and Actuators B*, **24-25** (1995) 465-468.
18. M. C. Horrillo, J. Gutiérrez, L. Arés, J. I. Robla., I. Sayago, J. Getino, and J. A. Agapito, *Sensors and Actuators B*, **24-25** (1995) 507-511.
19. G. Williams and G. S. V. Coles, *Sensors and Actuators B*, **24-25** (1995) 469-473.
20. K. S. Yoo, N. W. Cho, H. S. Song and H. J. Jung, *Sensors and Actuators B*, **24-25** (1995) 474-477.
21. G. Gaggiotti, A. Galdikas, S. Kaciulis, G. Mattogno, and a. Setkus, *Sensors and Actuators B*, **24-25** (1995) 516-519.
22. W. Göpel and K. D. Schierbaum, *Sensors and Actuators B*, **26-27** (1995)1-12.
23. J. Z. Jiang, R. Lin, S. Morup, K. Nielsen, F. W. Poulsen, F. J. Berry and R. Clasen, Physical Review B, Vol. 55, 1 (1997) 11-14.
24. J. Z. Jiang, R. Lin, K. Nielsen, S. Morup, K. Dam-Johansen and R. Clasen, *J. Phys. D: Appl. Phys.*, **30** (1997) 1459-1467.
25. Hong-Tao Sun, C. Cantalini, M. Faccio and M. Pelino, *Thin Solid Films*, **269** (1995) 97-101.
26. A. V. Chadwick, N. V. Russell, A. R. Whitham and A. Wilson, Sensors and Actuators B, 18-19 (1994) 99-102.
27. P. della Porta, " *Gettering"-An Integral Part of Vacuum Technology*, Technical Paper TP 202 (1992), SAES Getters S. p. A., 215 Via Gallarate, Milano, Italy.
28. R. L. Holtz, V. Provenzano, and C. Boffito, unpublished data.
29. R. L. Holtz, V. Provenzano and M. A. Imam, *NanoStructured Materials*, Vol. 7, **1-2** (1996) 259-264.
30. R. L. Holtz and M. A. Imam, *Journal of Materials Science*, **32** (1997) 2267-2274.
31. R. L. Holtz, M. A. Imam and D. A. Meyn, Proceedings of Symposium on *Synthesis/Processing of Lightweight Materials*, Edited by F. H. Froes, C. Suryanarayana and C. M. Ward-Close, The Minerals, Metals & Materials Society (1995) 339-346.
32. R. L. Holtz and M. A. Imam, Proceedings of Symposium on *Synthesis/Processing of Lightweight Metallic MaterialsII*, eds. C. M. Ward-Close, F. H. Froes, D. J. Chellman, and S. S. Cho (The minerals, Metals & Materials Society (1997) 313-320.

NANOCRYSTALLINE CERAMICS FOR STRUCTURAL APPLICATIONS: PROCESSING AND PROPERTIES

M.J. MAYO
*Dept. Materials Science & Eng., The Pennsylvania State University,
University Park, PA 16802 USA*

ABSTRACT. Several techniques for the consolidation and sintering of nanocrystalline ceramics are reviewed. For pressureless sintering, the presence of large, interagglomerate pores in the nanoceramic powder compact is shown to be the root cause of high sintering temperatures and slow densification rates. The strong dependence of densification rate on pore size is demonstrated, with smaller pores yielding faster densification rates and higher densities at a given grain size. In contrast, variations of the ceramic's sintering schedule are generally unproductive or counterproductive with respect to altering the density-grain size trajectory. Two consolidation methods which successfully produce fully dense, nanocrystalline ceramics include dry pressing at high (1 GPa) pressures prior to pressureless sintering, or sinter-forging. Both approaches greatly reduce or eliminate the large (interagglomerate) pores present in the starting compact. Wet processing techniques, though not yet widely used for nanoparticles, also show promise in this regard, due to the minimal force required to obtain efficient particle packing. Some properties and applications of nanocrystalline ceramics are also discussed; these include sintering temperature (in the absence of agglomeration), hardness, fracture toughness, superplasticity, thermal conductivity, and diffusion bonding ability.

1. Introduction

The field of nanocrystalline materials has evolved significantly since Herbert Gleiter first popularized it a decade ago. In the mid 1980's, the most significant problem facing scientists wishing to study the properties of solids at extremely fine length scales was the scarcity of starting materials, a dilemma reflected in the conclusion of the 1988 Panel Report on Clusters and Cluster-Assembled Materials [1]:

> Present methods for the synthesis of useful amounts of size-selected clusters, with surface chemical properties, purposefully controlled and/or modified, are almost non-existent, and these fundamentally limit our ability to explore the assembly of clusters into potentially novel materials . . . Progress on macroscopic synthetic methods for size-selected clusters of controlled surface properties is the most important immediate goal recognized by the Panel.

It should be noted that, in the context of the Panel report, the term "clusters" referred primarily, though not exclusively, to nanoparticles in the 1-100 nm range — i.e., those particles which comprise the powders from which nanocrystalline materials are typically made.

G.M. Chow and N.I. Noskova (eds.), Nanostructured Materials, 361–385.
© *1998 Kluwer Academic Publishers.*

Little could the Panel know, at that time, that the vast potential of nanocrystalline materials would so intrigue the scientific community that the problem of synthesizing ultrafine powders would solve itself. Independently, more and more scientists began to tackle the synthesis problem until today, in 1997, a vast number of synthesis techniques, many of them eminently scalable, have been developed for an extensive array of powder compositions. Summary tables showing a partial fraction of these techniques can be found across several review papers [2-4]). Ceramic compositions, in particular, are well represented, and commercial production of certain nanocrystalline ceramics in the tonnage size range is already in place [5]. Thus, the problem of fabricating a nanocrystalline ceramic does not, today, rely on one's ability to obtain a powder, but rather on one's ability to manipulate that powder into a dense ceramic with a nanocrystalline (<100 nm) grain size.

2. The Problem of Agglomeration

Nanocrystalline ceramics are made primarily by the compaction and sintering of ultrafine particles. (A few nanoceramics are made by the crystallization of glasses and/or gels). A brief look at the theoretical literature on the sintering of ultrafine powders tends to be very encouraging: rules-of-thumb such as the Herring scaling law [6] would predict that nanoparticles should sinter easily, at low temperatures. For instance, in comparing two powders with grain or crystallite sizes G_2 and G_1, respectively, the Herring scaling law predicts that the sintering temperature will be lowered from T_2 to T_1 according to the relation

$$\frac{RN}{Q} \ln\left(\frac{G_2}{G_1}\right) = \frac{1}{T_1} - \frac{1}{T_2}, \tag{1}$$

which is derived from the generalized relationship between densification rate and grain size,

$$\frac{d\rho}{dt} = \frac{1}{G^N} \exp(-Q/RT). \tag{2}$$

In the above expressions N is the grain size exponent (typically 3 if the densification mechanism is surface diffusion, 4 if the densification mechanism is grain boundary diffusion), R is the gas constant, ρ is the relative density (on a scale of 0 to 1), and t is time. Predictions cited in the literature [7] using this law include a 355°C reduction in sintering in sintering temperature for a ten-fold reduction in grain size for Al_2O_3, such that Al_2O_3 with a grain size of 60 nm would sinter at the remarkably low temperature of 1145°C.

In practice, such easy sintering of nanoceramic powders is rarely achieved. More commonly, the nanocrystalline powder begins to sinter at quite low temperatures, but full density is not reached until much higher temperatures or longer times, with the result that the grains grow well beyond their intended nanometric size. As an example, Figure 1 shows the results of sintering a nanocrystalline titania powder with a starting grain size of 10 nm and a green density (density of the powder compact prior to sintering) of 46% of theoretical [8]. The powder begins sintering around 610°C, but after 6.5 hours at 610°C, the grains have already grown to 123 nm and the density is still only 79% [8]. The cause of this poor sintering behavior is also evident in Fig. 1: though the individual crystallites in the powder may be nanosized, they are clumped together so inhomogeneously that the powder compact contains micron-sized pores that are very difficult to eliminate during sintering. At low magnification, these pores appear as the spongy, mottled white regions of Fig. 1a; at higher magnification (Fig. 1b), the individual pores can be resolved. The particular powder compact shown in Fig. 1 is an extreme example of a pervasive problem in the sintering of nanocrystalline ceramics: agglomeration.

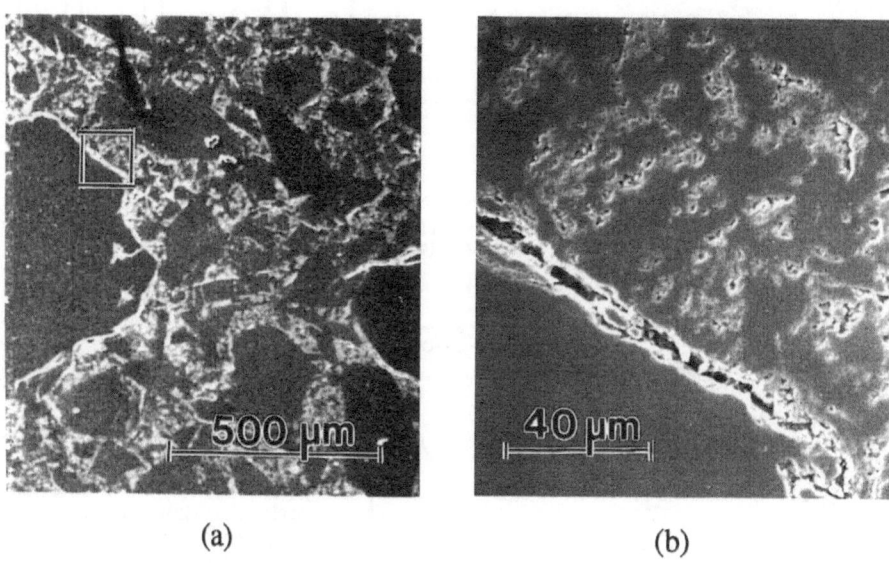

(a)   (b)

Figure 1. Scanning electron micrographs of a heavily agglomerated nanocrystalline TiO$_2$ powder after pressureless sintering at 610°C for 6.5 hours.

An agglomerate can be defined as a bonded mass or group of particles; an agglomerated powder contains many such masses. When an agglomerated powder is compacted, the hard agglomerates pack together as rigid entities, leaving behind pores on the scale of the agglomerate size (Fig. 2). If the agglomerates are micron-sized, one obtains the very large, micron-sized pores of Fig. 1 that, in turn, lead to the requirement of high sintering temperatures. Agglomerates can be detected in a powder or powder compact through a pore size distribution (Fig. 3), typically obtained by either nitrogen adsorption porosimetry or mercury porosimetry. These techniques detect the presence of open pore channels between powder particles. In a non-agglomerated powder, only one population of pores, that corresponding to the small, intercrystallite pores (also called *intra*agglomerate pores) exists. The presence of a secondary peak or multiple peaks in the pore size distribution, located at larger pore sizes, signals the presence of interagglomerate pores and hence agglomerates. It should be noted that standard porosimeter operation best reveals pores in the 1-100 nm range [9]; very large pores (and hence very large agglomerates) can be missed unless scanning electron microscopy or similar visual data are used to complement the porosimetry results.

364

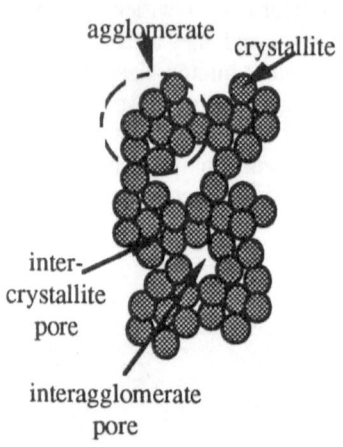

agglomerate

crystallite

inter-
crystallite
pore

interagglomerate
pore

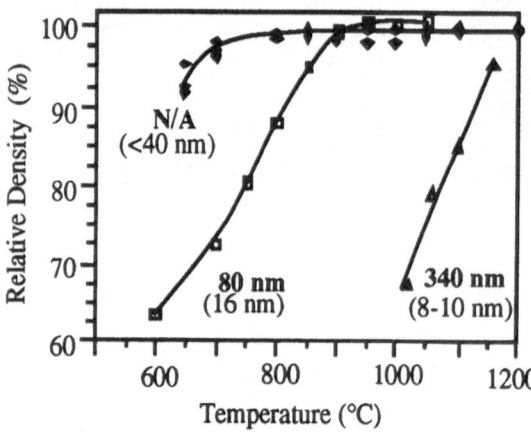

Pore Diameter (nm)

Figure 2. Schematic diagram of an
agglomerated powder

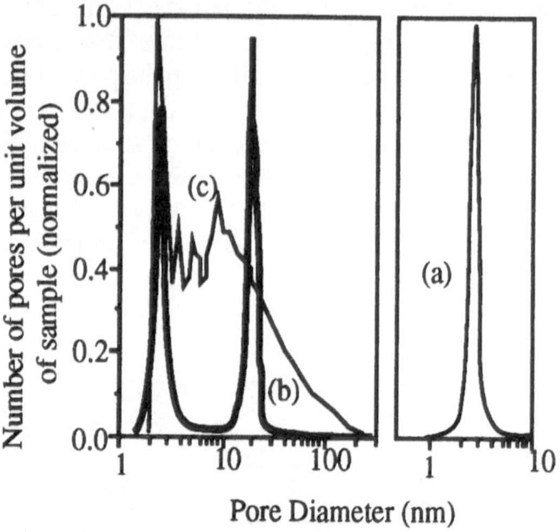

Figure 3. Schematic pore size distribution of powders that (a)
are non-agglomerated, (b) contain agglomerates of only one
size, and (c) contain agglomerates in a variety of sizes.

N/A
(<40 nm)

80 nm
(16 nm)

340 nm
(8-10 nm)

Figure 4. Densification behavior of
nanocrystalline TiO_2 (anatase
crystal structure) with three different
agglomerate sizes. Crystallite sizes
are shown in light face type;
agglomerate sizes in bold-faced
type. From L to R, the data are from
Refs. [10-12].

The extent to which agglomeration can impair sintering in nanocrystalline ceramic
powders is illustrated in Figure 4. Here, data from three nanocrystalline powders are
compared. Each of the powders is prepared by the same chemical technique (alkoxide
hydrolysis), is of the same chemical composition (TiO_2), and possesses the same
crystal structure (anatase). There are, however, large differences in the agglomerate size
of the powder, and hence, the sintering temperature. The sintering temperature increases
from about 700°C to 900°C to well over 1100°C as the agglomerate size is increased
from <40 nm (the size of the crystallites themselves, in a non-agglomerated or N/A
powder) to 80 nm to 340 nm. Note that crystallite size, shown in light type in Fig. 4,
has virtually no impact on the sintering temperatures of the powders. Thus, the most
important feature in determining sintering temperature of a nanocrystalline ceramic

powder is not the crystallite size (grain size), but rather, the agglomerate size. This distinction becomes important in evaluating nanocrystalline powders because, in practice, the term "particle" is used rather indiscriminately to refer to both agglomerates and crystallites, yet there is a large difference in sintering behavior between a powder composed of 40 nm agglomerates and one composed of 40 nm crystallites bonded together in 0.2 - 1 µm hard agglomerates.

The presence of agglomerates is one reason why the rosy predictions of the Herring Scaling Law [6] are never quite reached in practice. The Herring Scaling Law, like much of elementary sintering theory, does not take into account the presence of crystallites and agglomerates but rather assumes that all particles are perfect uniform entities. In this ideal scenario, scaling down the particle size causes all other microstructural features — notably pores — to scale down also.

The fact that pore size does not scale with grain size in agglomerated powders suggests that a refined sintering relation, which separately accounts for grain size and pore size effects, may be more appropriate for the sintering of nanocrystalline powders. There is limited evidence that this is the case; for example, Fig. 5 shows densification rate data for several lightly agglomerated nanocrystalline zirconias (white squares), compacted to different pressures and at different stages of sintering, alongside data for submicron zirconia (black diamonds) composed of 200 nm particles containing 30 nm grains. Despite the variety of powders and sintering conditions, these data are all successfully normalized to a single line once pore size and grain size are both taken into account. (It should be noted that Fig. 5 reports the instantaneous pore sizes and grain sizes measured during sintering, not the initial ones in the powder compact.)

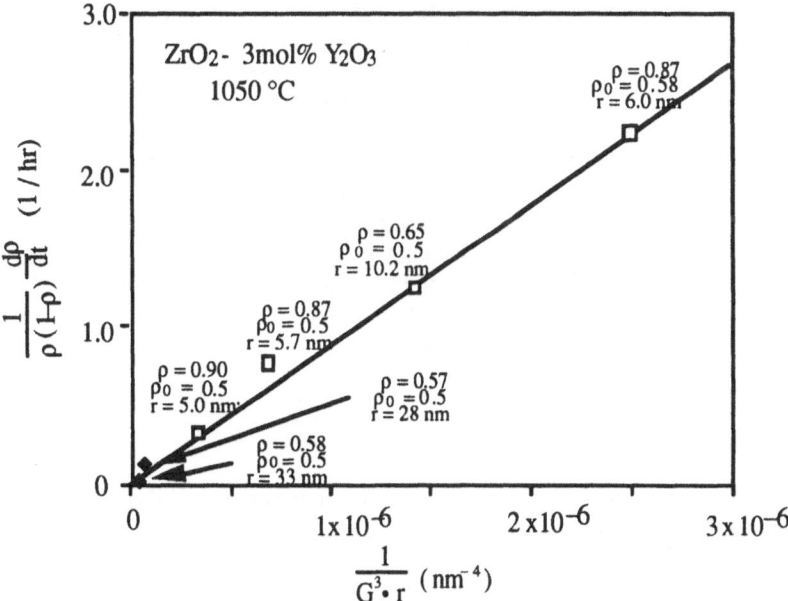

Figure 5. Correlation of the normalized densification rate with a combined grain size - pore size parameter. Data are for several nanocrystalline zirconia powders (white squares), with different green densities (ρ_0), taken at different stages of sintering; data area also shown for a commercial submicron zirconia powder (black diamonds).

If agglomeration is the primary source of sintering difficulty in nanocrystalline powders, the logical next step would be to determine the cause of agglomeration and eliminate it. Indeed, much progress has been made on this front. For example, agglomeration in aerosol processes has been attributed to the collisions of hot nanoparticles undergoing Brownian motion: when the particles are very hot, they coalesce much as molten droplets to form a single, larger particle; however, at somewhat lower temperatures, these particles are merely "sticky" and weld together to form complexes of particles, or agglomerates [13]. Suggested solutions to this problem include using a high temperature at the vapor source and quenching the vapor rapidly such that it passes quickly through the undesirable temperature region where welding of particles takes place [13].

When particles are made by liquid phase synthesis, agglomeration is largely the result of interactions involving the surface chemical species on the particle. These species can cause direct chemical bonding of the particles, as demonstrated in the case of zirconium hydroxide (a common precursor to ZrO_2): here, hydroxyl groups on adjacent particles bond to each other using the intervening water molecules as a bridge; subsequent removal of the water then leaves behind a direct oxygen bond between the particles themselves [14]. Another, more common scenario — especially in the case of nanoparticles — is that the surface of the particle reacts with the surrounding liquid to form a different species which is eminently soluble in the liquid. The surface of the particle then dissolves, and the dissolved cations and anions then reprecipitate out at interparticle contacts [15, 16]. In the case of oxide particles in water, the offending surface species is usually a hydroxide; unfortunately these hydroxides turn back into the corresponding oxide when the powder is heated. Thus, by immersing oxide particles in water under conditions which promote solubility of the surface hydroxides, one can inadvertently form extremely strong ceramic bridges between particles, once the powder is dried and calcined [16]. Not surprisingly, the large surface area associated with nanoparticles predisposes them to high solubility even at pH's where dissolution of larger particles is not a problem [15]. Also, the smaller radius of curvature associated with ultrafine particles leads to a larger driving force for the redeposition of the dissolved matter at interparticle contacts (Ostwald-Freundlich relation) [15]. In general, with oxide powders, one should try to avoid conditions which would cause the corresponding hydroxide to be soluble. Checking species stability diagrams (e.g., Ref. [16]) for pH's where the hydroxides are not soluble, using non-aqueous liquids for processing, avoiding repeated drying from solution and not storing the powder in a hot, humid environment are all reasonable precautions.

3.   Pressureless Sintering Behavior

For nanocrystalline ceramics, the principles governing sintering behavior do not appear to differ appreciably from those of conventional (submicron or micron-grained) materials. However, the extrapolation of those principles to ultrafine size scales can occasionally yield some surprising results. A case in point is rapid rate sintering. This technique, which employs ultrafast heating rates en route to the final sintering temperature, has enjoyed success in the production of dense, submicron-grained $BaTiO_3$ and related ceramics [17-19]. In theory, the fast heating rates allow one to quickly bypass an unproductive sintering regime, where densification rates are slow and grain growth rates are fast, in order to reach a more useful sintering regime where densification rates are at least on a par with grain growth rates [20]. For the case of nanocrystalline materials, one might expect favorable results from bypassing a low temperature surface diffusion regime (where grain growth occurs rapidly and densification not at all) in order to quickly reach a grain boundary or lattice-driven diffusion regimes at higher temperatures. When this approach was used in the sintering of nanocrystalline thin films, the results seemed generally positive [21]; however, application of this protocol to the sintering of bulk nanocrystalline ZrO_2-3mol% Y_2O_3 proved disastrous [22]. As Fig. 6 shows, the faster heating rates actually resulted in samples with dramatically *lower* densities than in the case of slower heating rates. Curiously, the same effect was not observed for submicron ZrO_2-3mol% Y_2O_3 (Fig. 7).

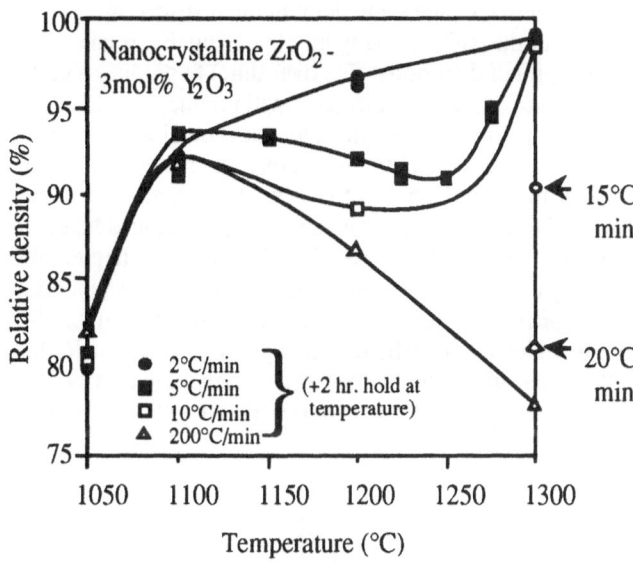

Figure 6. Densification results for the rapid rate sintering of nanocrystalline ZrO_2-3mol% Y_2O_3.

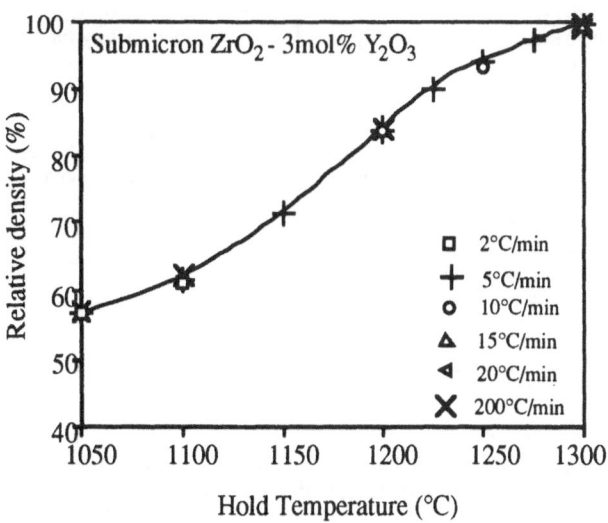

Figure 7. Densification results for the rapid rate sintering of submicron ZrO_2-3mol% Y_2O_3.

The problem was eventually traced, ironically, to the fast densification rates of the nanocrystalline material [22]. When the nanocrystalline sample experienced thermal gradients caused by the fast heating rates, the outside of the sample densified quickly and thoroughly. The rigid shell thus created then mechanically constrained the interior from shrinking, even once the interior reached temperature. The slower densification rates of the submicron sample, which were comparable to the rates of heat transfer into the sample, saved it from a similar fate.

Since accelerating the heating rate is not always a viable strategy, the question arises as to whether there are other ways to modify a sintering schedule in order to enhance sintering. Changing the combination of hold time and hold temperatures — using a long hold at low temperatures vs. a short hold at higher temperatures — might seem a reasonable approach. In practice, however, if the activation energies of densification and grain growth are the same, the same point in microstructural development is reached regardless of the path used (Figure 8): essentially, it takes a certain "amount" of diffusion (integral of diffusional flux over time) to produce a certain degree of grain growth, and that same "amount" of diffusion will produce a certain degree of pore shrinkage, or densification. Whether that amount is obtained by waiting a long time at low temperatures or a short time at higher temperatures does not matter. On a grain size-density plot, the sample will arrive at the same point in microstructural evolution, regardless (Figure 8). Two caveats, however, should be mentioned here. First, as a practical matter, it is much easier not to "overshoot" the targeted degree of microstructural development using longer time, lower temperature exposures; sintering times on the order of seconds are difficult to control. Secondly, the trend of Figure 8 applies only for the usual case in which the densification and grain growth kinetics have the same activation energy. If the densification rate has a larger activation energy than the grain growth rate, higher temperature - shorter time exposures should benefit the sample.

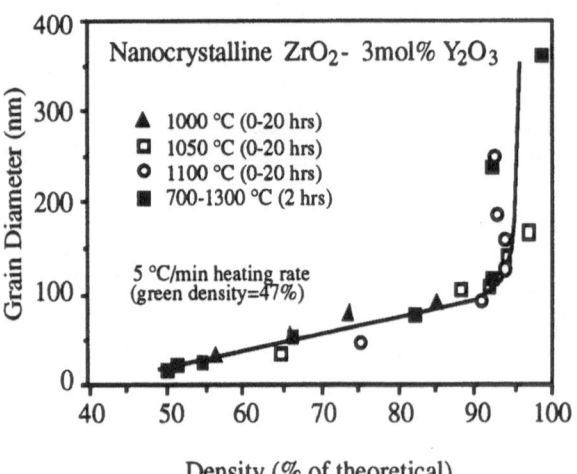

Figure 8. Grain-size density trajectory for a dry-pressed nanocrystalline ZrO_2-$3mol\%Y_2O_3$ compact.. Data for all time/ temperature combinations fall on the same curve.

The discussion so far suggests that, for a given sample, the microstructural development of that sample (its grain size - density curve) is more or less set, and is not easily influenced by the sintering schedule. This means that the only effective way to increase the sinterability of a nanocrystalline ceramic is to change the sample itself. By eliminating the large interagglomerate pores, or otherwise having the pores within the sample down at a reasonable size, it is possible to obtain a faster densification rate relative to the grain growth rate. The effectiveness of this strategy is shown in Figure 9. Here, the green (unsintered) samples have been compacted to higher and higher pressures to induce more agglomerate breakage and more particle rearrangement within the powder compact, ultimately producing pores of a smaller size and a more homogeneous (narrower) size distribution.

As can be seen in Figure 9, the increased compaction pressures prior to sintering cause the grain size-density curve to shift systematically further to the right, towards higher densities. Thus, with a high enough compaction pressure, it is possible to produce a nanocrystalline ceramic which will pressureless sinter to full density while retaining a grain size of less than 100 nm. It should be mentioned that it is not the

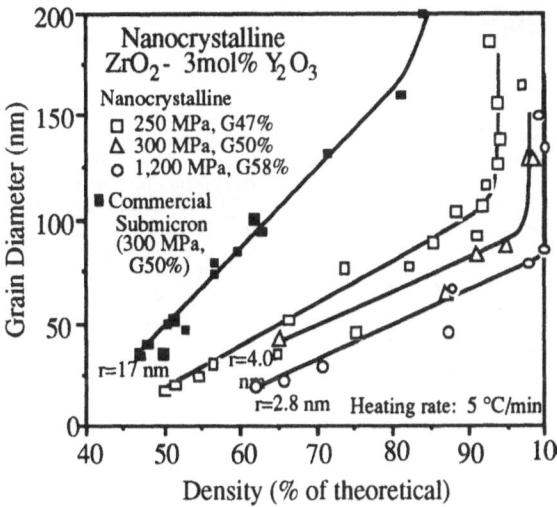

Figure 9. With smaller pores, the density grain size curves shift increasingly to the right, towards higher densities. Initial pore sizes are indicated as r values; initial (green) densities are denoted with G's (e.g., G50%), alongside the compaction pressures.

higher green density which accomplishes this feat, but rather the smaller pore size. The commercial submicron powder in Fig. 9 has the same green density, 50%, as the G50% nanocrystalline powder, and also has a similar grain size (30 nm vs. 15 nm) and grain growth behavior (Fig. 10). However, because the submicron powder is composed of 200 nm-sized hard aggregates (dense agglomerates with no intercrystallite pores), its pore size is 17 nm, vs. the nanocrystalline powder's 4 nm, and the result of that difference is reflected in the very different grain size-density trajectories of Fig. 9.

The shift of the grain size-density curves as a result of increased compaction pressures is clearly due to densification effects alone. Grain growth is largely unaffected by the different compaction pressures or the correspondingly different green densities of the samples (Fig. 10). However, pore size and pore size evolution — which translate directly to differences in densification behavior — are both notably altered (Fig. 11).

Given that densification is the process of pore shrinkage and elimination, it is not surprising that the G58% sample, which begins with smaller pores than the G50% sample (2.8 vs. 4.0 nm), and keeps those pores smaller throughout sintering, has the faster densification kinetics. Recall from Fig. 5 that the instantaneous densification rate

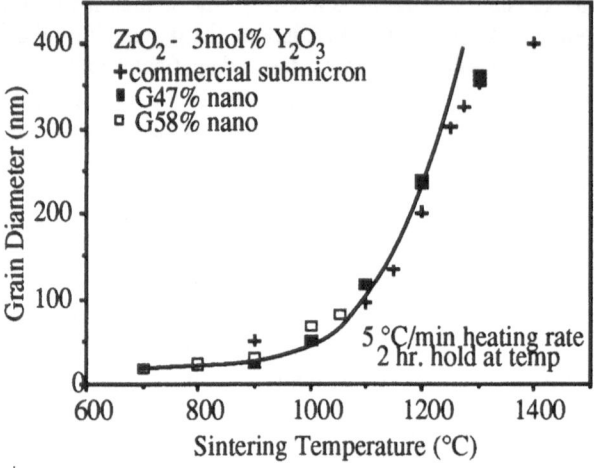

Figure 10. Grain growth behavior for several of the zirconias in Figure 9. It can be seen that, for this system, grain growth is largely a function of time and temperature, with virtually no differences resulting from variations in green density, pore size, etc.

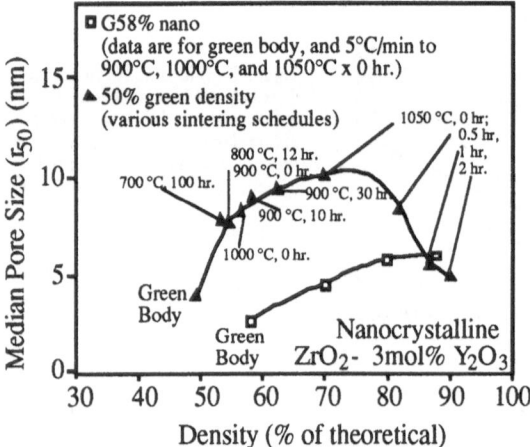

Figure 11. Changes in the microstructural arrangements of particles brought about by different compaction pressures not only result in different initial pore sizes, but change the trends in pore size evolution throughout sintering. It should be noted that, for any one starting microstructure, the path of pore size evolution is fixed. Changes in sintering schedule (long times at low temperatures vs. short times at high temperatures) do not affect the pore size- density trajectory for a given sample.

is inversely proportional to the instantaneous pore size. Thus, The G58% sample reaches full density in 5 hours at 1050 °C; the G50% sample, in comparison, takes over 20 hours. A close examination of the G50% data in Fig. 11 illustrates, once again, why simply changing the sintering schedule does not help promote better densification (at least in nanocrystalline zirconia): pore size develops according to a fixed microstructural path, regardless of the sintering schedule used. Thus, for a fixed starting microstructure, the growth and shrinkage of pores, and therefore the densification behavior of the sample, are also fixed.

Since the only way to affect the pressureless sintering behavior of the ceramic is to change its starting microstructure — preferably such that the starting pore size is reduced, and the pore size distribution narrowed — it is not surprising that this avenue has been the one most frequently exercised by experimentalists. Dry pressing to very high pressures (> 1 GPa) prior to sintering has been successfully used by a number of researchers to increase the sintered density of nanocrystalline ceramics with little adverse affect on grain growth [23-28]. The main limitation of this procedure is that the samples it produces are small. For a typical 100 kN press, obtaining 1 GPa of pressure requires that a disk-shaped sample have a diameter no greater than about a centimeter. In addition, the large compaction pressures promote extreme residual stresses and residual stress gradients [29], which tend to make the samples fracture or disintegrate upon ejection from the die. Minimizing these stresses requires that the sample height be small, typically no greater than 2 mm for most ceramics. Note that even if a coherent sample can be produced after compaction, the stress gradients induced during compaction also promote density gradients within the sample, and these can lead to differential densification and the creation of sintering flaws when the sample is heated.

4.   Sinter-Forging

The work of many researchers [30-40] has shown that samples that do not respond well to pressureless sintering, even after compaction at high pressures, can sometimes be made successfully by sinter-forging. Sinter-forging is an operation in which the sample is compressed at elevated temperatures such that it deforms and densifies simultaneously. It differs from hot pressing in that a die is not used, so that there is a higher ratio of deviatoric (shear) to hydrostatic stress. There appear to be at least two densification mechanisms which operate during sinter-forging [41, 42]. The first, stress-assisted diffusion, is similar to the diffusional densification mechanism in pressureless sintering. The difference is that the applied compressive stress lowers the vacancy concentration at grain boundaries so that the vacancies now move more quickly from the pores (where the vacancy concentration is high) to the grain boundaries, in accordance with Fick's first law. A faster pore shrinkage rate, and hence a faster densification rate, results.

The second mechanism of densification during sinter-forging is the closure of pores by plastic flow of the surrounding matrix [41, 42]. In this mechanism, matrix material is physically pushed into the pores to fill them as the bulk material is deformed. By vitue of plastic deformation — a property unique to fine-grained ceramics — it is possible to close pores which are much, much larger than those that can be closed by diffusion. As an example, the sample with many large pores shown in Fig. 1, which could not be densified at 610° by pressureless sintering, saw all of those large pores eliminated in 45 minutes of sinter-forging at the same temperatures (Fig. 12).

Figure 12. Nanocrystalline titania samples after sinter-forging for 45 minutes at 610 °C. Sample is almost completely featureless in SEM; no pores are visible at this or higher magnifications. Starting powder is identical to that used in the pressureless sintering experiment of Figure 1.

Theoretically, there is no limit to how large a pore can be closed by plastic flow; however, it should be noted that a significant amount of plasticity is required (strains of 0.62 by one theoretical calculation [43], but more is often needed in practice [41]). Hence, only those ceramics which are superplastic, i.e., capable of deforming to true strains of a hundred percent or greater, are able to take advantage of this mechanism.

Despite the advantages of being able to readily close large, interagglomerate pores, sinter-forging does offer some disadvantages. A significant one is the fact that the applied strain causes an additional type of grain growth, termed dynamic grain growth. In dynamic grain growth, the grains enlarge in direct proportion to the amount of applied strain [44]. Thus, the very large strains needed to close pores also induce a correspondingly large degree of grain growth.

A final difficulty with sinter-forging is that die friction effects can prevent plastic strain — and hence densification — from being evenly distributed within the sample. Figure 13 shows the complex stress state which exists in a uniaxially deformed sample when die friction is significant. The nearly hydrostatic stress state in Region I cannot give rise to macroscopic plastic strain; this requires shear stresses, which are dominant in Region II. Not surprisingly, the sinter-forged material in Region II is denser than that in Region I [8]. Region III suffers from significant tensile hoop stresses which not only discourages the closure of pores but also causes cracks to form on the outer surface of the sample.

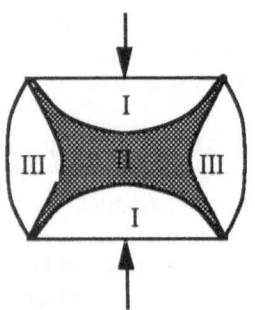

Figure 13. Different flow zones in a sample sinter-forged under conditions of substantial friction at the sample-platen interface (i.e., at top and bottom of specimen).

I: Dead zone; stress state has large hydrostatic component, relatively undeformed material
II: High deviatoric (shear) stress component; material deforms extensively
III: Compressive axial stresses and tensile hoop stresses; radial cracks form

5. Wet Processing

The two consolidation techniques discussed so far — dry pressing + pressureless sintering, and sinter-forging, both have severe drawbacks in terms of commercial application. The former is limited to small samples (due to the high pressure requirement for most nanoceramic powders), and the latter is limited to simple geometries and, furthermore, is time, equipment, and energy intensive. A solution to both these limitations may be found in the realm of wet consolidation. Because particles rearrange much more easily in a liquid medium than in a dry state, it is possible to get superior packing with a minimum of force. Thus, wet processing is able to provide samples that are as good as dry-pressed samples, but at a fraction of the pressure. This capability is demonstrated in Figure 14, which compares the grain size - density curves for dry pressed ZrO_2 - 3 mol% Y_2O_3 and ZrO_2 - 3 mol% Y_2O_3 prepared by centrifugation from a suspension. Note that despite starting out at a lower overall green density, the efficient particle packing and lack of large pores/flaws in the wet-processed sample allow it to achieve a density - grain size combination normally reserved for samples compacted under 1 GPa of pressure or more. The equivalent applied stress in the centrifuged case, however, is less than 0.06 MPa.

For oxide particles, the surface charge on the particles can easily be modified by changing the pH of the suspension in which they are immersed. Control of the surface charge, in turn, allows one to control particle packing in a way that is not possible in dry pressing. Highly charged particles will all attempt to repel one another and seek to maximize their interparticle distance; this results in a highly ordered arrangement of particles in the suspension, where each particle maintains a fixed distance from its neighbor. Upon drying, such a suspension yields a well-ordered compact with a small interparticle pore size. At pH's where little or no charge exists on the particles, the particles will come together in loose groups (flocculate), causing a looser and less homogeneous packing arrangement with a larger interparticle pore size. As in the case of dry pressed powders, those compacts which have the smaller pore sizes — pH 2 compacts in the case of ZrO_2 - 3 mol% Y_2O_3 — achieve faster densification per unit grain growth and ultimately achieve the densest samples at a given grain size (Fig. 15). For centrifuged ZrO_2 - 3 mol% Y_2O_3 samples prepared at pH 2, 100% density is reached while the sample still has a true grain diameter of 80 nm (mean linear intercept grain size of 51 nm).

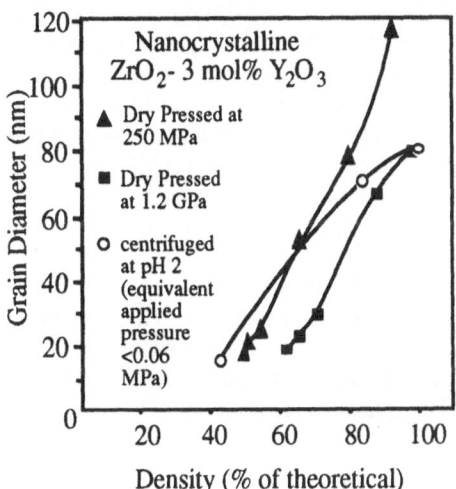

Figure 14. Grain size-density trajectories for wet processed vs. dry pressed green bodies upon sintering. Sintering time for the 250 MPa and 1.2 GPa dry-pressed samples is 2 hours, for the centrifuged samples, 1 hr.

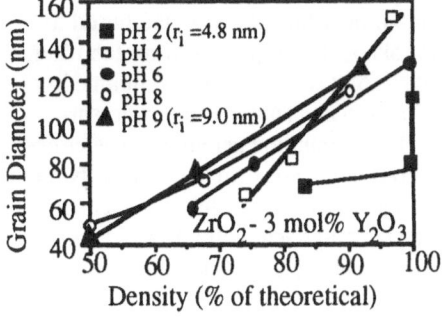

Figure 15. Grain size - density trajectories of centrifuged samples consolidated from suspensions prepared at different pH's. Values of r_i indicate initial pore sizes.

5.1 WET PROCESSING ROUTES ON THE DRAWING BOARD

There are a number of wet processing techniques currently being tested for the fabrication of nanocrystalline ceramics. Most have not quite reached the stage of producing publishable results (in some cases this is more due to industrial proprietary concerns than any deficiency of the technique). Since this is an area in which the future of nanocrystalline ceramics lies, it is useful to look at some of those techniques here.

Figure 16.
Osmotic
Consolidation

5.1.1 *Osmotic Consolidation.*
Osmotic consolidation is a technique long used in the biological sciences to separate proteins from the suspension containing them. The same technique was pioneered for the use of small ceramic particles by Miller and Zukoski [45]. Briefly, an aqueous suspension containing the ceramic particles is put inside a bag made of a semipermeable membrane (e.g., cellulose acetate, cellulose nitrate). The bag of suspension is then placed in a container of water + some soluble species. The soluble agent can be any soluble macromolecule such as a polymer or large carbohydrate. Because the chemical potential of the water inside the bag is quite high (the water is pure, with nothing dissolved in it) compared to the water outside the bag, there is a driving force for water to leave the bag, dilute the suspension on the outside, and thereby equalize the two chemical potentials. The act of the water leaving the bag, however, serves to consolidate the particulate suspension inside. Miller and Zukoski estimate that, under suitable conditions, it is possible to extract water from the saturated green body to the point that air or vapor gaps open up between the water-logged particles [45]. This has been successfully achieved in the case of nanocrystalline hydrous zirconia particles; unfortunately, no attempt was made to sinter the resulting samples, in part because they were not bona fide oxides [45, 46].

A major advantage of osmotic consolidation is that forcibly extracts water from a suspension without applying force to the particles in that suspension. The osmotic force acts on only the water component of the system. Because the particles see no applied force (other than gravity) they do not experience the drawbacks associated with forced compaction. Stress gradients and density gradients are minimal.

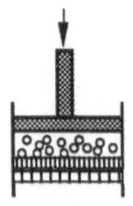

Figure 17.
Pressure Filtration

5.1.2. *Pressure Filtration.*
Pressure filtration is an operation in which a suspension of particles is mechanically forced through a filter, thereby forming a cake of particles on top of the filter while allowing the liquid to pass through the filter. The cake is then separated from the filter, dried, and sintered to produce a solid ceramic. This technique has been used successfully with micron and submicron powders to produce high strength / low flaw population ceramics. The stress requirements for pressure filtration are not stringent, typically about 12 MPa (???). Because biological filters (e.g., cellulose acetate, cellulose nitrate, Nafion) are available in a pore size range that will allow the filtration of nanoparticles, pressure filtration experiments with nanoparticles are now being attempted.

Figure 18.
Tape Casting

5.1.3. *Tape Casting.*

In tape casting, a suspension of ceramic particles and dissolved polymers is spread as a thin, wet film onto a conveyor belt and allowed to dry. The result is a flexible, thin "tape" of (mostly) ceramic material that can be wound on collection spools for easy storage but which will become a fully dense, rigid, thin ceramic sheet when unwound and

sintered. Tape casting is one of the primary production routes for electronic ceramic components; extending this operation to the nanoparticle realm would open up substantial commercial possibilities. The suspension chemistries in tape casting are often complex; there is an additional degree of complexity in the nanoparticle realm, since it is not clear how to scale the size and amount of polymer additions (binders, stabilizers, dispersants, etc.) If one assumes that the amount of polymer should increase in proportion to the surface area the polymer is modifying, the volume of polymer additions in a nanoparticle suspension would quickly reach massive proportions. Reports of successful nanopowder tape casting operations indicates that this is probably not the case [47], though published data are virtually non-existent.

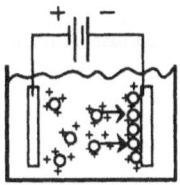

Figure 19.
Electrophoretic
Deposition

5.1.4. *Electrophoretic Deposition.*

Electrophoretic deposition is yet another wet processing technique which has been successfully applied to submicron and larger particles, but which is only now being applied to nanometer-sized particles. In this operation, dense particulate coatings are created by forcing charged particles in suspension to migrate to an electrode submerged in that suspension. The requisite surface charge on the particles develops naturally, through surface dissolution or reaction with the surrounding liquid, and it can be modified via pH changes. The surface charge can also be artificially contrived through the addition of ionic surfactants to the suspension. These

then adsorb on the particle surface, lending their own charge to the particle. The advantages of electrophoretic deposition include the simplicity of the equipment required, the rapidity of the deposition (0.1 to 10 microns/sec, depending on experimental conditions), the wide range of coating thicknesses that can be achieved (1 μm to 1 cm; Refs.[48, 49]), the variety of shapes that can be made (everything from micron-resolution, patterned thin films on CRT screens [50] to bulk sanitary ware [49]), and the highly dense nature of the as-deposited coatings. Results with nanoparticles appear to be available but not published [51].

6. Properties of Nanocrystalline Ceramics

Once the nanocrystalline ceramic is made, by whatever route — dry press & sinter, sinter-forge, one of the wet processes — properties can be measured. Some of the properties relevant to structural applications are summarized below.

6.1 REDUCED SINTERING TEMPERATURE

The most obvious property of nanocrystalline ceramics is that their sintering temperature is significantly lower than their submicron and micron-grained counterparts, once agglomeration problems are dealt with. Figure 20 shows up to a 250°C reduction in sintering temperature in going from a commercial ZrO_2-3mol%Y_2O_3 to a nanocrystalline version. In many cases, such a reduction allows for the use of less expensive furnaces, or for the ceramic to be cofired with certain metals that otherwise would have melted.

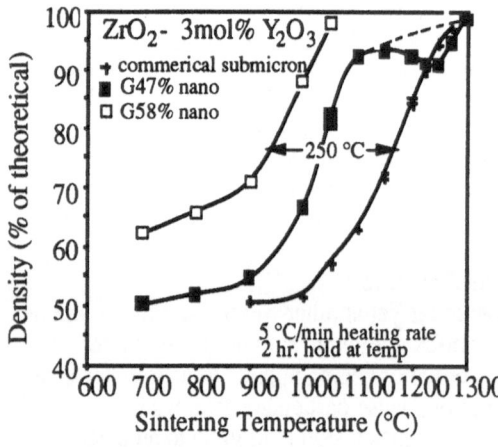

Figure 20. Reduction in sintering temperature through the use of nanocrystalline grain sizes. "G" values indicate green densities. Note that the dip in the G47% curve is due to rapid rate heating effects peculiar to nanocrystalline zirconia[22] and disappears when slower (2 °C/min) heating rates are used (dashed line).

6.2. HARDNESS

The hardness of several nanocrystalline ceramics has also been measured; in many cases the hardness appears to be less than the reported (or measured) hardnesses for large-grained ceramics; however, when comparisons are made between samples with the same porosity levels, this difference almost always disappears (Fig. 21).

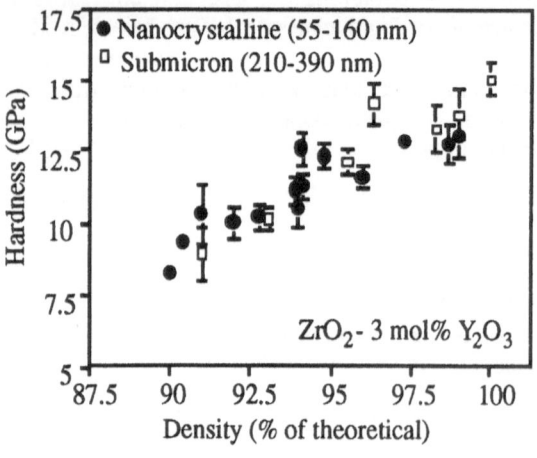

Figure 21. Hardness as a function of density for nanocrystalline and submicron-grained zirconias.

6.3. FRACTURE TOUGHNESS

With respect to fracture toughness, there is no indication at this point in time that there should be a significant difference between the fracture toughness of most nanocrystalline ceramics and their large-grained counterparts. The one striking, and important, exception is in the case of transformation-toughened ceramics (ZrO_2-based ceramics with one or more stabilizing dopants added). In this class of materials, a phase transformation occurs in the vicinity of a propagating crack, due to the tensile stress created by the crack. The existing tetragonal phase is transformed to a monoclinic phase of larger specific volume, and this additional volume is accommodated by pushing the crack shut [52]. The forcible closure of the crack as the crack attempts to propagate then leads to fracture toughnesses on the order of 8-12 $MPa \cdot m^{1/2}$, as compared to the 2-3 $MPa \cdot m^{1/2}$ of non-transformation-toughened ceramics.

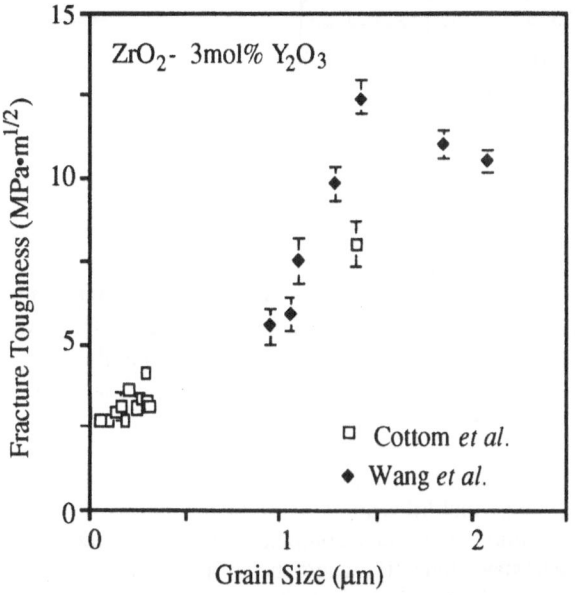

Figure 22. Fracture toughness vs. grain size for ZrO_2-$3mol\%Y_2O_3$. Data are from Ref. [53], Cottom et al., and Ref. [54], Wang et al..

For nanocrystalline ZrO_2-$3mol\%Y_2O_3$, fracture toughness is severely compromised by going to a nanocrystalline grain size, as shown in Fig. 22. However, the results are not unexpected. As the grain size becomes smaller, the tetragonal phase becomes increasingly stable with respect to the monoclinic phase. It therefore is less likely to transform (requires a higher applied tensile stress for transformation) as a crack passes through [54]. Without the transformation, there is no crack closure mechanism, and a low toughness results. Note that the nanocrystalline data of Fig. 22 are completely consistent with the trend of decreasing fracture toughness with decreasing grain size.

To counter the overstabilization of the tetragonal phase brought about by the ultrafine grain sizes, other contributions to the phase stabilization must be reduced. If this is done, e.g., by reducing the Y_2O_3 content in the zirconia ceramic, the high fracture toughness associated with transformation toughening can be regained (Fig. 23). Current work suggests it is possible to raise the fracture toughness to 16 $MPa \cdot m^{1/2}$

378

through a precise pairing of grain size with yttria concentration which puts the ceramic just on the verge of transformation prior to crack introduction [55].

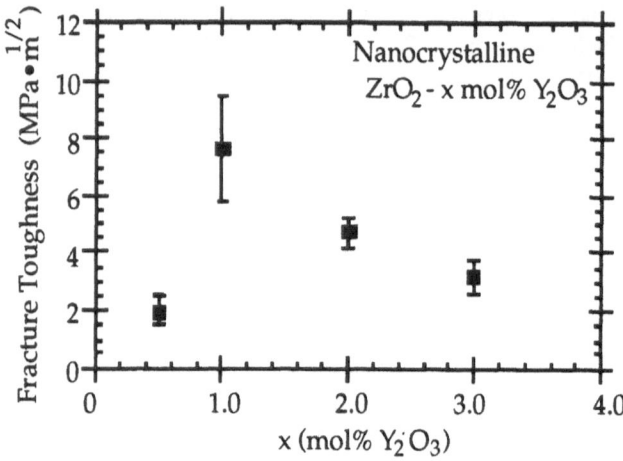

Fig. 23. Fracture toughness for nanocrystalline (80-90 nm) yttria-stabilized zirconia [56]. Data were obtained from indentation tests using the half-penny crack assumption for calculating toughness [57] (same method as for Cottom data in Fig. 22).

6.4 SUPERPLASTICITY

A fourth property of nanocrystalline ceramics, and one of great popular interest, is superplasticity. Superplasticity can be defined as the ability of a crystalline material to experience large elongations to failure. For metals, superplastic behavior has been well known since the 1930's [58]; for ceramics, the phenomenon was observed in the early 1980's (for a review of this early work, see Ref. [59]) but was not widely recognized until 1986, when the first tensile test results were published [60]. In metals, superplastic deformations of up to 8000% have been achieved [61, 62], for ceramics, tensile elongations to failure have reached 800% [63]. (In comparison, the ductility of steel is about 30%). These phenomenal ductilities allow complex parts to be manufactured in simple, one-step forming operations; commercially, the superplastic metals have found a home in the aerospace industry, where the ability to form lightweight, integrated parts is highly valued. Superplasticity in ceramics is not commercially accepted, despite the obvious potential, in large part because the forming temperatures are still quite high, and the forming rates are distressingly slow. Nanocrystalline ceramics have the opportunity to overcome this barrier, because the relationship between grain size and forming rate and temperature is as follows:

$$\dot{\varepsilon} = A\sigma^n d^{-p} D_0 \exp(-Q/RT) \tag{3}$$

In this equation, $\dot{\varepsilon}$ is the strain rate (forming rate), A is a material constant, σ is the applied stress, d is the grain size, and $D_0 \exp(-Q/RT)$ is the diffusivity, which depends on Q, the activation energy for diffusion, R, the gas constant, and T, the temperature in degrees Kelvin. Exponents of interest include p, the grain size exponent, which has a value between 2 and 3, and n, the stress exponent, which has a value between 1 and 3. It should be noted that superplasticians often refer to m, the strain rate sensitivity, which is the inverse of the stress exponent (m=1/n) and which, for ceramics, has a value between 0.3 and 1. According to Eq. 3, a ten-fold reduction in grain size should accelerate forming rates by a factor of 100 to 1000, and decrease the forming temperatures by about 300 °C (assuming an arbitrary but typical Q value of 450 kJ/mol and a reference temperature of 1400 °C).

Experimental data for the deformation of nanocrystalline ceramics support both the possibility of faster strain rate deformation and the possibility of lower temperature deformation. Strain rate comparisons of submicron (0.3 μm) and nanocrystalline (80 nm) ZrO_2-3mol%Y_2O_3 show the latter to deform 34 times faster than the former (Fig. 24). Extrapolating to a 20 nm grain size would push the deformation stresses and strain rates into the stress-strain rate range in which the superplastic forming of metals is currently conducted. The lower temperature possibilities are illustrated by Figure 25 for titania, which shows how nanocrystalline titania deforms at 250-300°C lower temperatures than micron-grained titania, when equivalent stress/strain rate conditions are imposed. It is interesting to note that the 691°C deformation temperature of the nanocrystalline titania places it well below the 950°C temperatures used for the commercial superplastic forming of titanium metal.

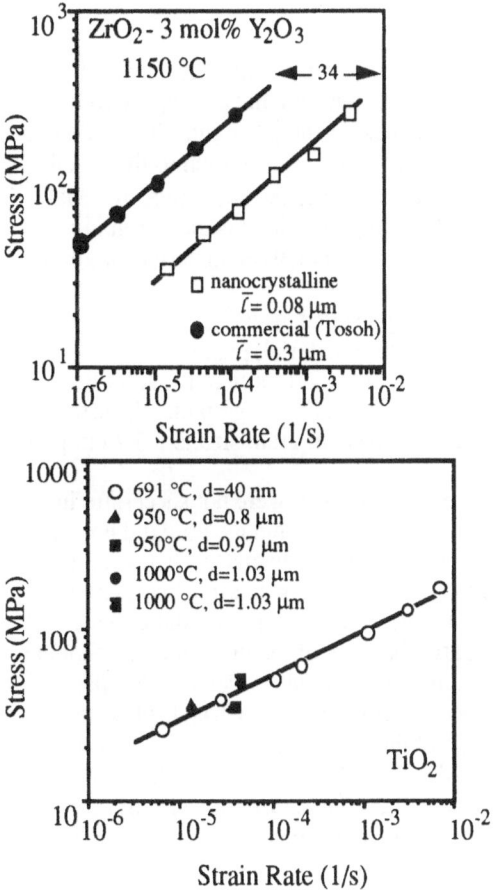

Figure 24. Stress-strain rate data for nanocrystalline (80 nm) and submicron (300 nm) yttria-stabilized zirconia, deformed in compression. Data from Ref. [64].

Figure 25. Stress-strain rate data for nanocrystalline (40 nm) titania deformed at 691 °C and micron-grained titania deformed at 950 and 1000 °C, in compression. Nanocrystalline titania is able to achieve the same forming conditions at 250-300°C lower temperatures. Nanocrystalline data from Ref. [65]; micron-grained data from Ref. [66].

Evidence of superplastic formability in nanocrystalline ceramics comes not only from compression tests, but also from bulge forming tests [67] and from tests in which micron-level surface detail of a die was almost perfectly replicated in the deformed ceramic [68]. Still, there has yet (to the author's knowledge) to be a published account of a conventional tensile test performed on a fully dense nanocrystalline ceramic. This limitation is almost entirely due to the large sample fabrication difficulties discussed in the first portion of this paper.

Though one tends to think of superplasticity as a method for near net-shape forming large objects, many applications may be more subtle. For instance, the ability of the ceramic to deform, rather than break, in response to a tensile stress should allow it to survive the thermal mismatch stresses generated in bonded components where the different materials have widely different thermal expansion coefficients — e.g., metal-ceramic joints, ceramic-metal laminates, or ceramic-coated metal substrates. The successful application of this technology will depend on the ability to match the stress relaxation time of the ceramic (dictated by its strain rate at the temperature and stress in question) with the rate at which stress is being applied (dictated by the heating rate of the laminate or joint).

6.5 THERMAL CONDUCTIVITY

One coating-related technology in which nanocrystalline ceramics may find application is in the area of thermal barrier coatings, the ceramic coatings used to protect metal superalloy turbine blades from the heat of the jet engine. These coatings are typically ZrO_2 with 6-8 wt% (=3.4-4.5 mol%) Y_2O_3. In addition to the possibility (not yet proven) that a nanocrystalline, equiaxed grain structure could relax residual stresses formed upon fabrication (and possibly during use) there is the intriguing possibility that the thermal conductivity of these nanocrystalline materials might be modified towards more desirable lower values.

There are data in the literature [69] which suggest that, as the grain size approaches the phonon mean free path length in a (non-metallic) material, phonons begin to scatter off of the grain boundaries, thereby rendering heat transport through the solid less efficient and lowering the thermal conductivity of the solid. In graphite, the effect is well over an order of magnitude, in going from a 105 nm grain size to a 3.7 nm grain size[69].

Experiments with yttria-stabilized zirconias, however, were not successful in resolving this effect, in large part because the processing technique used did not allow full density samples to be produced with grain sizes (grain diameters) below 63 nm. The data, shown in Fig. 26, exhibit no effect of grain size on thermal conductivity within the 78 nm-400 grain size range, for a fixed composition The flat curves, indicative of phonon scattering by defects, arise because of the interaction of phonons with oxygen vacancy defects in the ceramic, not grain boundary defects. Supporting this view is the fact that nanocrystalline zirconia with no added yttria displays the monotonically decreasing curve characteristic of the temperature dependence of thermal conductivity when no defects are present. Also, samples with higher yttria content show decreased thermal conductivity relative to the samples with lower yttria content, thereby suggesting again that yttria-induced defects (oxygen vacancies) are responsible for the scattering.

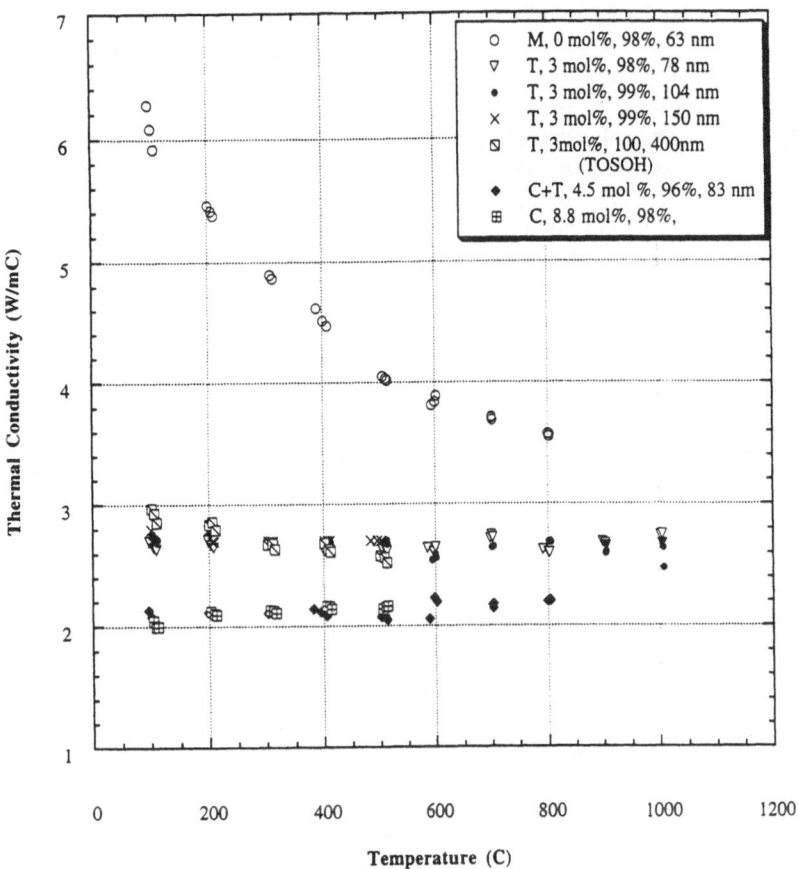

Figure 26. Thermal conductivity data for various yttria-stabilized zirconias. From Ref. [70].

Key to legend: M=monoclinic; T=tetragonal; C=cubic; mol%=mol% Y_2O_3; %=density in percent of theoretical; nm=grain diameter in nm, as measured by x-ray line broadening

6.6 DIFFUSION BONDING

As a final example of the applications to be found for nanocrystalline ceramic structural materials, one can find several research groups which are pursuing diffusion bonding [71, 72]. In diffusion bonding, two ceramic pieces are joined together directly using moderate temperature and applied stress to allow diffusion to occur across the joint line. This process is not particularly successful in micron-grained ceramics, but the many grain boundaries present in nanocrystalline ceramics increase the total diffusional flux across the bond line, and the nanoceramic's superplasticity additionally helps to fill topographical disparities between mating interfaces by plastic flow.

Early work [71] showed that nanocrystalline zirconias could be directly and seamlessly bonded to each other in this manner at surprisingly low temperatures and stresses (see Fig. 27), but the same study also concluded that using a nanocrystalline powder interlayer to join larger-grained ceramics was problematic because the drying and sintering of the powder slurry used to form the interlayer left behind large voids which severely compromised the strength of the joint. Sintered interlayers, which did not have the residual porosity problem, did work quite well [71]. A subsequent advance was made in the study of Ferkel and Riehemann [72], who abandoned the slurry interlayer approach and used a thin, pre-pressed dry nanocrystalline powder compact as an interlayer instead. The nanocrystalline interlayer in this case not only successfully bonded two large grained alumina pieces at 1100 °C, but the joint appeared to have far fewer pores and cracks than the parent material being bonded. Further advances in this area await a means of depositing a dense layer of particles on a surface which does not have to be perfectly flat; electrophoretic deposition (mentioned in Section 5.1.4) is one possibility.

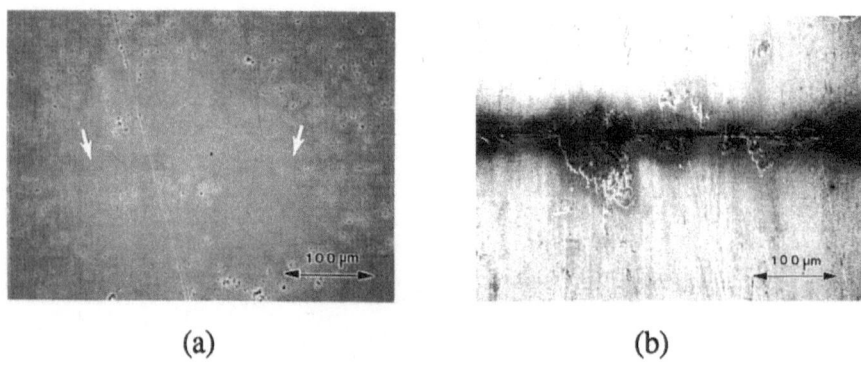

(a)   (b)

Figure 27.   (a) Diffusion bonding of nanocrystalline ZrO$_2$-3mol% Y$_2$O$_3$ blocks to each other using 10 MPa pressure at 1090°C. Joint line is indicated by arrows.   (b) Diffusion bonding of submicron ZrO$_2$-3mol% Y$_2$O$_3$ blocks to each other using 10 MPa pressure at 1090°C. Joint line is evident.

7. Conclusions

The wide variety of techniques available for the synthesis of ultrafine ceramic powders is now enabling the production of ceramics with grain sizes on the nanometer scale. Because of their high surface area and strong tendency to agglomerate, however, the nanocrystalline powders are far more difficult to process than their macrocrystalline counterparts. To achieve full density and still retain an ultrafine grain size often requires that large, interagglomerate pores be removed from the powder. This point is underscored by data showing pore size's significant influence on densification rate — vs. data which show little impact of altering sintering conditions. Successful methods of removing large pores include dry pressing at high (>1 GPa) pressures prior to compaction, or sinter-forging. The former eliminates large pores by crushing agglomerates and forcibly rearranging them; the latter fills large pores by plastic flow of the surrounding matrix during high temperature deformation. Because of the ability of particles to rearrange easily and therefore pack efficiently in a liquid medium, wet processing techniques also hold promise for the production of nanocrystalline ceramics. Centrifugation, pressure filtration, osmotic consolidation, and electrophoretic deposition have all been attempted.

   Fabrication problems aside, nanocrystalline ceramics have a number of rather useful properties. Sintering temperatures are typically several hundred degrees lower than for

383

more conventionally-sized ceramics. Hardness, though often lower as well, is usually an artifact of porosity present in the final product. Fracture toughness in transformation-toughened nanoceramics suffers somewhat due to a disruption in the balance between the stabilities of the monoclinic and tetragonal phases; superior toughness can be restored, however, by adjusting the chemical stabilizer content to compensate for the grain size effect on phase stability. Superplasticity in oxide nanoceramics is well documented; it has the added feature of occurring at faster strain rates (forming rates) and lower temperatures than in submicron and micron-grained ceramics. Theoretical work and limited experiments on graphite indicate it should also be possible to adjust the thermal conductivity of a ceramic through grain size refinement down to the nanometer range; however, data presented for yttria-stabilized zirconia do not appear to have entered this realm. A final application, diffusion bonding, has shown somewhat more success to date: with a minimum of pressure and modest temperatures, nanocrystalline ceramics have been successfully joined to each other and/or have been used as interlayers to join larger-grained ceramics.

8. References

[1]   DOE Council on Materials Science (1988). Department of Energy, Monterey, CA. Same article is reprinted in J. Mater. Res. **4** (1989) 704.
[2]   R. A. Andrievski (1994). "Review: Nanocrystalline High Melting Point Compound-Based Materials," *J. Mater. Sci.* **29**, 614-631.
[3]   M. J. Mayo (1996). "Processing of Nanocrystalline Ceramics from Ultrafine Particles," *International Materials Reviews* **41**, 85-115.
[4]   P. Matteazzi, D. Basset, F. Miani, and G. L. Caër (1993). "Mechanosynthesis of Nanophase Materials," *Nanostructured Materials* **2**, 217-229.
[5]   J. Parker (1995). Nanophase Technologies, personal communication.
[6]   C. Herring (1950). "Effect of Change of Scale on Sintering Phenomena," *J. Appl. Phys.* **21**, 85-87.
[7]   G. L. Messing and M. Kumagai (1994). "Low Temperature Sintering of α-Alumina-Seeded Boehmite Gels," *Am. Ceram. Soc. Bull.* **73**, 88-91.
[8]   D. C. Hague (1995). "Sinter-Forging of Nanocrystalline Ceramics," Ph.D. Thesis. The Pennsylvania State University.
[9]   S. Lowell and J. E. Shields (1979). *Powder Surface Area and Porosity.* Chapman and Hall, New York.
[10] M. F. Yan and W. W. Rhodes (1983). "Low Temperature Sintering of TiO_2," *Mater. Sci and Eng.* **61**, 59-66.
[11] D. C. Hague (1992). "Chemical Precipitation, Densification, and Grain Growth in Nanocrystalline Titania Systems," M.S. Thesis. The Pennsylvania State University.
[12] E. A. Barringer, R. Brook, and H. K. Bowen (1984). "The Sintering of Monodisperse TiO_2," in *Sintering and Heterogeneous Catalysis*, G. C. Kuczynski, A. E. Miller, and G. A. Sargent, Eds. Plenum Press, New York. pp. 1-21.
[13] R. C. Flagan and M. M. Lunden (1995). "Particle Structure Control in Nanoparticle Synthesis from the Vapor Phase," *Mater. Sci. and Eng.* **A204**, 113-124.
[14] M. S. Kaliszewski and A. H. Heuer (1990). "Alcohol Interaction with Zirconia Powders," *J. Am. Ceram. Soc.* **73**, 1504-9.
[15] S. Kwon and G. L. Messing (1997). "The Effect of Particle Solubility on the Strength of Nanocrystalline Agglomerates: Boehmite," *Nanostructured Materials* **8**, 399-409.
[16] J. H. Adair, H. G. Krarup, S. Venigalla, and T. Tsukada (1997). "A Review of the Aqueous Chemistry of the Zirconium-Water System," in *Aqueous Chemistry and Geochemistry of Oxides, Oxyhydroxides, and Related Materials* (Mater. Res. Soc. Symp. Proc. **432**), J. A. Voigt, T. E. Wood, B. C. Bunker, W. H. Casey, and L. J. Crossey, Eds. Materials Research Society, Pittsburgh, PA. .
[17] C. E. Baumgartner (1988). "Fast Firing and Conventional Sintering of Lead Zirconate Titanate Ceramic," *J. Am. Ceram. Soc.* **71**, 350-353.
[18] H. Mostaghaci and R. J. Brook (1983). "Production of Dense and Fine Grain Size $BaTiO_3$ by Fast Firing," *Trans. Brit. Ceram. Soc.* **82**, 167-70.
[19] H. Mostaghaci and R. J. Brook (1986). "Microstructure Development and Dielectric Properties of Fast-Fired BaTiO3 Ceramics," *J. Mater. Sci.* **21**, 3575-3580.
[20] M. P. Harmer and R. J. Brook (1981). "Fast Firing— Microstructural Benefits," *J. Brit. Ceram. Soc.* **80**, 147-149.

384

[21] P. Vergnon, M. Astier, and S. J. Teichner (1974). "Initial Stage for the Sintering of Ultrafine Particles (TiO2 and Al2O3)," in *Fine Particles*, W. E. Kuhn, Ed. Electrochemical Society, Princeton, NJ. pp. 299-307.

[22] D.-J. Chen and M. J. Mayo (1996). "Rapid Rate Sintering of Nanocrystalline ZrO2-3mol%Y2O3," *J. Am. Ceram. Soc.* **79**, 906-12.

[23] D.-J. Chen and M. J. Mayo (1993). "Densification and Grain Growth of Ultrafine 3 mol% Y2O3 - ZrO2 Ceramics," *Nanostructured Mater.* **2**, 469-478.

[24] R. W. Siegel, S. Ramasamy, H.Hahn, Z. Zonghuan, and L. Ting (1988). "Synthesis, Characterization, and Properties of Nanophase TiO2," *J. Mater Res.* **3**, 1367.

[25] H. Hahn, J. Logas, and R. S. Averback (1990). "Sintering Characteristics of Nanocrystalline TiO2," *J. Mater. Res.* **5**, 609-614.

[26] (no author) (1991). "Preparation and Sintering of Ultrafine SiC Particles," *Progress in Materials Science* **35**, 66-70.

[27] A. Pechenik, G. J. Piermarini, and S. C. Danforth (1992). "Fabrication of Transparent Silicon Nitride from Nanosize Particles," *J. Am. Ceram. Soc.* **75**, 3283-88.

[28] R. A. Andrievski (1994). "Compaction and Sintering of Ultrafine Powders," *Intl. J. Powder Metall.* **30**, 59-66.

[29] D. Train (1957). "Transmission of Forces Through A Powder Mass During the Process of Pelleting," *Trans. Instn. Chem. Engrs.* **35**, 258-266.

[30] D. C. Hague and M. J. Mayo (1993). "Sinter-Forging of Chemically Precipitated Nanocrystalline TiO2," in *Mechanical Properties and Deformation Behavior of Materials Having Ultrafine Microstructures*, M. Nastasi, D. Parkin, and H. Gleiter, Eds. Klewer, Dordrecht, The Netherlands. pp. 539-545.

[31] D. C. Hague and M. J. Mayo (1993). "The Effect of Crystallization and a Phase Transformation on the Grain Growth of Nanocrystalline Titania," *Nanostructured Mater.* **3**, 61-7.

[32] M. J. Mayo, D. C. Hague, and D.-J. Chen (1993). "Processing Nanocrystalline Ceramics for Applications in Superplasticity," *Mater. Sci. & Eng.* **A166**, 145-159.

[33] R. S. Averback, H. J. Höfler, and R. Tao (1993). "Processing of Nano-Grained Materials," *Mater. Sci. and Engineering* **A166**, 169-177.

[34] M. M. R. Boutz, A. J. A. Winnubst, A. J. Burggraaf, M. Nauer, and C. Carry (1993). "Low Temperature Sinter Forging of Nanostructured Y-TZP," in *Science and Technology of Zirconia V*, S. Badwal, J. Bannister, and R. Hannink, Eds. Technomic Pub. Co., Lancaster, PA. pp. 275-283.

[35] A. J. A. Winnubst, Y. J. He, P. M. V. Bakker, R. J. M. O. Scholtenhuis, and A. J. Burggraaf (1993). "Sinter Forging as a Tool for Improving the Microstructure and Mechanical Properties of Zirconia Toughened Alumina," in *Science and Technology of Zirconia V*, S. Badwal, J. Bannister, and R. Hannink, Eds. Technomic Pub. Co., Lancaster, PA. pp. 284-291.

[36] O.-H. Kwon, C. S. Nordahl, and G. L. Messing (1995). "Submicrometer Transparent Alumina by Sinter-Forging Seeded γ-Al2O3 Powders," *J. Am. Ceram. Soc.* **78**, 491-94.

[37] G. Skandan, H. Hahn, B. H. Kear, M. Roddy, and W. R. Cannon (1994). "Processing of Nanostructured Zirconia Ceramics," in *Molecularly Designed Ultrafine/Nanostructured Materials*, (Mat. Res. Soc. Symp. Proc. **351**), K. E. Gonsalves, G.-M. Chow, T. D. Xiao, and R. C. Cammarata, Eds. Materials Research Society, Pittsburgh, PA. pp. 207-12.

[38] M. J. Mayo and D. C. Hague (1994). "Superplastic Sinter-Forging of Nanocrystalline Ceramics," in *Superplasticity in Advanced Materials ICSAM-94*, (Materials Science Forum **170-172**). Trans Tech Publications, Switzerland. pp. 141-146.

[39] D. M. Owen and A. H. Chokshi (1993). "An Evaluation of the Densification Characteristics of Nanocrystalline Materials," *Nanostructued Mater.* **2**, 181-7.

[40] M. Uchic, H. J. Hofler, W. J. Flick, R. Tao, P. Kurath, and R. S. Averback (1992). "Sinter-Forging of Nanophase TiO2," *Scripta Metall. et Mater.* **26**, 791-6.

[41] D. C. Hague and M. J. Mayo (1995). "Modelling Densification During Sinter-Forging of Yttria-Partially-Stabilized Zirconia," *Mater. Sci. and Eng.* **A204**, 83-9.

[42] D. C. Hague and M. J. Mayo (1997). "Sinter-Forging of Nanocrystalline Zirconia: I. Experimental," *J. Am. Ceram. Soc.* **80**, 149-156.

[43] B. Budiansky, J. W. Hutchinson, and S. Slutsky (1982). "Void Growth and Collapse in Viscous Solids," in *Mechanics of Solids, the Rodney Hill 60th Anniversary Volume*, H. G. Hopkins and M. J. Sewell, Eds. Pergamon Press, Oxford. pp. 13-45.

[44] D. S. Wilkinson and C. H. Caceres (1984). "On the Mechanism of Strain-Enhanced Grain Growth in Microduplex and Second Phase Dispersed Alloys"," *Acta Metall.* **32**, 1335-1345.

[45] K. T. Miller and C. F. Zukoski (1994). "Osmotic Consolidation of Suspensions and Gels," *J. Am. Ceram. Soc.* **77**, 2473-8.

[46] K. T. Miller, R. M. Melant, and C. F. Zukoski (1996). "Comparison of the Compressive Yield Response of Aggregated Suspensions: Pressure Filtration, Centrifugation, and Osmotic Consolidation," *J. Am. Ceram. Soc.*, 2545-2556.

[47] J. H. Adair, personal communication

[48] Z. Surowiak (1973). "On the Technology of Deposition of Polycrystalline Thin Films of Ferro- and Anti-Ferroelectrics on Metallic Substrates," *Acta Physical Polonica* **A34**.

[49] W. Ryan, E. Massoud, and C. T. S. B. Perera (1979). "Electrophoretic Deposition Could Speed Up Ceramic Casting," *Interceram* **2**, 117-119.

[50] J. Miziguchi, K. Sumi, and T. Muchi (1993). "A Highly Stable Nonaqueous Suspension for the Electrophoretic Deposition of Powdered Substances," *J. Electrochem. Soc.* **130**, 1819-1825.

[51] U. Eisele and C. Randall (1997), personal communication.

[52] D. J. Green, R. Hannik, and M. Swain (1988). *Transformation Toughening of Ceramics*. CRC Press, Boca Raton.

[53] B. A. Cottom and M. J. Mayo (1996). "Fracture Toughness of Nanocrystalline ZrO_2-3mol%Y_2O_3 Determined by Vickers Indentation," *Scripta Met. et Mater.* **34**, 809-814.

[54] J. Wang, M. Rainforth, and R. Stevens (1989). "The Grain Size Dependence of the Mechanical Properties in TZP Ceramics," *Br. Ceram. Trans. J.* **88**, 1-6.

[55] A. Bravo-Leon (1997), unpublished work.

[56] Y. Morikawa (1996), unpublished work.

[57] G. R. Anstis et al. (1981). "A Critical Evaluation of Indentation Techniques for Measuring Fracture Toughness. Parts I and II.," *J. Am. Ceram. Soc.* **64**, 533-543.

[58] C. E. Pearson (1934). "The Viscous Properties of Extruded Eutectic Alloys of Lead-Tin and Bismuth-Tin," *J. Inst. Met.* **54**, 111.

[59] C. Carry and M. Mocellin (1985). "High Ductilities in Fine Grained Ceramics," in *Superplasticity/Superplasticité*, B. Baudelet and M. Suery, Eds. Centre National de la Recherche Scientifique, Paris. pp. 16.1-16.19.

[60] F. Wakai, S. Sakaguchi, and Y. Matsuno (1986). "Superplasticity of Yttria-Stabilized Tetragonal ZrO_2 Polycrystals," *Adv. Ceram. Mater.* **1**, 259-263.

[61] O. D. Sherby and J. Wadsworth (1990). "Observations on Historical and Contemporary Developments in Superplasticity," in *Superplasticity in Metals, Ceramics, and Intermetallics* (MRS. Symp. Soc. Proc. **196**), M. J. Mayo, M. Kobayashi, and J. Wadsworth, Eds. MRS, Pittsburgh, PA. pp. 3-14.

[62] Y. Nakatani, T. Ohnishi, and K. Higashi (1984). *Jpn. Inst. Met.* **48**, 113.

[63] T. G. Nieh, C. M. McNally, and J. Wadsworth (1989). "Superplasticity in Intermetallic Alloys and Ceramics," *JOM* **41**, 31-35.

[64] M. Çiftçioglu and M. J. Mayo (1990). "Processing of Nanocrystalline Ceramics," in *Superplasticity in Metals, Ceramics, and Intermetallics*, (Mater Res. Soc. Symp. Proc. **196**), M. J. Mayo, M. Kobayashi, and J. Wadsworth, Eds. MRS, Pittsburgh, PA. pp. 77-86.

[65] H. Hahn and R. S. Averback (1991). "Low-Temperature Creep of Nanocrystalline Titanium(IV) Oxide," *J. Am. Ceram. Soc.* **74**, 2918-21.

[66] C. Carry and A. Mocellin (1987). "Structural Superplasticity in Single Phase Crystalline Ceramics," *Ceram. Intl.* **13**, 89-98.

[67] Z. Cui and H. Hahn (1992). "Tensile Deformation of Nanostructured TiO_2 at Low Temperatures," *Nanostructured Mater.* **1**, 419-425.

[68] J. Karch and R. Birringer (1990). "Nanocrystalline Ceramics: Possible Candidates for Net-Shape Forming," *Ceram. Intl.* **16**, 291-4.

[69] J. G. Castle, T. Beime, and J. M. Hutchen, "Proceedings of the First and Second Conferences on Carbon, 1953 and 1955," ,University of Buffalo, Buffalo, NY, 1956. As presented in W.E. Kuhn, "Consolidation of Ultrafine Particles," *Ultrafine Particles—Proceedings of a Symposium Sponsored by the Electrothermics and Metallurgy Division of the Electrochemical Society.* W.E. Kuhn, Ed. New York: Wiley and Sons, 1963, pp. 41-103.

[70] S. Raghavan, R. Dinwiddie, W. Porter, H. Wang, and M. Mayo (1997). "The Effect of Grain Size, Porosity and Yttria Content on the Thermal Conductivity of Nanocrystalline Zirconia," *Scripta Met et Mater.*, submitted.

[71] T. H. Cross and M. J. Mayo (1994). "Ceramic-Ceramic Diffusion Bonding Using Nanocrystalline Interlayers," *Nanostructured Materials* **3**, 163-8.

[72] H. Ferkel and W. Riehemann (1996). "Bonding of Alumina Ceramics with Nanoscaled Alumina Powders," *Nanostructured Mater.* **7**, 835-845.

FEATURES OF NANOCRYSTALLINE STRUCTURE FORMATION ON SINTERING OF ULTRA-FINE POWDERS

V. V. SKOROKHOD, A. V. RAGULYA

Frantzevich Institute for Problems of Material Science NAS,
3, Krzhizhanovsky str., 252680 Kiev, Ukraine.

1. Abstract

This paper discusses the non-isothermal sintering of crystalline nanosize particles of metals, oxides and refractory compounds.

The processes of pore and particle coalescence by surface self-diffusion at the initial sintering stage are considered. Local non-uniform densification, and the evolution of pore morphology at the intermediate stage as well as local non-uniform diffusion viscous flow and grain growth at the final stage of sintering are theoretically described.

Special attention is paid to non-isothermal sintering methods, particularly the rate-controlled sintering (RCS). Major mechanisms of mass transport are analyzed from the view point of their dependence on temperature and size of particles. The influence of the local densification rate fluctuation on tensile stresses and, therefore, low density region formation is considered, as well as possible interpretation of the maximum safe rate, the key RCS parameter, is proposed. Grain structure development results from the pore size distribution at the late stage of rate-controlled sintering. Several experimental results are presented to describe the sintering of ultrafine single-phase systems such as: nickel, barium titanate and titanium nitride as well as some nano-composites under different temperature regimes (isothermal, constant rate of heating, rate-controlled sintering). Final parameters such as density, grain structure, and mechanical properties are discussed.

2. Introduction

The development of novel non-isothermal methods of sintering is based on two motivations. First of them is a practical one, since current research interests shifted to the realm of nanocrystalline and submicrongrained materials manufactured from ultrafine powders which are known to possess high sintering activity and the main part of their densification occurs on heating. The second one is connected with clear theoretical understanding of the main mass transfer mechanisms defining densification pore and grain growth, their concurrence and dependence on

G.M. Chow and N.I. Noskova (eds.), Nanostructured Materials, 387–404.

sintering schedule. Nevertheless, many researchers use isothermal or constant heating rate sintering modes, although there are no clear physical reasons to fix the temperature or heating rate. Nanocrystalline powders show very rapid densification kinetics on heating. This process is often accompanied by the formation of large stable pores and rapid local coarsening of the grain structure. The mentioned peculiarities of sintering demand unconventional sintering methods, favorable to accelerate pore elimination and to retard grain growth.

In the past, the rapid heating rate sintering, especially for the first sintering stage, was proposed to overcome the undesirable effect of pore coarsening through surface diffusion at low and moderate temperatures [1-3]. However, the fast firing does not always give a benefit because of extreme tendency of the agglomerated nanocrystalline powders to the interagglomerate crack-like voids which hinder further densification under fast heating rates [4,5]. Thus, the search of the best sintering mode for nanocrystalline materials is in progress. Bellow, are listed the novel nonisothermal sintering techniques suitable for consolidation of nano-size powders:

- rate-controlled sintering (RCS) [6-9];
- constant heating rate sintering (CHRS)[2-5];
- ultra-rapid sintering (URS) [1,10]
- microwave sintering (MWS) [11];
- selective laser sintering (SLS) [12] etc.

Some useful combinations of these methods can be widely applied to summarize the advantages of each of the combined methods.

One of the contemporary trends in unconventional sintering techniques is connected with rate-controlled sintering (RCS) as a morphology oriented control concept. This method of non-isothermal, non-linear sintering schedule attracts the attention of ceramists due to its wide possibilities of structural control. Specimens sintered under the RCS conditions are characterized by finer grain sizes and pore-free structures. The concept of rate controlled sintering was first introduced by Palmour and Johnson, who postulated that a feedback has to be established between the sintering temperature and the instantaneous density to develop a finer-grained microstructure in a nearly dense sample [6]. Thus, during RCS, the temperature becomes a dependent variable, in direct role reversal from conventional sintering. Rate controlled sintering has been used successfully to sinter different ceramic powders, particularly, alumina ceramic powders, to a final grain size of 1 μm, as opposed to 2 μm by conventional (linear heating rate + isothermal hold) sintering. On the micron scale, this improvement may seem insignificant, however, a reduction of grain size by a factor of two on the nanometer scale (e.g., from 70 to 35 nm, or 30 to 15 nm), leads to a significant increase in the absolute number of grain boundaries present, and hence a much stronger expression of grain boundary-dependent properties. Authors of the RCS concept proposed the heuristic idea [7]: there is a maximum safe rate of densification for each level of fractional density. This fastest rate allowed, usually occuring only in the last part of the initial stage of sintering, is necessary to control both pore and grain structure evolution. Nevertheless, the maximum safe rate still has no thorough theoretical explanation and can not be calculated beforehand.

This paper aims at contributing to the research of the non-isothermal sintering and rate controlled sintering through, the development of both theoretical explanation and experimental verification of RCS validity to obtain dense and nano-grained sintered materials with improved properties. From the beginning, we are going to discuss, why the sintering schedule of nanocrystalline powders should be non-isothermal, nonlinear and rather rate-controlled, to give the possible interpretation of the maximum safe rate and then show experimental proofs of the advantages of the RCS.

3. Theory

A dispersive system tends to decrease its free surface energy, which serves as the driving force for sintering (a thermodynamically irreversible process). The process of densification is of essentially rheological character, and its rate is determined by the effective viscosity of the dispersive system as a whole. The other process of free energy decrease has a character of a diffusion coalescence. The two processes are in concurrence which can be controlled through the temperature schedule. Since the technological aim of sintering is usually the attainment of a high relative density of a material, it is reasonable to make the first process intensive, while the second one should be retarded or completely suppressed. Consider this competition in detail.

Define the specific surface area as the total surface area of a powder, or pores in a porous body diviled by total volume of matter $s_s = S/V_m$ (here s_s is in $cm^2/cm^3 = cm^{-1}$). Thus, the specific surface area is proportional to the ratio: s_s ~ A/L, where A is a numerical constant $(3 \div 4)$ and L is the structural parameter (average size of particles for powder, or average diameter of pores in the porous body, for the cylinder pores $L = 2R_{cyl}$). If the volume V of a body and L are independent variables, then the total differential of s_s can be written as in [13]:

$$ds_s = \frac{\partial s_s}{\partial V}\bigg|_{L=const} dV + \frac{\partial s_s}{\partial L}\bigg|_{V=const} dL \cdot \quad (1)$$

In the rheological theory of sintering where the statistical model of a porous body is used, the second term has been ignored in the assumption that the surface modification is exclusively due to the densification of the sintered body. Such an approximation is justified if the grain size is greater than 10^{-3} cm. In the case of nano-particle ensemble, the surface area variation occurs mainly through the variation of structural parameter.

At low and moderate temperatures, the presintering of nanosize particles occurs by the surface self-diffusion mechanism. In this case, the kinetic equation can be easily deduced for isometric variation of the structural parameter using a modified equation of diffusion coalescence [13]:

$$L^4 = L_0^4 + \frac{32\sigma D_s \delta^4}{kT}t \qquad (2)$$

where D_s is a surface self-diffusivity, σ is a surface tension, δ is a surface thickness, L_0 is the initial value of structural parameter, k is Boltzmann constant, and T is temperature. Equation (2) is suitable to describe the coalescence in nanostructured systems such as pore ensemble or particle ensemble. At low temperatures, the coalescence of pores prevails over the coalescence of particles. Thus, surface relaxation by pore coalescence leads to the increase in mean pore size. In the real nano-size systems, it appears to be impossible to retard this process. However, it seems to be reasonable to minimize this contribution using the competition between pore coalescence and densification at the initial and intermediate stages of non-isothermal sintering. This competition is graphically illustrated in Fig. 1.

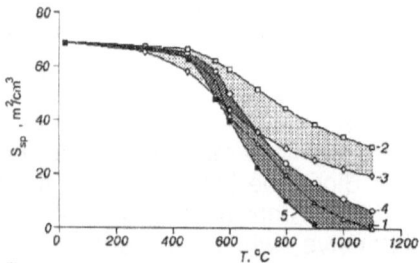

Fig. 1. Temperature dependence of the specific surface area of pores for nano-Ni powder (heating rates are of 1,2,4 - 0.44 °C/s; 3,5 - 0.18 °C/s: 1- experimental data, 2-5 - calculated data (2,3 for coalescence and 4,5 for densification) [4].

To develop this plot, the coalescence of pores (the field between curves 2 and 3) and densification (curves 4 and 5) were calculated and compared with the experimental data (curve 1) [4]. The higher is the heating rate, the lower is the contribution of pore coalescence to the decrease in the general surface area. At low heating rates, (curves 3 and 5) there is a temperature range where coalescence prevails over densification. At higher heating rates, the densification predominates over coalescence in the whole temperature interval, and the pore growth becomes impeded.

Thus, when nanocrystalline powders are sintered, there exists significant instability of geometrical structure affecting the level of the driving force for sintering. For nanocrystalline powders where the mean characteristic size of the structure elements grows by the mechanism of surface diffusion according to Eq.(2), it is reasonable to assume the possibility of a linear viscous flow by the interparticle sliding mechanism [14] under the action of capillary forces. In this case, the system viscosity is inversely proportional to the cube of the linear structure element size [14]:

$$\frac{1}{\eta} = 100\frac{\Omega b D_b}{kTL^3}, \qquad (3)$$

where Ω is an atomic volume, D_b is grain boundary diffusion coefficient, b is a grain boundary thickness.

The influence of heating rate on sintering is quite complex, therefore, the linear heating rate regime can not be considered optimal for the best structure development path. It is easy to show why the temperature-time path should be non-linear. The system of kinetic equations describing densification and grain growth at the intermediate sintering stage is considered. In this case, the pore surface area decrease is assumed to occur by the surface self-diffusion. The locally-inhomogeneous diffusion-viscous flow, accompanying by thermally activated sliding along grain boundaries, is responsible for the macroscopic densification. As nano-particles are often monocrystals, the single structural parameter of the system becomes equal to the mean particle size a [13]. Thus, the main differential equation for densification kinetics can be written as:

$$\frac{d\theta}{\theta dt} = -A\frac{\sigma D_b \delta^4}{a^4 kT},$$ (4)

where θ is a porosity and A is a numerical constant of 10^3. Mean size of particles is grown due to coalescence, described by the relation similar to (2). So, taking into consideration this kinetics of growth, the integral in Eq.(4) in the case of isothermal sintering can be presented as:

$$\ln\left(\frac{\theta}{\theta_0}\right) = -\frac{A\,D_b}{B\,D_s}\ln\left(1 + \frac{BD_s\sigma\delta^4}{a_0^4 kT}t\right),$$ (5)

or

$$\theta = \theta_0\left(1 + \frac{BD_s\sigma\delta^4}{a_0^4 kT}t\right)^{-\frac{AD_b}{BD_s}}.$$ (6)

To analyze the simplest case of non-isothermal sintering with constant rate of heating, it is necessary to obtain the equation of non-isothermal coalescence. Having accomplished simple transformations in relation (2), one can deduce [15]:

$$a^4 = a_0^4 + \frac{B}{HE_s}\sigma\delta^4 TD_{s0}\exp\left(-\frac{E_s}{kT}\right),$$ (7)

where D_{so}, E_s are the preexponent and activation energy of surface self-diffusion, respectively. The differential equation describing densification under constant heating rate sintering is:

$$\frac{d\theta}{\theta dT} = -\frac{A\sigma\delta^4 D_{b0}\exp(-E_b/kT)}{HkT\left(a_0^4 + \frac{B}{HE_s}\sigma\delta^4 TD_{s0}\exp(-E_s/kT)\right)}.$$ (8)

For nanocrystalline systems where $a_0 \leq 10^{-5}$ cm, at moderate temperatures and heating rates, the second term in the denominator of (8) is considerably greater than 10^{-20} cm^4 so, it would be correct to neglect by the addend a_0^4. Thus, the densification rate does not depend on heating rate and initial particle size [15]:

$$\frac{d\theta}{\theta dT} = -\frac{AE_s}{BkT^2}\frac{D_{bo}}{D_{so}}\exp\left(-\frac{E_b - E_s}{kT}\right).$$ (9)

Using the procedure of graphical differentiation of the experimentally determined curves θ vs. T, and plotting the function $\ln T^2 \Delta\theta/\theta\Delta T \div 1/T$, it appears possible to calculate the difference between activation energies of grain boundary and surface diffusion. The linearity of such function confirms the postulated sintering mechanism at the intermediate sintering stage —

grain boundary diffusion-viscous flow going with coalescence-controlled grain growth by the surface self-diffusion. Thus, nano-sized monocrystal particles appear to be unique model objects for basic sintering study.

By integrating equation (9), one can obtain the universal dependence of densification against the final temperature of non-isothermal sintering with any constant heating rate. This formula contains basic kinetic parameters of the substance only, i.e. the coefficients of grain boundary and surface self-diffusion and their activation energies:

$$\ln\left(\frac{\theta_0}{\theta}\right) = \frac{A}{B} \frac{E_s}{E_b - E_s} \frac{D_{b0}}{D_{s0}} \exp\left(-\frac{E_b - E_s}{kT}\right). \tag{10}$$

These kinetic parameters depend on chemical bonding in sintered substances, so, the equation (10) allows the prediction of sinterability for metals, high melting point compounds, oxides and covalent crystals in nanocrystalline state.

To generalize the presented theoretical results, the kinetic equations will be rewritten as follows:

$$f(\theta)\frac{d\theta}{dt} \approx \frac{A\sigma D}{a^m kT}, \tag{11}$$

$$a^n = a_0^n + \frac{B\sigma D'}{kT}t, \tag{12}$$

where t is time, m and n are power indexes, D and D′ are the effective diffusivities defining densification and grain growth respectively. Combining this equations and taking into consideration the relation T=T₀+Ht, one can obtain the differential equation for the non-isothermal linear heating rate regime:

$$f(\theta)\frac{d\theta}{dT} \approx \frac{A' \exp\left(-\dfrac{Q}{RT}\right)}{HT\left(ka_0^n + \dfrac{B}{H}T\exp\left(-\dfrac{Q'}{RT}\right)\right)^{m/n}}, \tag{13}$$

where Q and Q′ are the effective activation energies of densification and grain growth, respectively. Neglecting the addend ka_0^n, one obtains:

$$\frac{d\theta}{dT} \approx const \frac{\exp\left(-\dfrac{Q - m/n Q'}{RT}\right)}{H^{1-\frac{m}{n}}}. \tag{14}$$

This consideration permits us to carry out the general analysis of densification dependence on the heating rate. First of all, it should be mentioned that densification does not depend on heating rate if m=n.

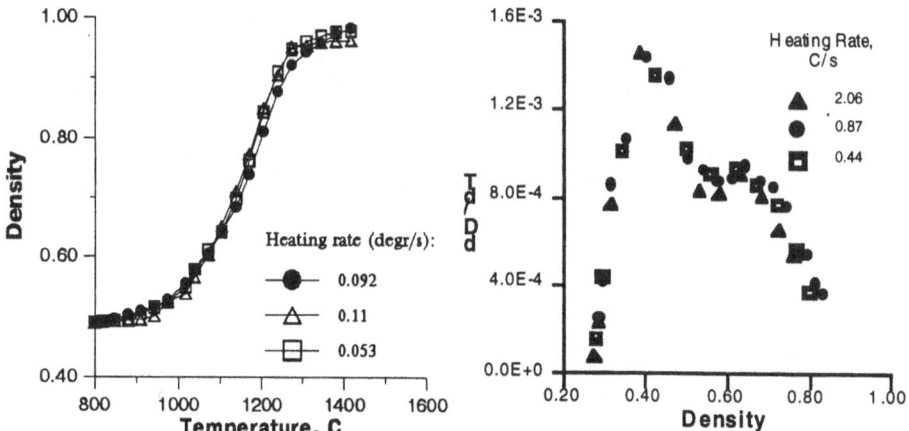

Fig. 2. Density of nanocrystalline TiN powder as a function of instantaneous temperature of sintering with different heating rates.

Fig. 3. Temperature derivative of relative density versus density of nanocrystalline nickel powder sintered with different heating rates.

If grain boundary diffusion predominates over volume diffusion at the initial and intermediate stages the power index of densification will be m=4. Meanwhile, the power index of grain growth is n=4 if ultrafine particles coalesce on the same stages through the action of surface diffusion. Indeed, it is a common case for nanosize powders when temperature derivative of density does not depend upon heating rate. For nanocrystalline titanium nitride, nano-nickel powders as well as composition B_4C-TiB_2 [15] this fact has been experimentally confirmed. Here these results are presented in Figs. 2,3,4.

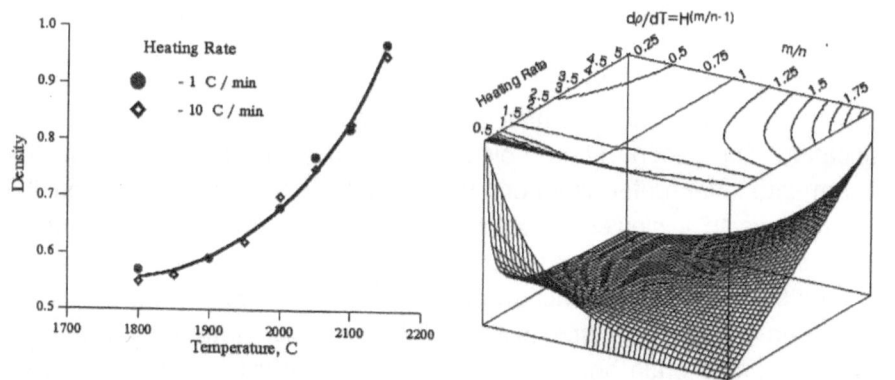

Fig. 5. Temperature derivative of fractional density vs. heating rate and ratio m/n in the eq. 14.

Fig. 4. Temperature dependence of relative density of the B_4C-TiB_2 composite.

If m>n, the temperature derivative of density (or porosity) is directly proportional to the heating rate and, vise versa, at m<n, it is inversely proportional to the heating rate. The both cases seem to be eventual at the late intermediate and final stages at high temperatures, i.e., where the contribution of volume diffusion overlaps that of grain boundary diffusion

and the surface mass transfer is diminished. The temperature - time path of the optimized rate-controlled sintering (see bellow) always includes the final part of heating rate increase. The singular point, as heating rate tends to zero, is rather an artifact of evaluations (Fig. 4). There is a range of heating rates (around one) where densification is not depends upon the ratio m/n. If the contribution of different mass transfer mechanisms to the integral densification kinetics varies with density and temperature, the ratio m/n will be also variable. Thus, the heating rate should be considered as a variable to optimize the densification process as a whole and for every instantaneous fractional density. This statement is a basis for rate - controlled sintering technique which is most available to prepare the dense and nano-grained materials.

4. Rate Controlled Sintering

The main peculiarities of sintering under linear heating conditions have been analyzed above. Authors [4] supposed that different micromechanisms of structural evolution (coalescence of pores, their shrinkage and grain growth) occurred concurrently and because of this competition the sintering process could be optimized through manipulation.

The typical RCS schedule uses a multizoned path included the fastest rates at the initial stage, slower rates at the intermediate stage, and then a regime of continually decreasing rates for the final stage of sintering [6]. In the first step, a fast and nearly linear densification rate is imposed, until the sample reaches a density of approximately 72.5% of theoretical. The purpose of this stage is to use fast heating rates to quickly bypass the surface diffusion regime, in which pore coarsening takes place with little densification. The second stage holds the densification rate constant at the maximum safe level (determined empirically) until 75% of theoretical density, then transitions to a slower densification rate form 75% to 77.5% density. The third stage of RCS imposes a considerably slower densification rate, up to about 85% density, to encourage pores within the microstructure to remain open as long as possible, because in this state they are very effective at pinning grain boundaries. From 85% to 100% density, the fourth stage of sintering is utilized; here, a still slower densification rate is imposed, which decreases in a logarithmic manner with increasing density. This stage permits the pores, now closed, sufficient time to disappear by diffusion without becoming detached from grain boundaries (an event which often provokes excessive grain growth).

Palmour [6,7] has postulated that to attain the uniform development of structure and finally minimal porosity, the maximum safe shrinkage rate should not be exceeded. Such a rule was deduced from numerous experimental observations. However, a rigorous criterion should be obtained theoretically from the competition between different mechanisms defining pore and grain structure evolution. Besides the above stated microscale mechanisms, the mesoscale phenomenon of local de-sintering occurs in agglomerated nanosize powders (the term is differential sintering [16]). This effect is a sequence of spatial heterogeneities, first of all, particle agglomerates. Therefore, the maximum safe rate can be determined by the deviation of local values of structural parameters from average ones. Physical interpretation was clearly formulated by Sñherer [17]: "the effective shrinkage rate of the heterogeneous porous body is a superposition of the uniform shrinkage rate and it's spatial variation". Consider a theoretical explanation of maximum safe rate of densification.

The general approach is considered here to estimate the principles of structural evolution in the course of sintering. It is based on the variation principle providing dissipation of free surface energy of a dispersed system. There are at least three paths for free surface energy decrease on sintering: surface smoothening, free surface area decrease by pore shrinkage, and boundary surface area decrease by grain growth. Thus, the free energy variation can be written:

$$dF = (\sigma ds_s + \gamma ds_b) v_m , \qquad (15)$$

where s_s and s_b are the specific surface areas (defined as above) of pores and grain boundaries respectively, v_m is the volume of material, σ and γ are the surface and grain boundary energies. The specific surface area is a function of porosity and mean curvature of pores was described by Skorokhod in the scope of the statistical model of porous body as follows [18]:

$$s_s = \frac{A}{a} \varphi(\theta, a, r), \quad where \ \varphi(\theta, a, r) = \frac{a\theta}{a\theta + (1-\theta)r} , \qquad (16)$$

and a is the mean grain size, r is the mean pore size, θ is porosity, and A is a geometric constant. This formula has been deduced by taking into account the proportion between interface area and general surface area of each phase in the statistical two phase mixture:

$$s_a = A/a, \quad s_\theta = A\theta/r(1-\theta), \quad s_s = s_a \cdot s_\theta/(s_a + s_\theta) \qquad (16a)$$

The function $\varphi(\theta, a, r)$ satisfies the following conditions: it is monotonous, equals to zero at $\theta = 0$ and is finite at $\theta \to 1$. The latter condition corresponds to the upper limit of the surface area which is proportional to the mean particle size of powder A/a. Therefore, the grain boundary surface area can be presented as:

$$s_b = \frac{1}{2}\left(\frac{A}{a} - s_s\right) = \frac{A(1-\theta)r}{2a(a\theta + (1-\theta)r)} . \tag{17}$$

By taking differentials of s_s and s_b and substituting them into Eq. (15) one can find the total differential of the free energy of a porous body:

$$dF = (2\sigma - \gamma)\frac{Av_m}{2(a\theta + r(1-\theta))^2}\left[rd\theta - \theta^2 da - \theta(1-\theta)dr\right] - \gamma\frac{Av_m}{2a^2}da \tag{18}$$

The energy dissipation rate is known to equal to the double value of the dissipation function describing the irreversible deformation of the porous body under the action of the surface tension:

$$-\frac{dF}{dt} = -\sigma\frac{dS}{dt} = 2\int_V \Psi dV \tag{19}$$

For the uniform densification, the dissipation function aquires the form:

$$\Psi = \frac{\zeta}{2}\dot{u}_{ll}^2; \quad \dot{u}_{ll} = \frac{dV}{Vdt} = \frac{\theta}{1-\theta}, \tag{20}$$

where θ is the mean porosity, $\dot{u}_{ll}$ is the tensor of strain rate, ζ is a volume viscosity. Dissipation function consists of two components corresponding to as macroscopic energy dissipation which is defined by porosity decrease as local energy dissipation during the local mass redistribution. So, the energy dissipation rate is:

$$\dot{H} = 2\int_V \Psi\left(\dot{u}_{ll}^2\right)dV = \frac{6\eta_0}{3 - \theta(1-\theta)}\dot{\theta} + 2\eta_0\int_{V_m}\left(\dot{u}^2 - \bar{\dot{u}}^2\right)dV_m, \tag{21}$$

where η_0 is the viscosity of pore-free sample and dots over symbols denote time derivatives, the first addend in (21) corresponds to uniform densification rate of porous body, and the second addend is responsible for local densification rate variation. Neglecting the energy dissipation of grain growth and minimizing the variation of the integral, the energy balance in the system can be rewritten for the uniform densification:

$$\dot{\theta}^2 + \Phi r\dot{\theta} - \Phi\theta(1-\theta)\dot{r} = \Phi\theta^2\dot{a} + K(3 - \theta(1-\theta))\dot{a}, \tag{22}$$

where Φ, K are functions:

$$\Phi = \frac{(2\sigma - \gamma)A(3 - \theta(1-\theta))}{12\eta_0[a\theta + r(1-\theta)]^2}; \quad K = \gamma\frac{A}{12a^2\eta_0} \tag{23}$$

To analyze Eq. (22), it is appropriate to find a solution if its right-hand part is equal to zero (the rate of grain growth is equal to zero). Therefore, there is a quadratic equation with respect to the densification rate. The roots of this equation represent upper and lower limits of the densification rate at instantaneous porosity and determine the range within which grain growth is minimal (Figs. 6,7).

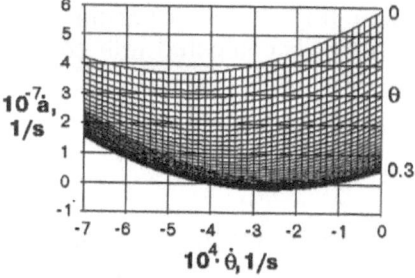

Fig. 6. Grain growth rate vs. densification rate and porosity (one projection): to the determination of the sintering safe rates minimizing the rate of grain growth.

Fig. 7. Grain growth rate vs. densification rate and local pore shrinkage rate (one projection). Current porosity 45%.

Both plots have been calculated for nanocrystalline Ni powder using tabulated values of diffusion coefficients for Ni. If one considers $\sim 4 \cdot 10^{-4}$ 1/s (Fig. 6) as the maximum safe rate allowing minimal grain growth at given porosity, then there is very good coincidence between the results of calculations and experimental results presented bellow (the experimental value for 33% porosity is around $2 \div 5 \cdot 10^{-4}$ 1/s [19]).

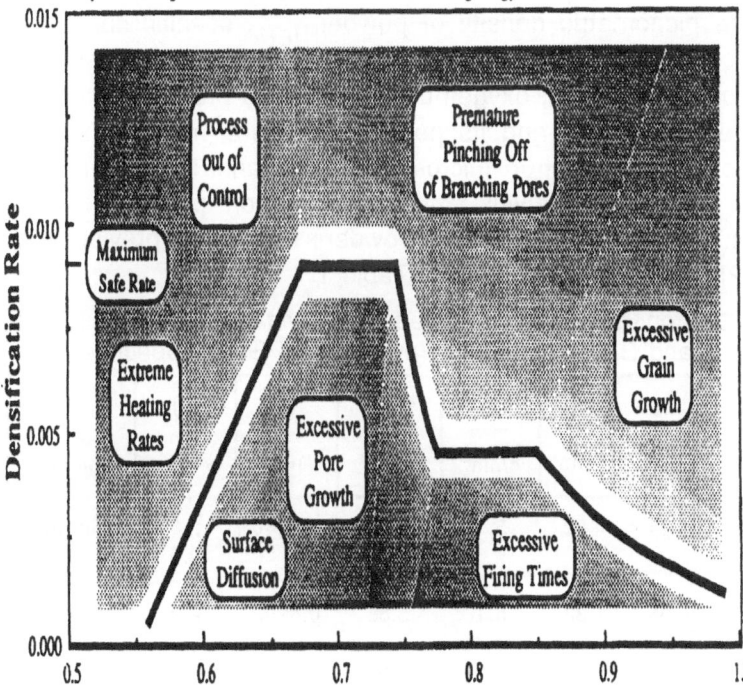

Fig. 8. Typical RCS profile, expressed as dD/dt vs. D, annotated to indicate some practical constraints as well as some underlying science issues that can - and should -influence path selection [8].

The scheme presents a rather qualitative explanation of maximum safe rate than quantitative procedure permitting the best sintering schedule prediction. However, it allows us to introduce the minimum safe densificatior rate, bellow which the microstructure coarsening occurs as it was shown above and according to Palmour's scheme (Fig. 8) [8]. Consider some practical examples illustrating advantages of RCS compared with constant heating rate sintering.

5. Experiments

To date, the methodology of rate-controlled sintering has been mainly applied to obtain dense nano-grained ceramic [20,21], and metallic materials [19,22]. Presented below are several examples of rate-controlled sintering application to develop dense materials with nanograined structure. The nanocrystalline (NC) powders of substances with different type of chemical bonding, namely, metallic NC-Ni, semi-metallic NC-TiN, and ionic NC-BaTiO$_3$ were selected to show the validity of RCS. All powders of different origination were produced by evaporation-condensation in vacuum, plasma synthesis and rate-controlled decomposition of organic precursors, respectively.

The picnometric density of powder $\gamma_{picn.}$, specific surface area S_{sp}, particle size distribution (average particle size R), theoretical density $\gamma_{theor.}$ and lattice parameter a, oxygen content [O] and pressability (green density of a specimen ρ_{green} and its porosity θ) were measured by nitrogen absorption (BET), electron microscopy (SEM and TEM), X-ray diffraction, neutron activation analysis, mercury intrusion porosimetry and other methods. These properties of the powders have been previously described in [19-22] and summarized here in Table 1.

TABLE 1. Properties of as-Prepared Nanocrystalline Titanium Nitride, Nickel and Barium Titanate Powders and Green Specimens

Powder	S_{sp}, m^2/g	R, nm	$\gamma_{picn.}$, g/ñm^3	$\gamma_{theor.}$, g/ñm^3	a, nm	[O] % mass	ρ_{green}, g/ñm^3	θ %
TiN(P)	14	82	5.1	5.35	0.4240	2.5 0.5	2.68±0.05	50.0
TiN(S)	35	32	5.1	5.35	0.4233	2.2	2.62±0.02	49.0
Ni	20.6	33	8.86	8.90	3.521	6.25 0.6-0.8	2.94±0.02	67.0
BaTiO$_3$	38	26	6.0	6.04	4.0050	-	2.80±0.03	53.5

The most important precondition for sintering optimization of NC powders is that the heating rate dependence of final density has a maximum. Figs. 9,10 show the details of the densification behavior for Ni and BaTiO$_3$ powders. Such a dependence allows one to carry out the rapid search of the best temperature-time path for sintering, and to deduce the important relations connecting the densification kinetics and structural evolution.

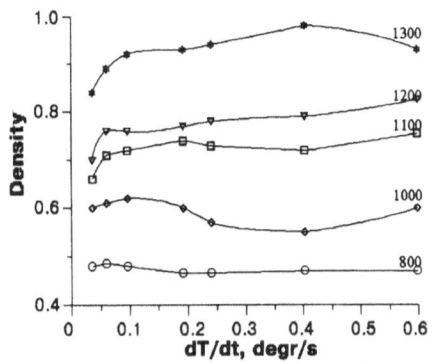

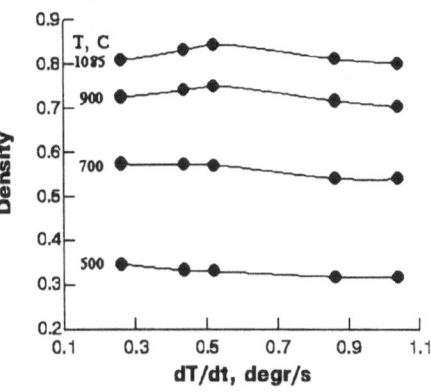

Fig. 10. Densification against heating rate for nano-Ni powder Best density 0.922 was achieved at heating rate ~0.5 degr/s.

Fig. 9. Densification against heating rate for nano-BaTiO$_3$ powder. Best density 0.98 was achieved at heating rate ~0.4 degr/s.

This extremum is the result of the concurrence between different mechanisms of structure evolution during sintering. By following the general methodology of RCS, the data base of kinetics responses was formed using preliminary kinetic experiments on sintering under constant heating rate. The temperature-time paths were developed to sinter nano-grained nickel, titanium nitride and barium titanate powders (Figs. 11,12). Figs. 11 illustrate the optimized sintering profiles, corresponding to curves of Figs. 12, which are the best temperature time paths.

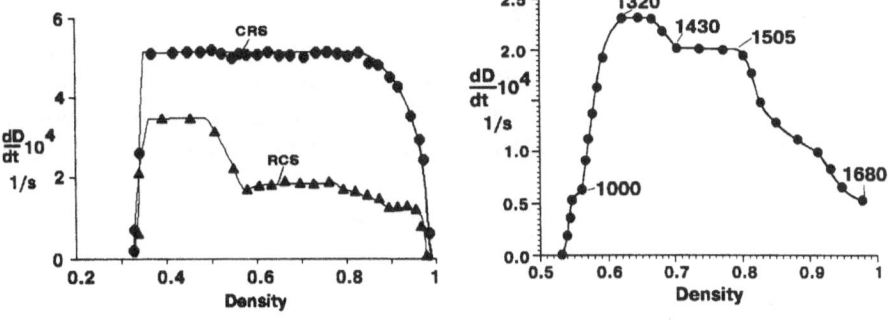

A   B

Fig. 11 Optimized rate-controlled sintering paths for Ni (A) and TiN(P) (B) RCS-rate-controlled sintering, CRS- constant rate of sintering

One can point out that every substances studied which have beer undergone RCS achieved high final densities exceeding 0.985 for NC-N₁ and 0.992 for both NC-TiN powders. The sintering under rate controlled conditions allowed the grain size to remain around 80 nm and 150-200 nm for Ni and TiN, respectively, at final density. As a result of the rate-controlled sintering, barium titanate also densified up to 99.9 % of theoretical density, and an average grain size of 0,1 - 0,3 μm was obtained. The final temperature of sintering of BaTiO₃ was about 1285 ± 5 °Ñ which is 50-70 °C lower than that for conventional powders.

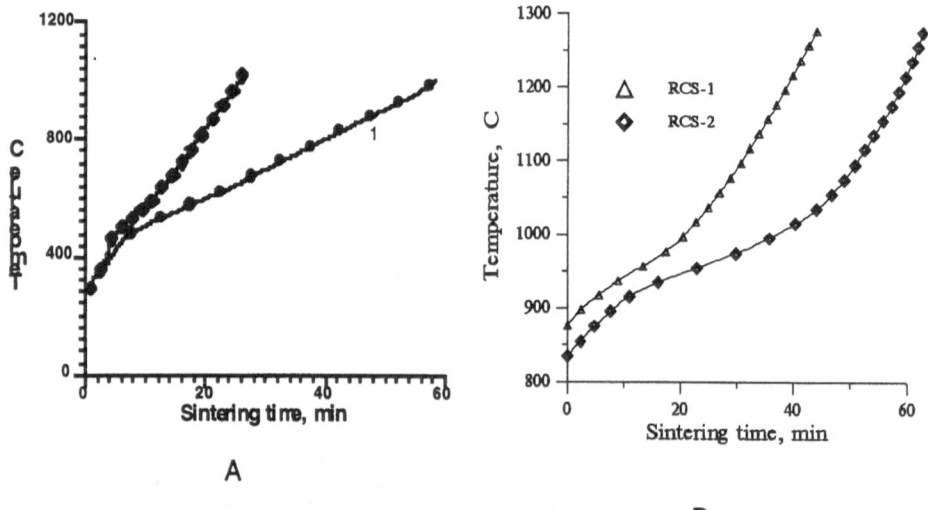

A

B

Fig. 12. Best temperature-time paths for the rate-controlled sintering of the nanocrystalline nickel powder (A) and nanocrystalline barium titanate powder (B).

6. Evolution of porous structure under non-isothermal sintering.

The difference in final residual porosity at various heating rates is explained by path for the pore structure evolution as it is shown for nanosize Ni powder (Fig.13,14) [4]. For nano-powders of TiN(P) and BaTiO₃ the similar behavior was found as well in [20-22]. Figs. 13 A,B for Ni and 14 A,B present the pore size distribution for conventional sintering mode A (constant rate of heating - CHRS) and controlled sintering process B respectively.

After comparing these pore size distributions one can conclude that the initial densification stages are similar: the average pore diameter remains approximately constant and the number of small pores decreases. During CHRS process at higher temperatures pore coarsening predominates over pore shrinkage and the coarse pores become stable. On the contrary, during the rate - controlled processes, a uniform decrease of pore volume occurs in the whole density and temperature ranges and there is an approximate balance between pore growth and shrinkage. The influence of heating rate on pore size distribution is of twofold character. When slowly heated, the ripening of pores and coarse pore stabilization occurs due to coalescence by surface diffusion at low temperatures. The effect of pore coarsening is also observed in the case of excess rapid heating, due to the effect of local intraagglomerate densification (differential densification). The latter can be exacerbated by the further isothermal hold. Thus, a different range of heating rates where pore coarsening is sufficiently compensated by pore shrinkage (see above the case n=m) should be found to optimize the heating mode.

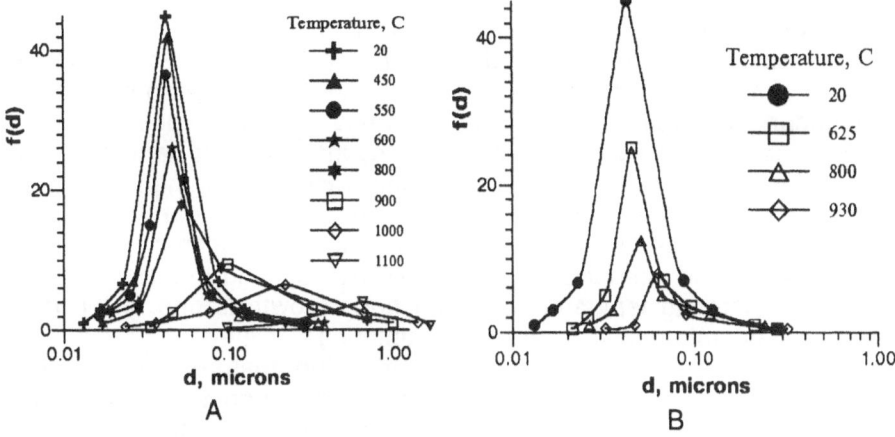

Fig. 13. Pore size (d) distribution in the sintered NC-Ni after heating up to different temperatures at Constant heating rate of 0.44 °C/s (A) and Rate-controlled sintering (B)

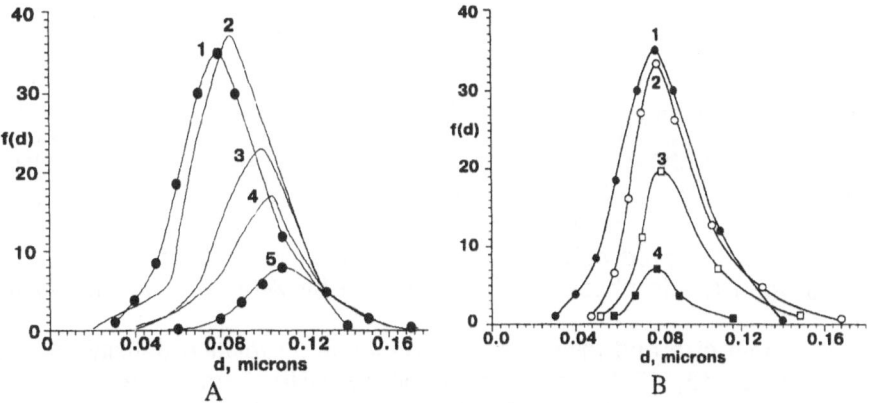

Fig. 14 Pore size (d) distribution in the sintered NC-TiN(P) after heating up to different temperatures at Constant heating rate of 0.44 °C/s (A) and Rate-controlled sintering (B)

Thereby, the maximum safe rate should not be exceeded mainly at the initial and intermediate stages of sintering. A density level close to theoretical one was reached at temperatures less than those for the constant heating rate mode in the preliminary experiments. It is a favorable fact to prevent the grain growth. During rate-controlled sintering there is no noticeable pore coalescence at lower temperatures and average pore size monotonously decreases at high temperatures (Figs. 13B, 14B). It is worth pointing out that closed porosity appears during controlled sintering at a porosity level less than 14 % and stays negligible (less than 1.3-1.6 %) until full densification, whereas under conventional constant heating rate sintering, pore isolation occurs approximately at 17-18% of general porosity and leads to closed porosity increase up to 5-6%.

7. Properties

Material properties, such as strength, hardness, fracture toughness, superplasticity, coercive force, and dielectric constant all depend on the grain size and the grain boundary state (e.g., crystallographic orientation of the boundary, boundary energy, the presence of foreign phases at the boundary, and impurity segregation at the boundary). In a nanocrystalline material, for which up to 50% of the atoms reside on or near the grain boundaries, the details of the grain boundary state can completely determine the material properties. High energy grain boundaries, which lead to large surface and interfacial tensions, also directly lead to faster sintering rates (lower sintering temperatures). The grain growth is one of the main grain boundary state-defining process on sintering.

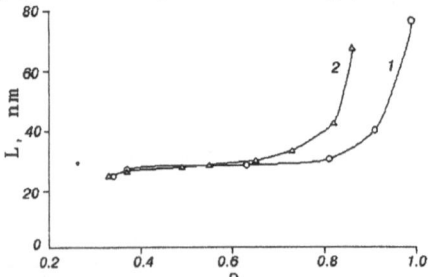

Fig. 15. Density dependence of the coherent scattering field for sintered NC-Ni powder: 1 - RCS, 2 - CHRS

Consider the features of the grain growth during non-isothermal sintering. The experimental data were obtained for sintered nanocrystalline Ni and TiN. The increase of the grain size is compared for both linear heating rate sintering and rate-controlled sintering (Fig. 15). Very close final grain sizes were observed in the two regimes but the residual porosity differed by 7-8%. An additional isothermal hold after CHRS results in relative density increase up to 98.6% and grain size increase up to 0.9 μm. Despite high densities of the samples (~99%), a reduction in hardness was observed because of grain boundary contamination [19] by excess oxygen (oxide phase was not detected by X-ray but was found by AES).

The grain size and microhardness depends on relative density as it is shown in Figs. 16, 17 for both conventional (CHRS) and rate-controlled sintering modes. Sintered nanocrystalline TiN(P) with high density and fine-grain structure has the highest hardness. Microhardness was measured using the loading of 50 g. At the same density, both sintering modes (RCS and CHRS) result in different microhardness due to different pore and grain size (Figs. 16,17). The hardness of RCS-sintered nano-grained titanium nitride around 23.5 GPa is higher than the hardness after the CHRS (16 GPa), is close to the best results of hot pressing (24 GPa), and is lower than one for TiN thin films (28-35 GPa) [23,24].

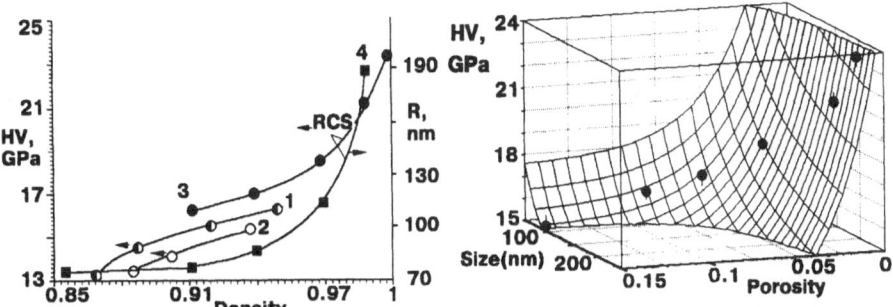

Fig. 16. Hardness (curves 1 - 3) and grain size (curve 4) versus porosity: linear heating rates of 1 - 0.3, 2 - 0.1 °C/s, 3, 4 - RCS.

Fig. 17. Hardness versus porosity and grain size (approximated 3-d plot with experimental points).

According to Hall-Petch type empirical equation $H=H_0+kL^{-0.5}$, the hardness should increase in pore-free titanium nitride from 20 GPa at grain size $L=10^3$ nm (experimental value) to 33 GPa at $L=70$ nm (expected value) from the best approximation of experimental data.

The best hardness at RCS in comparison with CHRS can be explained by higher density and finer grain size at RCS. Grain growth and density increase results in decrease of hardness. The three-dimension surface approximating the dependence of hardness on porosity (θ) and grain size (Fig. 17) was calculated as a sum of an exponential decay function and a Hall-Petch function:

$$H=H_0+10^{-4}/L^{1/2} + 10 \exp(-40 \cdot \theta), \tag{24}$$

where $H_0=15.5$ GPa is the initial microhardness. This plot shows that the hardness depends on the porosity stronger than on the grain size especially at a porosity of less than 5%.

8. Conclusion

The sintering of ultrafine powders is a perspective technological process for manufacturing of dense nanocrystalline single-phase and multiphase materials. However, this technological process should be strictly controllable. Isothermal sintering of ultrafine powders is not efficient because of rapid termination of densification at large residual porosity and the intensive pore and grain coarsening can not be prevented.

The processes of nonisothermal sintering seem to be more perspective in view of simultaneous achievement of both full densification and microstructural refinement which are sensitive to heating rate. The optimization of temperature-time path of rate-controlled sintering (RCS) permits the best combination of density and grain size to be obtained, and thus, influences the grain growth on the late stage of sintering. The set of contemporary sights on the mechanism and kinetics of mass transfer in ultrafine systems allows the scientific basis to be given for an empirically found optimum mode of sintering at RCS.

At present, sintered materials have been already manufactured on the basis of nanocrystalline powders of metals, refractory nitrides, carbides, borides and oxides. Enhanced mechanical and electrophysical properties were achieved due to high density and nanocrystalline structure.

9. References

1. Johnson, D. L. and Rizzo, R.A. (1980) Plasma Sintering of β-alumina, *J.Amer. Ceram Soc. Bull.* **59**, 467-468.
2. Mostaghaci, H. and Brook, R.J. (1981) Fast Firing of Non-Stoichiometric BaTiO3 *J.Brit.Ceram.Soc.* **80**, 148-149.
3. Baumgartner, C.E. (1988) Fast Firing and Conventional Sintering of Lead Zirconate Titanate Ceramics. *J.Amer.Ceram.Soc.* **71**, C350-C353.
4. Skorokhod, V. V. and Ragulya, A.V. (1995) Evolution of the Porous Structure During Non Isothermal Sintering of Fine Powders, *Science of Sintering*, **27**, 89-98.
5. Chen, D.-J. and Mayo, M.J. (1996) Rapid Rate Sintering of Nanocrystalline ZrO2-3 mol% Y2O3, *J.Amer.Ceram.Soc.* **79**, 906-912.
6. Palmour III, H. and Johnson, D.R. (1967) Phenomenological Model for Rate-Controlled Sintering. *Sintering and related phenomena*, Gordon & Breach publishers, New-York 779.
7. Palmour III, H. and Hare, T.M. (1987) Rate-Controlled Sintering Revisited. *Sintering'85* Plenum Press, New-York, 16-34.
8. Palmour III, H. (1989) Rate Controlled Sintering for Ceramics and Selected Powder Metals. *Sintering'89*, Plenum Press, New-York 337.
9. Palmour III, H., Huckabee, M.L., and Hare, T.M. (1979) Rate-Controlled Sintering: Principles and Practice. *Material Science Monographs*, **4**, Elsevier, Amsterdam, 46-56.
10. Johnson, D. L. (1984) Ultra Rapid Sintering. *Sintering and Heterogeneous Catalysis*, G.C Kuczynski, A.E. Miller, G.A.Sargent, eds., Plenum Publ. Corp., New-York, 243-252.
11. I-Nan Lin, Horng-Yi Chang (1996) Microwave Sintering of Semiconducting Perovskite and Their Electrical Properties, *Sintering Technology*, Marcel Dekker, Inc. New-York, 439-447
12. Bourell, D. L., Marcus, H. L., Barlow, J. W., and Beaman, J.J. (1992) Selective laser sintering of metals and ceramics. *Int. J.Powder Metallurgy*, **28**, 369-382.
13. Skorokhod, V. V. (1987) Surface Relaxation, Dynamics of Geometric Structure and Macrokinetics of Densification During Sintering of Ultrafine Powders. *Sintering'85*, Plenum Press, New-York, 81-87.
14. Geguzin, Ya. E. (1984) *Physics of Sintering*, Nauka, Moscow.
15. Skorokhod, Vl. V., Vlajic, M.V., and Krstic, V.D. (1997) Sintering and microstructural development of the B4C-TiB2 ceramic composites. *Proc. of Yugoslav Conference on New Mater. YUGOMAT'97*, to be published.
16. Shueh, C.H., Evans, A. G., Cannon, R.M., and Brook, R.J. (1986) Viscoelastic stresses and sintering damage in heterogeneous powder compacts, *Acta Metallurgica* **34**, 927-936.
17. Scherer, G. W. (1984) Viscous Sintering of a Bimodal Pore Size Distribution, *J.of the Amer.Ceram.Soc.* **67**, 709-715.
18. Skorokhod, V.V. (1972) *Rheological Basis of Theory of Sintering*, Naukova Dumka, Kiev.
19. Ragulya, A.V., Skorokhod, V.V. and Andrievski, R.A. (1997), *NanoStructured Materials*, to be published
20. Ragulya, A.V., Vasylkiv, O.O., Skorokhod, V.V. (1997) *Powder Metallurgy(Rus)* in print
21. Ragulya, A.V. and Skorokhod, V. V. (1997) Validity of Rate-Controlled Sintering Method for Consolidation of Dense Nanocrystalline Materials, *14 Plansee Seminar*, **2**, 735-744.
22. Ragulya, A.V. and Skorokhod, V. V. (1995) Rate-controlled Sintering of Ultrafine Nickel Powder, *NanoStructured Materials* **7**, 835-844.
23. Andrievski, R.A., Kalinnikov, G.V., Potafeev, A.F. and Urbanovich, V.S. (1995) *NanoStructured Materials* **7**, 353.
24. Andrievski, R. A. (1994) Nanocrystalline high melting point compound- based materials (Review). *J.Mater. Sci.* **29**, 614-431.

CONSOLIDATION OF NANOCRYSTALLINE MATERIALS AT HIGH PRESSURES

V.S. URBANOVICH
Institute of Solid State and Semiconductor Physics
National Academy of Sciences of Belarus
17, P.Brovka Str., Minsk, 220 072, Belarus

1. Introduction

The main problem of consolidation of particulate nanostructured materials is to provide a complete compaction with retention of the nanocrystalline structure, i.e. maintaining the grain size characteristic of the initial ultrafine powder (UFP). Conventional sintering and hot pressing methods are not acceptable due to intensive recrystallization process. Currently, high energy consolidation techniques - various static and dynamic techniques of high pressures (conventional techniques of high pressures and high temperatures, hot forging and hot extrusion, explosive compaction and shock wave sintering), as well as electric discharge compacting are the most promising routes to fabricate powder nanocrystalline materials [1,2].

Among the above mentioned methods the static high pressures and shock wave sintering has gained the widest acceptance in the research and development practices and in the industry [3-6]. A successful use of these methods for production of diamond and cubic boron nitride-based superhard materials is widely known [6,7]. Application of high pressures for production of high-melting ceramic materials is extremely efficient to achieve high density without using additives. It provides a maximum realization of potentially high physical and mechanical properties of initial substances in the sintered sample [8].

The first experiments in high pressure consolidation of UFP revealed many important aspects and considerations [19-16]. Though the information regarding the use of these methods for production of nanocrystalline materials (NM) is still rather scarce, the number of publications on this topic is increasing year after year. The reviews [1,2,17,18] discussing the production and properties of high-melting compound-based NM contain a part of the information on the UFP high pressure consolidation. Most works are related to dynamic compaction methods of the UFP [19-27]. Recently, however, a number of interesting results have been obtained using static high pressures. The production of high-melting NM with increased hardness has been demonstrated [28-32]. In this paper the consolidation methods and mechanisms of forming a dense nanocrystalline structure of the material under high pressure are considered.

G.M. Chow and N.I. Noskova (eds.), Nanostructured Materials, 405–424.
© 1998 *Kluwer Academic Publishers.*

2. High Pressure Techniques and the Methods of Consolidation of Ultrafine Powders

The principal schemes of application of dynamic and static high pressures for consolidation of UFP are considered: explosive compaction or sintering in shock waves, magnetic impulse compaction, sintering at high static pressures and temperatures, as well as plastic deformation by torsion under high pressures.

2.1. DYNAMIC HIGH PRESSURES

Figure 1 provides a schematic diagram of material compression with shock waves in a cylindrical (a) and plane (b) capsules [3].

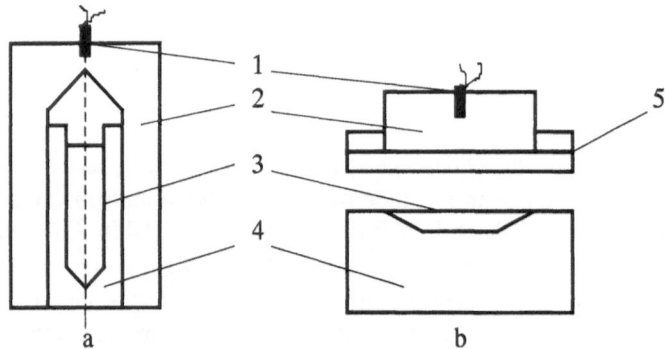

Figure 1. Schematic diagram of material compression with shock waves in a cylindrical (a) and plane (b) capsules: 1 - detonator, 2 - explosive charge, 3 - substance being compressed, 4 -capsule, 5 - flyer plate [3].

Method of retaining metal and non-metal materials in the cylindrical capsule (Fig.1a) after the shock wave effect was proposed first by Yu.N. Ryabinin [33]. The shock-loading-induced change of the substance state may be characterized by shock adiabat. A part of the shock wave energy is used for heating the material under adiabatic compression conditions. It has been determined that the higher pressure applied, the higher the proportion of the thermal component. With increasing pressure, for example, from 50 to 100 GPa on the Ni sample, its temperature increased from approximately 280° to 830 °C [4]. In the processes of the explosive compaction of porous materials an additional heating is induced not only due to the plastic deformation of particulates, but also due to their friction, while they are being compacted [4]. A more simple plane loading system is used commonly in the experiments to investigate the powder compacting in shock waves [12,19,25].

Characteristic properties of the shock loading compaction assemblies include the possibilities to apply a wide range of pressures from 2.5-100 GPa, a high loading velocity of 1-3 km/s and an exceptionally short time of pressure applied on the sample, i.e. units of microseconds. This condition provides a greater depth of solid phase processes at high loading velocities compared to the static pressures. Pressure and temperature distribution on end surfaces of the explosive-treated diamond powder sample is shown in Fig.2. [20].

It is clear from Fig. 2 that the time of the pressure and temperature effect is strongly dependent on the sample composition and is as low as 0.25-1 microsecond at 1000-1200 °C. Intensive labor is required to recover the treated sample from the steel capsule, which is a disadvantage of the shock compaction technique.

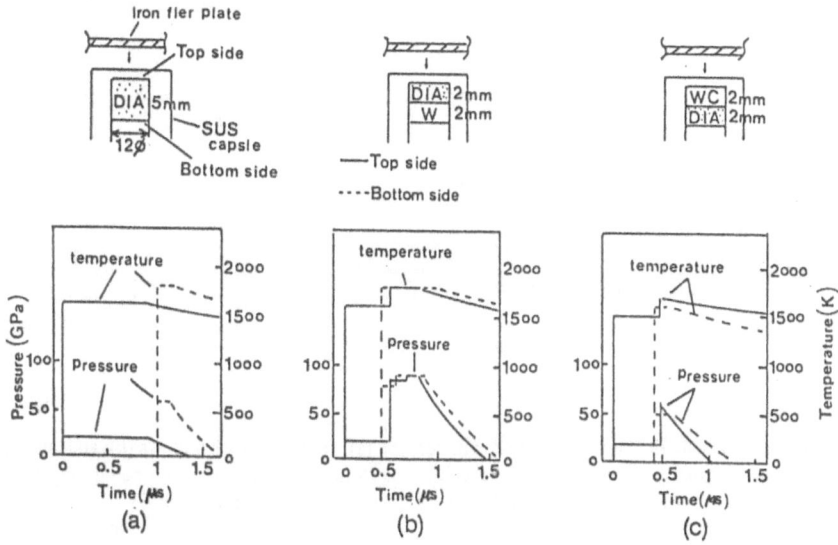

Figure 2. Cross sections of various assemblies and profiles of shock pressures and continuum temperatures at the top (solid line) and the bottom (broken line) sides for samples (a), (b) and (c). The release state due to the rarefaction wave is exaggerated [20].

Recently a magnetic impulse pressing method has been used for fabrication of high-density compacts from ceramic nanopowders, in addition to the shock wave technique [23,24,26]. Its assembly comprises a mould with two cylindrical WC-Co hard alloy punches in a vacuum chamber. A mechanical impulse compacting a powder in the mould is transmitted to one of the punches through a concentrator which is pushed from the region of the impulse magnetic field of the inducer due to a diamagnetic effect.

The magnetic impulse pressing technique provides compaction of ceramic nano-powders in vacuum at pressures up to 10 GPa. In this case the pressure pulse duration is about 250 μs [26], that is the time of the pressure applied on the powder is approximately two orders of magnitude longer compared to the explosive pressing. The method provides loading with preheating of the powder sample to 600 °C.

2.2. STATIC HIGH PRESSURES

The schematics of high static pressure assembly using a wide "technological" pressure and temperature range (P=1-15 GPa, $T_{max} \sim 3000$ °C) are shown in Fig.3. They a cylinder type, an anvil with hollows and a multipunch type apparatuses [34]. Solid substances are used in these devices as the media for transmitting pressure to the sample, thus com-

commonly referred to high-pressure solid state apparatuses.

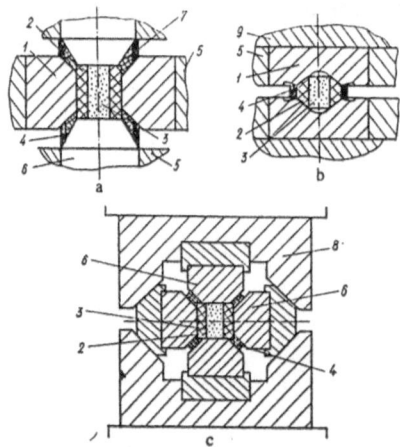

Figure 3. Schematic diagram of HPA: cylinder type (a); anvil type with hollows (b); and cubic (c):
1 - matrix; 2 - container; 3 - reaction composition; 4 - coupling; 5 - binding rings;
6 - punch; 7 - sealing gasket; 8 - pressure plate; 9 - support plate [34].

The cylinder type high-pressure apparatus (HPA) (Fig.3a) consists of a matrix fastened by binding rings provided with an axial hole for disposing the reaction cell with a sample and sealing gaskets; two punches having working end faces in a form of a truncated cone. The apparatus features a greater compression stroke due to the deforming seal (gasket) positioned at an angle to the punch motion direction. Various modifications of the device of this type are known as "belt" and "girdle" and their description is extensively provided in the literature [35,36]. Devices of this type are mainly used by the western researchers.

High pressure apparatuses of the anvil-and-hollow type (Fig.3b) are devices with the improved Bridgman anvil [37]. The apparatus comprises two matrices provided with hollows on working end faces to accommodate a container with a sample to be treated. The pressure in the reaction zone is produced by compressing the container and tightening it due to friction forces both in the container material and at a container-matrices interface. Soviet scientists have reached the most significant progress in developing the above devices. As a result a number of simple-in-design, cost-effective and easy-to-use high-pressure apparatuses of this type have been developed, which are different from each other by the relief of a matrix taper surface around the hollows [38-43].

The anvil-and-hollow apparatuses produce pressure over 10 GPa in the working volume up to 25 cm^3 [34]. They are most simple in manufacture and widely used both for industrial synthesis of superhard materials and research. They provide the maximum efficiency and cost-effectiveness in production of powders of superhard materials, especially during short synthesis cycles. They feature lower specific tungsten consumption and significantly lower production cost of various superhard materials compared to those of the high pressure cylinder apparatuses [6]. Negligeable value of the integral lateral force exerted on the chamber walls with respect to the axial force is a significant advantage of the anvil-and-hollow type over the "belt" type devices. This results in smaller device dimensions. These devices are better suited to be used in the press units developing force from 5-20 MN.

The cylinder type HPA, however, are superior to the anvil-and-hollow apparatuses in some technical characteristics. They produce a higher working pressure in a considerably larger volume compared to the high-pressure anvil type apparatuses [6]. They are more efficient and durable and provide more uniform pressure and temperature conditions.

Multipunch apparatuses (Fig.3c) are used mainly for research purposes [44]. Compared to the uniaxal compression apparatuses, they require much more sophistication in manufacture and assembly. Despite the lower durability, they provide a more uniform pressure distribution over the working chamber volume and a higher working pressure in a larger working volume. They are promising for large volume applications [6].

A characteristic feature of static high pressures is the possibility to control the temperature in the sintering process. This is an important factor in a number of cases for producing the required nanocrystalline structure. The sample size, however, determines the time of establishing the isothermic sintering conditions in the reaction cell and it may require tens of second [45]. Temperature and pressure may be maintained for several days in some types of high-static pressure solid phase apparatuses. An easy recovery of the sintered sample from the cell is an additional advantage compared to the shock wave techniques.

2.3. METHOD OF INTENSIVE PLASTIC DEFORMATION

Another effective method for consolidation of NM is intensive plastic deformation [46]. This method has two variations (Fig.4): equal-channel angular pressing and torsion deformation under a high pressure. The method of equal-channel angular pressing (Fig. 4a) was developed by V.M. Segal in 1972 [47,48]. It includes a standard pressing process in a mould provided with an axial hole turned at an angle of 45-90°. The pressing pressure may reach several GPa. The required size reduction of a microstructure is obtained by multiply passing the material through the matrix channels at a room temperature. Usually this technique is used to produce NM by deforming rod-shape metal blanks. The main advantage that this method provides is the possibility to fabricate massive porousfree samples.

The torsion strain method (Fig. 4b) produces an axial compression of the powder under high pressures followed by the rotation of the punch on its axis. Usually the punch makes several rotations. This results in the required shearing strain in a thin 0.1-0.2 mm layer being pressed. The first experiments using high pressures combined with a shearing strain to investigate an oxide were described by Bridgman [49] and Vereshchagin et al. [50]. Pressures may reach 50 GPa in the above experiments [51].

410

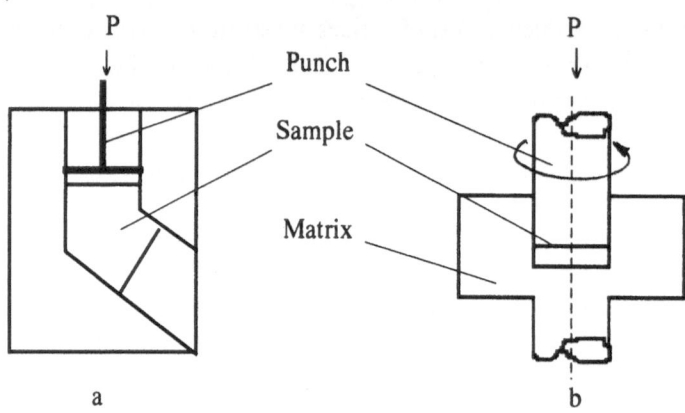

P   P
↓   ↓

Punch

Sample

Matrix

a   b

Figure 4. Method of intensive plastic deformation [46]: a - equal-channel
angular pressing, b - torsion straining .

3. Properties of Ultrafine Powders

UFP have many specific features. They differ from conventional powders in morphology, composition of surface layers, sinterability, high content of adsorbed gases and the behavior in the process of compaction and sintering.

Maximum size of UFP particulates is not more than 100 μm. The amount of surface atoms in the UPF particulates may reach several tens of percent. For example, according to data [52,53] approximately 30 % of atoms are located on the particulate surface in UPF with ~10 nm particulates resulting in the symmetry break of the force and mass distribution compared to space. The surface atoms with neighbors only on one side, their equilibrium is disturbed, resulting in a structural relaxation leading to displacement of interatomic distance in the 2-3 nm thick layer. Thus the surface particulate layers are stretched, while the inner layers are compressed [54]. The values of specific surface usually covers the range from several units to several hundreds of square meters per gram.

UFP particulate morphology may be different depending on the techniques of their fabrication. Most particulates have structures which are face-centered cubic and close-packed hexagonal lattices, since these close-packed structures have a minimal surface energy [52]. It has been found by using X-ray/neutron and electron diffraction that many UPF compounds (for example, Si_3N_4, Al_2O_3, SiC, TiB_2 and others) may be in the amorphous state. The lattice constant and microdeformation (distortion) are effected by the particulate size [1].

Ultrafine particulates are characterized by an increased value of mean square atom dislocations relative to positions in lattice sites compared to the values observed in bulk crystals. This results in a decrease of melting temperature and an increase in an atom diffusion due to lower diffusion activation energy [55]. Reduction of lattice constant has been reported for many UFP particulates for both metal and high-melting compounds. This is related to the occurrence of capillary compression which may be rather significant. The capillary pressure may reach approximately 1 GPa, for example, for particulates of ~10 nm at a surface tension of ~2 N/m, and this may result in deformation of approximately 10^3 nm according to [1]. Undoubtedly the capillary compression will considerably effect the mechanism of UFP consolidation, especially under high pressures.

The difference of a stoichiometric composition of the UFP is one more important feature. A lower atomic ratio [N]/[B] and a lower concentration of electrons in the volume unit [1,56] is characteristic, for example, of the UFP of a graphite-like BN. In general, ultrafine particles posses considerable excess of surface energy and lattice defect energy.

4. Consolidation of Ultrafine Powders at High Pressures without Heating

Some reports [57-59] showed that consolidation of conventional powders under high pressures results in an intensive fracture of their particulates and grains. The slipping of partriculates and positioning of finer fragmentation products in interparticulate pores contribute to the increase of the compact density. These processes are extremely characteristic to brittle ceramic powders. The consolidation of metal powders takes place to a large extent due to plastic deformation of particles [60].

According to [61] a noticeable plasticity in brittle materials may occur even under pressures over 10^{-3} K at normal temperatures where K is a compressibility modulus. For example, considering that the K =280.17 GPa [62] for TiN, plastic deformation may be expected even under pressures of about 0.3 GPa. For diamond, these values are 442.43 GPa [62] and 0.4 GPa, respectively.

In reference [63], a linear relationship between the lattice distortion and pressure has been established at a static compression (P=1.5-5.5 GPa) for an AlN powder with the particulate distribution from submicron to hundreds of microns size. Extrapolation of a straight line to a zero value of the lattice strain leads to a pressure value of 0.6 GPa, showing good agreement with brittleness-plasticity transition for AlN. It has been shown in [63] that the lattice distortions are significantly higher at a dynamic compression than in the static case for the same density, and they increase with increase of the loading rate. As a result, the crystallite size after the impact compression is ~100 nm.

The consolidation of conventional powders at high pressures is accompanied by a brittle and plastics deformation of the particulates, resulting in a finer system. Regarding the case of high-pressure compaction of UFP the information [58,64] is not widely available.

The main characteristics of the UFP consolidation are defined by the properties of their particulates, the large amount of surface energy related to a high specific surface of the powder particulates.

In [1,16,18,52,65] using Si_3N_4 and TiN, as an example it has been shown that as the powder particulate size decreases, the compactibility significantly decreases on pressing (Fig.5,6). The influence of the range of particle dimensions manifests itself within a wide range of pressures up to several GPa. A theoretical density is not reached even at high pressures of 7-10 GPa. In the best case the maximum value of the relative density is about 0.85 for TiN UFP.

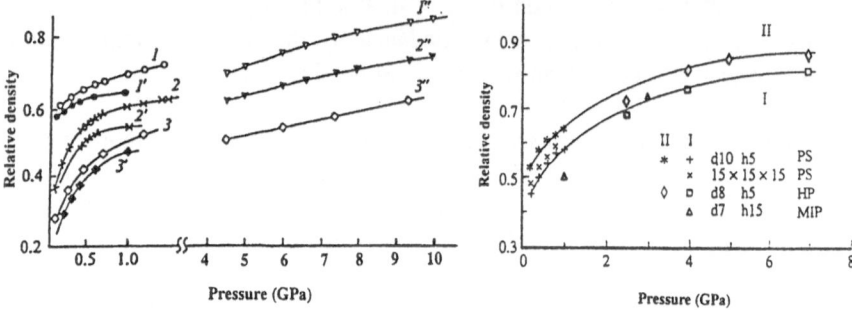

Figure 5. Relative density dependence of compacting pressure of Si₃N₄ powders having different graininess. 1-3 - compaction in a hydrostat ; 1'-3' - compaction in conventional molds; 1"-3"- compaction in high pressure apparatuses; d_s, nm: 1, 1', 1" - 1000; 2, 2', 2" - 50; 3, 3', 3" - 17 [1,16,18,52].

Figure 6. Compaction pressure effect on relative density of the ultrafine TiN powder: 1 - d_s~70 nm, II - d_s~80 nm; PS - pressureless sintering, HP - high pressures, MIP - magnetic impulse pressing. Sample dimensions are given in millimeters [1,65].

The dependence of the compact density on the UFP particulate size is observed for metallic powders as well. According to [16,52], for example, the relative density of Ni powder samples compacted under 1 GPa was 0.49 and 0.58 for the particulate sizes of 15 and 50 nm , respectively. In hydrostatic pressing under the same pressure, these values were 0.61 and 0.68, respectively. It has been found in [11] that increasing pressing pressure of Ni UFP with 60 nm particulate size over 4 GPa did not result in a marked increase in density. The maximum relative density of the compact was not more than 0.9.

Vacuum compaction under high pressures (3-5 GPa) was also used [10,13] for producing metallic NM. The samples were produced in the unit combining a condensing and vacuum pressing processes for UFP production that provided samples free from impurity. The condensation method combined with the high pressure compaction of powders was used for production of NM from titanium dioxide [15,66-68]. They, however, failed to prepare high density samples. According to [1,69] low UFP compactibilty was caused by a high level of interparticulate friction characteristic to powders with a highly developed surface.

In consolidation of UFP under high pressures, particulate surfaces transform into the grain boundaries of the material being consolidated. The boundaries feature a low atomic density and high atomic energy [70]. The state of boundary atoms considerably affects the structure and properties of nanocrystalline materials. High-pressure treatment of nanoscale powders of brittle materials results in a finer structure, as in conventional powders [57,59,71]. According to [64], for example, after the ultrafine $La_{0.7}Sr_{0.3}Mn_{0.99}Fe_{0.1}O_3$ powder was compacted at 4.5 GPa pressure, the average crystallite size decreased from 30 to 18 nm for orientations [110] and [111] and from 24 and 20 nm to 10 nm for orientations [100] and [211], respectively. Two situations emerge in this case: a large particulate fragments into multiple finer structures or many crystallites merge into a small particle.

The density of samples compacted from nanopowders under high pressures is also dependent of the loading rate and holding time under pressure. The density continues to grow at a constant pressure when it is fixed. It was observed [72] in both metallic (Fe, Ni) and ceramic (Si_3N_4, TiN) powders with a particulate size of 15-20 nm. The relative change in density within 3-hour holding under pressure may reach 5 % (Si_3N_4) at 0.5 GPa pressure. The density increment under pressure decreases to 0.8 % with pressure increasing to 5 GPa. It has been found that the rate of the abnormal density relaxation of such powder compact under high pressure is not dependent of the powder material. This agrees with the conclusions regarding that the behavior of powder bodies as they are pressed is determined predominantly by the particulate dispersity and morphology, and the role of the chemical bond nature is reduced to a large extent [69]. The time dependent change in the powder compact density of TiN (20 nm) has been investigated at different loading rates of dP/dt=0.1 and 0.01 GPa/min. As a result it has been found that the density increment of the compact for a high loading rate was higher after the pressure was set. A higher compression rate in the process of magnetic and impulse pressing also conditions a higher density of the compact from Al_2O_3 UFP compared to the stationary pressing [26]. The same law is also observed for the conventional powders [73].

Intensive plastic shearing deformation under high pressure is a promising candidate for fabrication of high-density NM. It may be used to produce NM by consolidating not only UFP, but more coarse-grained powders. In the work [74], for example, porous-free NM with a grain size 80 and 65 nm, respectively, were fabricated using this method from Ti powder with an average particulate size ~20 μm and a powder mixture with 5% SiC (d_s = 20 μm) addition at 5 GPa pressure. The microhardness of the powder Ti compact exceeds two-fold that of the cold plastic strain-treated bulk Ti samples, and the thermal stability.

The compaction process of the UFP under high pressure without heating is accompanied by a plastic deformation as in the case of conventional powders. The pressed samples are characterized by high microstresses, finer microstructure (decrease of the crystal size). Lower compactibility of the UFP is caused by a higher degree of mechanical friction. The role of the chemical bond nature is reduced to a large extent [69]. The influence of the range of particle dimensions manifests itself within a wide range of pressures up to several GPa (Fig. 5,6). The pressing pressure increase over 4-5 GPa did not result in density increase over 0.85-0.9 [11,65].

In some cases, pore-free nanostructured samples (< 1mm) were produced by shearing deformation of coarse-grain powders at high pressures [74]. In general, high density of compacts has not been produced by subjecting the UFP to compression only at a high pressure without heating.

5. Sintering in Shock Waves

Dynamic high pressures or sintering in shock waves has long been used for the production of dense compacts from conventional powders [4,5,60,75]. However, this consolidation method for production of NM from the UFP has not been well studied. Most works are related to ceramic diamond, BN, Al_2O_3, SiC, Si_3N_4 powders [1,12,17,19,21 22,25,76].

The phase transformations are specific to these compounds; their mechanisms under different pressure and temperature conditions have not yet been completely understood However, they play a critical role in forming the material structure under high pressure consolidation conditions.

The consolidation of Si_3N_4 UFP, SN-E10 brand, with the crystallite size of 70 nm containing more than 95% of α-Si_3N_4 has been studied [22]. A nanocrystalline ceramics of α-Si_3N_4 with a relative density of 0.99 and a microhardness of $H_V \sim 23$ GPa have been produced at the 40 GPa pressure and 930 °C temperature. The analysis of the micro-structure using the transmission electronic microscope showed the presence of plasti-cally deformed grains with a size of 50-60 nm and amorphous grains. It has been also found that the pressure increase and, to a larger extent, temperature increase, resulted in the formation of β - Si_3N_4 with accelerated grain growth, leading to low density and mi-crostresses.

The compaction process of β-SiC UFP with particle size of 280 nm within a wide range of pressures (2.4-29.9 GPa), temperatures (700-3130 °C) and loading rates (1.5-3 km/s) has been studied [12]. A strong dependence of the compact density and micro-hardness on initial green density (30, 50, 70 %) has been pointed out. It is found that maximum values of these parameters for green denser blanks were obtained at higher pressures, temperatures and loading rates.

The ceramic samples produced under optimal conditions (P = 22.2 GPa and T = 1510 °C) had a relative density of 0.97 and microhardness of $H_V \sim 27$ GPa [12]. The structure of the ceramic comprised submicron grains bonded with a fused material. The samples were characterized by a small size of crystallites (~ 100 nm) and a high level of micro stresses (~ $8.2 \cdot 10^{-3}$).

Nanocrystal ceramic material obtained by dynamic high-pressure consolidation tends to fracture. This is caused by strong internal strains occurring in the process and after the shock compaction [19,20].

It has been found [19] that preheating compacts from α-Al_2O_3 powders with 500 nm and 200 µm particulate sizes to 1200 °C allows the shock effect energy to be lowered and microfractures to be avoided. As a result the compact hardness may be increased from 12 to 14 GPa (Knoop, 10 N load). The ceramic samples produced had a relative density of 0.94, the crystallite size less than 10 nm and the microstress value ~3×10^{-3} only about half that of the coarse-grain powder samples (d=200 µm). It has been found that consolidation of a fine powder takes place through regrouping and a plastic defor-mation of particulates without a considerable change of their size compared to a coarse grain powder fracturing even under hot compaction [19]. A higher density of the com-pact (0.98) is obtained for the Ni-63 mol % TiN powder mixture with the particulate size of 4-6 nm under shock loading (39 GPa) with preheating up to 580 °C [27].

The work [20] discusses the issue of the possibility of producing high-dense nanoc-rystal materials by consolidating the diamond UFP in shock waves. The first attempts were not a success [20,25]. The relative density and microhardness of compacts of the diamond powder with the particle size of 4-6 nm were 0.90 and $H_V \sim 16$ GPa [25], re-spectively, though the grain growth and graphitization were not observed. Plastic deformation of nanoparticulates compacted into a homogenous mixture and a low strength of their bond are noted, though double compaction at pressures 23 and 54 GPa was applied to increase density [25]. Adding a submicrocrystalline diamond powder (d = 0.25 - 1μm) in proportion 1:3 did not result in a significant increase of the diamond compact characteristics (ρ = 0.93, H_v ~ 25 GPa), though it has been possible to produce a denser and stronger ceramics (ρ = 0.94, H_v ~ 56-65 GPa, at 10 N load) using a pure submicrocrysalline diamond powder without additives using the conditions of P = 90 GPa, T = 1550 °C, v ~ 2.6 km/s [20]. It has been found that polycrystalline powder with smaller particulate size (0.25 - 0.75 μm) was compacted worse than the monocrystalline powder (0.25 - 1μm).

The major requirements for production of the dense nanoceramics by a shock compaction for all materials are discussed in [20]: 1)Use of a fine powder with a short time of thermal relaxation susceptible to plastic deformation within the shock wave front, 2)Use of monosize powders or powders having the narrowest particulate size distribution, 3)Decrease of the pore size and their distribution by finely dividing the powder particulates, 4)Thorough control over the shock loading conditions and increase of the impact time for attaining the plastic deformation of particulates and preventing their fracturing.

The work [26] shows that, in principle, it is possible to produce hard ceramics with a nanosize structure (d=16-18 nm) by using a magnetic impulse pressing of the preheated (600 °C) TiO_2, Al_2O_3, TiN UFP at pressures (up to 10 GPa). The relative density and hardness of ceramic samples were 0.89-0.94 and 11-20 GPa, respectively.

Thus, it has not been possible to attain a complete compaction of powders and high hardness in the most cases even though the nanocrystalline structure of the material has been preserved. The main mechanism of consolidation of UFP in shock waves is plastic deformation with surface melting of particles.

6. Sintering Under High Static Pressures

Significant densification may be obtained while maintaining the nanocrystalline structure by using high pressures combined with heating up to the temperatures not resulting in intensive recrystallization. Only limited data, however, have been obtained to-date. A part of the information on this problem is provided in the reviews [1,2]. One of the first results in this direction have been obtained in the works [9,11,16,77].

The difference between densification of coarse-grain powders and UFP manifests itself during heating under high pressures as well. In this case, the densification of the UFP is worse and incomplete. For instance, it was shown [11] in Ni powder densification at temperatures up to 900 °C, the relative density of particulates having the size of 15 μm and 60 nm after sintering (P = 5 GPa, t = 15 s) was 0.99 and 0.94, respectively (Fig. 7). Prior to compaction the powder samples were put under vacuum to remove the adsorbed gas molecules. Despite a lower density, nearly 2-fold hardness increase has been observed in Ni UFP compared to that of coarse-grain powders (Fig.7). The hardness increase was caused by the fine grain size.

416

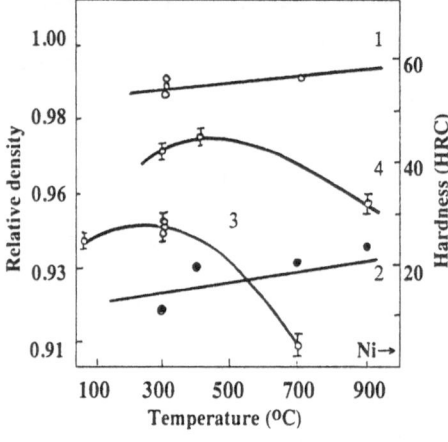

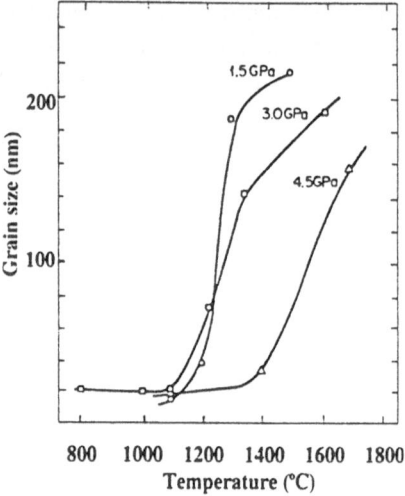

Retardation of Ni UFP recrystallization at high quasi-hydrostatic 5 GPa pressure and temperatures from 600-900 °C has been reported in [14]. The resulting porosity of samples produced under high pressure is 3 times lower compared to that obtained under standard sintering of the same powder without pressure applied.

Using the model of the boundary motion due to sliding of grain-boundary dislocations, the authors [14] associate the crystallization retardation with the increase of energy of athermanous vacancy formation from the surface of the moving stages of sliding grain boundary dislocations.

Interesting results have been obtained in the work [9] in sintering TaC UFP with 28 nm particulate size under high pressures. The pressures from 1.5 to 4.5 GPa were applied to decrease the diffusion rate in sintering and retard grain growth. The procedure of sample preparation included a slow pressure loading, followed by a rapid heating, holding for 5 - 15 min at 1100 -1400 °C, cooling down to 700 °C and off-loading at a rate of 5 t/min and, finally, complete cooling. The delamination of the produced samples is one of the problems of high-pressure sintering, due to high internal stress. That is why off-loading at 700 °C was used [9] to prevent such separation. Fig.8. shows the results of the grain structure of the samples produced using the transmission electron microscope.

As shown in Fig.8, grain growth decreases with increasing pressure. The authors [9] managed to produce high-density nanocrystal TaC samples with 20 nm grain size and 22.5 GPa hardness (1N load), which is higher than that of a single-crystal.

The retardation effect of recrystallization processes in high-pressure sintering is observed not only for metals [11,14], but also for high-melting compounds [9]. In this respect investigating the consolidation of the UFP and other high-melting compounds under high pressures and temperatures is of interest. Given the potential of developing novel superhard materials with a high level of physical and chemical properties based on the nanocrystal high-melting compounds, it is of critical importance [17].

The work [65] describes that at high pressures (5-7 GPa) even a slight heating (800 °C) is sufficient to increase the relative density of the TiN UFP from 0.85 to 0.95. The consolidation of TiN UFP and TiN-TiB$_2$ composition has been investigated in works [28-32] as a continuation of these investigations. It has been demonstrated that application of pressures about 4 GPa at temperatures 1200-1400 °C allows production of low-porosity samples (1-2 %) with a grain size of about 50 nm and hardness nearly 30 GPa, that is 1.5-2 times higher than in the case of a standard sintering and hot pressing [2].

Fig. 9 shows the relative density relationship of hardness and elastic modulus of nanocrystalline TiN as a function of sintering temperature at pressures of 4 GPa and 7.7GPa [29]. The data for pressureless sintering of the same TiN UFP [78] were also demonstrated for comparison. One can see that using the compacting pressure of 4 GPa results in a decrease of the active densification temperature to about 600-800 °C. In the case of the pressureless sintering the densification begins only at $T>1050$ °C as determined by the specific surface area measurements. However, it is also evident that the further pressure increase has little effect on the relative density values higher than 0.98.

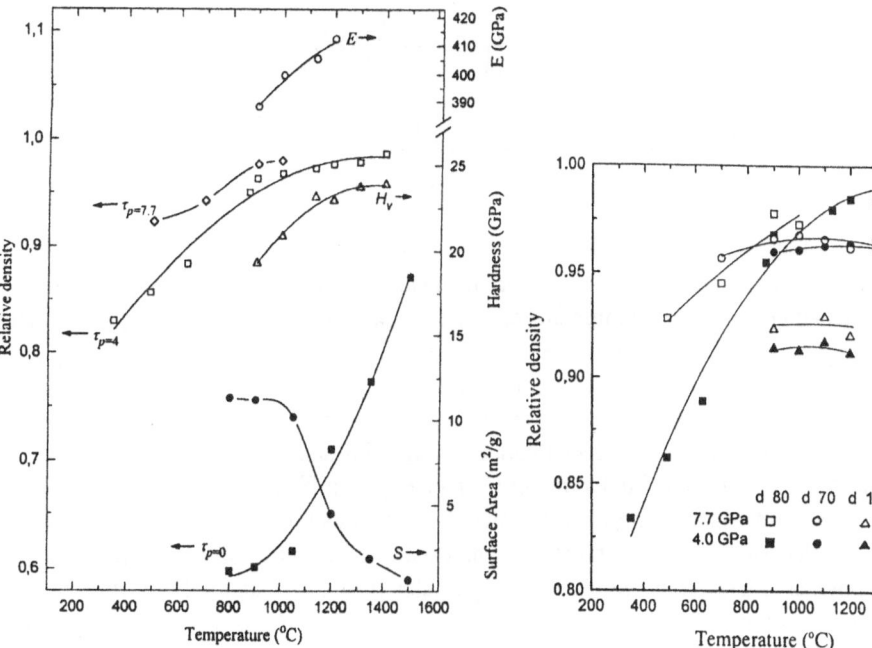

Figure 9. Effect of temperature on relative density ($\tau_{P=4}$ and $\tau_{P=7.7}$), microhardness (H_V), and elastic modulus (E) at high pressure sintering and on relative density ($\tau_{P=0}$) and spesific surface area (S) at pressureless sintering [29].

Figure 10. Effect of temperatures, pressure and grain size of particles on relative density of UFP TiN [30].

The fact that the hardness was not changed in the interval from 1130 °C to 1400 °C may be explained by the possible competition between the porosity decrease and the re-crystallization progress.

Change of the relative density samples versus particle size of TiN UFP and temperature of sintering (duration of exposure was 5 minutes) at pressures 4 and 7.7 GPa is shown in Fig.10 [30]. As evident, pressure increase in the range of these values of density has insignificant effect on the compaction. However, the initial particle size rather considerably effects the change in density. This has been already noted for silicon nitride (T = 20 °C and T = 1100 - 2000 °C, pressure - up to 9 GPa) [16] and for titanium nitride when treated under high pressures at a room temperature (Fig.6). The role of interparticle friction manifesting itself in the decreased compaction is also significant in high temperature consolidation of UFP under high pressures. It should be noted that this fact is indicative of a considerable difference of compaction mechanisms under the so-called pressureless sintering (i.e. without applying an external pressure) when, as is well known [30] particle size decrease contributes to shrinkage. In a high pressure sintering, interparticle movement and interparticle friction significantly increases.

The following formulae (1), (2) for the initial ($\rho \leq 0.9$) and final ($\rho \geq 0.9$) diffusion stages, respectively, have been proposed in the work [79] to describe the densification process of Si_3N_4 powder in sintering under high pressures assuming the predominant role of the boundary diffusion:

$$\rho = \rho_o + \frac{1-\rho_o}{0.36}\left(16.72\frac{\delta D_b}{kTr^3}\Omega P t\right)^{1/3} \qquad (1)$$

$$\rho = 1 - \frac{1-\rho_o}{0.36}\left[0.3162 - 135\frac{\delta D_b}{kTr^3}\Omega P(t - t_o)\right]^2 \qquad (2)$$

where: ρ - relative density, ρ_o - initial value, δ - boundary width (thickness of the layer in which diffusion occurs), D_b - boundary diffusion coefficient, Ω - molecular space, k - Boltzmann constant, T - temperature, r - particulate radius, t_o - time required for transition to the final stage (obtaining density of 0.9) and equaling $1.05 \times 10^{-3} kTr^3/(\delta D_b \Omega P)$, P -pressure. A satisfactory agreement of experimental and calculated data for powders with 1.1 and 0.75 μm particulate size has been pointed out by the authors [79].

The mechanism of powder densification strongly influences the compaction process and the properties of the final product. For example, differences in the deformation mechanisms under static high pressures and temperatures [76] are observed in the diamond powders of the impact and static synthesis. The densification of the impact synthesis powder is caused by fracture of the particles themselves and pulverization at the particulate corners, where the stresses are concentrated at the contacting points with other particulates. This results in increase of a free surface not subjected to a high pressure and, hence, to graphitization. In addition, fragmentation of the particulate areas conditions their additional heating facilitating the graphitization process [20]. In contrast to the above, the plastic deformation of particulates is characteristic to the static synthesis diamond powder. This provides a higher density of compacts. But characteristics of such diamond compacts (ρ=0.85, H_v=29 GPa) produced under high static pressures (P= 5.8 GPa, T=1520 °C, t = 30 min) are inferior to those of the impact compaction samples (ρ=0.94, H_v= 64 ± 12 GPa) [20].

7. Conclusion

Therefore, the information available in the literature suggests that, in principle, production of high-hardness NM by consolidating UFP at high pressures is possible. A high-density powder compact may be produced only by the combined action of the high pressure and temperature.

Contrary to pressureless sintering, in UPF sintering under high static pressures the interparticle friction, manifesting itself in the decreased densification with the reduced particle size, plays a significant role. Depending on the UPF characteristics the densification at high static pressures may be reached both due to a brittle fracture of particles and their plastic deformation. The characteristic feature is the retardation of recrystallization processes. The grain growth slows down with the increased pressure. The main mechanism of UPF shock wave consolidation is the plastic deformation and surface melting of particles.

An advantageous utilization of high pressures for producing NM with improved properties may be determined both by expansion of possibilities to control the temperature regime of a high pressure sintering process and the investigation of structural features of UPF plastic deformation at high pressures related, for example, to the contribution of strengthening, the role of defects of the crystal structure, intergrain and interparticle sliding [52].

Acknowledgment

The author would like to express his gratitude to Prof. R.A. Andrievski for support and for providing publication prints on the subject of the review.

References

1. Andrievski, R.A. (1994) Fabrication and Properties of Nanocrystalline High-Melting Compounds, *Successes of Chemistry* **63**, 431-448 (in Russian).
2. Andrievski, R.A. (1997) State-of-Art and Perspectives in the Fild of Particulate Nanostructured Materials, *Int. Symp."New Materials and Technologies in Powder Metallurgy", March 19-20, 1997, Minsk, Belarus* (in press.).
3. Krupin, A.V., Solovyov, V.Ya., Sheftel, N.I., Kobelev A.G. (1975) *Explosive Deformation of Metals*, Metallurgia, Moskow (in Russian).
4. Altshuler, L.V. (1965) Application of Shock Waves in High Pressure Physics, *Successes of Physical Sciences* **85**, 197-258 (in Russian).
5. Sawaoka, A.B. (1988) Shock Compaction and Consolidation of Non-Oxide Ceramic Powders as a Manufacturing Process, in L.E. Murr (ed.), *Shock Waves for Industrial Applications*, Oregon Graduate Center, Beaverton, New-Jersey, USA, pp. 380-405.

420

6. Novikov N.V. et al. (1986) *Synthetic Superhard Materials,* Naukova Dumka, Kiev (in Russian).
7. Frantsevich, I.N., Gnesin, G.G., Kurdumov, A.V. et al. (1980) *Superhard Materials*, Naukova Dumka, Kiev (in Russian).
8. Urbanovich, V.S., Shipilo, V.B. (1997) Physicomechanical and Thermophysical Properties of High-Melting Compounds-Based Ceramics Sintered at High Pressures in P. Abelard et al. (eds.), *Euro Ceramics V,* Trans Tech Publications, Switzerland **2**, pp. 1027-1030.
9. Yohe, W.C. and Ruoff, A.L. (1978) Ultrafine-Grain Tantalum Carbide by High Pressure Hot Pressing, *Am. Ceram. Soc. Bull.* **57**, 1123-1125, 1130.
10. Gleiter, H. (1981) in N.Hansen, T.Leffers, and H.Lilholt (eds.) *"Deformation of Polycrystals:Mechanisms and Microstructures"* Riso Nat.Labor., Roskilde, p. 15.
11. Jakovlev, E.N., Grjaznov, G.M.,.Serbin, V.I. et al. (1983) Fabfication of Nickel Polycrystals with Higher Hardness by Ultrafine Powders Pressing, *SURFACE. Physics, Chemistry, Mechanics* **4**, 138-141 (in Russian).
12. Kondo, K., Soga, S., Sawaoka, A., and Araki, M. (1985) Shock Compaction of Silicon Carbide Powder, *J. of Material Science* **20**, 1033-1048.
13. Birringer, R., Herr, U., Gleiter, H. (1986) *Trans. Jpn. Inst. Met. Suppl.,* **27**, 43.
14. Novikov, V.I., Ganelin, V.Ya, Trusov, L.I. et al. (1986) Retardation of Recrystallization of Ni Ultrafine Powder under High Hydrostatic Pressure, *Metal- Physics* **8**, 111-113 (in Russian).
15. Siegel, R.W., Hahn, H. (1987), in M.Yussouff (ed.), *Current Trends in the Physics of Materials*, World Scientific, Singapore, p.403.
16. Andrievski, R.A., Zeer, S.E., Leontiev, M.A. (1987) Peculiarities of Pressing and Sintering of Nickel and Silicon Nitride Ultrafine Powders in I.V. Tananaev (ed.), *Physics-Chemistry of Ultradispersed Media,* Nauka, Moscow, pp. 197-203 (in Russian).
17. Andrievski, R.A. (1993) Properties of Nanocrystalline High-Melting Compounds (Review), *Powder Metallurgy* **11/12**, 85-91 (in Russian).
18. Andrievski, R.A. (1995) Silicon Nitride - Synthesis and Properties, *Russ. Chem. Rev.*, **64**, 311-329 (in Russian).
19. Taniguchi, T. and Kondo, K. (1988) Hot Shock Compaction of α-Alumina Powder, *Advanced Ceramic Materials* **3**, 399-402.
20. Kondo, K. and Sawai, S. (1990) Fabricating Nanocrystalline Diamond Ceramics by a Shock Compaction Method, *J. Am. Ceram. Soc.* **73**, 1983-1991.
21. Kovtun, V.I., Kurdyumov, A.V., Zelyavski, V.B. (1992) Phase and Structural Transformation of Wurtzite Boron Nitride during the Sintering in Shock Waves, *Powder Metallurgy* **12**, 38-43 (in Russian).
22. Hirai, H. and Kondo K. (1994) Shock-Compacted Si_3N_4 Nanocrystalline Ceramics: Mechanisms of Consolidation and of Transition from α- to β-form, *J. Am. Ceram. Soc.* **77**, 487-492.
23. Ivanov, V.V., Kotov, Yu.A., Samatov, O.N. et al. (1995) Synthesis and Dynamic Compaction of Ceramic Nano Powders by Techniques based on Electric Pulsed Power, *Nanostructured Materials* **6**, 287-290.

24. Andrievski, R.A., Vikhrev, A.N., Ivanov, V.V. et al. (1996) Compacting Ultrafine Titanium Nitride Using a Magnetic Impulse Method and under a Shearing Strain and Higher Pressures, *Physics of Metals and Metal Science* **81**, 137-145 (in Russian).

25. Kondo, K and Hirai, H. (1996) Shock-Compaction of Nano-Sized Diamond Powder, as Examined by Microstructural Analysis, *J. Am. Ceram. Soc.* **79**, 97-101.

26. Ivanov, V.V., Paranin, S.N., Vikhrev, A.N. (1997) Compacting Nanosize Powders of Hard Materials Using a Magnetic Impulse Method in G.G. Taluts and N.I. Noskova (eds.), *Structure, Phase Transformations and Properties of Nanocrystalline Alloys*, UD RAS, Ekaterinburg, pp. 46-56 (in Russian).

27. Ogino, Y., Yamasaki, T. and Shen, B.L. (1997) Indentantion Creep in Nanocrystalline Fe-TiN and Ni-TiN Alloys Prepared by Mechanical Alloying, *Metallurgical and Materials Transactions B* **28B**, 299-306.

28. Andrievski, R.A., Kalinnikov, G.V., Potafeev, A.F., and Urbanovich, V.S. (1995) Synthesis, Structure and Properties of Nanocrystalline Nitrides and Borides, *Nanostructured Materials* **6**, 353-356.

29. Andrievski, R.A., Urbanovich, V.S., Kobelev, N.P., and Kuchinski, V.M. (1995) Structure, Density and Properties Evolution of Titanium Nitride Ultrafine Powders under High Pressures and High Temperatures, in A. Bellosi (ed.), *Fourth Euro Ceramics, Basic Sciences - Trends in Emerging Materials and Applications*, Gruppo Edit. Faenza, Printed in Italy **4**, pp. 307-312.

30. Andrievski, R.A., Urbanovich, V.S., and Kobelev N.P. (1997) High-Temperature Consolidation and Physical and Mechanical Properties of Nanocrystalline Titanium Nitride, *Reports of Russian Academy of Sciences*, (in Russian), (in press).

31. Andrievski, R.A., Kalinnikov, G.V., and Urbanovich, V.S., Consolidation and Evolution of Physical and Mechanical Properties of Nanocomposite Materials Based on High-Melting Compounds, *Nanophase and Nanocomposite Materials II*, **457**, MRS, Pittsburgh (in press).

32. Andrievski, R.A. (1997) Properties and Structure of Nanocrystalline and Multilayer of Titanium Nitrides and Borides, in G.G. Taluts and N.I. Noskova (eds.), *Structure, Phase Transformations and Properties of Nanocrystalline Alloys*, UD RAS, Ekaterinburg, pp. 37-46 (in Russian).

33. Ryabinin, Yu.N. (1956) About Some Experiences for Dynamic Compacting of Substances, *Technical Physics Review* **26**, 2661-2666 (in Russian).

34. Gerasimovich, A.V. (1987) General Problems of Engineering High Pressure Apparatuses in B.I. Beresnev (ed), *Effect of High Pressures on Substance*, Naukova Dumka, Kiev, **2**, pp. 88-98 (in Russian).

35. Bradley, C.C. (1969) *High Pressure Methods in Solid State Research*, Butterworths, London.

36. Tsiklis, D.S. (1976) *Techniques of Physical and Chemical Investigations at High and Superhigh Pressures*, Khimiya, Moskow (in Russian).

37. Bridgman, P.W. (1952) *Proc. Amer. Acad. Arts Sci.*, **81**, 165.

38. Pat. 1360281, Greate Britain, ICI B 01 J 3/00. *High Pressure and High Temperature Apparatus,* Vereschagin, L.F., Bakul, V.N., Semertchan, A.A. et al. - Publ. 17.07.74.
39. Pat. 3695797, USA, ICI B 30 B 11/32. *Apparatus for Creation of High Pressure,* Bakul, V.N., Prihna, A.I., Shuljenko, A.A. and Gerasimovich, A.V. - Publ. 03.10.74.
40. Pat. 3790322, USA, ICI B 30 B 11/32. *Apparatus for Creation of High Pressures and High Temperatures,* Sirota, N.N., Mazurenko, A.M. and Strukov, N.A. - Publ. 05.02.74.
41. Pat. 1392, Belarus, ICI B 01 J 3/06. *Apparatus for Creation of High Pressure,* Shipilo, V.B. - Prioritet 08.07.94 (in Russian).
42. Mazurenko, A.M., Urbanovich, V.S., and Kuchinski V.M. (1994) A High Pressure Apparatus for Sintering of High-Melting Compounds Based Ceramics, *Vestsi of Belarus. Acad. Sci., ser. phys.-techn. sci.* **1**, 42-45 (in Russian).
43. Urbanovich, V.S. (1994) On account for the Geometrical Form of Hard-Alloy dies when Designing HPA of the Anvil-with Receeses Type, *Physics and Technique of High Pressures, FTINT of Ukr. Acad. Sci. Kharkov,* **4**, 66-69 (in Russian).
44. Platen, B. (1962) Multi-Piston High Pressure and High Temperature Apparatus in R.H.Wentorf (ed.) *Modern Very High Pressure Techniques,* Schenectady, London, pp. 191-216.
45. Urbanovich, V.S. (1997) The Diamond and Hard Alloy-Based Composite Material in M.A.Prelas et al. (eds.) *Diamond Based Composites and Related Materials,* Kluwer Academic Publishers, Dordrecht, pp. 53-62.
46. Alexandrov, I.V., Kilmametov, A.R., Mishlyaev, M.M., Baliev, R.Z. (1997) Peculiarities of Nanocrystalline Material Structure Produced Using an Intensive Plastic Strain in G.G. Taluts and N.I. Noskova (eds.) *The Structure, Phase Transformations and Properties of Nanocrystalline Alloys,* UD RAS, Ekatirinburg, pp. 57-69.
47. Segal, V.M., Reznikov, V.I., Kopilov, V.I. et al. (1994) *Processes of Plastic Structure Formation In Metals,* Nauka i Tekhnika, Minsk (in Russian).
48. Segal, V.M (1974) *Methods for Investigating the Stressed-Strained State on the Processes of Plastic Deformation,* Doctoral Dissertation of Eng., Minsk.
49. Bridgman, P.V. (1955) *Investigation of Severe Plastic Deformations and Raptures,* Izdatelstvo inostrannoi literaturi, Moscow (in Russian).
50. Vereshchagin, L.F., Zubova, E.V., Shapochkin, V.A. (1960) Equipment and Methods of measuring shear in solids under high pressures, *Instrumentation and Equipment of the Experiment* **5**, 89-93 (in Russian).
51. Vereshchagin, L.F., Zubova, E.V., Burdina, K.P., Aparnikov, G.L. (1982) Behavior of Oxides under the Action of High Pressures with Simultaneous Application of the Shearing Stress in I.S. Gladkaja et al.(eds.) *Vereshchagin, L.F. Synthetic Diamonds and Hydroextrusion,* Nauka, Moscow, pp. 319-321 (in Russian).
52. Andrievski, R.A. (1991) *Particulate Materials Science,* Metallurgia, Moscow (in Russian).
53. Morokhov, I.D., Trusov, L.I., and Lapovok, V.N. (1984) *Physical Phenomena in Ultrafine Media,* Energoatomizdat, Moscow (in Russian).

54. Kisly, P.S., and Kuzenkova, M.A. (1980) *Sintering of Refractory Compounds*, Naukova Dumka, Kiev (in Russian).

55. Chigik, S.P., Gladkikh, N.T., Grogorieva, L.K. and Kuklin, R.N. (1984) Dimensional Dependence of Diffusion in Small Particulates, in Physics and Chemistry and Technology of Fine Powders, IPM AS USSR, Kiev, pp.121-124 (in Russian).

56. Shulga, Yu.M., Moravskaya, T.M., Gurov, S.V. (1990) Investigation of the Ultrafine Boron Nitride Using X-Ray Photoelectron Spectroscopy and Spectroscopy of the Characteristic Energy Losses Techniques, *SURFACE. Physics, Chemistry, Mechanics*, **10**, 155-157 (in Russian).

57. Djamarov, S.S., Kurdyumov, A.V., Oleinik, G.S. et al. (1982) Specifics of Forming the Sinter Microstructure Based on Wurtcit BN (hexanite - P), *Powder Metallurgy* **8**, 32-37 (in Russian).

58. Djamarov, S.S., Pavlenko, N.P., Bozhko A.V., Kornienko, P.A. (1982) Specifics of Cold Compaction of Wurtcit Boron Nitride under High Pressures, *Powder Metallurgy* **10**, 6-10 (in Russian).

59. Mazurenko, A.M., Urbanovich, V.S., Leonovich, T.I. (1987) Physical and Mechanical Properties of Metal Diborides of IVa, Va Groups Sintered under High Pressures, *Powder Metallurgy* **7**, 37-40 (in Russian).

60. Meyers, M.A., Thadhani, N.N. and Yu, L.H. (1988) Explosive Shock Wave Consolidation of Metal and Ceramic Powders, in L.E.Murr (ed.), *Shock Waves for Industrial Applications*, Oregon Graduate Center, Beaverton, New-Jersey, USA, pp. 265-334.

61. Leontieva, A.V., Streltsov, V.A., Feldman, E.P. (1986) Brittle and Plastic Transition in Crystals under the Hydrostatic Pressure, *Physics and Techniques of High Pressures* **22**, 16-30 (in Russian).

62. Frantsevich, I.N., Voronov, F.F. and Bakuta, C.A. (1982) *Elastic Constants and Elasticity Moduli of Metals and Non-Metals, Reference Book*, Naukova Dumka, Kiev (in Russian).

63. Akaishi, M., Fukunaga, O., Horie, Y. Et al. (1984) Effects of Dynamic and Isostatic Compaction on the Microstructure and Mechanical Behaviour of AlN, TiB$_2$ and TiC in *High Pressure Science and Technology - Proc. IX AIRAPT Int. High Pressure Conf., Albany*, N.Y. July 24-29, 1983, New York e.a. **3**, pp.159-162.

64. Su, W., Sui, Yu., Xu, D., and Zheng, F. (1996) High Pressure Research on Nanocrystalline Solid Materials, in W.A.Trzeciakowski (ed.) *High Pressure Science and Technology*, World Scientific Publishing Co. Pte. Ltd, Singapore, pp. 203-207.

65. Andrievski, R.A., Grebtsova, O.M., Domachneva, E.P. et al. (1993) Consolidation of Ultrafine Titanium Nitride at High Pressures, *Rus. Reports Acad. Sci.* **331** (3), 306-307 (in Russian).

66. Gleiter, H. (1992) *Nanostructured Materials* **1**, 1.

67. Siegel, R.W. (1993) *Nanostructured Materials* **3**, 1.

68. Averback, R.S., Hofler, H.J., Hahn, H. Et al. (1992) Sintering and grain growth in nanocrystalline ceramics, *Nanostructured Materials* **1** , 173-178.

69. Andrievski, R.A. (1988) Role of the Chemical Bond Nature and Dispersity in Formation of Particulate Materials, *Powder Metallurgy* **8**, 40-47 (in Russian).

424

70. Gleiter, H. (1995) Nanostructured Materials: State of the Art and Perspectives, *Z.Metallkd.* **86**, 78-83.

71. Mazurenko, A.M., Urbanovich, V.S., Olekhnovich, A.I., Voitenko, A.A. (1987) Fine Crystalline Structure of Niobium and Tantalum Diborides Sintered under High Pressures, *Superhard Materials* **6**, 34-36 (in Russian).

72. Tsiok, O.B., Sidorov, V.A., Bredikhin, V.V. et al. (1997) Dynamics of Powder Systems under High Pressures, in G.G.Taluts and N.I.Noskova (eds.), *Structure, Phase Transformations and Properties of Nanocrystalline Alloys*, UD RAS, Ekaterinburg, pp. 79-72 (in Russian).

73. Mazurenko, A.M., Urbanovich, V.S., Olekhnovich, A.I. et al. (1990) Investigation of Effect of Cold Pressing on the Aluminum Nitride Properties in A.M.Mazurenko and V.M.Dobryanski (eds.), *High Pressures Techniques and Technologies*, Uradjai, Minsk, pp. 139-143 (in Russian).

74. Sergeeva, A.V., Islangaliev, R., and Valiev, R.Z. (1997) Microstructure and Thermal Stability in Metal-Ceramic Titanium Based Nanocomposites, *Programme and Abstracts of NATO ASI "Nanostructured Materials: Science and Technology", August 10-20, 1997, St.-Petersburg* , Poster P13-12.

75. Gorobtsov,V.G., Furs, V.Ya., Shevchenok, A.A., Bondarenko S.N. (1991) *Application of Impulsive Loading for Fabricating Superconducting Ceramic Materials.*, BelNIINTI, Minsk (in Russian).

76. Kondo, K., Sawai, S., Akaishi, M., and Yamaoka, S. (1993) Deformation behaviour of shock-synthesized diamond powder under high pressures and high temperatures, *J. of Materials Science Letters* **12**, 1383-1385.

77. Andrievski, R.A., Konyaev, Yu.S., Leontiev, M.A., and Pivovarov, G.I. (1989) The Influence of High Pressures on Structure and Properties of Silicon Nitride in N.V.Novikov (ed.), *High Pressure Science and Technology*, **2**, Naukova Dumka, Kiev, p. 170-173 (in Russian).

78. Andrievski, R.A., Ljutikov, R.A., Torbova, O.D. et al. (1993) Gassing and Porosity at Sintering of Ultrafine Titanium Nitride, *Inorganic Materials* **29**, 1641- 1644 (in Russian).

79. Andrievski, R.A., Voldman, G.M., Leontiev, M.A. (1991) About Boundary Parameters of Diffusion in Silicon Nitride, *Inorganic Materials* **27**, 729-732 (in Russian).

MAGNETIC STATE, TRANSPORT PROPERTIES AND STRUCTURE OF GRANULAR NANOPHASED SYSTEMS

1. *Mechanically Alloyed Cu-20%Co System*
2. *Hydrogenated Pr(Cu,Co)$_5$ intermetallics*

A.YE.YERMAKOV, M.A.UIMIN, N.V.MUSHNIKOV, N.K.ZAJKOV,
V.V. SERIKOV, A.YU.KOROBEJNIKOV AND N.M.KLEINERMAN
*Institute for Metal Physics, Ural Branch of Russia Academy of
Sciences, 620219, Ekaterinburg, Russia*
A.K. SHTOLZ,
Ural State Technical University, 620002, Ekaterinburg, Russia

Abstract

In this paper we report the investigation on the formation, if any, of supersaturated solid solution by mechanical alloying of Cu-20%Co and by hydrogenation of Pr(Co,Cu)$_5$ intermetallics. A comparative analysis of structure and magnetic properties of above mentioned systems are performed.

The structural state of Cu$_{80}$Co$_{20}$ system after mechanical alloying and a subsequent thermal treatment has been studied by NMR-spectroscopy and X-ray diffraction. Analysis of the data obtained as well as magnetic and magnetoresistive properties make it possible to assume the formation of ultradispersed cobalt clusters in a copper matrix. A partial dissolving of cobalt in copper is likely to take place in defect regions. Such a structure is responsible for superparamagnetism of Cu$_{80}$Co$_{20}$ system at T > 150 K and for a considerable giant magnetoresistance (GMR) effect (12% at T = 77 K). A shift of X-ray peaks of a copper matrix can partially be due to a coherence of Cu and Co phases.

Quasi-binary Pr(Co$_{1-x}$Cu$_x$)$_5$ intermetallics with $0 \leq x \leq 1$ were hydrogenated at elevated temperatures to precipitate Co and Cu and to study their mutual solubility. Low temperature hydrogenation was found to form a CaCu$_5$-type hydride containing about 1.6 hydrogen atoms per formula unit. Above 500°C the sample decomposed into PrH$_{2.6}$, Cu and HCP-Co. In the temperature range 330 - 450°C the CaCu$_5$-type hydride coexisted with the decomposed phases. Structural and magnetic measurements indicated that no solid solution was formed in Co-Cu decomposed phases. The magnetoresistance on both parent and hydrogenated samples did not exceed 0.5 %.

Keywords: mechanical alloying, hydrogenation, crystal structure, decomposition, Cu-Co, Pr(Co$_{1-x}$Cu$_x$)$_5$, magnetic properties, magnetoresistance.

G.M. Chow and N.I. Noskova (eds.), Nanostructured Materials, 425–440.
© *1998 Kluwer Academic Publishers.*

1. Mechanically Alloyed Cu-20%Co System

1.1 INTRODUCTION

The development of new methods for the preparation of solids such as magnetron sputtering, melt spinning, mechanical alloying, molecular beam epitaxy opens up new possibilities for the synthesis of non-equilibrium phases, supersaturated solid solutions, amorphous and nanocrystalline states with new physical properties. Systems of magnetic elements in a non-magnetic conducting matrix (for example, Fe-Cu, Co-Cu, Fe-Cr) are of special interest. Multilayer structures of these components exhibit a giant magnetoresistance effect (GMR) [1]. It was revealed later that GMR effect is also observed in some nanophase granular systems.This effect together with a number of other interesting physical properties result from the structural peculiarities of nanophased materials. When the volume of a ferromagnetic particle decreases to a certain value (an order of several nm), the probability of thermal fluctuations of the particle magnetic moment orientation becomes considerable. This phenomenon is known as superparamagnetism. For every measurement time t characteristic of the given experiment T_b (the blocking temperature) below which these fluctuations will not be significant can be evaluated. T_b depends on magnetocrystalline anisotropy constant and mean particle volume. At temperatures higher than T_b (but below Curie temperature) the dependence of the magnetization of the particle system on the field is described by the Langevin function. In this case using the experimental data it is possible to determine mean volume of a particle and its magnetic moment. Unfortunately, this is not applied to granular nanophase systems with a high magnetic phase concentration due to the bipolar interaction and/or exchange coupling of the magnetic particles. This interaction affects both T_b and the system behaviour in a magnetic field. In adddition the structure of granular systems can have a number of features related to their methods of fabrication (the form of particles, parameters of the interface layer and so on). However, it should be mentioned that even if experimental results are not accurately described by Langevin-type functions, approximate estimates of the particle sizes based on these measurements can be useful.

The microstructure details mentioned above dramatically influence also the transport properties, i.e. electric resistance and GMR effect. The GMR effect is due to a spin-dependent contribution to conduction electron scattering on structure elements with a varying magnetic moment orientation. In nanophase granular materials ferromagnetic particles can be such elements due to the disorientation of their magnetic moments in the absence of the external field. Furthemore, interface can have inhomogeneous magnetic structure and also give some contribution to scattering. In any case careful analysis of the structure of these objects is critical for explaining their magnetic and transport properties. Unfortunately conventional structure analysis method such as X-ray diffraction turns out uneffective for detecting the components of nanoscale sizes. Moreover transmission electron microscopy can also prove a failure when the system under investigation consists of the phases having the same type of crystal structure and close lattice parameters. In this case, the detection of superparamagnetic features or significant magnetoresistance can turn out to be more sensitive indication of the dispersed state of the object.

Similar situation had occured in the research of mechanical alloying in Cu-Co and Cu-Fe systems. Authors [5] believe that during mechanical treatment of the Cu-Co system a homogeneous solid solution forms and then upon annealing, phase separation takes place which leads to the onset of magnetoresistance. The conclusion on the formation of supersaturated solid solutions in the Cu-Co system upon mechanical treatment is reported in other works as well [6,7]. On the other hand, several authors point out that X-ray data used in most works as a proof of dissolving (a shifting of matrix peaks and disappearance of the second element peaks) can be interpreted alternatively: similar results would be observed in a dispersed (particles + matrix) coherent system as a result of either strains [8,9] or interference effect [10]. Stacking faults can also give rise to the displacement of diffraction peaks. Huang et al. [11] obtained more direct evidence of element mixing at an atomic level in mechanically alloyed Cu-Fe system. The analysis of particle composition with a microprobe showed that mixing really takes place as particle composition varied from pure copper to pure iron. As far as we know, for the Cu-Co system such research has not been carried out. In this case NMR method can be used which allows one to determine the number of non-magnetic copper atoms in the nearest cobalt neighborhood [12]. The system $Cu_{80}Co_{20}$ prepared by mechanical alloying possesses a considerable GMR effect and magnetic properties characteristic of cluster magnetics [13]. However, it is not clear whether these properties are the result of the composition fluctuation in the Cu-Co solid solution or of the presence of the cobalt particles as suggested for melt spun Cu-Co systems. In this paper we report our investigation of this Cu-Co system.

1.2. EXPERIMENTAL DETAILS

Powders of pure Cu, Co elements (20:80 at% composition) were milled in a centrifugal -planetary Fritsch mill in steel vials with powder to ball mass ratio 1:2 under argon atmosphere. To reduce contamination of the powders by the material of the mortar a preliminary treatment of similar powder that was later removed was carried out. The remaining material on the walls of vials and balls greatly reduced the possibility of contamination of subsequent powders prepared in the same mortar. The Fe impurity content did not exceed 1% after treatment.

The X-ray diffraction analysis was carried out with a DRON-4 diffractometer with Cu K_{α} - radiation. The extrapolation method along stronger lines was used for determination of the FCC lattice parameter. Magnetic measurements were carried out with a vibrating sample magnetometer in fields up to 6.4 MA/m at temperature from 4.2 to 293 K. Temperature dependence of the magnetic susceptibility was measured in a small alternating field at a frequency of about 100 Hz. Magnetoresistance was measured by a traditional four-probe method at T=77 K in fields up to 1.6 MA/m.

428

The samples for resistance measurement were obtained by compacting the powder by means of shear under pressure [13]. As a result of this treatment perfect density was attained at the surface of a sample while, in the center, powder packing was not so high, which resulted in some increase of the integral value of specific electric resistance.

NMR spectra were studied at 4.2 K in the range 150-350 MHz by a two-pulse spin echo technique with equal pulses having a duration of 2 ms. The interval between pulses was 15-20 ms. The value of radiofrequency pulses was chosen for obtaining the maximum signal value. Calibration of radiofrequency channel was performed at every point with a standard signal generator. The spectra were corrected taking into account sensitivity changes with respect to frequency.

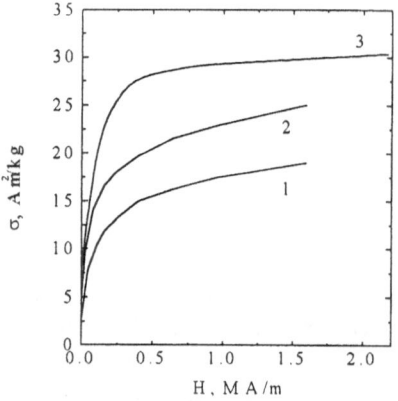

Figure 1. Magnetization curves of Cu-20%Co alloy samples after milling for 20 hrs. T = 293 K (1), 77 K (2), 4.2 K (3).

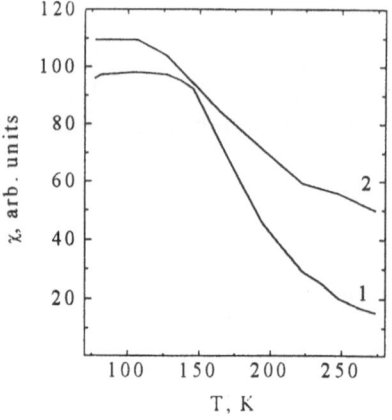

Figure2. Temperature dependence of ac susceptibility after milling for 20 hrs (1) and annealing for 2 hrs at 400 °C (2).

1.3. RESULTS

1.3.1. *Magnetic Properties*

Magnetic properties and magnetoresistance turned out to be close to those presented earlier [13]. After 20-hour milling magnetization curves at 293 K have a superparamagnetic nature (Fig. 1, curve 1). The magnetization variation in high fields can be due to a great number of particles with magnetic moment of the order of several dozens of μ_B . The susceptibility increases smoothly at cooling from 250 to 150 K as a result of freezing the particle magnetic moments (Fig. 2).

At 4.2 K the samples are magnetized almost to saturation in fields 4 MA/m and more, however, the magnetization value seems to be smaller than in the initial state by 25%. Thus, after milling about one quarter of Co atoms are nonmagnetic. Magnetioresistance $\Delta\rho$ attains its maximum after milling (Fig.3). Low-temperature annealing (400°C) considerably decreases ρ_0 (resistance at H = 0) from 170 to 37 $\mu\Omega$ cm owing to annealing of deformation defects and, probably, to a better contact between particles in the central part of the sample. At

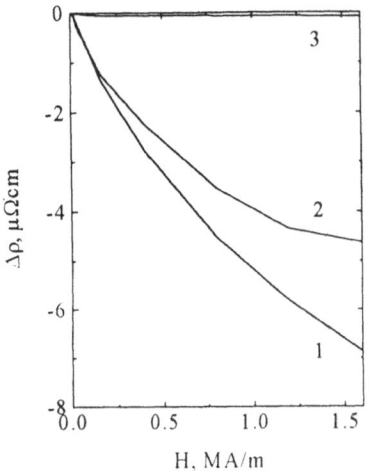

Figure 3. Field dependencies of magnetic resistance at T = 77 K; 1 - after milling for 20 hrs, 2 - after annealing for 2 hrs at T = 400 °C, 3 - after annealing for 0.5 hrs at 600 °C.

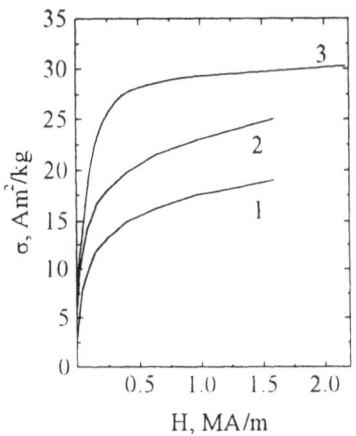

Figure 4. Magnetization curves of samples after milling for 20 hrs and annealing for 2 hrs at 400 °C (1,2); for 0.5 hrs at 600 °C (3). Temperature of measurements: 293 K (1,3); 77 K (2).

the same time $\Delta\rho$ value decreases only by 25%; as a result value $\Delta\rho/\rho_0$ increases to 12% at T=77 K. Such annealing results in a considerable magnetization increase but some superparamagnetic contribution to magnetization process preserves (Fig. 4).

Susceptibility decreases remarkably at heating from 77 K (Fig. 2) which alongside with the high mag-netoresistance value indicates the presence of the disperse magnetic structure. After annealing at 600°C the magnetic resistance decreases to a negligible value and magnetization curves look like the curves in the initial state, probably, due to a considerable enlargement of the particles. The saturation magnetization value is equal to the corresponding value 30 Am^2/kg for a mixture of components, which can be considered as the proof of low powder contamination by Fe from the mortars and balls.

1.3.2. *X-RAY DIFFRACTION DATA*

After 12 hour of milling Cu lines broaden and Co lines almost disappear. Tabl. 1 shows the lattice parameters calculated by the position of different peaks after 20 hrs of milling (1) and after 2 hrs of annealing at T=400 °C (2) and after 0.5 hrs at 600 °C (3). The half-width of lines (β) is also indicated. The error of calculated lattice parameters does not exceed 0.0002 nm. In Fig.5. the lattice parameters as function of cos(⁻) ctg(⁻) are given, which is usually used for a more accurate determination of the lattice parameter by extrapolation of

Table 1. X-ray diffraction data (lattice parameter a, half-width β) for $Cu_{80}Co_{20}$ after 20 hrs of milling (1) and following heat treatments: 2 hrs of annealing at T = 400 °C (2) and 0.5 hrs of annealing at 600 °C (3).

treatment	hkl	a, nm	β, deg.
1	111	0.3607	0.65
2	111	0.3611	0.60
3	111	0.3613	0.58
1	200	0.3611	1.01
2	200	0.3615	0.96
3	200	0.3612	0.79
1	220	0.3608	1.06
2	220	0.3608	1.12
3	220	0.3611	0.83
1	311	0.3612	1.50
2	311	0.3613	1.58
3	311	0.3612	1.09
1	222	0.3610	1.05
2	222	0.3611	1.13
3	222	0.3612	0.91
1	400	0.3605	1.47
2	400	0.3607	1.04
3	400	0.3613	1.24
1	331	0.3609	2.20
2	331	0.3610	2.35
3	331	0.3612	1.92
1	420	0.3610	2.60
2	420	0.3611	2.55
3	420	0.3613	2.08

the curves to Θ = 90°. The considerable variation of thelattice parameter values determined from different peaks is observed. Relative shift of peaks (111) and (200) as well as the shifts of other lines with respect to the straight approximation line match in sign to the shifts appearing when stacking faults are available in the FCC structure. A remarkable broadening of peaks (311) and (331) in comparison to others can be the indication of stacking faults. However, the density of the stacking faults calculated from a relative shift of the peaks (111) and (200) appeared to be rather small (0.8%), which does not correspond to greater shifts of other peaks, especially, the peak (400). Probably, in this case the shift of the peaks is due not to stacking faults but to other reasons, for example, a coherenct conjugation of the particles of

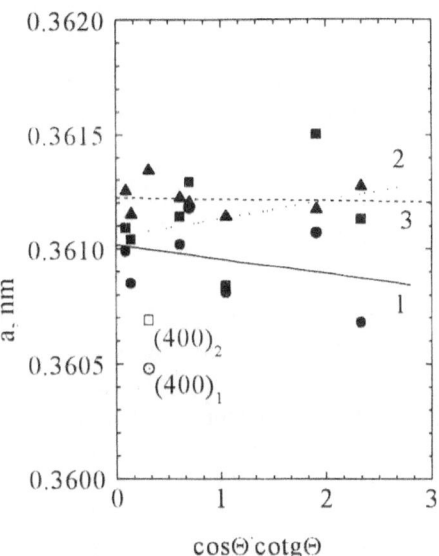

different phases. A more detailed analysis is difficult because of the error in the determination of the line positions.

Furthermore, when determining the lattice parameter we didn't try to correct the calculated parameter values but limited ourselves to the linear extrapolation along the seven most strong peaks taking no account of the position of peaks (400) (Fig. 5). Such extrapolation gives practically equal parameter values before and after annealing at T = 400 °C - 0.3610 nm, which is a bit lower than the copper lattice parameter (0.3615 nm). Even after annealing at T=600 °C peaks of the copper matrix remain shifted in respect to the initial position. The possible reasons of this fact will be discussed later.

Figure5. FCC-phase lattice parameter determination for Cu-20%Co samples. 1- 20 hr milling; 2 - annealing at 400°C. 2 hr; 3 - annealing at 600°C. 0.5 hr.

1.3.3. NMR -SPECTRA

Fig. 6 gives NMR-spectra for the samples with different milling time. At the formation of a solid solution satellite lines in the frequency range 180-200 MHz corresponding to appearance of one, two etc. atoms of copper in the first coordination sphere of Co atoms should be observed. As seen from Fig. 6, a signal is observed in this frequency range but its value does not vary depending on the milling time and corresponds to the initial state of samples. All variations of the spectrum take place in the range 200-230 MHz and are related to the increase of the amount of the FCC-phase and stacking faults with milling time.

The influence of annealing on the structural state was studied on samples after 20 hrs of milling. NMR-spectra are given in Fig. 7. As seen from the Figure, the main variations of spectra after annealing are related to the decrease of the amount of defects and the increase of intensity of the FCC-phase line. Besides, annealing at 400 °C results in a signal increase near 200 MHz, which can be interpreted as the appearance of one impurity atom in the first coordination shell of cobalt. Supposing the binomial distribution one can estimate the impurity content about 1%. A similar result was noticed earlier in Cu-Co films [12]. The growth of copper solubility in cobalt at T = 400 °C may be related to the increasing amount of vacancies in the cobalt lattice in the vicinity of the polymorphous HCP - FCC transformation or facilitating the copper diffusion in defect cobalt matrix. After

432

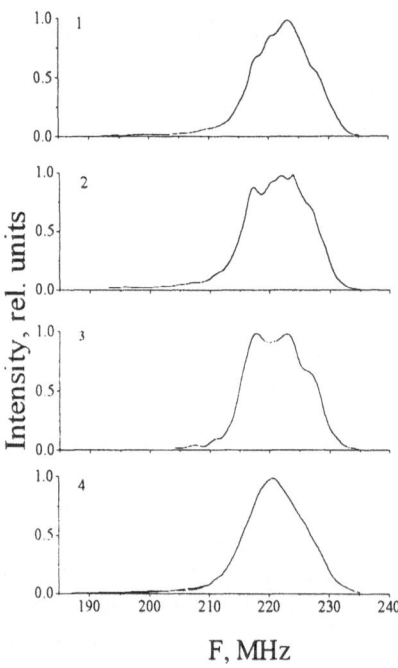

Figure 6. NMR-spectra of samples after different milling duration: (1), (2) and (3) - 6, 12 and 20 hrs respectively. Pure cobalt spectrum (4) is given for comparison.

change the lattice parameter for copper matrix. Even after annealing at 600 °C, the lattice parameter of copper matrix does not attain the value of pure copper. It is suggested that the dissolution of cobalt occurs in the copper matrix defect regions which do not form the X-ray peaks, such as dislocation nuclei or grain boundaries. Then the shift of X-ray peaks may be due to the coherency of Cu and Co phases. Such coherency was observed in our TEM studies of the samples. It should be mentioned that in papers [14], [15], even greater discrepancy in the kinetics of

annealing at 600 °C the shoulder that appeared after annealing at 400 °C disappears and the NMR-spectrum is presented by a narrow resonance line corresponding to FCC-Co (Fig. 7). At 600 °C annealing of defects is likely to take place both in the Cu matrix and the Co particles, resulting in a drastic decrease of copper content in the latter.

Let us discuss the possible reasons for the decrease of the copper matrix lattice parameter as well as the magnetization of the Cu-Co-system after mechanical treatment. Both results can be due to a partial Co dissolving into the copper matrix - the first corresponds to the Vegard's law. the second, to the absence of a magnetic moment for cobalt atoms in diluted Cu-Co solutions. However, annealing at 400 °C, though resulting in the magnetization increase, does not

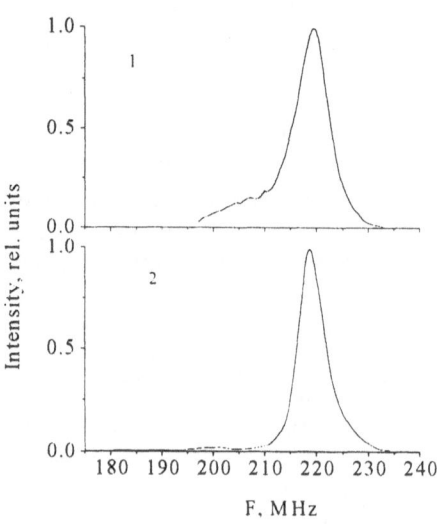

Figure 7. NMR-spectra of samples after milling for 20 hrs and different annealing: 1 - for 2 hrs at 400 °C, 2 - for 0.5 hrs at 600 °C.

variations of the copper matrix lattice parameter and the magnetization of the system was observed upon annealing of melt spun $Cu_{85}Co_{15}$ alloys. Just after melt spinning, the saturation magnetization (per g-Co) was a little more than half of the cobalt magnetization. After annealing at 600 °C the magnetization was 84% of the Co magnetization, (i.e. there remained not more than 2.5 % Co dissolved in Cu), while the matrix lattice parameter increased from 0.3605 nm to only 0.3608 nm. The shifting of X-ray peaks due to the Co and Cu coherency is likely to take place here.

2. Hydrogenated $Pr(Co_{1-x}Cu_x)_5$ Intermetallics

2.1. INTRODUCTION

To date, it is unclear, if a solid solution in this system can be formed using non-equilibrium methods, e.g., mechanical alloying [16,6,17], or Cu and Co will form a two-phase mixture with a characteristic grain size of the order of a few nanometers [13].

To prepare a solid solution in an immiscible system, one can start by using pure components or their mixture as it was above mentioned. Alternatively, a quasi-binary rare-earth intermetallic compound can be used as a starting material where Cu and Co are mixed at an atomic level in the framework of one sublattice. Destroying this intermetallic phase by the reaction of rare-earth (R) with a chemically active element (H, O, N or other), it is possible to select Cu and Co and to investigate their mutual solubility.

Among binary R-Co and R-Cu intermetallics common stoichiometries are 1:2 and 1:5, but only in the case of 1:5 and for light R atoms (from La up to Gd) the compounds RCo_5 and RCu_5 crystallize in the same structural type - $CaCu_5$. The system $Sm(Co,Cu)_5$ was investigated in detail, as it was widely used for permanent magnets. The mutual solubility of Co and Cu in this system was found to be only partial. A spinodal decomposition was observed giving a solid solution 1:5 enriched by Co and the isostructural solid solution enriched by Cu [18]. In the same systems with R=La and Pr no decomposition was observed [19,20].

In the present work we have studied the structure and magnetic properties of phases arising at interaction of $Pr(Co,Cu)_5$ with hydrogen, to determine whether it is possible to form a solid solution in Co-Cu system.

2.2. EXPERIMENTAL DETAILS

The alloys of $Pr(Co_{1-x}Cu_x)_5$ composition with $0 \leq x \leq 1$ were melted in an induction furnace in an argon atmosphere in alumina crucibles. After annealing at 850°C during 4 days the amount of extraneous phases in samples did not exceed 3 %.

Hydrogenation of alloys was carried out using pure hydrogen gas obtained at decomposition of $LaNi_5H_x$ hydride, at hydrogen pressure 7.5 atm, in the temperature range 300 - 650°C. The amount of absorbed hydrogen was determined by weighing of a sample. Contrary to most of other intermetallics, hydrogenation of the alloys given does not provide destruction of the sample, which has allowed to investigate both powdered and bulk samples.

X-ray diffraction patterns were obtained using Cu K_α-radiation on powdered samples at room temperature. The average crystallite size was estimated from half-width of lines, recorded in a mode of step-by-step scanning. A pure annealed copper was used as a reference sample for determination of physical width of lines.

Magnetic properties were studied on a vibrating sample magnetometer in magnetic fields up to 2 T in the temperature range 77 - 300 K. Magnetoresistance was measured by a 4-probe method on the rectangular samples of $1.3 \times 1.3 \times 6$ mm size at 77 and 290 K in a magnetic field up to 2 T.

2.3. EXPERIMENTAL RESULTS AND THEIR DISCUSSION

The initial $Pr (Co_{1-x}Cu_x)_5$ intermetallics for all x possess a hexagonal $CaCu_5$-type crystal lattice. The measured lattice parameters for $x=0$ and $x=1$ are in good agreement with the previous data [21-22]. The concentration dependencies of crystal lattice parameters a and c (Fig. 8) are monotonous indicating the solubility in 3d-sublattice.

After hydrogen treatment at the temperature 500°C and higher the samples contain 2.5 - 3 atoms of hydrogen per formula unit and represent a mixture of phases $PrH_{2.6}$ (FCC CaF_2 - type), Cu and HCP-Co, the lines of the latter are extremely broadened.

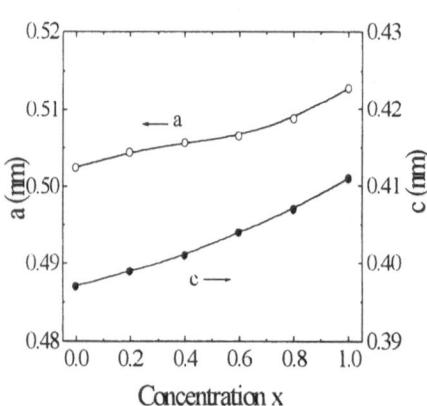

The only ferromagnetic phase after such decomposition is Co, and if it will not form a solid solution with Cu, the magnetization of such multiphase mixture should be equal to the magnetization of pure Co multiplied on a weight percent of Co in a sample. As can be seen from Fig. 9, this condition is valid for most of concentrations. Hence, after hydrogen treatment at 500°C and higher, both Co and Cu are separate phases.

Figure 8. Composition dependence of lattice constants for $Pr(Co_{1-x}Cu_x)_5$ intermetallics.

Giant magnetoresistance in binary Co-Cu system is known to be observed in the range of compositions 5 - 30 % Co [8]. Therefore, for more detailed researches we have chosen the composition $PrCo_{1.5}Cu_{3.5}$. Fig. 10 shows X-ray diffraction patterns of this alloy after hydrogen treatment during 5 hours at different temperatures. It can be seen that hydrogenation at 300°C leads to the formation of isostructural with a parent compound $PrCo_{1.5}Cu_{3.5}H_{1.6}$ hydride, which represents a

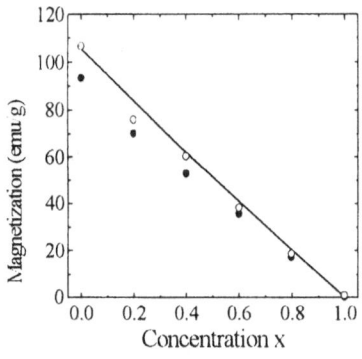

Figure 9. Composition dependence of the saturation magnetization at 77 K (open circles) and 293 K (solid circles) for $Pr(Co_{1-x}Cu_x)_5$ hydrognated at 500°C. Solid line is a calculation for completely decomposed system.

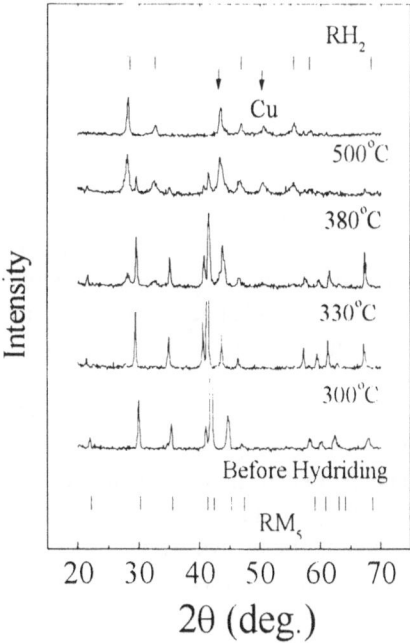

Figure 10. X-ray diffraction patterns of $PrCo_{1.5}Cu_{3.5}$ compound hydrogenated for 5 h at different temperatures.

solid solution of hydrogen in $CaCu_5$ lattice. As the treatment temperature increases, the structure is gradually transformed to a mixture of phases $PrH_{2.6}$ and Cu. The lines of cobalt are not detected up to temperatures 500°C.

Changes of magnetic properties of the samples after hydrogenation at different temperatures (Fig.11) correlate with structural transformations in an alloy. The parent $PrCo_{1.5}Cu_{3.5}$ intermetallic is a ferromagnet. however, its temperature dependence of the magnetization (curve (a) on Fig. 11) differs from that for classical magnetically ordered

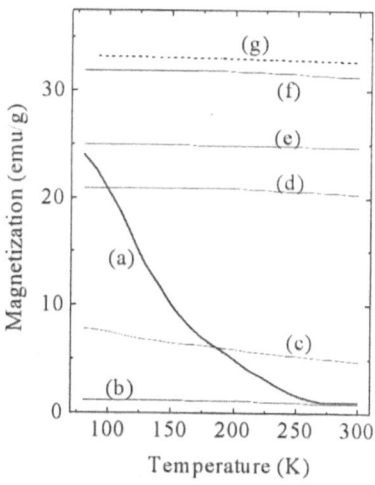

Figure 11. Temperature dependencies of the magnetization of $PrCo_{1.5}Cu_{3.5}$ in the field 0.6 T: (a) before hydrogenation; (b) after hydrogenation for 5 h at 300°C; (c) at 330°C; (d) at 380°C; (e) at 420°C; (f) at 500°C; (g) calculated for completely decomposed system.

substances.

If we take into account that $PrCu_5$ is a paramagnet [23], whereas $PrCo_5$ is a ferromagnet with the Curie temperature

912 K [21], a smooth reduction of a magnetization of quasi-binary intermetallic with the temperature increase can be explained by a spinodal-like decomposition or structural clusters having various relative Co and Cu concentration (rather than on $PrCo_5$ and $PrCu_5$ phases). As mentioned above, such effects were found earlier for $Sm(Co,Cu)_5$ [18]. Hydride $PrCo_{1.5}Cu_{3.5}H_{1.6}$ (curve (b), Fig. 11) has no spontaneous magnetic moment down to 77 K. After hydrogen treatment at 500°C temperature dependence of the magnetization (curve (f)) is close to that calculated in assumption of completely decomposed system, whereas all other dependencies shown in Fig. 11 can be interpreted in the first approximation as superpositions of curves (b) and (f) with weight factors, proportional to the amount of the appropriate phases.

The estimate of the grain sizes of copper phase, listed in the Table 2, shows continuos growth of Cu grains with the increase in the treatment temperature. It is

TABLE 2. Mean grain size of Cu precipitations, determined from Cu [200] line width after hydriding at different temperatures

Hydriding temperature (°C)	Mean grain size (nm)
330	7
350	10
380	16
500	21
650	38

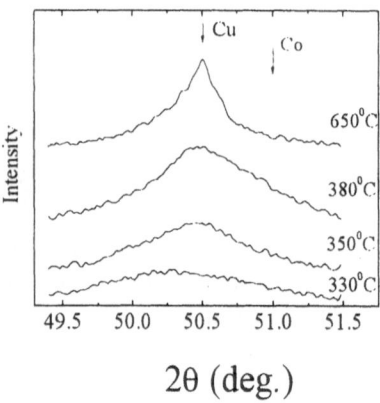

2θ (deg.)

Figure 12. Shape of the Bragg reflection (200) of Cu precipitated from $PrCo_{1.5}Cu_{3.5}$ after hydrogenation for 5 h at different temperatures. The arrows indicate the positions of (200) reflection for pure Cu and FCC-Co.

evident by the reduction of half-width of X-ray diffraction line [220] of copper (Fig. 12). The average grain size of other phases remains practically unchanged for different treatments and makes 15 - 30 nm for $PrH_{2.6}$ and 70-90 nm for hydride $PrCo_{1.5}Cu_{3.5}H_{1.6}$. It is necessary to note that two last grain sizes can be underestimated, as both of hydrogen-containing phases are non-stoichiometric by hydrogen.

The phase transition $PrCo_{1.5}Cu_{3.5}H_{1.6} \rightarrow PrH_{2.6} + (Co + Cu)$ is a thermally activated process, and, hence, it develops not only with the temperature growth, but also with increase in the time of hydrogenation, as seen from Fig.13. The relative

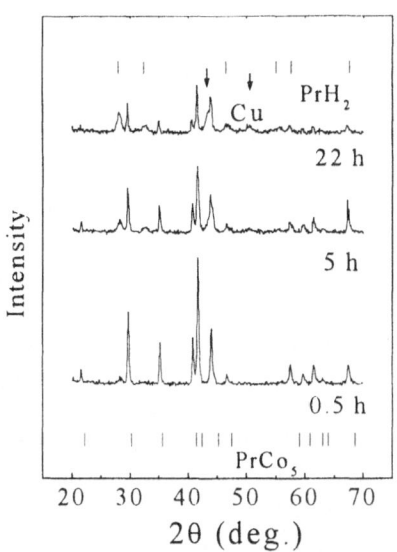

Figure 13. X-ray diffraction patterns of PrCo$_{1.5}$Cu$_{3.5}$ hydrogenated at 330°C for different time.

amount (d) of PrH$_{2.6}$ phase calculated from the intensities of X-ray diffraction patterns follows the law $d^n = at$, where t is the time of hydrogen treatment, a is a characteristic parameter of the diffusion process, with the power factor n ≈ 2.3. This indicates that the phase precipitation rate is mainly controlled by the grain boundaries [24]. It is reasonable to assume that the energy of grain boundaries of nano-sized Co and Cu phases will increase the free energy of the system, which will stabilize the multiphase mixture and reduce the decomposition rate.

Fig. 14 shows temperature dependencies of the magnetization of the samples after different times of hydrogen treatment at 330°C together with the calculated one using the quantitative phase analysis of X-ray diffraction patterns (Fig.13). If a solid solution is formed in Co-Cu system after decomposition of the parent quasi-binary intermetallic, we can expect the reduction of a Co magnetic moment owing to the change of the density of states in a d-zone. This will result in reduction of the magnetization at low temperatures. However, the experimental and calculated magnetization values appear to be nearly the same at low temperatures, and only with an increase in temperature the mismatch enlarged. Such non-Brillouin behavior of temperature dependencies of the magnetization indicates that the decomposition leads to the formation

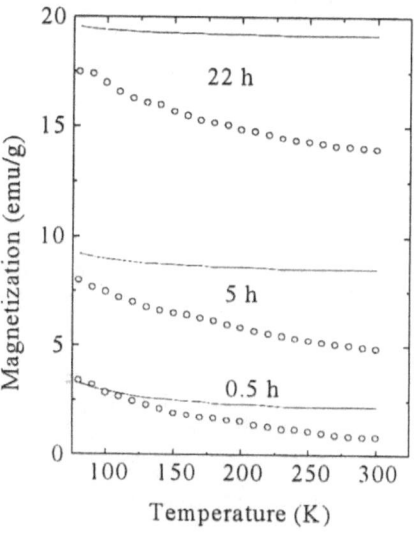

Figure14. Temperature dependencies of the magnetization in the field 0.6 T of PrCo$_{1.5}$Cu$_{3.5}$ hydrogenated at 330°C for different time (open circles). Solid lines are calculated dependencies.

of a nanocrystalline mixture of Co and Cu, where Co particles have different sizes and different transition temperatures into a superparamagnetic state.

In the case of a small grain size Co can crystallize into a FCC-structure, and then its X-ray diffraction line [200] ($2\Theta=51.0$ deg.) should be located near the same line of copper. As it can be seen from Fig. 12, such line of Co is not observed. The small asymmetry of a Cu [200] line at 380°C can be attributed to the superposition with an extremely broadened line of FCC-Co. At the same time, no shift of the maximum of Cu line is observed, which could have been expected at the formation of a solid solution. Hence, Co forms an extremely small-size or even amorphous separate grains.

Measurements of a conductivity showed that the magnetoresistance effect in the field up to 2 T does not exceed 0.5 % in an initial alloy. Practically the same value of the magnetoresistance is observed on samples hydrogenated at 380 and 400°C, on which the decomposition is nearly complete. Probably, the sizes of Co and Cu phases precipitated at such temperatures, and also phase structure of the sample are not optimal for appearance of giant magnetoresistance. We expect that a smaller grain size of Co can be obtained at a similar treatment of R(Co,Cu)$_2$ system, since, firstly, hydrogen-induced decomposition of this system should take place at a lower temperature and, secondly, as it is known, hydrogen treatment allows to completely amorphize Laves phases [25].

3. Summary

In this paper we have shown that:

1. The mechanical alloying of Cu and Co powders can result in an ultradisperse structure consisting of magnetic clusters in a non-magnetic matrix. Magnetic clusters in the initial (after milling) state responsible for superparamagnetic properties of such a system at elevated (> 150 K) temperatures and for a considerable GMR effect are the cobalt particles. After annealing at T = 400 °C, copper content in these particles according to NMR data does not exceed 1%.

2. The shift of X-ray peaks of the copper matrix after mechanical alloying is due to, at least partially, the coherency of Cu and Co phases. Probably a larger portion of dissolved cobalt atoms is concentrated in the matrix defect regions.

3. As a result of hydrogen treatment of Pr(Co,Cu)$_5$ system both a solid solution of hydrogen based on the CaCu$_5$-type structure and a decomposition to PrH$_{2.6}$ hydride and (Co + Cu) can be obtained depending on the hydriding conditions. No hydrogen-induced amorphization was found in this system.

4. There exists a wide range of temperatures and times of hydriding, in the limits of which a multiphase microcrystalline mixture is stable, including the hydride Pr(Co,Cu)$_5$H$_x$, PrH$_{2.6}$ and (Co + Cu).

5. The solid solution of Co and Cu, available in the initial compound, is not maintained during decomposition of an intermetallic, and probably, precipitations of Co and Cu with the typical grain size of the order of 10-20 nm are formed. The giant magnetoresistance effect is not observed in such a multiphase mixture.

Acknowledgments

Authors are grateful to Dr.A.G.Kuchin and Dr. V.S.Gaviko for helpful discussion.

4. References

1. Baibich, M.N., Broto, J.M., Fert A., Nguyen, Van Dan F., Petroff, F., Eitenne, P., Creuzet, G., Freiderich, A. and Chazelas, J. (1988) Giant magnetoresistance of (001)Fe/(001)Cr magnetic superlattices, *Phys. Rev. Lett.* **61**, 2472-2475.

2. Chien, C.L, Xiao, J.Q., and Jiang, J.S. (1993) Giant magnetoresistance in granular ferromagnetic systems, *J. Appl. Phys.* **73**, 5309-5314.

3. Howson, M.A., Musa, S.A., Walker, M.J., Hickey, B.J., Cochrane, R. and Stevens, R. (1994) Giant magnetoresistance in melt-spun $Cu_{87}Co_{13}$. *J. Appl. Phys.* **75**, 6546-6547.

4. Ounadiela, K., Herr, A., Poinsot, R., Coey, J.M.D., Fagan, A., Staddon, C.R., Daniel, D., Gregg, J.F., Thompson. S.M., O'Grady, K. and Grieves, S. (1994) Giant magnetoresistance and induced exchange anisotropy in machanically alloyed $Co_{30}Ag_{70}$, *J. Appl. Phys.* **75**, 6921-6923.

5. Mahon, S.W., Song, X., Howson, M.A., Hickey, B.J. and Cochrane, R. (1996) GMR in $Cu_{90}Co_{10}$ Alloy Produced by Mechanical Alloying, *Mater. Sci. Forum* **225-227**, 157-162.

6. Baricco, M., Battezzati, L., Enzo, S., Soletta, I. and Cocco, G. (1993) X-Ray absorption spectroscopy and diffraction study of miscible and immiscible binary metallic systems prepared by ball milling , *Spectrochimica Acta* **49A**, 1331-1344..

7. Elkalkouli, R., Chartier, P. and Dinhut, J.-F. (1995) Structure and thermal stability of CuCo and CuFe alloys prepared by mechanical milling. *Mater. Sci. Forum* **179-181**, 267-272.

8. Dieny, B., Chamberod, A., Cowache, C., Genin, J.B., Teixeira, S.R., Ferre, R. and Barbara, B. (1994) Giant magnetoresistance in melt-spun metallic ribbons . *J. Magn. Magn. Mater.* **135**, 191-199.

9. Harris, V.G., Kemner, K.M., Das, B.N., Koon, N.C., Ehrlich, A.E., Kirkland, J.P., Woicik, J.C., Crespo, P., Hernando, A. and Garcia Escorial, A. (1996) Near-neigbor mixing and bond dilation in mechanically alloyed Cu-Fe, *Phys. Rev. B* **54**, 6929-6940.

10. Michaelsen, C. (1995) On the structure and homogeneity of the solid solutions: the limits of conventional X-ray diffraction, *Phil. Mag. A* **72**, 813-828.

11. Huang, J.Y., Yu, Y.D., Wu, Y.K., Li, D.X. and Ye, H.Q. (1997). Microstructure and nanoscale composition analysis of the mechanical alloying of Fe_xCu_{100-x} (x=16,60), *Acta Mater.* **45**, 113-124.

12. Meny, C., Panissod, P., Loloee, R. (1992) Structural study of cobolt-cupper multilayers by NMR, *Phys. Rev. B* **45**, 12269-12277.

13. Yermakov, A.Y., Uimin, M.A., Shangurov, A.V., Zarubin, A.V., Chechetkin, Y.V., Shtolz, A.K., Kondratyev, V.V., Konygin, G.N, Yelsukov, Y.P., Enzo, S., Macri, P.P., Frattini, R. and Cowlam, N. (1996) Structure and

magnetoresistance of Cu-Co and Cu-Fe immiscible systems prepared by mechanical alloying , *Mater. Sci. Forum* **225-227**, 147-156.

14. Yu, R.H., Zhang, X.X., Tejada, J., Knobel, M., Tiberto, P. and Allia, P. (1995) Magnetic properties and giant magnetoresistance in melt-spun $Co_{15}Cu_{85}$ alloys, *J. Phys.: Condens. Matter* **7**, 4081-4093.

15. Yu, R.H., Zhang, X.X., Tejada, J., Knobel, M., Tiberto, P. and Allia, P. (1996) Giant magnetoresistance in magnetic granular $Co_{15}Cu_{85}$ alloys annealed by direct-current Joule heating, *J. Magn. Magn. Mater.* **164**, 99-104.

16. Ueda, Y. and Ikeda, S. (1995) Magnetoresistance in Co-Cu alloys prepared by the mechanical alloying, *Materials Transactions, JIM* **36**, 384-388.

17. Cabanas-Moreno, J.G.and Lopez-Hirata, V.M. (1995) Copper and cobalt alloys made by mechanical alloying, *Materials Transactions, JIM* **36**, 218-227.

18. Hofer, F. (1970) Physical metallurgy and magnetic measurements of $SmCo_5$ - $SmCu_5$ alloys, *IEEE Trans. Magn.* **6**, 221-224.

19. Brouha, M. and Buschow, K.H.J. (1975) Magnetic properties and pressure dependence of the Curie temperature of $LaCo_{5x}Cu_{5-5x}$, *J. Appl. Phys.* **46**, 1355-1358.

20. Maeda, H. (1973) Temperature dependence of the magnetic easy direction of $Pr(Co_{1-x}Cu_x)_5$ intermetallic compounds, *Jap. J. Appl. Phys.*, **12**, 1825-1826.

21. Buschow, K.H.J. (1977) Intermetallic compounds of rare-earth and 3d transition metals, *Rep. Progr. Phys.* **40**, 1179-1256.

22. Buschow, K.H.J.and Van der Goot, A.S. (1971) Composition and crystal structure of hexagonal Cu-rich rare earth - copper compounds. *Acta Crystallogr. B* **27**, 1085-1088.

23. Andres, K., Bucher, E., Schmidt, P.H., Maita, J.P.and Darack, S. (1975) Nuclear-induced ferromagnetism below 50 mK in the Van Vleck paramagnet $PrCu_5$, *Phys. Rev. B* **11**, 4364-4372.

24. Exner, H.E. (1983) *Physical Metallography*, by R.W.Cahn, P.Haasen (eds.), North-Holland, **ch.2**, p.50.

25. Aoki, K. and Masumoto, T. (1993) Solid state amorphization of intermetallic compounds by hydrogenation, *J. Alloys and Compounds* **194**, 251-261.

METALLIC SUPERLATTICES
WITH GOVERNED NON-COLLINEAR MAGNETIC ORDERING:
ATOMIC STRUCTURE, INTERLAYER EXCHANGE AND
MAGNETOTRANSPORT PROPERTIES

V.V.USTINOV and E.A.KRAVTSOV
Institute of Metal Physics
18, Kovalevskaya St., Ekaterinburg, 620219, Russia

1. Introduction

Over the last decade there have been significant advances in studying magnetic multilayered structures which, being a new kind of artificial magnetic materials, display a wide array of fascinating properties. These structures consist of thin alternating layers of magnetic and non-magnetic materials with close lattice parameters. Typical examples of magnetic multilayers are Fe/Cr, Co/Cu, and Ni/Ag structures. Magnetic multilayers were found to exhibit numerous intriguing phenomena having no parallel in the bulk properties of the multilayer constituents. It is possible wherewith to influence properties of magnetic multilayers by varying such parameters as material, thickness and morphology of layers, preparation conditions, etc. Thus, one can engineer magnetic multilayers with desirable characteristics. All the above makes magnetic multilayers a challenging object both from a fundamental point of view and for their great potential for technological and engineering applications.

2. Bilinear exchange in magnetic multilayers

2.1. EXPERIMENTAL EVIDENCE

Magnetic multilayers have drawn the attention initially due to their unusual magnetic behaviour. In 1986, P.Grunberg at al. [1] have discovered that in Fe/Cr trilayers the magnetisation vectors of the neighbouring iron layers were aligned antiparallel to each other when the Cr spacer was thin enough. In 1990, after having probed systematically a number of multilayered systems (Co/Cr, Fe/Cr, Co/Ru) S.S.P. Parkin et al. [2] have found and described the effect received the name oscillating interlayer coupling.

This effect implies a magnetic coupling between the magnetic layers mediated by the nonmagnetic spacer, with successive magnetic layers arranging themselves with

G.M. Chow and N.I. Noskova (eds.), Nanostructured Materials, 441–456.
© 1998 *Kluwer Academic Publishers.*

442

their magnetisation either antiparallel (antiferromagnetic coupling, Figure 1a) or parallel (ferromagnetic coupling, Figure 1b).

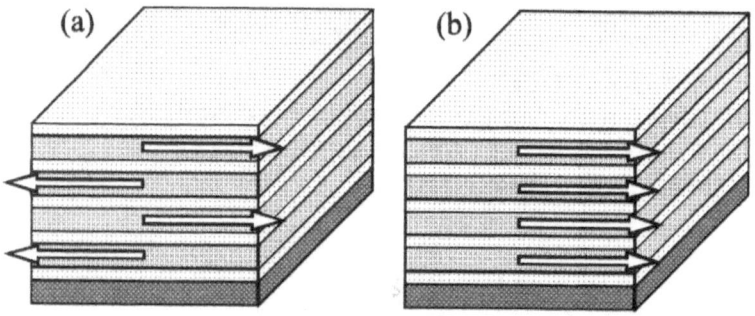

Figure 1. Schematic representation of the antiparallel (a) and parallel (b) arrangement in magnetic multilayers.

The arrangement is a function of the material and thickness of the spacer; for a given system, it oscillates periodically between parallel and antiparallel depending on the spacer thickness.

In interpreting experiments, it is convenient to use the phenomenological expression for the interlayer coupling energy W:

$$W = -J_1 \cos \Theta. \tag{1}$$

Here Θ is the angle between the two adjacent magnetisations and J_1 is the bilinear coupling parameter which is positive ($J_1 > 0$) for ferromagnetic coupling and negative ($J_1 < 0$) for antiferromagnetic. Generally, J_1 is a periodical function of the spacer thickness. The oscillation period d is usually about 10 angstroms for most spacers and about 18 angstroms for Cr. The typical behaviour of J_1 via the spacer thickness in Fe/Cr multilayers is reproduced in Figure 2

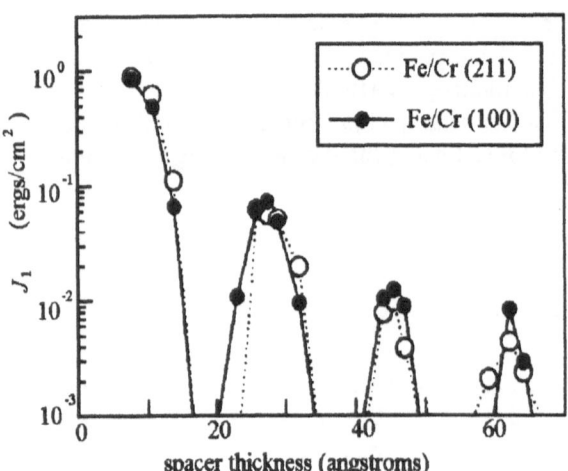

Figure 2. Dependence of bilinear coupling parameter J_1 on the spacer thickness for (211) and (100)-orientated Fe(14Å)/Cr(t_{Cr}) superlattices at room temperature (after [3]).

More recently, for extremely good samples J.Unguris et al. have found that there are shorter oscillations to be about two atomic layers for Fe/Cr [4], Fe/Ag [5] and Fe/Au [6]. In Figure 3 are shown the coupling periods measured by using scanning electron microscopy with polarization analysis (SEMPA).

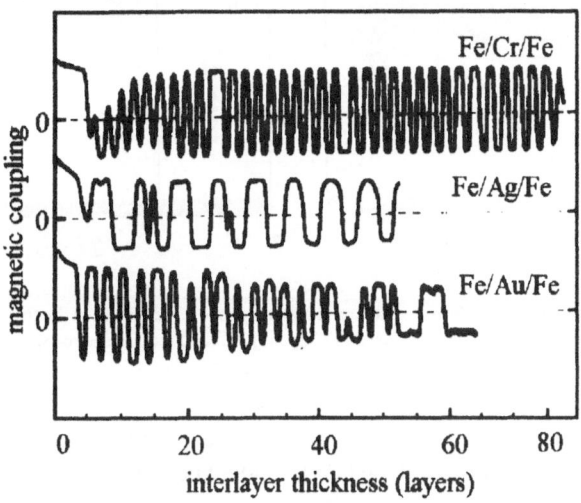

Figure 3. Short-period magnetic coupling in Fe/Cr, Fe/Ag and Fe/Au trilayers as a function of the number of monolayers in the spacer (after [7]).

2.2. SIMPLE MODEL OF OSCILLATING COUPLING: INFINITE QUANTUM WELLS

The oscillating interlayer coupling is intensively investigated since 1986 and a substantial body of literature has evolved which traces numerous aspects of this phenomenon in different systems. It will suffice to mention the last comprehensive reviews in this field [8-9]. There exist a number of sophisticated theoretical approaches to account for the oscillating coupling. Here we are not going to discuss these theories but we will try to clarify the qualitative physics concerned with the oscillations. With this in mind, we exploit the simplest model of infinite quantum wells [10] which enables the underlying physical mechanisms to be explained without going into detail.

Let us consider a trilayer consisting of two magnetic layers of thickness L separated by a non-magnetic spacer of thickness L_0. We assume that the electronic states near the Fermi level in the magnetic slabs are occupied fully for the spin-up projection on the direction of the layer magnetisation and free for the spin-down one, with those in the non-magnetic spacer being occupied for both the spin states. For the ferromagnetic configuration, the electrons appear to be confined within the quantum well of width L_0+2L for spin-up configuration and within that of width L_0 for the spin-down one; for the antiferromagnetic case both the wells have the same width $L+L_0$ (see Figure 4).

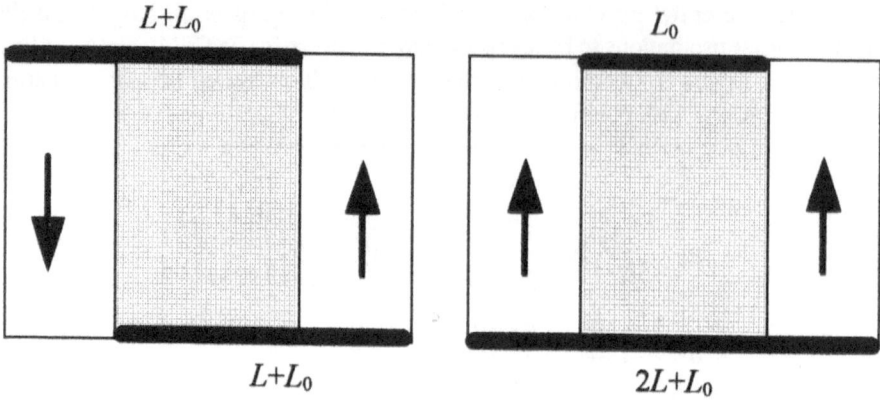

$L+L_0$ L_0

$L+L_0$ $2L+L_0$

Figure 4. Quantum wells for antiferromagnetic and ferromagnetic arrangement in the trilayered system (after [10]). In bold black are shown the regions permitted for the electron moving.

Let us analyze the electron moving in a separated one-dimensional well. For every separated well of l in width the energy should be the form

$$E_n = \frac{\pi^2 \hbar^2}{2ml^2} n^2 \qquad (2)$$

where n=1,2,3,...
The density of states per unit of area can be written as

$$\rho(E) = \frac{m}{2\pi\hbar^2} \sum_n \eta(E - E_n), \qquad (3)$$

with $\eta(x)$ standing for the step function.

By using (2) and (3) one finds successively the number of particles $N(l)$, the total energy $E(l)$ and the thermodynamical potential $\Omega(l)$:

$$N(l) = \frac{m}{2\pi\hbar^2} \varepsilon_F v_l \left\{ 1 - \frac{(1+v_l)(1+2v_l)}{6\xi_l^2} \right\}, \qquad (4)$$

$$E(l) = \frac{m}{4\pi\hbar^2} \varepsilon_F v_l \left\{ 1 - \varepsilon_F \frac{(1+v_l)(1+2v_l)(3v_l^2 + 3v_l - 1)}{30\xi_l^4} \right\}, \qquad (5)$$

$$\Omega(l) = \frac{m}{4\pi\hbar^2} \varepsilon_F^2 v_l \left\{ -1 + \frac{(1+v_l)(1+2v_l)}{\xi_l^2}(1 - \frac{3v_l^2 + 3v_l - 1}{10\xi_l^2}) \right\}. \qquad (6)$$

Here we introduced the value $\xi_l = \frac{l\sqrt{2m\varepsilon_F}}{\pi\hbar}$ and its integral part $v_l = \text{Int}[\xi_l]$. In fact, v_l is the number of discrete energy levels under the Fermi level in the well.

Let us apply the above formulae to the trilayer system of Figure 4. The Fermi level will be concerned to be fixed and therefore independent of the spacer thickness. Then the coupling strength per unit area can be defined as

$$J(L_0)= \Omega(L+L_0)-\{\Omega(2L+L_0)-\Omega(L_0)\}/2. \qquad (7)$$

In Figure 5 is shown the calculated dependence of coupling strength J on the value $\xi_0 = \dfrac{\sqrt{2m\varepsilon_F}}{\pi\hbar} L_0$.

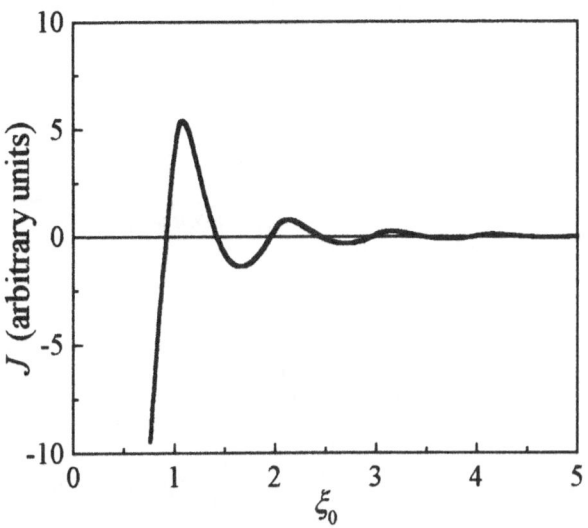

Figure 5. Calculated coupling strength J as function of ξ_0 (on setting $\xi_L =1$)

The changes in the sign of J in Figure 5 are associated with the oscillations between ferromagnetic and antiferromagnetic coupling. From the above consideration, we deduce that these oscillations are governed by filling quantum electron levels near the Fermi surface.

Our consideration provides insight only into the qualitative picture of oscillation coupling. To obtain the quantitative results on oscillations in specific multilayered structures, it is necessary to use more realistic and sophisticated models in which the account is taken of the real form of the Fermi surface [11-14].

Take as an example Fe/Cr structure. There the short-period oscillatory coupling arises from the 'nested' parts of the Cr Fermi surface. 'Nested' refers to regions of the Fermi surface that are parallel to each other over an extended area in reciprocal space. As for the long-period oscillations, most analyses relate its origin to critical spanning vectors of the Cr Fermi surface. These critical vectors may be associated with the 'lens' of the Fermi surface [14], or spanning vectors across the N-centered ellipsoids in the Cr Fermi surface [15], or a zone centered spanning vector that is aliased by the antiferromagnetic order in the Cr [16]. Since our consideration gives no way of analyzing this point in more detail, we refer an interested reader to the original works.

3. Giant magnetoresistance

3.1. FEATURES OF GIANT MAGNETORESISTANCE

Magnetic multilayers with antiferromagnetic coupling were found to exhibit new physical effects. One of the most intriguing phenomena, being of both significant physical interest and great practical importance, is the giant magnetoresistance effect. The first evidence of this effect was obtained in Fe/Cr by Baibich et al [17] who observed halving resistance under the applied magnetic field (see Figure 6).

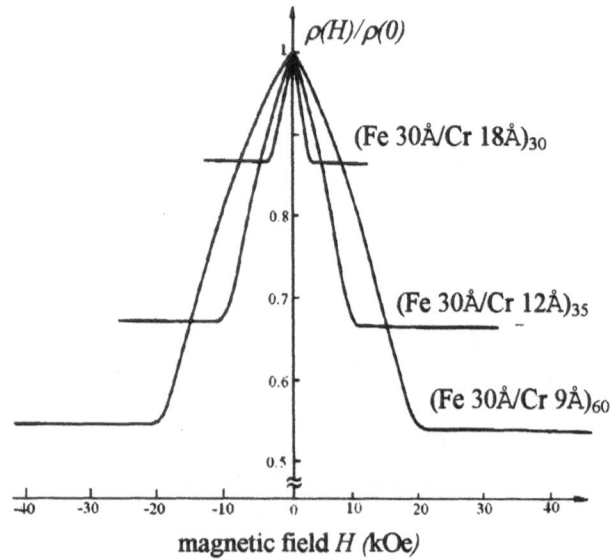

Figure 6. Magnetic field dependence of normalized magnetoresistance for antiparallel ordered Fe/Cr multilayers at T=4K (after [17]).

More recently, the effect was found to occur also in other magnetic multilayered systems (see for reference review [18]). Typical of the giant magnetoresistance effect are its negative sign (the magnetoresistance decreases with magnetic field), its large value (up to 65% at room temperature in Co/Cu), and its dependence on the magnetic arrangement in the multilayer.

The last point is quite important for understanding the physical background of the phenomenon. In Figure 7 is shown the dependence of the magnetoresistance ratio $\Delta\rho/\rho$ on the spacer thickness in Fe/Cr. Note that the only regions where $\Delta\rho/\rho$ is large correspond to the regions for which there is the antiferromagnetic coupling. By comparing Figures 2 and 7 one can see that the magnetoresistance ratio $\Delta\rho/\rho$ oscillates with the spacer thickness and mirrors the oscillations of the magnetic coupling.

The effect originates from the fact that the resistance of a magnetic multilayer structure is much higher in the antiferromagnetic configuration of the magnetic layer magnetisations than in the ferromagnetic one. It is known that the magnetic order of the multilayer is strongly influenced by an applied magnetic field. The magnetic field applied to a multilayer changes the magnetic arrangement from antiferromagnetic to ferromagnetic. When the magnetic field is strong enough, all the magnetisations are forced to lie in the same direction. Simultaneously, the resistance of the multilayer decreases drastically.

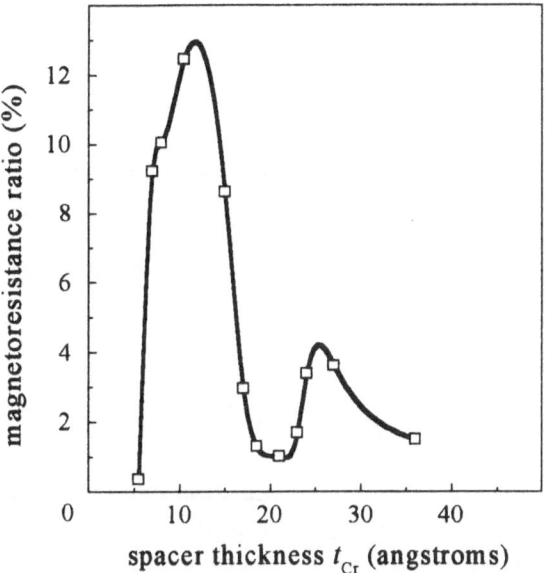

Figure 7. The dependence of magnetoresistance ratio on the spacer thickness in Fe(11Å)/Cr(t_{Cr}) multilayers

3.2. PHYSICAL BACKGROWND: TWO-CURRENT MODEL

Here we will consider the simplest two-current model to give only the basic physical picture of the effect. We assume that the electrons in every ferromagnetic layer form two groups according to the projection of their spins along the local magnetisation. The electrons with opposite spin projections having different scattering rates within the layer, we introduce the resistances ρ_+ and ρ_- for electrons of different groups.

Let us consider the case when the electron mean-free paths are much longer than the layer thickness. Then an electron of a given spin traveling in the whole multilayer sees regions of different resistivities. The distribution of these regions is one for the ferromagnetic configuration and another the antiferromagnetic one.

The total resistance resulting from two independent spin-current channels can be described in terms of an equivalent scheme [19] as shown in Figure 8. At this scheme, we introduced a 'mixing' resistance ρ_{mix} to account for the dependence of the multilayer magnetoresistance on the magnetic configuration. When the applied magnetic field rotates the layers magnetisations from antiferromagnetic to ferromagnetic configuration, the 'mixing' resistance falls from infinity to zero, with the spin channels being switched.

448

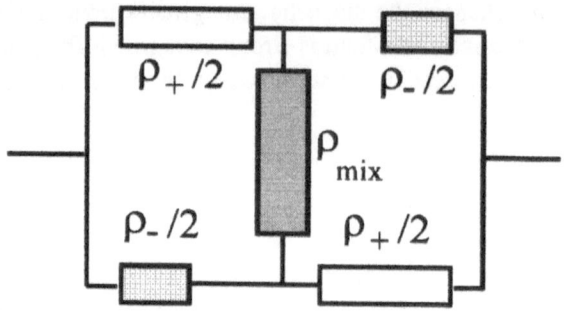

Figure 8. The equivalent current scheme for the multilayer resistance.

The resistivity becomes $\rho_f = \rho_+\rho_-/(\rho_+ + \rho_-)$ in the ferromagnetic configuration and $\rho_{af} = (\rho_+ + \rho_-)/4$ in the antiferromagnetic one. The magnetoresistance ratio $\Delta\rho/\rho$ is therefore given by

$$\Delta\rho/\rho = (\rho_f - \rho_{af})/\rho_{af} = -\left(\frac{\rho_+ - \rho_-}{\rho_+ + \rho_-}\right)^2 \tag{8}$$

According to (8), for the giant magnetoresistance effect to appear, the resistivities ρ_+ and ρ_- should be different. Hence, the electron scattering responsible for the giant magnetoresistance is spin-dependent either in the bulk of the ferromagnetic layers or at the interfaces. The relative importance of the bulk and interface scattering in giant magnetoresistance varies from system to system.

4. Biquadratic interlayer coupling and non-collinear magnetic ordering

4.1 EXPERIMENTAL FACTS

The discovery of the oscillating interlayer coupling and the giant magnetoresistance effect generated interest in magnetic multilayers in the late 80s. Magnetic multilayers having been investigated by various scientific groups in detail, the physical nature and microscopic origin of the above effects appear to be quite clear. It turns out that the peculiarity of physics of magnetic multilayers extends much further the above effects.

The significant discovery has been made in 1991 by Ruhrig et al [20] who observed that in Fe/Cr trilayers there is a non-collinear magnetic arrangement, with this ordering appearing in the transition regions between the ferromagnetic and antiferromagnetic collinear alignment. These regions correspond changing in sign of the bilinear parameter J_1 in Figure 2.

In the phenomenological description, it means that one should add a new biquadratic term into the expression (1) . Then the coupling energy will be written in the form

$$W = -J_1 \cos \Theta - J_2 \cos^2 \Theta \,.$$ (9)

For non-collinear ordering to exist the biquadratic term should be large and negative $(-2J_2 > |J_1|)$.

The first experimental evidence for biquadratic coupling was obtained on magnetic trilayered structures and gave very small values J_2 near the zero-transitions of J_1. The results obtained on multilayered systems show that the above situation is not common and significant biquadratic coupling may coexist with large non-zero bilinear one, leading to the non-collinear arrangement far from zeros of J_1. A. Schreyer at al [21] have observed the coupling angle of 50° between the magnetisations of the neighbouring iron layers in Fe/Cr multilayers by spin-polarized neutron reflectometry. Later the non-collinear magnetic ordering in Fe/Cr was studied by magnetooptical methods and it has been demonstrated that the coupling angle Θ_0 (see Figure 9) varies with the Cr thickness [22].

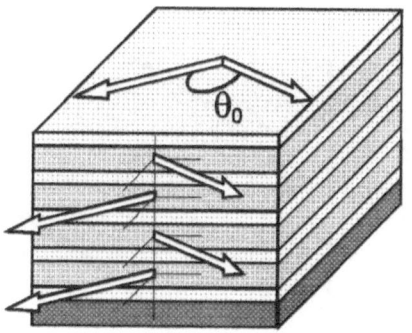

Figure 9. Non-collinear arrangement in magnetic multilayers. The coupling angle Θ_0 takes a value between 0 and π.

The last point means that the magnetic coupling in multilayers is governed by the spacer thickness, so that one can prepare multilayers with desirable magnetic ordering. The structures with significant biquadratic coupling were also observed in trilayers [23-25] and multilayers [26] of other constituents. It should be pointed out also that there exist systems where the biquadratic coupling is large but not enough to overcome the bilinear one [27].

The non-collinearity significantly modifies the magnetizing processes. The multilayer magnetisation is nonzero in the nearest vicinity of zero magnetic field (**H**=0). By analyzing the expression (9), one can see that the relative multilayer magnetisation in zero magnetic field $\mu_0 = \cos\Theta_0/2$ takes the form

$$\mu_0 = \sqrt{\frac{J_1 - 2J_2}{4|J_2|}}$$ (10)

450

An applied magnetic field **H** changes the angle Θ between the neighbouring layer magnetisations. Note that if the magnetic field applied at a non-zero angle Φ relative to the layer planes, the magnetisation curves would have peculiarities associated with the non-collinearity [28].

In Figure 10 are shown typical magnetisation curves for the magnetic field applied in ($\Phi = 0$) and perpendicular to ($\Phi = \pi/2$) the layer planes. If **H** lies in the film plane, the magnetization curve (solid circles) is smooth and convex upward. It is to be noted that in a weak field, $H < 150$ Oe, the magnetization is determined by the domain walls displacement and by the in-plane anisotropy, therefore this region is excluded from our consideration. If **H** is perpendicular to the plane, the magnetization curve (solid squares) has a weak peculiarity near 6 kOe; one can easily confirm the existence of the peculiarity by plotting the curve $\partial\mu/\partial H$ versus H. No jumps on a magnetization curve were observed.

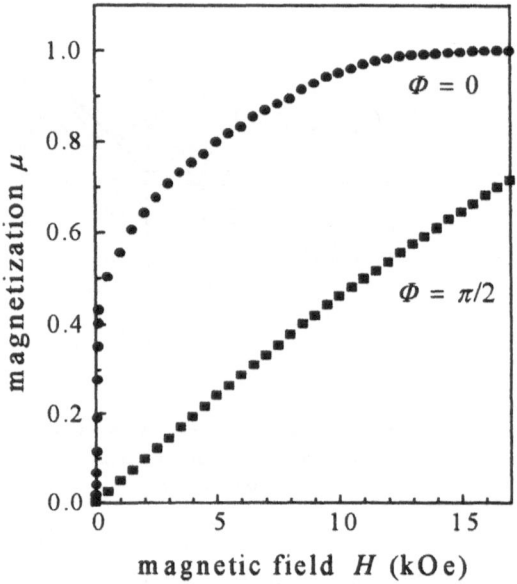

Figure 10. Magnetisation curves in Fe(23 Å)/Cr(8 Å) multilayers for the magnetic field applied in ($\Phi=0$) and perpendicular ($\Phi = \pi/2$) to the layers plane (after [28]).

4.2. MAGNETORESISTANCE IN NON-COLLINEAR STRUCTURES

Of particular interest are magnetoresistive properties of magnetic multilayers with non-collinear ordering [28]. In Figure 11 are shown the magnetoresistance curves in Fe/Cr for magnetic field applied in and perpendicular to the layers plane.

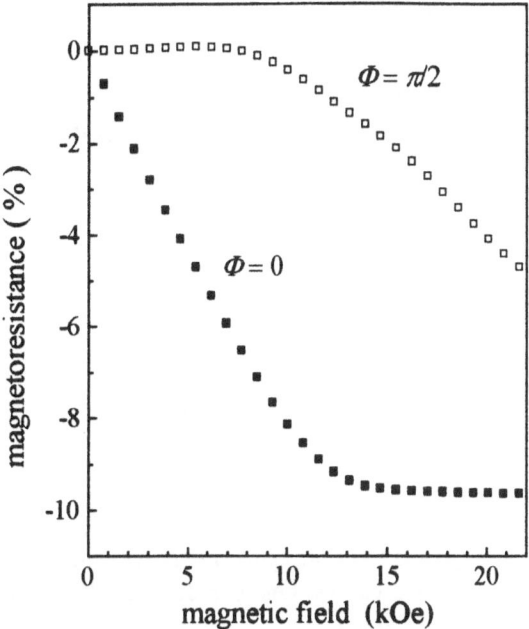

Note the following features characteristic of magnetoresistance in non-collinear structures.

(i) The in-plane magnetoresistance is negative and decreases linearly over a field range from zero to a value close to saturation. In contrast, for collinear structures the magnetoresistance is quadratic in weak magnetic fields (see Figure 6).

(ii) As a rule, the perpendicular-to-plane magnetoresistance does not change noticeably and is positive in a wide field region up to several kOe.

It is significant that magnetoresistance curves for various angles Φ between the applied magnetic field and the layers plane are not independent. It is possible to calculate the magnetoresistance curve for arbitrary Φ if the curves for some certain Φ's are known. Consider this point in more detail.

Let us fix the value r of magnetoresistance ρ. Our basic assumption is that magnetoresistance $\rho(H,\Phi)$ depends on its arguments only through the angle between the neighbouring magnetisations $\Theta = \Theta(H,\Phi)$.

Then the relation

$$\rho(\Theta(H,\Phi)) = r \qquad (11)$$

defines the implicit function $H_r(\Phi)$. By differentiating (11) on Φ we obtain the equation

$$\frac{\partial\Theta}{\partial H}\frac{\partial H_r(\varPhi)}{\partial\varPhi}+\frac{\partial\Theta}{\partial\varPhi}=0. \tag{12}$$

Finding $\partial\Theta/\partial\varPhi$ and $\partial\Theta/\partial H$ by making use of the expression for the free energy, one can obtain a differential equation for $H_r(\varPhi)$ and hence find this function. The equation for $H_r(\varPhi)$ has the form

$$\frac{\partial}{\partial\varPhi}\left(\frac{1}{\sin(2\varPhi)}\frac{\partial}{\partial\varPhi}\frac{1}{H_r^2(\varPhi)}\right)=0. \tag{13}$$

The result of solving (13) can be written as

$$H_r(\varPhi)=\left(\frac{\cos^2\varPhi}{H_r^2(0)}+\frac{\sin^2\varPhi}{H_r^2(\pi/2)}\right)^{-\frac{1}{2}}. \tag{14}$$

The solution contains two arbitrary constants which is the values of the function $H_r(\varPhi)$ at $\varPhi=0$ and $\varPhi=\pi/2$. The equation (14) can be applied to interpretation of the experimental data. In Figure 12 are presented the curves calculated from (14) and the experimental magnetoresistive curves. One can see that the theoretical curves are in good agreement with the experiment.

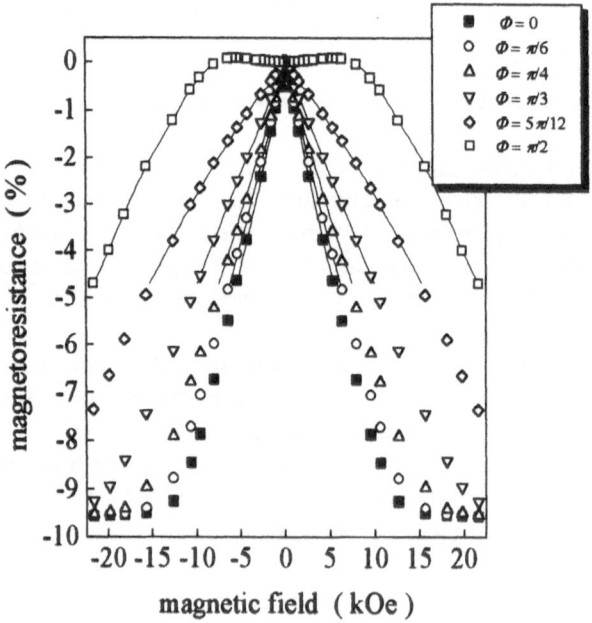

Figure 12. Magnetoresistance of samples Fe(23 Å)/Cr(8 Å) measured (symbols) and calculated from (14) (lines) for some angles $\varPhi$ (after [28]).

4.3. MICROSCOPIC ORIGIN OF BIQUADRATIC COUPLING: BASIC APPROACHES

The problem of microscopic origin of the biquadratic coupling is not resolved yet and remains a subject of much controversy. It is not clear if the biquadratic coupling would be attributed to intrinsic mechanisms arising from the electron structure of the spacer or it is due to extrinsic factors. The last ones may be imperfections in samples such as fluctuations in the spacer thickness, superparamagnetic impurities within the spacer ('loose spins'), or rough interfaces resulting in dipolar fields. Here we will concern ourselves with the basic extrinsic mechanisms proposed for non-collinear coupling.

At first glance the biquadratic coupling in (9) is easily obtainable because it can be treated simply as a second order term relative to the bilinear one. Actually, the biquadratic term would appear intrinsically within the framework of any indirect mechanism if it were to consider higher order perturbation theory. However, the concrete calculations [29] made it apparent that the intrinsic mechanisms are too weak to account for the experimentally observed biquadratic coupling. The intrinsic explanations having collapsed, J. Slonczewski suggested that the biquadratic coupling would be not fundamental but secondary extrinsic effect. He proposed two possible mechanisms for the effect.

J.Slonczewski associated the first mechanism [30] with spatial fluctuations of the bilinear coupling that are due to surface roughness at the interfaces. The roughness is modeled by terraces that, being infinitely long in one direction, have a characteristic width L (see Figure 13a).

Figure 13. The thickness-fluctuation (a) and loose-spin (b) models for biquadratic coupling (after [31]).

The local bilinear coupling $J_1(x)$ between the faces of the magnets is assumed to have the mean J_1 and steps by amount $\pm 2\Delta J$ at the edges of terraces. Then the sum of interface coupling and intralayer ferromagnetic stiffness exchange leads to static spin-wave fluctuations which penetrate exponentially a distance of the order L into the ferromagnets. The energy of these fluctuations is minimized when the mean magnetisations of the neighbouring ferromagnets are orthogonal. If one wants to speak about the coupling in terms of mean magnetisations, ignoring the nonhomogeneity, he has to write the mean energy in the form (9) where Θ is the angle between the *averaged* neighbouring magnetisations. The Slonczewski's model gives J_2 as

$$J_2 = -\frac{4L(\Delta J)^2}{\pi^3 A}\coth\left(\frac{\pi D}{L}\right) \tag{15}$$

where A is the exchange stiffness within the ferromagnetic layers of thickness D. It is notable that the fluctuation model predicts increasing J_2 with L, i.e., with increasing specimen perfection. The fluctuation model was found consistent with the data on Fe/Cr [20], Fe/Ag [27] and Fe/Cu [24] systems. Its significance was confirmed recently by extended analysis and numerical investigations.

The second Slonczewski's mechanism [32] implies the existence of localized electron states with unpaired spins located inside or at the interfaces of the spacer. Exchange coupling of these spins to the ferromagnetic layers contributes to the mutual coupling between the layers. Let us consider the model of Figure 13b, where a loose spin with momentum operator $\hbar S$ lies at position z between the two ferromagnets having magnetisation vectors $\mathbf{m}_1$ and $\mathbf{m}_2$. The ferromagnets induce the exchange-coupling field $\mathbf{U}(z)=U_1(z)\mathbf{m}_1+U_2(z)\mathbf{m}_2$ to be interact with the loose spin. The energy levels of the spin defect are $\varepsilon_m = -U(\Theta)m/S$ with $m= -S,-S+1,\dots S$ and $U(\Theta)=\sqrt{U_1^2+U_2^2+2U_1U_2\cos\Theta}$. From the conventional statistics, the free energy per loose spin is

$$f(T,\Theta) = -kT\ln(\frac{\sinh\{[1+(2S)^{-1}]U(\Theta)/kT\}}{\sinh[U(\Theta)/2SkT]}) . \tag{16}$$

The non-analytic loose-spin energy has not the form of equation (9) but also leads to the non-collinear arrangement and it is expandable in the form (9) in a number of limiting cases. The loose spin model was found consistent with the value and temperature dependence of the biquadratic coupling in such systems as Fe/Al [23].

5. Conclusion

In conclusion, the magnetic multilayers comprise an exiting kind of nanostructures and in our overview we concerned ourselves with only some of interesting phenomena characteristic of magnetic multilayers. We would like to mention other properties, such as anomalous elastic properties, enhanced magneto-optic Kerr rotation, magnetic anisotropy perpendicular to the layers plane. The physics of magnetic multilayered structures is considered now as being promising part of solid state physics and its study is proceeding vigorously.

6. Acknowledgments

This publication was made possible by Grant No. 95-02-04813 from the Russian Foundation of Basic Researches and Grant-in-Aid for Research Program from the State Committee of Russian Federation for Science and Technology.

7. References

1. Grunberg, P., Schreiber, R., Pang, Y., Brodsky, M.B., Sowers, H. (1986) Layered magnetic structures: evidence for antiferromagnetic coupling of Fe layers across Cr interlayers. *Physical Review Letters* **57**, 2442-2445.
2. Parkin, S.S.P., More, N., Roche, K.P. (1990) Oscillations in exchange coupling and magnetoresistance in metallic superlattice structures: Co/Ru, Co/Cr and Fe/Cr. *Physical Review Letters* **64**, 2304-2307.
3. Fullerton, E.E., Conover, M.J., Mattson, C.H., Sowers, C.H., Bader, S.D. (1993) Oscillatory interlayer coupling and giant magnetoresistance in epitaxial Fe/Cr(211) and (100) superlattices. *Physical Review B* **48**, 15755-15763.
4. Unguris, J., Celotta, R.J., Pierce, D.T. (1991) Observation of two different periods in the exchange coupling of Fe/Cr/Fe (100). *Physical Review Letters* **67**, 140-143.
5. Unguris, J., Celotta, R.J., Pierce, D.T. (1993) Oscillatory magnetic coupling in Fe/Ag/Fe(100) sandwich structures. *J. Magnetism and Magnetic Materials* **127**, 205-213.
6. Unguris, J., Celotta, R.J., Pierce, D.T. (1994) Oscillatory exchange coupling in Fe/Au/Fe(100). *J. Applied Physics* **75**, 6437-6439.
7. Celotta, R.J., Pierce, D.T., Unguris, J. (1995). SEMPA studies of exchange coupling in magnetic multilayers. *MRS Bulletin* **10**, 30-33.
8. Heinrich, B. and Bland, A., eds. (1994) *Ultrathin Magnetic Structures*, Springer, Berlin.
9. Grunberg, P., Wolf, J.A., Schafer, R. (1996) Long-range exchange interactions in epitaxial layered magnetic structures. *Physica B* **221**, 357-365.
10. Uzdin, V.M. and Yartseva, N.S. (1996) Quantum wells in trilayers: dependence of the properties on the thickness of magnetic and nonmagnetic layers. *J. Magnetism and Magnetic Materials* **156**, 193-194.
11. Bruno, P. (1995) Theory of interlayer magnetic coupling. *Physical Review B* **52**, 411-439.
12. Erickson, R.P, Hathaway, K.B., Cullen, J.B. (1993) Mechanism for non-Heisenberg-exchange coupling between ferromagnetic layers. *Physical Review B* **47**, 2626-2635.
13. Wang, Y., Levy, P.M., Fry, J.L. (1990) Interlayer magnetic coupling in Fe/Cr multilayered structures. *Physical Review Letters* **65**, 2732-2735.
14. Stiles, M.D. (1993) Exchange coupling in magnetic geterostructures. *Physical Review B* **48**, 7238-7256.
15. Stiles, M.D. (1996) Oscillatory exchange coupling in Fe/Cr multilayers. *Physical Review B* **54**, 14679-14685.
16. Mirbt, S., Niklasson, M.N., Johansson, B., Skriver, H.L. (1996) Calculated oscillation periods of the interlayer coupling in Fe/Cr/Fe and Fe/Mo/Fe sandwiches. *Physical Review B* **54**, 6382-6392.
17. Baibich, M.N., Broto, J.M., Fert, A., Nguyen Van Dau, F., Petroff, F., Etienne, P., Creuzet, G., Friederich, A., Chazelas, J. (1988) Giant magnetoresistance of (001)Fe / (001)Cr magnetic superlattices. *Physical Review Letters* **61**, 2472-2475.
18. Levy, P.M. (1994) Giant magnetoresistance in magnetic layered and granular materials. *Solid State Physics*, **47**, 367-462.
19. Ustinov, V.V. and Kravtsov, E.A. (1995) A unified semiclassical theory of parallel and perpendicular giant magnetoresistance in metallic superlattices. *J. Physics: Condensed Matter* **7**, 3471-3484.
20. Ruhrig, M., Schafer, R., Hubert, A., Mosler, R., Wolf, J.A., Demokritov, S., Grunberg, P. (1991) Domain observations in Fe-Cr-Fe layered structures. *Physica Status Solidi (a)* **125**, 635-656.
21. Schreyer, A., Ankner, J.F., Zeidler, Th., Zabel, H., Scafer, M., Wolf, J.A., Grunberg, P., Majkrzak, C.F. (1995) Non-collinear and collinear magnetic structures in exchange coupled Fe/Cr(001) superlattices. *Physical Review B* **52**, 16066-16083.
22. Ustinov, V.V., Kirillova, M.M., Lobov, I.D., Mayevskii, V.M., Makhnev, A.A., Minin, V.I., Romashev, L.N., Del, A.R., Semerikov, A.V. (1996) Optical, magnetooptical properties and giant magnetoresistance of Fe/Cr superlattices with non-collinear arrangement of iron layers. *J. Experimental and Theoretical Physics* **109**, 477-494.
23. Filipkowski, M.E., Gutierrez, C.J., Krebs, J.J., Prinz, G.A. (1993) Temperature-dependence of the 90-degrees coupling in Fe/Al/Fe(001) magnetic trilayers. *J. Applied Physics* **73**, 5963-5965.
24. Heinrich, B., Celinski,Z., Cochran, J.F., Arrott, A.S., Myrtle, K., Purcell, S.T. (1993) Bilinear and biquadratic exchange coupling in bcc Fe/Cu/Fe trilayers - ferromagnetic-resonance and surface magnetooptical Kerr-effect studies. *Physical Review B* **47**, 5077-5089.
25. Filipkowski, M.E., Krebs, J.J., Prinz, G.A., Gutierrez, C.J. (1995) Giant near – 90° coupling in epitaxial CoFe/Mn/CoFe sandwich structures. *Physical Review Letters* **75**, 3172-2475.

456

26. Rodmacq, B., Dumesnil, K., Mangin, P., Hennion, M. (1993) Biquadratic magnetic coupling in NiFe/Ag multilayers. *Physical Review B* **48**, 3556-3559.

27. Celinski, Z., Heinrich, B., Cochran, J.F. (1993) Analysis of bilinear and biquadratic exchange coupling in Fe/Ag/Fe(001) trilayers. *J. Applied Physics* **73**, 5966-5968.

28. Ustinov, V.V., Bebenin, N.G., Romashev, L.N., Minin, V.I., Milyaev, A.R., Del, A.R., Semerikov, A.V (1996) Magnetoresistance and magnetization of Fe/Cr(001) superlattices with non-collinear magnetic ordering. *Physical Review B* **54**, 15958-15966.

29. Edwards, D.M., Ward, J.M., Mathon, J. (1993) Intrinsic and secondary machanesms for biquadratic exchange coupling in magnetic trilayers. *J. Magnetism and Magnetic Materials* **126**, 380-383.

30. Slonczewski, J.C. (1991) Fluctuation mechanism for biquadratic exchange coupling in magnetic multilayers. *Physical Review Letters* **67**, 3172-2475.

31. Slonczewski, J.C. (1995) Overview of interlayer exchange theory. *J. Magnetism and Magnetic Materials* **150**, 13-24.

32. Slonczewski, J.C. (1993) Origin of biquadratic exchange in magnetic multilayers. *J. Applied Physics* **73**, 5957-5961.

Subject Index

aerosol, 15
agglomeration, 1, 15, 31, 361
alloying, 93, 243
amorphous alloys, 143
applications, 71

ball milling, 47

coagulation, 15
coatings, 31, 283
consolidation, 405
cryogenic milling, 283
crystallization, 143, 163

defect, 163, 207
deformation, 243
devitrification, 143
diffusion bonding, 361
disclinations, 207
dislocation, 163, 183, 207, 243
disordered interfaces, 183

electrodeposition, 47
electroless deposition, 31
enhanced transformation, 319

FTIR spectrometry, 303

gas-getters, 335
gas-reactive applications, 335
gas-sensors, 335
grain boundaries, 207

high melting temperature
        compounds, 263
high pressure consolidation, 405

interfaces, 207
interlayer exchange, 441

kinetics, 319

laser ablation, 1

magnetic multilayer, 441
magnetic ordering, 441
magnetic properties, 93, 143, 425
magnetoresistance, 425, 441
mechanical alloying, 283, 425
mechanical milling, 47
mechanical properties, 93, 143, 163,
        207, 243, 361
melt quenching, 163
multiphase process approach, 71

non-isothermal sintering, 387
nucleation, 15

phase transformation, 243
porosity, 283

quasiperiodic grain boundaries, 183
quasiperiodic interfaces, 183
quenching, 93

radiation-induced effects, 243
rate-controlled sintering, 387
reactive milling, 47

severe plastic deformation, 121
shock compaction, 405
size distribution, 1, 15
solution chemistry, 31
stability, 283
sinter-forging, 361
sintering, 319, 361, 387, 405
surface characterization, 303
surface modification, 303
surfactant, 31

thermal conductivity, 361
thermal spray processing, 283
thermophoresis, 15
thin films, 31, 183, 207
transport properties, 425

vacancy, 163

world-wide perspective/
        technology, 71